Methods in Molecular Biology

Series Editor
John M. Walker
School of Life and Medical Sciences
University of Hertfordshire
Hatfield, Hertfordshire, UK

For further volumes:
http://www.springer.com/series/7651

For over 35 years, biological scientists have come to rely on the research protocols and methodologies in the critically acclaimed *Methods in Molecular Biology* series. The series was the first to introduce the step-by-step protocols approach that has become the standard in all biomedical protocol publishing. Each protocol is provided in readily-reproducible step-by-step fashion, opening with an introductory overview, a list of the materials and reagents needed to complete the experiment, and followed by a detailed procedure that is supported with a helpful notes section offering tips and tricks of the trade as well as troubleshooting advice. These hallmark features were introduced by series editor Dr. John Walker and constitute the key ingredient in each and every volume of the *Methods in Molecular Biology* series. Tested and trusted, comprehensive and reliable, all protocols from the series are indexed in PubMed.

Plant and Food Carotenoids

Methods and Protocols

Edited by

Manuel Rodríguez-Concepción

Centre for Research in Agricultural Genomics (CRAG), CSIC-IRTA-UAB-UB, Barcelona, Spain

Ralf Welsch

Institute for Biology II, University of Freiburg, Freiburg, Germany

Humana Press

Editors
Manuel Rodríguez-Concepción
Centre for Research in Agricultural Genomics (CRAG)
CSIC-IRTA-UAB-UB
Barcelona, Spain

Ralf Welsch
Institute for Biology II
University of Freiburg
Freiburg, Germany

ISSN 1064-3745 ISSN 1940-6029 (electronic)
Methods in Molecular Biology
ISBN 978-1-4939-9954-5 ISBN 978-1-4939-9952-1 (eBook)
https://doi.org/10.1007/978-1-4939-9952-1

This Humana imprint is published by the registered company Springer Science+Business Media, LLC, part of Springer Nature.
The registered company address is: 233 Spring Street, New York, NY 10013, U.S.A.

Preface

Carotenoids are a large group of isoprenoid metabolites covering more than 600 compounds. A typical carotenoid molecule is a C40 hydrocarbon and forms a chromophore of double bonds in conjugation which absorbs light in the blue light spectrum, thus appearing in colors ranging from yellow to orange and red. Although carotenoids are usually synthesized by photosynthetic organisms such as plants, animals including humans need to consume carotenoids with their diet. Carotenoids present in our food are an essential source of retinoids such as vitamin A and visual pigments, but they are also used to produce other biologically active molecules that prevent degenerative diseases and promote health. In plants, the functions of carotenoids cover almost all aspects in biological systems, including photosynthesis, protection against oxidative stresses, visual attraction, signaling, aroma composition, developmental regulation, and membrane stabilization.

Accordingly, research on carotenoids developed rapidly within the last years, supported by the development of methods suitable to analyze carotenoid cleavage products and also by numerous approaches succeeding in altering carotenoid amounts and patterns in various plant tissues and food sources. Associated with the development of these techniques, knowledge on the functions of carotenoids increased significantly and provided novel insights on how carotenoid biosynthesis is dynamically regulated in coordination with their storage and degradation in plant cells as well as the contribution of carotenoid cleavage products in numerous processes in plants and animals.

This collection of methods is intended to enable carotenoid research into these recently extended research areas and answer questions which now can be asked. The methods are also thought to support researchers approaching carotenoid functions from a different research area, who are thus new in the field or have only little experience. Accordingly, the methods are described with additional details of practical relevance usually not included in research publications, which are therefore sometimes difficult to translate into real lab life. We cover a wide spectrum of research aspects, including established protocols for carotenogenic and carotenoid-cleavage enzyme characterizations, a large chapter on analytical methods which meets the need of increasing knowledge on catabolic carotenoid pathways, chapters on how both anabolic and catabolic pathways can be considered as well as a chapter on imaging carotenoid storage and carotenogenic metabolon formation in living systems. Finally, we included practical instructions on how carotenoid metabolism can be altered in model organisms and how carotenoid function can be assessed in animal systems.

This method collection builds on the willingness of all contributing authors to provide their detailed lab protocols and make them available to a greater audience which we highly appreciate. These methods are intended to be used as reference methods to unify research methods and to allow carotenoid research to be extended into other fields of plant and animal research. We would like to thank John M. Walker for the invitation to create this volume of Methods in Molecular Biology and for his careful instructions on the way to finalize it.

Barcelona, Spain — ***Manuel Rodríguez-Concepción***
Freiburg, Germany — ***Ralf Welsch***

Contents

Part VI Biotechnology

Part VII Nutrition and Health

Contributors

OUSSAMA AHRAZEM • *Departamento de Ciencia y Tecnología Agroforestal y Genética, Instituto Botánico, Universidad de Castilla-La Mancha, Albacete, Spain*

YAGIZ ALAGOZ • *Hawkesbury Institute for the Environment, Western Sydney University, Penrith, NSW, Australia*

SALIM AL-BABILI • *The BioActives Lab, Biological and Environmental Sciences and Engineering Division, King Abdullah University of Science and Technology (KAUST), Thuwal, Kingdom of Saudi Arabia*

JAVIER AVALOS • *Departamento de Genética, Facultad de Biología, Universidad de Sevilla, Seville, Spain*

APARNA BALAKRISHNA • *The BioActives Lab, Biological and Environmental Sciences and Engineering Division, King Abdullah University of Science and Technology (KAUST), Thuwal, Kingdom of Saudi Arabia*

RAFAL BARANSKI • *Faculty of Biotechnology and Horticulture, Department of Plant Biology and Biotechnology, University of Agriculture in Krakow, Krakow, Poland*

M. VICTORIA BARJA • *Centre for Research in Agricultural Genomics (CRAG), CSIC-IRTA-UAB-UB, Barcelona, Spain*

JESÚS BELTRÁN • *Department of Biological Sciences, Lehman College, The City University of New York, Bronx, NY, USA; The Graduate School and University Center-CUNY, New York, NY, USA; Department of Botany and Plant Sciences, University of California, Riverside, CA, USA*

M. LUISA BONET • *Grup de Recerca Nutrigenòmica i Obesitat, Laboratori de Biologia Molecular, Nutrició i Biotecnologia (LBNB) of the Universitat de les Illes Balears, CIBER de Fisiopatología de la Obesidad y Nutrición (CIBERobn) and Institut d'Investigació Sanitària Illes Balears (IdISBa), Palma de Mallorca, Spain*

ALBERT BORONAT • *Center for Research in Agricultural Genomics (CSIC-IRTA-UAB-UB), Barcelona, Spain; Department of Biochemistry and Molecular Biomedicine, Faculty of Biology, University of Barcelona, Barcelona, Spain*

MAURIZIO CAMAGNA • *Graduate School of Bioagricultural Sciences, Nagoya University, Nagoya, Japan; Faculty of Biology, University of Freiburg, Freiburg, Germany*

CHRISTOPHER I. CAZZONELLI • *Hawkesbury Institute for the Environment, Western Sydney University, Penrith, NSW, Australia*

JOSÉ-ANTONIO DARÒS • *Instituto de Biología Molecular y Celular de Plantas (Consejo Superior de Investigaciones Científicas-Universitat Politècnica de València), Valencia, Spain*

RITA DE BRITO FRANCISCO • *Department of Plant and Microbial Biology, University of Zurich, Zurich, Switzerland*

OLIVIA COSTANTINA DEMURTAS • *ENEA, Italian National Agency for New Technologies, Energy and Sustainable Economic Development, C.R. Casaccia, Rome, Italy*

KAREL DE POURCQ • *Center for Research in Agricultural Genomics (CSIC-IRTA-UAB-UB), Barcelona, Spain; Department of Biochemistry and Molecular Biomedicine, Faculty of Biology, University of Barcelona, Barcelona, Spain*

NAMRAJ DHAMI • *Hawkesbury Institute for the Environment, Western Sydney University, Penrith, NSW, Australia*

GIANFRANCO DIRETTO • *Italian National Agency for New Technologies, Energy, and Sustainable Development, Casaccia Research Centre, Rome, Italy*

ALEXANDRA G. DURÁN • *Allelopathy Group, Department of Organic Chemistry, Institute of Biomolecules (INBIO), School of Sciences, University of Cadiz, Cadiz, Spain*

JAVIER FONTECHA • *Institute of Food Science Research (CSIC-UAM), Madrid, Spain*

JUAN M. GARCÍA • *Department of Soil Microbiology and Symbiotic Systems, Estación Experimental del Zaidín (CSIC), Granada, Spain*

SALVADOR GENOVÉS • *Cell Biology Laboratory, Food Biotechnology Department, Biópolis SL/Archer Daniels Midland, Paterna, Valencia, Spain*

JOSÉ VICENTE GIL • *Food Technology Area, Faculty of Pharmacy, University of Valencia, Burjassot, Valencia, Spain; Food Biotechnology Department, Instituto de Agroquímica y Tecnología de Alimentos (IATA), Consejo Superior de Investigaciones Científicas (CSIC), Paterna, Valencia, Spain*

DANIELE GIUFFRIDA • *Department BIOMORF, University of Messina, Messina, Italy*

GIOVANNI GIULIANO • *ENEA, Italian National Agency for New Technologies, Energy and Sustainable Economic Development, C.R. Casaccia, Rome, Italy*

LOURDES GÓMEZ-GÓMEZ • *Departamento de Ciencia y Tecnología Agroforestal y Genética, Instituto Botánico, Universidad de Castilla-La Mancha, Albacete, Spain*

CHRISTIAN GONZALEZ-CALQUIN • *Centro de Biología Molecular Vegetal, Departamento de Biología, Facultad de Cienicas, Universidad de Chile, Ñuñoa, Santiago, Chile*

ALAIN GOOSSENS • *Department of Plant Biotechnology and Bioinformatics, Ghent University, Ghent, Belgium; VIB Center for Plant Systems Biology, Ghent, Belgium*

ANTONIO GRANELL • *Instituto de Biología Molecular y Celular de Plantas, CSIC-Universidad Politécnica de Valencia, Valencia, Spain*

EWA GRZEBELUS • *Faculty of Biotechnology and Horticulture, Department of Plant Biology and Biotechnology, University of Agriculture in Krakow, Krakow, Poland*

FRANCISCO J. HEREDIA • *Food Colour and Quality Laboratory, Area of Nutrition and Food Science, Universidad de Sevilla, Seville, Spain*

DÁMASO HORNERO-MÉNDEZ • *Departament of Food Phytochemistry, Instituto de la Grasa (CSIC), Seville, Spain*

THOMAS JACOBS • *Department of Plant Biotechnology and Bioinformatics, Ghent University, Ghent, Belgium; VIB Center for Plant Systems Biology, Ghent, Belgium*

KUN-PENG JIA • *The BioActives Lab, Biological and Environmental Sciences and Engineering Division, King Abdullah University of Science and Technology (KAUST), Thuwal, Kingdom of Saudi Arabia*

JULIAN KOSCHMIEDER • *Faculty of Biology, University of Freiburg, Freiburg, Germany*

JOANNA LADO • *Instituto Nacional de Investigación Agropecuaria (INIA), Salto, Uruguay*

LI LI • *Robert W. Holley Center for Agriculture and Health, USDA-ARS, Cornell University, Ithaca, NY, USA; Plant Breeding and Genetics Section, School of Integrative Plant Science, Cornell University, Ithaca, NY, USA*

M. CARMEN LIMÓN • *Departamento de Genética, Facultad de Biología, Universidad de Sevilla, Seville, Spain*

SILVIA LLOPIS • *Cell Biology Laboratory, Food Biotechnology Department, Biópolis SL/Archer Daniels Midland, Paterna, Valencia, Spain*

JUAN A. LÓPEZ-RÁEZ • *Department of Soil Microbiology and Symbiotic Systems, Estación Experimental del Zaidín (CSIC), Granada, Spain*

FRANCISCO A. MACÍAS • *Allelopathy Group, Department of Organic Chemistry, Institute of Biomolecules (INBIO), School of Sciences, University of Cadiz, Cadiz, Spain*

PAULA MAPELLI-BRAHM • *Food Colour and Quality Laboratory, Area of Nutrition and Food Science, Universidad de Sevilla, Seville, Spain*
JULIA MARENTE • *Departamento de Genética, Facultad de Biología, Universidad de Sevilla, Seville, Spain*
ENRICO MARTINOIA • *Department of Plant and Microbial Biology, University of Zurich, Zurich, Switzerland*
PATRICIA MARTORELL • *Cell Biology Laboratory, Food Biotechnology Department, Biópolis SL/ Archer Daniels Midland, Paterna, Valencia, Spain*
SHIZUE MATSUBARA • *IBG-2: Plant Sciences, Forschungszentrum Jülich, Jülich, Germany*
ANTONIO J. MELÉNDEZ-MARTÍNEZ • *Food Colour and Quality Laboratory, Area of Nutrition and Food Science, Universidad de Sevilla, Seville, Spain*
JIANING MI • *The BioActives Lab, Biological and Environmental Sciences and Engineering Division, King Abdullah University of Science and Technology (KAUST), Thuwal, Kingdom of Saudi Arabia*
CHRIS MITCHELL • *Hawkesbury Institute for the Environment, Western Sydney University, Penrith, NSW, Australia*
JOSÉ M. G. MOLINILLO • *Allelopathy Group, Department of Organic Chemistry, Institute of Biomolecules (INBIO), School of Sciences, University of Cadiz, Cadiz, Spain*
LUIGI MONDELLO • *Department CHIBIOFARAM, University of Messina, Messina, Italy; Chromaleont s.r.l., at Department CHIBIOFARAM, Polo Annunziata, University of Messina, Messina, Italy; Department of Medicine, University Campus Bio-Medico of Rome, Rome, Italy*
TOMASZ OLESZKIEWICZ • *Faculty of Biotechnology and Horticulture, Department of Plant Biology and Biotechnology, University of Agriculture in Krakow, Krakow, Poland*
BEGOÑA OLMEDILLA-ALONSO • *Instituto de Ciencia y Tecnología de Alimentos y Nutrición (ICTAN-CSIC), Madrid, Spain*
PEDRO ORTEGA • *Departamento de Genética, Facultad de Biología, Universidad de Sevilla, Seville, Spain; Centro Andaluz de Biología Molecular y Medicina Regenerativa-CABIMER, Seville, Spain*
MARTA Z. PACIA • *Jagiellonian Centre for Experimental Therapeutics (JCET), Jagiellonian University, Krakow, Poland*
ANDREU PALOU • *Grup de Recerca Nutrigenòmica i Obesitat, Laboratori de Biologia Molecular, Nutrició i Biotecnologia (LBNB) of the Universitat de les Illes Balears, CIBER de Fisiopatología de la Obesidad y Nutrición (CIBERobn) and Institut d'Investigació Sanitària Illes Balears (IdISBa), Palma de Mallorca, Spain*
JAVIER PARDO-MEDINA • *Departamento de Genética, Facultad de Biología, Universidad de Sevilla, Seville, Spain*
LAURENS PAUWELS • *Department of Plant Biotechnology and Bioinformatics, Ghent University, Ghent, Belgium; VIB Center for Plant Systems Biology, Ghent, Belgium*
ANTONIO PÉREZ-GÁLVEZ • *Food Phytochemistry Department, Instituto de la Grasa (CSIC), Sevilla, Spain*
MARÍA J. POZO • *Department of Soil Microbiology and Symbiotic Systems, Estación Experimental del Zaidín (CSIC), Granada, Spain*
RENA F. QUINLAN • *Department of Biological Sciences, Lehman College, The City University of New York, Bronx, NY, USA*
JOSÉ L. RAMBLA • *Instituto de Biología Molecular y Celular de Plantas, CSIC-Universidad Politécnica de Valencia, Valencia, Spain; Department of Agricultural and Environmental Sciences, Jaume I University, Castellón de la Plana, Spain*

Srinivasagan Ramkumar • *Department of Pharmacology, School of Medicine, Case Western Reserve University, Cleveland, OH, USA*
Daniel Ramón • *Cell Biology Laboratory, Food Biotechnology Department, Biópolis SL/ Archer Daniels Midland, Paterna, Valencia, Spain*
Carlos Rial • *Allelopathy Group, Department of Organic Chemistry, Institute of Biomolecules (INBIO), School of Sciences, University of Cadiz, Cadiz, Spain*
Joan Ribot • *Grup de Recerca Nutrigenòmica i Obesitat, Laboratori de Biologia Molecular, Nutrició i Biotecnologia (LBNB) of the Universitat de les Illes Balears, CIBER de Fisiopatología de la Obesidad y Nutrición (CIBERobn) and Institut d'Investigació Sanitària Illes Balears (IdISBa), Palma de Mallorca, Spain*
María Roca • *Food Phytochemistry Department, Instituto de la Grasa (CSIC), Sevilla, Spain*
María Jesús Rodrigo • *Food Biotechnology Department, Instituto de Agroquímica y Tecnología de Alimentos (IATA), Consejo Superior de Investigaciones Científicas (CSIC), Paterna, Valencia, Spain*
Manuel Rodríguez-Concepción • *Centre for Research in Agricultural Genomics (CRAG), CSIC-IRTA-UAB-UB, Barcelona, Spain*
Francisco J. Rodríguez-Pulido • *Food Colour and Quality Laboratory, Area of Nutrition and Food Science, Universidad de Sevilla, Seville, Spain*
Maria Shumskaya • *SONS Biology, CNAHS, Kean University, Union, NJ, USA; Department of Biological Sciences, Lehman College, The City University of New York, Bronx, NY, USA*
Claudia Stange • *Departamento de Biología, Facultad de Cienicas, Centro de Biología Molecular Vegetal, Universidad de Chile, Ñuñoa, Santiago, Chile*
Carla M. Stinco • *Food Colour and Quality Laboratory, Area of Nutrition and Food Science, Universidad de Sevilla, Seville, Spain*
Bojan Stojnic • *Grup de Recerca Nutrigenòmica i Obesitat, Laboratori de Biologia Molecular, Nutrició i Biotecnologia (LBNB) of the Universitat de les Illes Balears, CIBER de Fisiopatología de la Obesidad y Nutrición (CIBERobn) and Institut d'Investigació Sanitària Illes Balears (IdISBa), Palma de Mallorca, Spain*
Tianhu Sun • *Robert W. Holley Center for Agriculture and Health, USDA-ARS, Cornell University, Ithaca, NY, USA; Plant Breeding and Genetics Section, School of Integrative Plant Science, Cornell University, Ithaca, NY, USA*
Gwen Swinnen • *Department of Plant Biotechnology and Bioinformatics, Ghent University, Ghent, Belgium; VIB Center for Plant Systems Biology, Ghent, Belgium*
Yaakov Tadmor • *Plant Science Institute, Israeli Agricultural Research Organization, Newe Yaar Research Center, Ramat Yishai, Israel*
Björn Thiele • *IBG-2: Plant Sciences, Forschungszentrum Jülich, Jülich, Germany; IBG-3: Agrosphere, Forschungszentrum Jülich, Jülich, Germany*
Linda Dora Thomas • *Department of Pharmacology, School of Medicine, Case Western Reserve University, Cleveland, OH, USA*
Rosa M. Varela • *Allelopathy Group, Department of Organic Chemistry, Institute of Biomolecules (INBIO), School of Sciences, University of Cadiz, Cadiz, Spain*
Isabel Viera • *Food Phytochemistry Department, Instituto de la Grasa (CSIC), Sevilla, Spain*
Johannes von Lintig • *Department of Pharmacology, School of Medicine, Case Western Reserve University, Cleveland, OH, USA*
Ralf Welsch • *Institute for Biology II, University of Freiburg, Freiburg, Germany*

ELEANORE T. WURTZEL • *Department of Biological Sciences, Lehman College, The City University of New York, Bronx, NY, USA; The Graduate School and University Center-CUNY, New York, NY, USA*
JAIME ZACARIAS • *Instituto de Agroquímica y Tecnología de Alimentos, Consejo Superior de Investigaciones Científicas (IATA-CSIC), Valencia, Spain*
LORENZO ZACARÍAS • *Food Biotechnology Department, Instituto de Agroquímica y Tecnología de Alimentos (IATA), Consejo Superior de Investigaciones Científicas (CSIC), Paterna, Valencia, Spain*
MARIOSIMONE ZOCCALI • *Department CHIBIOFARAM, University of Messina, Messina, Italy*

Part I

Introduction

Chapter 1

Pathways for Carotenoid Biosynthesis, Degradation, and Storage

Tianhu Sun, Yaakov Tadmor, and Li Li

Abstract

Carotenoids are isoprenoid compounds synthesized de novo in all photosynthetic organisms as well as in some nonphotosynthetic bacteria and fungi. In plants, carotenoids are essential for light harvesting and photoprotection. They contribute to the vivid color found in many plant organs. The cleavage of carotenoids produces small molecules (apocarotenoids) that serve as aroma compounds, as well as phytohormones and signals to affect plant growth and development. Since carotenoids provide valuable nutrition and health benefits for humans, understanding of carotenoid biosynthesis, catabolism and storage is important for biofortification of crops with improved nutritional quality. This chapter primarily introduces our current knowledge about carotenoid biosynthesis and degradation pathways as well as carotenoid storage in plants.

Key words Carotenoids, Biosynthesis, Degradation, Sequestration/Storage, Pathways

1 Introduction

Carotenoids are a group of isoprenoids synthesized de novo in all photosynthetic organisms of plants, algae and cyanobacteria, as well as in some nonphotosynthetic bacteria and fungi [1–3]. Animals, with very few exceptions (i.e., pea aphids, spider mites, and gall midges), are unable to synthesize carotenoids but accumulate in specific tissues or organs. Carotenoids exert multifaceted functions for plant survival [4]. They are also important for human nutrition and health [2].

Carotenoids are produced exclusively in plastids in plants [5]. The biochemical backbone of carotenoid biosynthesis pathway was established around 1960s by the labeling and inhibition studies along with mutant analysis [6]. Nearly all the major genes that encode enzymes to catalyze the core reactions of carotenogenesis were cloned by the end of last century [7]. A nonmevalonate pathway, that is, the plastidial 2-C-methyl-D-erythritol

Manuel Rodríguez-Concepción and Ralf Welsch (eds.), *Plant and Food Carotenoids: Methods and Protocols*, Methods in Molecular Biology, vol. 2083, https://doi.org/10.1007/978-1-4939-9952-1_1,

4-phosphate (MEP) pathway, was identified for the biosynthesis of prenyl diphosphate precursors of plastidic isoprenoids including carotenoids around the same period of time [8, 9]. A great deal is now known about the biochemistry of carotenoid biosynthesis [1, 2, 10]. Carotenoid degradation occurs via nonenzymatic oxidation and enzymatic oxidation. Carotenoid cleavage dioxygenases (CCDs) catalyze oxidative cleavage of carotenoids to form various apocarotenoids [11, 12]. Their roles in the biosynthesis of phytohormones abscisic acid (ABA) and strigolactones are well elucidated [13, 14]. Since carotenoids are synthesized and stored in plastids in plants, the plastid sink strength or sequestration capacity is critically important for carotenoid accumulation [5, 15]. The biosynthesis activity, degradation rate, and stable storage in plastids define the final carotenoid content in plant tissues.

2 Carotenoid Biosynthesis

2.1 Plastidial Precursors for Carotenoid Biosynthesis in Plants

Plant carotenoid biosynthesis occurs exclusively in plastids and the plastid-localized MEP pathway provides precursors for carotenoid production. The MEP pathway utilizes pyruvate and glyceraldehyde 3-phosphate (GAP) to produce 5-carbon building blocks isopentenyl diphosphate (IPP) and dimethylallyl diphosphate (DMAPP) (Fig. 1). Deoxy-D-xylulose 5-phosphate synthase (DXS) and reductoisomerase (DXR) are two major rate-determining enzymes in the MEP pathway [9, 16]. By addition of three IPP molecules to DMAPP, geranylgeranyl diphosphate (GGPP) is produced by GGPP synthase (GGPS). GGPP is the direct precursor for carotenoid biosynthesis, as well as for many other important plastidial isoprenoids, such as gibberellins, chlorophylls, tocopherols, and plastoquinones.

Various evidence support the impact of precursor supply from the MEP pathway on carotenoid biosynthesis. By overexpression of DXS and DXR, carotenoid levels increase in plants [17–19]. Null mutants in the MEP pathway cause albino phenotype due to lack of precursors for the biosynthesis of carotenoids and other plastidial isoprenoids [20]. While a recent work indicates that GGPS activity also influences carotenoid accumulation [21], overexpression of GGPS usually exerts minimal impact on total carotenoid levels, likely due to the fact that GGPP is the precursor for many plastidial isoprenoids. Through direct interactions with different enzymes, GGPS has been shown to channel GGPP to different plastidial isoprenoid biosynthetic pathways [22, 23]. A recent study shows that enzyme fusion of GGPS and phytoene synthase (PSY) can effectively direct metabolic flux into carotenogenesis [24].

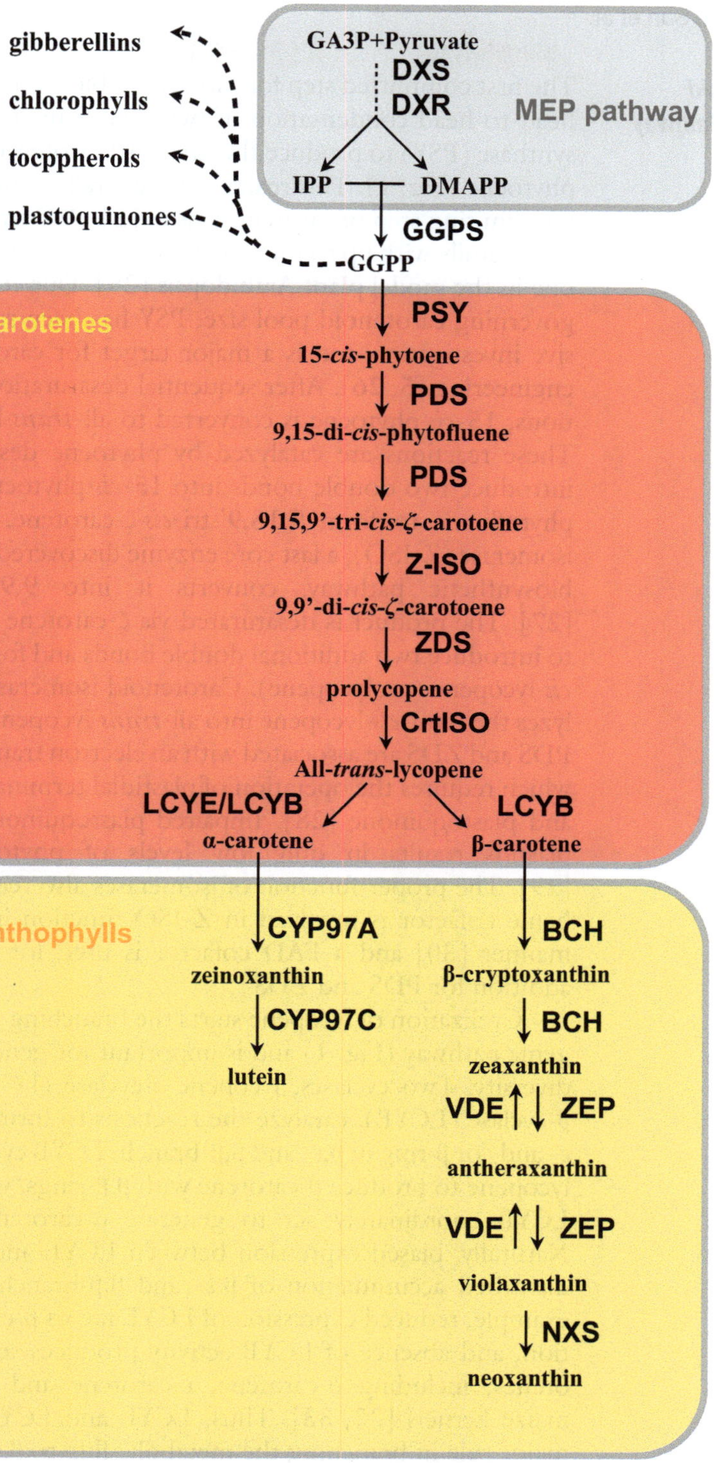

Fig. 1 Carotenoid biosynthesis pathway in plant. Carotene biosynthesis is shown in red box and xanthophyll biosynthesis in yellow box. Dashed lines represent multiple steps. Enzymes are in bold font. GA3P glyceraldehyde 3-phosphate, DXS 1-deoxy-D-xylulose 5-phosphate synthase, DXR 1-deoxy-D-xylulose 5-phosphate reductoisomerase, IPP isopentenyl diphosphate, DMAPP dimethylallyl diphosphate, GGPP geranylgeranyl diphosphate, GGPS geranylgeranyl pyrophosphate synthase, PSY phytoene synthase, PDS phytoene desaturase, Z-ISO ζ-carotene isomerase, CrtISO carotene isomerase, LCYB lycopene β-cyclase, LCYE lycopene ε-cyclase, CYP97A cytochrome P450 carotene β-hydroxylase, CYP97C cytochrome P450 carotene ε-hydroxylase, BCH β-carotene hydrolase, ZEP zeaxanthin epoxidase, VDE violaxanthin de-epoxidase, NXS neoxanthin synthase

2.2 Carotenoid Biosynthetic Pathway in Plants

The first committed step for carotenoid biosynthesis in plants is the head-to-head condensation of two GGPP molecules by phytoene synthase (PSY) to produce the first C40 carotenoid product 15-*cis*-phytoene (Fig. 1). This reaction is generally considered as a major rate-limiting step of carotenogenesis [1]. PSY typically comprises a small family with up to three members in many crops although only one in the model plant Arabidopsis [25]. Due to its crucial role in governing carotenoid pool size, PSY has been subjected to intensive investigation and is a major target for carotenoid metabolic engineering [5, 26]. After sequential desaturations and isomerizations, 15-*cis*-phytoene is converted to all-*trans*-lycopene (Fig. 1). These reactions are catalyzed by phytoene desaturase (PDS) to introduce two double bonds into 15-*cis*-phytoene via 9,15-di-*cis*-phytofluene to form 9,15,9′-tri-*cis*-ζ-carotene. Then ζ-carotene isomerase (Z-ISO), a last core enzyme discovered in the carotenoid biosynthetic pathway, converts it into 9,9′-di-*cis*-ζ-carotene [27]. The product is desaturated via ζ-carotene desaturase (ZDS) to introduce two additional double bonds and form 7,9,9′,7′-tetra-*cis*-lycopene (prolycopene). Carotenoid isomerase, CRTISO, catalyzes the tetra-*cis*-lycopene into all-*trans*-lycopene. The activities of PDS and ZDS are associated with an electron transport mechanism, which requires the operation of plastidial terminal oxidase (PTOX) and plastoquinone [28]. Impaired plastoquinone biosynthesis in plastids results in only low levels of phytoene desaturation [29]. The proper function of isomerases also requires cofactors. A heme cofactor is involved in Z-ISO function in redox-regulated manner [30] and a FAD cofactor is used for CRTISO [31] in addition for PDS and ZDS.

Cyclization of lycopene starts the branching point of carotenogenic pathway (Fig. 1) and is important for generating carotenoid diversity. Two cyclases, lycopene ε-cyclase (LCYE) and lycopene β-cyclase (LCYB), catalyze the reactions to form carotenoids with ε- and/or β-ring in β,ε- and β,β-branch. LCYB cyclizes both ends of lycopene to produce β-carotene with β,β-rings, whereas LCYE and LCYB coordinately act to generate α-carotene with β,ε-rings. Naturally, biased expression between LCYE and LCYB results in disturbed accumulation of β,ε- and β,β-branch carotenoids. For example, reduced expression of LCYE favors β-carotene accumulation, and absence of LCYB activity produces several unusual carotenes, including δ-carotene, ε-carotene and lactucaxanthin in maize kernels [32, 33]. Thus, LCYE and LCYB activities play a major role in branching the metabolic flux to β,ε- and β,β-branch. Their direct products β-carotene and α-carotene are the predominant provitamin A carotenoids found in fruits and vegetables. The production of β-carotene and α-carotene also represent the end of carotene biosynthesis (Fig. 1).

Subsequent sequential ring-specific hydroxylation of α-carotene and β-carotene by two different types of hydroxylases generates yellow xanthophylls of lutein and zeaxanthin (Fig. 1). The first type of hydroxylases includes two nonheme β-ring hydroxylases (BCH1 and BCH2) that primarily hydroxylate β-rings of β-carotene to form β-cryptoxanthin and then zeaxanthin. The second type of hydroxylases contains two heme-containing cytochrome P450 type hydroxylases (CYP97C and CYP97A) that mainly hydroxylates β- and ε-ring of α-carotene to synthesize lutein [34]. Lutein is the most abundant carotenoid pigment in photosynthetic tissues of plants. The synthesis of lutein represents the classic final product of the β,ε-branch in the carotenoid biosynthesis pathway.

In the β, β-branch, zeaxanthin is epoxidized by zeaxanthin epoxidase (ZEP) to sequentially produce antheraxanthin and then violaxanthin. Violaxanthin can be reversed back to zeaxanthin via light induced de-epoxidation by violaxanthin de-epoxidase (VDE). Interconversion of zeaxanthin and violaxanthin forms the xanthophyll cycle. The ubiquitous xanthophyll cycle along with the taxonomically restricted lutein epoxide cycle is an important mechanism to protect plants against photodamage [35]. The conversion of violaxanthin into neoxanthin by neoxanthin synthase (NXS) finalizes the core biosynthetic pathway [36].

Carotenoid biosynthetic enzymes implement carotenoid production. It is believed that carotenogenic enzymes are organized in multienzyme complexes to facilitate biosynthesis and metabolic channeling [7, 37]. Two large complexes containing PDS were found in daffodil chromoplasts and in Arabidopsis chloroplasts, and PDS in the small complex is enzymatic activity and membrane bound, and that in the large complex is inactive in stroma [38, 39]. Recently, GGPS was discovered to form carotenoid enzyme complex via physical interaction with PSY [22]. CYP97A and CYP97C also interact to facilitate the carotene hydroxylation of α-carotene [40]. In addition, other proteins like chaperones interact directly with carotenogenic proteins to assist enzyme functions. A J-protein J20 specifically interacts with inactive form of DXS and aids DXS destination either for activity or degradation [41, 42]. The protein homeostasis and activity of PSY are counterbalanced by ORANGE (OR) protein and Clp protease through direct physical associations [43–46].

2.3 Biosynthesis in Bacteria, Fungi, and Algae

Microalgae, photosynthetic bacteria, and some nonphotosynthetic bacteria and fungi also synthesize carotenoids [2, 3]. Unlike plants that synthesize mainly C40 carotenoids, some archaea and nonphotosynthetic bacteria also produce C30, C45 and C50 carotenoids. While most bacteria utilize MEP pathway as photosynthetic organisms to produce C5 IPP and DMAPP, fungi and some bacteria use mevalonate (MVA) pathway to provide the precursors for carotenoid biosynthesis [9]. In bacteria, the genes involved in carotenoid biosynthesis are normally organized in operons. The early biosynthetic pathway for C40 carotenoids is similar as in plants, which comprises of GGPP synthase (CrtE), phytoene synthase (CrtB), phytoene desaturase (CrtI), lycopene β-cyclase (CrtY), and β-carotene hydroxylase (CrtZ) to produce GGPP, phytoene, all-*trans*-lycopene, β-carotene, and zeaxanthin. A different biosynthetic pathway for C30 carotenoids is utilized in bacteria [3]. In fungi, CrtE, CrtI and CrtYB, a bifunctional enzyme with both phytoene synthase and lycopene cyclase activities, catalyze the synthesis of γ-carotene or β-carotene. In contrast to plants that utilize four enzymes to desaturate and isomerize phytoene into all-*trans*-lycopene, a single CrtI catalyzes all the desaturation and isomerization steps in bacteria and fungi. CrtZ and CrtW (carotene ketolase) in marine bacteria convert β-carotene into astaxanthin, whereas β-carotene ketolase (BKT) in green algae participates the formation of astaxanthin, the red carotenoid with strong antioxidant activity.

2.4 Special Carotenoid Products in Plants

Plants produce a large number and diverse secondary metabolites due to evolution for fitness and enzyme neofunctionalization. More than 1000 chemical structures of carotenoids have been documented in living organisms [47]. Besides the main carotenoid products from the core biosynthetic pathway, some crop species synthesize and accumulate high levels of unique carotenoid products, as exampled below.

Pepper (*Capsicum annuum*) is well-known for its unique carotenoid composition with 35% capsanthin, 6% capsorubin, and 4% cryptocapsin of total carotenoids in red ripen fruit [48]. Capsanthin-capsorubin synthase (CCS) catalyzes antheraxanthin and violaxanthin from the main pathway to generate capsanthin and capsorubin, respectively, which are also the signature pigments of tiger lily (*Lilium lancifolium*). CCS has high sequence similarity with lycopene cyclases. It is also a most abundant protein in pepper chromoplasts [49].

While most plants only have β,β-ring or β,ε-ring cyclization products, green lettuce (*Lactuca sativa*) produces a rare type of double ε-ring carotene lactucaxanthin by lettuce LCYE [50]. Several key amino acid residues of LCYE function as molecular switches to determine single or double ε-ring cyclization [51]. The large amount of lactucaxanthin in light-harvesting complexes of Photosystem II indicates its importance in lettuce [52].

A few *Adonis* species with blood-red flowers are the only known land plants to produce ketocarotenoid astaxanthin. Two enzymes carotenoid 4-hydroxy-β-ring 4-dehydrogenase (HBFD) and carotenoid β-ring 4-dehydrogenase (CBFD) catalyze three distinct reactions to convert β-carotene into astaxanthin in *Adonis*, utilizing a strikingly different way unlike in bacteria, cyanobacteria, and green algae [53]. In the later cases, astaxanthin is formed from β-carotene by 3-hydroxylation and 4-ketolation of β-rings catalyzed by BCH and BKT (CrtW), respectively. Ketolation is usually not found in plants [54, 55]. Since astaxanthin has strong antioxidant activity and is widely used in aquaculture supplements, great efforts have been done to metabolically engineer by introducing BKT with or without BCH for astaxanthin production in crops [56–59].

3 Carotenoid Degradation Pathways

3.1 Nonspecific Cleavage of Carotenoids

The electron-rich polyene chain of carotenoids makes them a group of unstable compounds. Carotenoids are susceptible to oxidative breakdown by nonspecific mechanism such as photochemical oxidation, and oxidation by nonspecific enzymes including lipoxygenases and peroxidases. The nonspecific oxidization of carotenoids leads to the generation of unspecific apocarotenoid products by random cleavage. Environment factors such as oxygen level, temperature, and light affect carotenoid oxidative degradation [60]. Nonenzymatic oxidation of carotenoids by reactive oxygen species oxidizes carotenoids to produce a variety of oxidized products, some of which such as β-cyclocitral act as stress signals in plants [61] (Fig. 2). Lipoxygenases are a family of nonheme iron-containing enzymes that mainly catalyze the oxidation of polyunsaturated fatty acids to yield hydroperoxides. In the presence of polyunsaturated fatty acids, lipoxygenases co-oxidize carotenoids by random attack of the carotenoid molecules to produce unspecific cleavage products [62].

3.2 Enzymatic Oxidative Cleavage of Carotenoids

Carotenoids also undergo specific enzymatic oxidative breakdown to produce apocarotenoids in living organisms [11, 63]. This process is catalyzed by a family of carotenoid cleavage dioxygenases (CCDs) or sometime referred to as carotenoid cleavage oxygenases (CCOs). CCDs are a group of nonheme iron-containing enzymes that cleave double bonds in the polyene chain of carotenoids. They are widespread not only in carotenogenic organisms of plants, algae, bacteria and fungi, but also in animals. In plants, members of the CCD family are typically named based on their sequence similarities to the Arabidopsis CCD enzymes. They are generally divided into two functionally different groups: 9-*cis*-epoxycarotenoid dioxygenases (NCEDs) that cleave 9-*cis*-violaxanthin and 9-*cis*-neoxanthin into xanthoxin for ABA biosynthesis [64, 65], and CCDs (CCD1, 2, 4, 7 and 8) that have different substrate specificities and catalyze different cleavage steps to produce a large number of apocarotenoids.

The NCEDs are exclusively involved in the production of ABA, an important hormone in regulating plant responses to various stresses and seed maturation. NCEDs normally exist as family

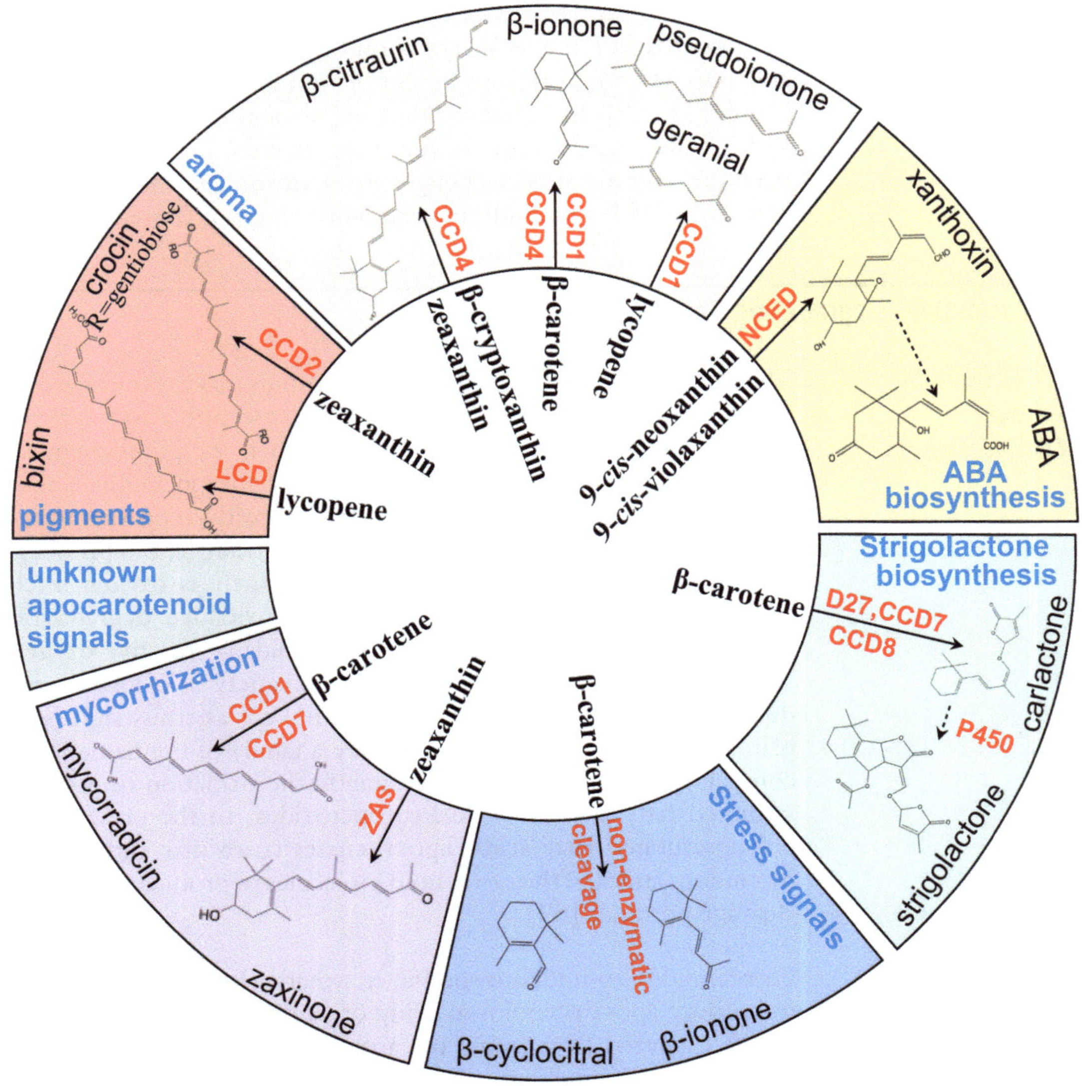

Fig. 2 Carotenoid degradation pathways in plant. The catabolic metabolites are grouped by their biological functions including plant aroma, pigments, signals, and phytohormones (ABA and strigolactone). Nonenzymatic cleavage and enzymatic catabolic processes are annotated in red. CCD carotenoid cleavage dioxygenase, NCED 9-*cis*-epoxycarotenoid dioxygenase, LCD lycopene cleave dioxygenase, ZAS zaxinone synthase

enzymes in plants, and five NCEDs (NCED2, 3, 5, 6, and 9) are identified in Arabidopsis [66]. The first *NCED* gene was cloned from the ABA-deficient maize mutant *viviparous 14* [67, 68]. It also represents the first *CCD* identified. The cleavage of carotenoids for ABA production starts with the formation of *cis*-isomers of violaxanthin and neoxanthin by an unknown isomerase. NCEDs catalyze the 9-*cis*-xanthophylls to produce C15 ABA precursor xanthoxin (Fig. 2). While the cleavage of *cis*-xanthophylls occurs in plastids, the xanthoxin product presumably migrates to cytosol, in where it is then converted by alcohol dehydrogenase to abscisic aldehyde and oxidized by abscisic aldehyde oxidase into ABA [69].

The enzymes CCD7 and CCD8 are involved in the production of strigolactones, the phytohormone inhibiting plant shoot branching, promoting germination of parasitic plants, and establishing mycorrhizal symbiosis. CCD7 and CCD8 cleave β-carotene sequentially to form carlactone (Fig. 2), the precursor for strigolactone biosynthesis [70]. β-carotene is initially isomerized by the carotenoid isomerase D27 to form 9-*cis*-β-carotene. CCD7 cleaves 9-*cis*-β-carotene to produce β-apo-10′-carotenal and β-ionone. The β-apo-10′-carotenal is then further catalyzed by CCD8 to form carlactone for subsequent synthesis of strigolactones [71].

Both CCD1 and CCD4 recognize a large number of carotenoids and cleave them at various double bond locations to produce a wide array of apocarotenoid products such as bixin, saffron, α-ionone, β-ionone, β-cyclocitral, and β-citraurin (Fig. 2). CCD1 catabolizes carotenoid such as ζ-carotene, pro-lycopene, lycopene, α-carotene, β-carotene, β-cryptoxanthin, and zeaxanthin, whereas CCD4 degrades α-carotene, lutein, β-carotene, β-cryptoxanthin, and zeaxanthin [2]. Interestingly, CCD1 has cytosolic localization and is suggested to act as scavenger of apocarotenoids. Indeed, apocarotenoids are substrates of several CCDs [72]. The newly discovered CCD2 cleaves zeaxanthin to form crocetin in the *Crocus* species [73, 74].

In contrast to the excentric cleavage of carotenoids by CCDs in plants, central cleavage of dietary carotenes at the 15, 15′ double bond by CCOs plays a critical role for vitamin A production in mammals including humans. Human genome encodes three CCOs. Two members β-carotene-15,15′-oxygenase (BCO1) and β-carotene-9′,10′-oxygenase (BCO2) are involved in vitamin A formation [75]. BCO1 cleaves β-carotene at central 15,15′ position to produce two molecules of vitamin A. BCO2 catabolizes carotenes such as β-carotene and β-cryptoxanthin at 9′,10′ and 9,10 double bond sites to produce C27 apocarotenoid, which can be further cleaved by BCO1 for vitamin A production. BCO2 also cleaves a range of carotenoids to form various apocarotenoid products and plays a critical role in controlling carotenoid homeostasis in animal tissues [76]. Since no CCDs with central cleavage activities are known from plants, plants do not specifically produce vitamin A.

3.3 Carotenoid Degradation and Carotenoid Content in Plants

While there is no evidence that the activities of NCEDs and CCD7/8 affect carotenoid homeostasis in plants, CCD1 and CCD4 have been shown to influence carotenoid content in various plant species [77–79]. The loss-of-function mutation of *CCD1* and *CCD4* in Arabidopsis leads to more carotenoid accumulation in seeds [63, 79]. A significant negative correlation between *CCD1* copy number and carotenoid content was also reported in a panel of maize inbred lines [80]. However, CCD1 activity is also known to have minimal impact on the carotenoid content in other plant species [81, 82]. Negative correlations between CCD4 activity and total carotenoid levels are clearly demonstrated in peach and in chrysanthemum [77, 78, 83].

Nonspecific oxidation of carotenoids also mediates carotenoid content and is a noticeable problem particularly during postharvesting storage. Both enzymatic and nonenzymatic degradation of carotenoids can prevail in living cells, and the nonenzymatic turnover is predominant in mature dry grains [84]. Indeed, investigation of Arabidopsis mutants defect in *CCD* genes reveals that the oxidized carotenoid metabolites β-cyclocitral and β-ionone are not reduced during photoxidative stress [85], indicating the contribution of nonspecific oxidation of carotenoids. In durum wheat, the lipoxygenase activity was found to be negatively correlated to total carotenoid concentration [86]. During postharvesting storage of carotenoid enriched grains, the rapid turnover of β-carotene with the occurrence of β-apo-14′-carotenal and β-apo-13-carotenone, which are not the cleavage products of known CCDs, indicates the prevalence and contribution of nonenzymatic destruction in controlling carotenoid levels [84].

3.4 Carotenoid Degradation and Plant Aroma, Pigmentation, and Signaling

Carotenoid degradation to produce apocarotenoid volatiles particularly the C13 derivatives contributes to scent and aroma of plants (Fig. 2). In rice, CCD1 cleaves lycopene to produce geranial and also other carotenoids to form ketone products [87]. The temporal expression of CCD1 is essential for the emission of apocarotenoid volatiles in grape berries and petunia flowers [88, 89]. Carotenoid turnover also provides a large number of volatiles, such as β-ionone, β-cyclocytral, geranial, neral and farnesylacetone in melon fruit [90], and *cis*-pseudoionone, geranial, and farnesylacetone in tomato fruits [81]. In saffron flower, CCD4 activity contributes to the emission of volatile β-ionone during stigma development [91].

Carotenoid degradation also affects fruit and flower pigmentation (Fig. 2). In the flavedo of some citrus fruit, CCD4 catalyzes the cleavage of β-cryptoxanthin and zeaxanthin to produce β-citraurin, which gives the fruit their characteristic color [92]. In peach, CCD4 was found to control flesh color. The expression of CCD4 promotes carotenoid turnover and results in white-fleshed peach, while lack of CCD4 activity reduces carotenoid degradation to produce yellow-flesh fruit [83]. In chrysanthemum, CCD4a was showed to express strongly in the petals of white flower but weakly in yellow petals [77]. Knockdown CCD4a activity by RNAi in 'Jimba', the most popular white-flowered chrysanthemum cultivar in Japan, changes the petal colors to yellow with significant carotenoid accumulation [93]. CCD2 cleaves zeaxanthin to give crocetin pigment in the *Crocus* species [73, 74]. Bixin, a diapocarotenoid from *Bixa orellana* seed coats, is an important commercial pigment. Lycopene cleave dioxygenase (LCD) along with bixin aldehyde dehydrogenase and norbixin carboxyltransferase convert lycopene into bixin [94].

Apart from ABA and strigolactones, emerging evidences have revealed that a class of apocarotenoids function as signal molecules (Fig. 2) to regulate plant growth and development [95]. Apocarotenoids β-cyclocitral and β-ionone derived from nonenzymatic cleavage of β-carotene act as photo-oxidative stress signals to mediate stress-related gene expression under high light [96, 97]. Recently, a novel apocarotenoid metabolite zaxinone was identified as a regulator of plant development and mycorrhization [98]. There are also some uncharacterized apocarotenoid signals involved in the regulation of different biological processes in plants. Examples include a signal found in a ζ-carotene desaturase mutant in regulating chloroplast and leaf development [99], a signal in mediating lateral root branching [100], and apocarotenoids derived from prolycopene or *cis*-neurosporene in controlling the transcription of *PSY1* to affect tomato fruit color [101]. In addition, two types of apocarotenoids, glycosylated C13 α-ionols derivatives previously called cyclohexenone derivatives and C14 yellow pigment mycorradicin, cleaved by CCD7 and CCD1, are participated in arbuscular mycorrhizal symbiosis like strigolactones [102–104].

3.5 Apocarotenoid Transportation

Carotenoids are mainly localized in plastids in plants. Notably some carotenoids, especially apocarotenoids and ketocarotenoids, are found and metabolized outside of plastids. In *Crocus sativus* flower stigma and Bixa seed arils, apocarotenoids like crocins or bixin are formed in plastids, but store in vacuole in significant amounts [105, 106]. Xanthoxin, the cleavage product of the 9-*cis*-xanthophylls, is produced in plastids and converted to ABA in cytosol [64]. Carlactone is another carotenoid cleavage product that is cleaved in plastids and catalyzed into strigolactone in cytosol [107, 108]. Similarly, the carotenoids or apocarotenoids cleaved by CCD1 likely are present outside of plastids as CCD1 is localized in cytosol. Moreover, several green algae accumulate ketocarotenoids, especially astaxanthin, in cytoplasmic vesicles [109]. These observations imply that some apocarotenoids or ketocarotenoids are transported or trafficked from their sites of synthesis in plastids to other subcellular locations by unknown mechanisms. Crocins in *Crocus sativus* are thought to be transported via ABC transporter family [110] and via endoplasmic reticulum and cytoplasm [111]. However, the transport processes remain poorly understood.

4 Carotenoid Sequstration and Storage in Plants

4.1 Site and Structure for Carotenoid Storage

Plastids are the organelles for carotenoid biosynthesis and storage in plant cells [5]. Chloroplasts and chromoplasts are two types of plastids that accumulate high levels of carotenoids, although almost all plastids including etioplasts and amyloplasts possess the ability for carotenoid biosynthesis [37, 112]. Chloroplasts primarily synthesize and accumulate lutein, β-carotene, violaxanthin, and neoxanthin, along with minor amounts of zeaxanthin and α-carotene in green tissues. The majority of these carotenoids are located in the thylakoid membranes, where they function in light harvesting and photoprotection processes. Amyloplasts produce and store mainly xanthophylls such as lutein, zeaxanthin and violaxanthin in many agronomically important starchy grain seeds, as well as in potato tubers and cassava roots. Among all types of plastids, chromoplasts are specialized for massive storage of a vast range of carotenoid pigments in different plant organs including colored flowers, fruits, and vegetables [5, 113–115].

As carotenoid-accumulating plastids, chromoplasts have various morphologies. Depending on internal substructures for carotenoid-sequestering, chromoplasts are classified into five major types as globular, tubular, crystalline, fibrillar and membranous [5, 114, 115]. More than one type of chromoplasts can coexist in a species [116]. Plastoglobules (lipid droplets) are the major site for carotenoid storage in globular chromoplasts with apolar and ester carotenoids in the core of plastoglobules and polar carotenoids associated with surrounding membrane [117]. In tubular chromoplasts, carotenoids are usually sequestered in lipid monolayer tubes branching extensively cross the stroma. Crystalline chromoplasts are typically found in tissues over-accumulating lycopene and β-carotene. Those carotenoids in crystalline chromoplasts are deposited as red or orange crystals surrounded by lipid bilayers. In fibrillar chromoplasts, carotenoids are sequestered in those spindle shaped fibrils. Membranous chromoplasts contain multilayers of condensed whirly membrane structures and the stacked membranes provide sites for carotenoid storage. The various carotenoid sequestering substructures in chromoplasts likely affect the kinds and levels of carotenoids accumulated in different plant tissues [5, 118].

4.2 Carotenoid Sequestration in Plastids

Plastid substructures are critically important for carotenoid accumulation because they affect both carotenoid biosynthesis and storage capacity [5, 118]. Various types of plastids are formed with distinctive functions in plants, and they have dramatically different sink strength or capacity to synthesize and sequester carotenoids. Amyloplasts as starch-storing plastids in general synthesize and store relatively low levels of carotenoids in plastid membranes. The absence of appropriate lipoprotein sequestering substructures within amyloplasts perhaps restricts their capacity to significantly synthesize and stably store carotenoids. Chloroplasts that define plants are the photosynthetic plastids in green tissues. Abundant carotenoids are metabolized during active photosynthesis [119]. The majority of carotenoids are as protein-bound in the chloroplast thylakoid membranes, which serve as the sequestering structures to promote carotenoid biosynthesis in plastid envelope membranes, where the carotenogenic enzymes are primarily localized [37]. However, as indispensable components of the photosynthetic apparatus, carotenoids are synthesized tightly in coordination with chlorophyll formation in chloroplasts. Such tight regulation constrains excess carotenoid sequestration to maintain proper carotenoid and chlorophyll pigment ratio for photosynthesis. By contrast, chromoplasts as carotenoid storing plastids do not have such a constraint. Sequestration in various carotenoid-lipoprotein substructures enables chromoplasts to have high sink strength or great capacity to synthesize and store carotenoids. These sequestering substructures function not only in sequestering the newly synthesized carotenoids inside chromoplast for stable storage, but also in promoting continuous biosynthesis by driving carotenoid products away from the biosynthesis sites [5, 114]. As a result, alterations of both chromoplast numbers and compartment sizes positively affect total carotenoid levels.

4.3 Chromoplast Formation

Chromoplast biogenesis is critical important for high levels of carotenoid accumulation [5]. However, the process is not understood. Chromoplasts can be derived from various types of plastids, including chloroplasts, amyloplasts, and proplastids [113, 114]. The direct conversion of chloroplasts into chromoplasts at initiation of carotenoid overproduction was observed during tomato fruit ripening [120]. However, the factors or proteins that control chromoplast differentiation and/or transition remain elusive.

The ORANGE (OR) protein, which was first identified from orange curd cauliflower mutant, represents the only known protein that acts as a molecular switch to initiate chromoplast differentiation [121, 122]. A single amino acid substitution in the OR protein is responsible for chromoplast formation and carotenoid deposition in orange melon fruit [45, 123]. The role of OR in chromoplast biogenesis is also proved by its ectopic expression in potato, tomato and Arabidopsis calli [124–126]. In the OR-induced chromoplasts, carotenoids are sequestrated mainly in components of massive, highly ordered sheets with membranous substructure as showed by transmission electron microscopy [121, 125, 127]. In addition, OR posttranslationally regulates PSY protein stability and activity to promote carotenoid biosynthesis [43, 44, 46]. The dual roles of OR in initiating chromoplast formation and regulating PSY activity not only provide sink structures that limit carotenoid turnover, but also facilitate carotenoid biosynthesis [45].

A number of proteins have been shown to be associated with chromoplast development or be involved in carotenoid sequestration. A small molecular chaperone HSP21 protein and a plastid encoded fatty acid biosynthetic enzyme accD function in chromoplast development during tomato fruit ripening [128, 129]. Fibrillin proteins participate in the formation of carotenoid-lipoprotein complexes for carotenoid sequestration, which include CHRB [130], PAP [131] in pepper and CHRC from cucumber [132]. A plastid fusion and/or translocation factor (Pftf) is involved in chromoplast membrane biogenesis in pepper fruit [133]. Clearly, much are needed to be learned for the proteins and factors that mediate the formation of chromoplasts with great carotenoid storage capacity.

5 Concluding Remarks

Carotenoid accumulation is a dynamic process and a net result of biosynthesis activity, degradation rate, and stable storage in plastids. Great progresses have been achieved in our understanding of carotenoid metabolic pathways. The main carotenoid biosynthesis pathway in plants along in bacteria, fungi, and algae is well established. Although the enzymatic oxidative degradation by CCDs has been determined and new CCDs for apocarotenoid production are continuously being identified, the nonspecific oxidative cleavage of carotenoids remains to be fully elucidated. Also the identities of some apocarotenoids and transportations are unknown. Chromoplasts are the specific plastids for massive carotenoid biosynthesis and storage in plants. In comparison with the carotenoid biosynthesis and degradation pathways, much is unknown for the proteins in chromoplast biogenesis pathway and sequestration processes for carotenoid accumulation. Clearly, the advance of our knowledge about carotenoid metabolism and storage would continuously contribute to crop nutritional quality improvement for better human nutrition and health.

Acknowledgments

This work was supported by Agriculture and Food Research Initiative competitive award (grant no. 2016-67013-24612) from the USDA National Institute of Food and Agriculture and by the United States-Israel Binational Agricultural Research and Development Fund (grant no. US-4918-16CR).

References

1. Nisar N, Li L, Lu S, Khin NC, Pogson BJ (2015) Carotenoid metabolism in plants. Mol Plant 8(1):68–82
2. Rodríguez-Concepción M, Avalos J, Bonet ML, Boronat A, Gomez-Gomez L, Hornero-Mendez D, Limon MC, Meléndez-Martínez AJ, Olmedilla-Alonso B, Palou A, Ribot J, Rodrigo MJ, Zacarias L, Zhu C (2018) A global perspective on carotenoids: metabolism, biotechnology, and benefits for nutrition and health. Prog Lipid Res 70:62–93
3. Liang MH, Zhu J, Jiang JG (2018) Carotenoids biosynthesis and cleavage related genes from bacteria to plants. Crit Rev Food Sci Nutr 58(14):2314–2333
4. Wurtzel ET (2019) Changing form and function through carotenoids and synthetic biology. Plant Physiol 179:830–843
5. Sun T, Yuan H, Cao H, Yazdani M, Tadmor Y, Li L (2018) Carotenoid metabolism in plants: the role of plastids. Mol Plant 11(1):58–74
6. Spurgeon SL, Porter lW (1983) Biosynthesis of carotenoids. In: Porter, lW. and Spurgeon, SL (eds), Biosynthesis of Isoprenoid Compounds, Vol. 2. Wiley, New York, pp I-122.
7. Cunningham FX Jr, Gantt E (1998) Genes and enzymes of carotenoid biosynthesis in plants. Annu Rev Plant Biol 49(1):557–583
8. Lichtenthaler HK (1999) The 1-deoxy-D-xylulose-5-phosphate pathway of isoprenoid biosynthesis in plants. Annu Rev Plant Biol 50(1):47–65
9. Rodríguez-Concepción M, Boronat A (2015) Breaking new ground in the regulation of the early steps of plant isoprenoid biosynthesis. Curr Opin Plant Biol 25:17–22
10. Moise AR, Al-Babili S, Wurtzel ET (2013) Mechanistic aspects of carotenoid biosynthesis. Chem Rev 114(1):164–193
11. Giuliano G, Al-Babili S, von Lintig J (2003) Carotenoid oxygenases: cleave it or leave it. Trends Plant Sci 8(4):145–149
12. Walter MH, Strack D (2011) Carotenoids and their cleavage products: biosynthesis and functions. Nat Prod Rep 28(4):663–692
13. Jia KP, Baz L, Al-Babili S (2017) From carotenoids to strigolactones. J Exp Bot 69 (9):2189–2204
14. Hauser F, Waadt R, Schroeder JI (2011) Evolution of abscisic acid synthesis and signaling mechanisms. Curr Biol 21(9):R346–R355
15. Li L, Van Eck J (2007) Metabolic engineering of carotenoid accumulation by creating a metabolic sink. Transgenic Res 16(5):581–585
16. Cordoba E, Salmi M, León P (2009) Unravelling the regulatory mechanisms that modulate the MEP pathway in higher plants. J Exp Bot 60(10):2933–2943
17. Estévez JM, Cantero A, Reindl A, Reichler S, León P (2001) 1-Deoxy-D-xylulose-5-phosphate synthase, a limiting enzyme for plastidic isoprenoid biosynthesis in plants. J Biol Chem 276(25):22901–22909
18. Carretero-Paulet L, Cairo A, Botella-Pavía P, Besumbes O, Campos N, Boronat A, Rodríguez-Concepción M (2006) Enhanced flux through the methylerythritol 4-phosphate pathway in Arabidopsis plants overexpressing deoxyxylulose 5-phosphate reductoisomerase. Plant Mol Biol 62(4–5):683–695
19. Simpson K, Quiroz LF, Rodriguez-Concepción M, Stange CR (2016) Differential contribution of the first two enzymes of the MEP pathway to the supply of metabolic precursors for carotenoid and chlorophyll biosynthesis in carrot (Daucus carota). Front Plant Sci 7:1344
20. Chen N, Wang P, Li C, Wang Q, Pan J, Xiao F, Wang Y, Zhang K, Li C, Yang B, Sun C, Deng X (2018) A single nucleotide mutation of the IspE gene participating in the MEP pathway for Isoprenoid biosynthesis causes a green-revertible yellow leaf phenotype in Rice. Plant Cell Physiol 59 (9):1905–1917
21. Oleszkiewicz T, Klimek-Chodacka M, Milewska-Hendel A, Zubko M, Stróż D, Kurczyńska E, Boba A, Szopa J, Baranski R (2018) Unique chromoplast organisation and carotenoid gene expression in carotenoid-rich carrot callus. Planta 248(6):1455–1471
22. Ruiz-Sola MÁ, Coman D, Beck G, Barja M, Colinas M, Graf A, Welsch R, Rütimann P, Bühlmann P, Bigler L, Gruissem W, Rodriguez-Concepcion M, Vranová E (2016) Arabidopsis GERANYLGERANYL DIPHOSPHATE SYNTHASE 11 is a hub isozyme required for the production of most photosynthesis-related isoprenoids. New Phytol 209(1):252–264
23. Zhou F, Wang C-Y, Gutensohn M, Jiang L, Zhang P, Zhang D, Dudareva N, Lu S (2017) A recruiting protein of geranylgeranyl diphosphate synthase controls metabolic flux toward chlorophyll biosynthesis in rice. Proc Natl Acad Sci U S A 114(26):6866–6871

24. Camagna M, Grundmann A, Bär C, Koschmieder J, Beyer P, Welsch R (2019) Enzyme fusion removes competition for geranylgeranyl diphosphate in carotenogenesis. Plant Physiol 179(3):1013–1027
25. Li F, Tzfadia O, Wurtzel ET (2009) The phytoene synthase gene family in the grasses: subfunctionalization provides tissue-specific control of carotenogenesis. Plant Signal Behav 4(3):208–211
26. Giuliano G (2017) Provitamin a biofortification of crop plants: a gold rush with many miners. Curr Opin Biotechnol 44:169–180
27. Chen Y, Li F, Wurtzel ET (2010) Isolation and characterization of the Z-ISO gene encoding a missing component of carotenoid biosynthesis in plants. Plant Physiol 153 (1):66–79
28. Carol P, Kuntz M (2001) A plastid terminal oxidase comes to light: implications for carotenoid biosynthesis and chlororespiration. Trends Plant Sci 6(1):31–36
29. Hunter CT, Saunders JW, Magallanes-Lundback M, Christensen SA, Willett D, Stinard PS, Li QB, Lee K, DellaPenna D, Koch KE (2018) Maize w3 disrupts homogentisate solanesyl transferase (ZmHst) and reveals a plastoquinone-9 independent path for phytoene desaturation and tocopherol accumulation in kernels. Plant J 93(5):799–813
30. Beltrán J, Kloss B, Hosler JP, Geng J, Liu A, Modi A, Dawson JH, Sono M, Shumskaya M, Ampomah-Dwamena C, Love JD, Wurtzel ET (2015) Control of carotenoid biosynthesis through a heme-based cis-trans isomerase. Nat Chem Biol 11(8):598–605
31. Yu Q, Ghisla S, Hirschberg J, Mann V, Beyer P (2011) Plant carotene cis-trans isomerase crtiso a new member of the fadred-dependent flavoproteins catalyzing non-redox reactions. J Biol Chem 286(10):8666–8676
32. Bai L, Kim EH, DellaPenna D, Brutnell TP (2009) Novel lycopene epsilon cyclase activities in maize revealed through perturbation of carotenoid biosynthesis. Plant J 59 (4):588–599
33. Harjes CE, Rocheford TR, Bai L, Brutnell TP, Kandianis CB, Sowinski SG, Stapleton AE, Vallabhaneni R, Williams M, Wurtzel ET (2008) Natural genetic variation in lycopene epsilon cyclase tapped for maize biofortification. Science 319(5861):330–333
34. Kim J, Smith JJ, Tian L, DellaPenna D (2009) The evolution and function of carotenoid hydroxylases in Arabidopsis. Plant Cell Physiol 50(3):463–479
35. Jahns P, Holzwarth AR (2012) The role of the xanthophyll cycle and of lutein in photoprotection of photosystem II. Biochim Biophys Acta 1817(1):182–193
36. Neuman H, Galpaz N, Cunningham FX, Zamir D, Hirschberg J (2014) The tomato mutation nxd1 reveals a gene necessary for neoxanthin biosynthesis and demonstrates that violaxanthin is a sufficient precursor for abscisic acid biosynthesis. Plant J 78(1):80–93
37. Ruiz-Sola MÁ, Rodríguez-Concepción M (2012) Carotenoid biosynthesis in Arabidopsis: a colorful pathway. Arabidopsis Book 10: e0158
38. Al-Babili S, Jv L, Haubruck H, Beyer P (1996) A novel, soluble form of phytoene desaturase from Narcissus pseudonarcissus chromoplasts is Hsp70-complexed and competent for flavinylation, membrane association and enzymatic activation. Plant J 9 (5):601–612
39. Lopez AB, Yang Y, Thannhauser TW, Li L (2008) Phytoene desaturase is present in a large protein complex in the plastid membrane. Physiol Plant 133(2):190–198
40. Quinlan RF, Shumskaya M, Bradbury LMT, Beltrán J, Ma C, Kennelly EJ, Wurtzel ET (2012) Synergistic interactions between carotene ring hydroxylases drive lutein formation in plant carotenoid biosynthesis. Plant Physiol 160(1):204–214
41. Pulido P, Toledo-Ortiz G, Phillips MA, Wright LP, Rodríguez-Concepción M (2013) Arabidopsis J-protein J20 delivers the first enzyme of the plastidial isoprenoid pathway to protein quality control. Plant Cell 25 (10):4183–4194
42. Pulido P, Llamas E, Llorente B, Ventura S, Wright LP, Rodríguez-Concepción M (2016) Specific Hsp100 chaperones determine the fate of the first enzyme of the plastidial isoprenoid pathway for either refolding or degradation by the stromal Clp protease in Arabidopsis. PLoS Genet 12(1):e1005824
43. Zhou X, Welsch R, Yang Y, Álvarez D, Riediger M, Yuan H, Fish T, Liu J, Thannhauser TW, Li L (2015) Arabidopsis OR proteins are the major posttranscriptional regulators of phytoene synthase in controlling carotenoid biosynthesis. Proc Natl Acad Sci U S A 112(11):3558–3563
44. Park S, Kim HS, Jung YJ, Kim SH, Ji CY, Wang Z, Jeong JC, Lee H-S, Lee SY, Kwak S-S (2016) Orange protein has a role in phytoene synthase stabilization in sweetpotato. Sci Rep 6

45. Chayut N, Yuan H, Ohali S, Meir A, Sa'ar U, Tzuri G, Zheng Y, Mazourek M, Gepstein S, Zhou X, Portnoy V, Lewinsohn E, Schaffer AA, Katzir N, Fei Z, Welsch R, Li L, Burger J, Tadmor Y (2017) Distinct mechanisms of the ORANGE protein in controlling carotenoid flux. Plant Physiol 173(1):376–389

46. Welsch R, Zhou X, Yuan H, Álvarez D, Sun T, Schlossarek D, Yang Y, Shen G, Zhang H, Rodriguez-Concepcion M, Thannhauser TW, Li L (2018) Clp protease and OR directly control the proteostasis of phytoene synthase, the crucial enzyme for carotenoid biosynthesis in Arabidopsis. Mol Plant 11(1):149–162

47. Yabuzaki J (2017) Carotenoids database: structures, chemical fingerprints and distribution among organisms. Database 2017

48. Curl AL (1962) Red pepper carotenoids, the carotenoids of red bell peppers. J Agric Food Chem 10(6):504–509

49. Wang Y-Q, Yang Y, Fei Z, Yuan H, Fish T, Thannhauser TW, Mazourek M, Kochian LV, Wang X, Li L (2013) Proteomic analysis of chromoplasts from six crop species reveals insights into chromoplast function and development. J Exp Bot 64(4):949–961

50. Siefermann-Harms D, Hertzberg S, Borch G, Liaaen-Jensen S (1981) Lactucaxanthin, an ε, ε-carotene-3, 3'-diol from Lactuca sativa. Phytochemistry 20(1):85–88

51. Cunningham FX, Gantt E (2001) One ring or two? Determination of ring number in carotenoids by lycopene ε-cyclases. Proc Natl Acad Sci U S A 98(5):2905–2910

52. Phillip D, Young AJ (1995) Occurrence of the carotenoid lactucaxanthin in higher plant LHC II. Photosynth Res 43(3):273–282

53. Cunningham FX, Gantt E (2011) Elucidation of the pathway to astaxanthin in the flowers of Adonis aestivalis. Plant Cell 23 (8):3055–3069

54. Misawa N (2009) Pathway engineering of plants toward astaxanthin production. Plant Biotechnol 26(1):93–99

55. Zhu C, Naqvi S, Capell T, Christou P (2009) Metabolic engineering of ketocarotenoid biosynthesis in higher plants. Arch Biochem Biophys 483(2):182–190

56. Zhu Q, Zeng D, Yu S, Cui C, Li J, Li H, Chen J, Zhang R, Zhao X, Chen L, Liu Y-G (2018) From golden rice to aSTARice: bioengineering Astaxanthin biosynthesis in rice endosperm. Mol Plant 11(12):1440–1448

57. Campbell R, Morris WL, Mortimer CL, Misawa N, Ducreux LJM, Morris JA, Hedley PE, Fraser PD, Taylor MA (2015) Optimising ketocarotenoid production in potato tubers: effect of genetic background, transgene combinations and environment. Plant Sci 234:27–37

58. Farré G, Perez-Fons L, Decourcelle M, Breitenbach J, Hem S, Zhu C, Capell T, Christou P, Fraser PD, Sandmann G (2016) Metabolic engineering of astaxanthin biosynthesis in maize endosperm and characterization of a prototype high oil hybrid. Transgenic Res 25(4):477–489

59. Nogueira M, Enfissi EMA, Valenzuela MEM, Menard GN, Driller RL, Eastmond PJ, Schuch W, Sandmann G, Fraser PD (2017) Engineering of tomato for the sustainable production of ketocarotenoids and its evaluation in aquaculture feed. Proc Natl Acad Sci U S A 114(41):10876–10881

60. Boon CS, McClements DJ, Weiss J, Decker EA (2010) Factors influencing the chemical stability of carotenoids in foods. Crit Rev Food Sci Nutr 50(6):515–532

61. Havaux M (2014) Carotenoid oxidation products as stress signals in plants. Plant J 79 (4):597–606

62. Chedea VS, Jisaka M (2013) Lipoxygenase and carotenoids: a co-oxidation story. Afr J Biotechnol 12(20):2786–2791

63. Auldridge ME, McCarty DR, Klee HJ (2006) Plant carotenoid cleavage oxygenases and their apocarotenoid products. Curr Opin Plant Biol 9(3):315–321

64. Seo M, Koshiba T (2002) Complex regulation of ABA biosynthesis in plants. Trends Plant Sci 7(1):41–48

65. Walter MH, Floss DS, Strack D (2010) Apocarotenoids: hormones, mycorrhizal metabolites and aroma volatiles. Planta 232(1):1–17

66. Tan BC, Joseph LM, Deng WT, Liu L, Li QB, Cline K, McCarty DR (2003) Molecular characterization of the Arabidopsis 9-cis epoxycarotenoid dioxygenase gene family. Plant J 35 (1):44–56

67. Schwartz SH, Tan BC, Gage DA, Zeevaart JAD, McCarty DR (1997) Specific oxidative cleavage of carotenoids by VP14 of maize. Science 276(5320):1872–1874

68. Tan BC, Schwartz SH, Zeevaart JAD, McCarty DR (1997) Genetic control of abscisic acid biosynthesis in maize. Proc Natl Acad Sci U S A 94(22):12235–12240

69. Finkelstein R (2013) Abscisic acid synthesis and response. The Arabidopsis book. Am Society Plant Biologists 11:e0166

70. Alder A, Jamil M, Marzorati M, Bruno M, Vermathen M, Bigler P, Ghisla S, Bouwmeester H, Beyer P, Al-Babili S (2012) The path from β-carotene to carlactone, a

strigolactone-like plant hormone. Science 335(6074):1348–1351

71. Al-Babili S, Bouwmeester HJ (2015) Strigolactones, a novel carotenoid-derived plant hormone. Annu Rev Plant Biol 66:161–186
72. Alder A, Holdermann I, Beyer P, Al-Babili S (2008) Carotenoid oxygenases involved in plant branching catalyse a highly specific conserved apocarotenoid cleavage reaction. Biochem J 416(2):289–296
73. Frusciante S, Diretto G, Bruno M, Ferrante P, Pietrella M, Prado-Cabrero A, Rubio-Moraga A, Beyer P, Gomez-Gomez L, Al-Babili S (2014) Novel carotenoid cleavage dioxygenase catalyzes the first dedicated step in saffron crocin biosynthesis. Proc Natl Acad Sci U S A 111(33):12246–12251
74. Ahrazem O, Rubio-Moraga A, Berman J, Capell T, Christou P, Zhu C, Gómez-Gómez L (2016) The carotenoid cleavage dioxygenase CCD2 catalysing the synthesis of crocetin in spring crocuses and saffron is a plastidial enzyme. New Phytol 209(2):650–663
75. Amengual J, Widjaja-Adhi MAK, Rodriguez-Santiago S, Hessel S, Golczak M, Palczewski K, von Lintig J (2013) Two carotenoid oxygenases contribute to mammalian provitamin a metabolism. J Biol Chem 288 (47):34081–34096
76. Babino D, Palczewski G, Widjaja-Adhi MAK, Kiser PD, Golczak M, von Lintig J (2015) Characterization of the role of β-carotene 9, 10-dioxygenase in macular pigment metabolism. J Biol Chem 290(41):24844–24857
77. Ohmiya A, Kishimoto S, Aida R, Yoshioka S, Sumitomo K (2006) Carotenoid cleavage dioxygenase (CmCCD4a) contributes to white color formation in chrysanthemum petals. Plant Physiol 142(3):1193–1201
78. Falchi R, Vendramin E, Zanon L, Scalabrin S, Cipriani G, Verde I, Vizzotto G, Morgante M (2013) Three distinct mutational mechanisms acting on a single gene underpin the origin of yellow flesh in peach. Plant J 76(2):175–187
79. Gonzalez-Jorge S, Ha S-H, Magallanes-Lundback M, Gilliland LU, Zhou A, Lipka AE, Nguyen Y-N, Angelovici R, Lin H, Cepela J, Little H, Buell CR, Gore MA, DellaPenna D (2013) Carotenoid cleavage dioxygenase4 is a negative regulator of β-carotene content in Arabidopsis seeds. Plant Cell 25 (12):4812–4826
80. Vallabhaneni R, Bradbury LMT, Wurtzel ET (2010) The carotenoid dioxygenase gene family in maize, sorghum, and rice. Arch Biochem Biophys 504(1):104–111
81. Ilg A, Bruno M, Beyer P, Al-Babili S (2014) Tomato carotenoid cleavage dioxygenases 1A and 1B: relaxed double bond specificity leads to a plenitude of dialdehydes, mono-apocarotenoids and isoprenoid volatiles. FEBS Open Bio 4:584–593
82. Ilg A, Yu Q, Schaub P, Beyer P, Al-Babili S (2010) Overexpression of the rice carotenoid cleavage dioxygenase 1 gene in Golden Rice endosperm suggests apocarotenoids as substrates in planta. Planta 232(3):691–699
83. Adami M, De Franceschi P, Brandi F, Liverani A, Giovannini D, Rosati C, Dondini L, Tartarini S (2013) Identifying a carotenoid cleavage dioxygenase (ccd4) gene controlling yellow/white fruit flesh color of peach. Plant Mol Biol Rep 31(5):1166–1175
84. Schaub P, Wuest F, Koschmieder J, Yu Q, Virk P, Tohme J, Beyer P (2017) Non-enzymatic β-carotene degradation in (Provitamin A-biofortified) crop plants. J Agric Food Chem 65(31):6588–6598
85. Ramel F, Mialoundama AS, Havaux M (2012) Nonenzymic carotenoid oxidation and photooxidative stress signalling in plants. J Exp Bot 64(3):799–805
86. Leenhardt F, Lyan B, Rock E, Boussard A, Potus J, Chanliaud E, Remesy C (2006) Genetic variability of carotenoid concentration, and lipoxygenase and peroxidase activities among cultivated wheat species and bread wheat varieties. Eur J Agron 25(2):170–176
87. Ilg A, Beyer P, Al-Babili S (2009) Characterization of the rice carotenoid cleavage dioxygenase 1 reveals a novel route for geranial biosynthesis. FEBS J 276(3):736–747
88. Holloway GM, Gainer JL (1988) The carotenoid crocetin enhances pulmonary oxygenation. J Appl Physiol 65(2):683–686
89. Simkin AJ, Underwood BA, Auldridge M, Loucas HM, Shibuya K, Schmelz E, Clark DG, Klee HJ (2004) Circadian regulation of the PhCCD1 carotenoid cleavage dioxygenase controls emission of β-ionone, a fragrance volatile of petunia flowers. Plant Physiol 136 (3):3504–3514
90. Ibdah M, Azulay Y, Portnoy V, Wasserman B, Bar E, Meir A, Burger Y, Hirschberg J, Schaffer AA, Katzir N (2006) Functional characterization of CmCCD1, a carotenoid cleavage dioxygenase from melon. Phytochemistry 67 (15):1579–1589
91. Rubio A, Rambla JL, Santaella M, Gómez MD, Orzaez D, Granell A, Gómez-Gómez L (2008) Cytosolic and plastoglobule-targeted carotenoid dioxygenases from Crocus sativus are both involved in β-ionone release. J Biol Chem 283(36):24816–24825
92. Ma G, Zhang L, Matsuta A, Matsutani K, Yamawaki K, Yahata M, Wahyudi A,

Motohashi R, Kato M (2013) Enzymatic formation of β-citraurin from β-cryptoxanthin and zeaxanthin by carotenoid cleavage dioxygenase4 in the flavedo of citrus fruit. Plant Physiol 163(2):682–695
93. Ohmiya A, Sumitomo K, Aida R (2009) "Yellow Jimba": suppression of carotenoid cleavage dioxygenase (CmCCD4a) expression turns white chrysanthemum petals yellow. J JPN SocHortic Sci 78(4):450–455
94. Rivera-Madrid R, Aguilar-Espinosa M, Cárdenas-Conejo Y, Garza-Caligaris LE (2016) Carotenoid derivates in achiote (Bixa orellana) seeds: synthesis and health promoting properties. Front Plant Sci 7:1406
95. Hou X, Rivers J, León P, McQuinn RP, Pogson BJ (2016) Synthesis and function of apocarotenoid signals in plants. Trends Plant Sci 21(9):792–803
96. Ramel F, Birtic S, Ginies C, Soubigou-Taconnat L, Triantaphylides C, Havaux M (2012) Carotenoid oxidation products are stress signals that mediate gene responses to singlet oxygen in plants. Proc Natl Acad Sci U S A 109(14):5535–5540
97. Shumbe L, Bott R, Havaux M (2014) Dihydroactinidiolide, a high light-induced beta-carotene derivative that can regulate gene expression and photoacclimation in Arabidopsis. Mol Plant 7(7):1248–1251
98. Wang JY, Haider I, Jamil M, Fiorilli V, Saito Y, Mi J, Baz L, Kountche BA, Jia K-P, Guo X, Balakrishna A, Ntui VO, Reinke B, Volpe V, Gojobori T, Blilou I, Lanfranco L, Bonfante P, Al-Babili S (2019) The apocarotenoid metabolite zaxinone regulates growth and strigolactone biosynthesis in rice. Nat Commun 10(1):810
99. Avendano-Vazquez AO, Cordoba E, Llamas E, San Roman C, Nisar N, De la Torre S, Ramos-Vega M, Gutierrez-Nava MD, Cazzonelli CI, Pogson BJ, Leon P (2014) An uncharacterized Apocarotenoid-derived signal generated in zeta-carotene Desaturase mutants regulates leaf development and the expression of chloroplast and nuclear genes in Arabidopsis. Plant Cell 26 (6):2524–2537
100. Van Norman JM, Zhang J, Cazzonelli CI, Pogson BJ, Harrison PJ, Bugg TDH, Chan KX, Thompson AJ, Benfey PN (2014) Periodic root branching in Arabidopsis requires synthesis of an uncharacterized carotenoid derivative. Proc Natl Acad Sci U S A 111 (13):1300–1309
101. Kachanovsky DE, Filler S, Isaacson T, Hirschberg J (2012) Epistasis in tomato color mutations involves regulation of phytoene synthase 1 expression by cis-carotenoids. Proc Natl Acad Sci U S A 109(46):19021–19026
102. Walter MH, Stauder R, Tissier A (2015) Evolution of root-specific carotenoid precursor pathways for apocarotenoid signal biogenesis. Plant Sci 233:1–10
103. Vogel JT, Walter MH, Giavalisco P, Lytovchenko A, Kohlen W, Charnikhova T, Simkin AJ, Goulet C, Strack D, Bouwmeester HJ, Fernie AR, Klee HJ (2010) SlCCD7 controls strigolactone biosynthesis, shoot branching and mycorrhiza-induced apocarotenoid formation in tomato. Plant J 61(2):300–311
104. Akiyama K, K-i M, Hayashi H (2005) Plant sesquiterpenes induce hyphal branching in arbuscular mycorrhizal fungi. Nature 435 (7043):824
105. Gómez-Gómez L, Parra-Vega V, Rivas-Sendra A, Seguí-Simarro JM, Molina RV, Pallotti C, Rubio-Moraga Á, Diretto G, Prieto A, Ahrazem O (2017) Unraveling massive Crocins transport and accumulation through proteome and microscopy tools during the development of saffron stigma. Int J Mol Sci 18(1):76
106. Louro RP, Santiago LJ (2016) Development of carotenoid storage cells in Bixa orellana L. seed arils. Protoplasma 253(1):77–86
107. Abe S, Sado A, Tanaka K, Kisugi T, Asami K, Ota S, Kim HI, Yoneyama K, Xie X, Ohnishi T, Seto Y, Yamaguchi S, Akiyamab K, Yoneyama K, Nomura T (2014) Carlactone is converted to carlactonoic acid by MAX1 in Arabidopsis and its methyl ester can directly interact with AtD14 in vitro. Proc Natl Acad Sci U S A 111 (50):18084–18089
108. Seto Y, Sado A, Asami K, Hanada A, Umehara M, Akiyama K, Yamaguchi S (2014) Carlactone is an endogenous biosynthetic precursor for strigolactones. Proc Natl Acad Sci U S A 111(4):1640–1645
109. Grunewald K, Hagen C (2000) Extrusion of secondary carotenoid containing vesicles from flagellates of Haematococcus pluvialis (Volvocales; Chlorophyceae). JApplBotany-Angewandte Botanik 74(3–4):141–144
110. Mohiuddin T, Baba SA, Ashraf N (2018) Identification, phylogenetic analysis and expression profiling of ABC transporter family of Crocus sativus L: a step towards understanding apocarotenoid transport. Plant Gene 14:1–6
111. Demurtas OC, Frusciante S, Ferrante P, Diretto G, Azad NH, Pietrella M, Aprea G, Taddei AR, Romano E, Mi J, Al-Babili S, Frigerio L, Giuliano G (2018) Candidate

enzymes for saffron crocin biosynthesis are localized in multiple cellular compartments. Plant Physiol 177(3):990–1006
112. Cazzonelli CI, Pogson BJ (2010) Source to sink: regulation of carotenoid biosynthesis in plants. Trends Plant Sci 15(5):266–274
113. Egea I, Barsan C, Bian W, Purgatto E, Latché A, Chervin C, Bouzayen M, Pech J-C (2010) Chromoplast differentiation: current status and perspectives. Plant Cell Physiol 51 (10):1601–1611
114. Li L, Yuan H (2013) Chromoplast biogenesis and carotenoid accumulation. Arch Biochem Biophys 539(2):102–109
115. Schweiggert RM, Carle R (2015) Carotenoid deposition in plant and animal foods and its impact on bioavailability. Crit Rev Food Sci Nutr 57(9):1807–1830
116. Lado J, Zacarías L, Gurrea A, Page A, Stead A, Rodrigo MJ (2015) Exploring the diversity in Citrus fruit colouration to decipher the relationship between plastid ultrastructure and carotenoid composition. Planta 242 (3):645–661
117. Hempel J, Amrehn E, Quesada S, Esquivel P, Jimenez VM, Heller A, Carle R, Schweiggert RM (2014) Lipid-dissolved gamma-carotene, beta-carotene, and lycopene in globular chromoplasts of peach palm (Bactris gasipaes Kunth) fruits. Planta 240(5):1037–1050
118. Yuan H, Zhang J, Nageswaran D, Li L (2015) Carotenoid metabolism and regulation in horticultural crops. Hortic Res 2:15036
119. Beisel KG, Jahnke S, Hofmann D, Köppchen S, Schurr U, Matsubara S (2010) Continuous turnover of carotenes and chlorophyll a in mature leaves of Arabidopsis revealed by 14CO2 pulse-chase labeling. Plant Physiol 152(4):2188–2199
120. Egea I, Bian W, Barsan C, Jauneau A, Pech J-C, Latché A, Li Z, Chervin C (2011) Chloroplast to chromoplast transition in tomato fruit: spectral confocal microscopy analyses of carotenoids and chlorophylls in isolated plastids and time-lapse recording on intact live tissue. Ann Bot 108(2):291–297
121. Li L, Paolillo DJ, Parthasarathy MV, DiMuzio EM, Garvin DF (2001) A novel gene mutation that confers abnormal patterns of β-carotene accumulation in cauliflower (Brassica oleracea var. botrytis). Plant J 26 (1):59–67
122. Lu S, Van Eck J, Zhou X, Lopez AB, O'Halloran DM, Cosman KM, Conlin BJ, Paolillo DJ, Garvin DF, Vrebalov J, Kochian LV, Küpper H, Earle ED, Cao J, Li L (2006) The cauliflower or gene encodes a DnaJ cysteine-rich domain-containing protein that mediates high levels of β-carotene accumulation. Plant Cell 18(12):3594–3605
123. Tzuri G, Zhou X, Chayut N, Yuan H, Portnoy V, Meir A, Sa'ar U, Baumkoler F, Mazourek M, Lewinsohn E, Fei Z, Schaffer AA, Li L, Burger J, Katzir N, Tadmor Y (2015) A 'golden'SNP in CmOr governs the fruit flesh color of melon (Cucumis melo). Plant J 82(2):267–279
124. Lopez AB, Van Eck J, Conlin BJ, Paolillo DJ, O'Neill J, Li L (2008) Effect of the cauliflower or transgene on carotenoid accumulation and chromoplast formation in transgenic potato tubers. J Exp Bot 59(2):213–223
125. Yuan H, Owsiany K, Sheeja TE, Zhou X, Rodriguez C, Li Y, Welsch R, Chayut N, Yang Y, Thannhauser TW, Parthasarathy MV, Xu Q, Deng X, Fei Z, Ari Schaffer NK, Burger J, Tadmor Y, Li L (2015) A single amino acid substitution in an ORANGE protein promotes carotenoid overaccumulation in Arabidopsis. Plant Physiol 169 (1):421–431
126. Yazdani M, Sun Z, Yuan H, Zeng S, Thannhauser TW, Vrebalov J, Ma Q, Xu Y, Fei Z, Van Eck J, Tian S, Tadmor Y, Giovannoni JJ, Li L (2019) Ectopic expression of ORANGE promotes carotenoid accumulation and fruit development in tomato. Plant Biotechnol J 17(1):33–49
127. Paolillo DJ, Garvin DF, Parthasarathy MV (2004) The chromoplasts of or mutants of cauliflower (Brassica oleracea L. var. botrytis). Protoplasma 224(3):245–253
128. Neta-Sharir I, Isaacson T, Lurie S, Weiss D (2005) Dual role for tomato heat shock protein 21: protecting photosystem II from oxidative stress and promoting color changes during fruit maturation. Plant Cell 17 (6):1829–1838
129. Kahlau S, Bock R (2008) Plastid transcriptomics and translatomics of tomato fruit development and chloroplast-to-chromoplast differentiation: chromoplast gene expression largely serves the production of a single protein. Plant Cell 20(4):856–874
130. Newman LA, Hadjeb N, Price CA (1989) Synthesis of two chromoplast-specific proteins during fruit development in Capsicum annuum. Plant Physiol 91(2):455–458
131. Pozueta-Romero J, Rafia F, Houlne G, Cheniclet C, Carde J-P, Schantz M-L, Schantz R (1997) A ubiquitous plant housekeeping gene, PAP, encodes a major protein component of bell pepper chromoplasts. Plant Physiol 115(3):1185–1194

132. Vishnevetsky M, Ovadis M, Itzhaki H, Levy M, Libal-Weksler Y, Adam Z, Vainstein A (1996) Molecular cloning of a carotenoid-associated protein from Cucumis sativus corollas: homologous genes involved in carotenoid sequestration in chromoplasts. Plant J 10 (6):1111–1118

133. Hugueney P, Bouvier F, Badillo A, d'Harlingue A, Kuntz M, Camara B (1995) Identification of a plastid protein involved in vesicle fusion and/or membrane protein translocation. Proc Natl Acad Sci U S A 92 (12):5630–5634

Part II

Activities

Chapter 2

A Simple In Vitro Assay to Measure the Activity of Geranylgeranyl Diphosphate Synthase and Other Short-Chain Prenyltransferases

M. Victoria Barja and Manuel Rodríguez-Concepción

Abstract

Most carotenoids are C40 metabolites produced from C20 geranylgeranyl diphosphate (GGPP). The enzymes that produce this precursor, GGPP synthases (GGPPS), are members of the short-chain prenyltransferase (SC-PT) family. SC-PTs are enzymes that catalyze the sequential head-to-tail addition of one or more C5 molecules of isopentenyl diphosphate (IPP) to dimethylallyl diphosphate (DMAPP) with the concomitant release of pyrophosphate (PPi). SC-PTs produce linear isoprenyl diphosphates of up to C20 (GGPP) that serve as precursors for many groups of isoprenoids with a wide range of essential biological functions in Eucarya, Bacteria, and Archaea. Enzymatic analysis of SC-PT activity normally requires complex, laborious and expensive methods such as radioactivity-based assays or liquid chromatography-–mass spectrometry (LC-MS). Here we describe a fast and inexpensive spectrophotometric protocol for determining the kinetic parameters of SC-PTs in purified enzyme preparations, using an adapted assay for PPi quantification. We developed the method using the *Arabidopsis thaliana* GGPPS11 enzyme, which produces geranylgeranyl diphosphate for the synthesis of carotenoids in the chloroplast.

Key words Short-chain prenyltransferases, Isoprenyl diphosphate synthases, Enzymatic activity, Kinetic parameters, EnzChek kit, Geranylgeranyl diphosphate synthase, GGPP, Isoprenoids, Carotenoids

1 Introduction

Isoprenoids (also known as terpenoids) are a widely diverse group of natural metabolites, and many of them are essential in all-living organisms. Particularly, plant isoprenoids show an enormous structural and functional diversity. Plants produce a myriad of specialized isoprenoids involved in their interaction with the environment, but also a number of essential isoprenoids that participate in photosynthesis (chlorophylls, carotenoids, tocopherols, prenylquinones), respiration (ubiquinone), or growth regulation (cytokinins, brassinosteroids, gibberellins, abscisic acid, strigolactones) [1]. All

Manuel Rodríguez-Concepción and Ralf Welsch (eds.), *Plant and Food Carotenoids: Methods and Protocols*,
Methods in Molecular Biology, vol. 2083, https://doi.org/10.1007/978-1-4939-9952-1_2,

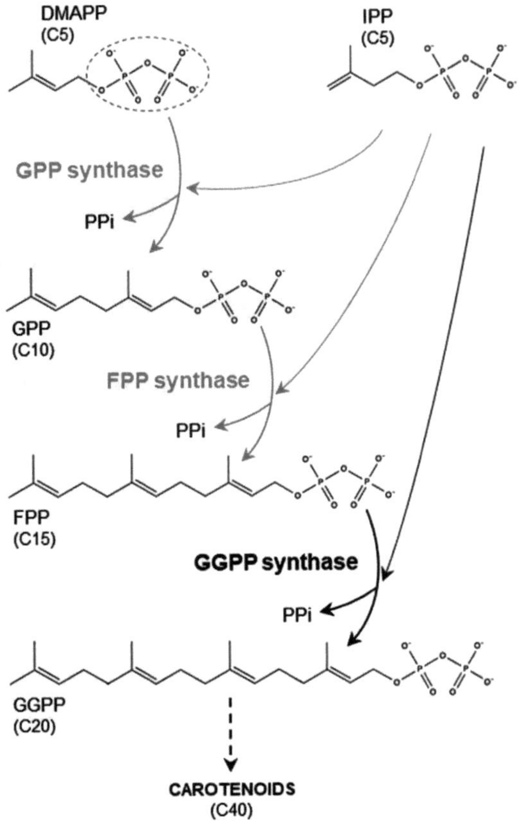

Fig. 1 Biosynthesis of short-chain prenyl diphosphates by short-chain prenyltransferases. Each IPP condensation involves the elongation of the prenyl diphosphate molecule with the subsequent release of one PPi molecule (dashed circle). Solid arrows represent one enzymatic step and dashed arrows indicate multiple enzymatic steps. Abbreviations: *DMAPP* dimethylallyl diphosphate, *IPP* isopentenyl diphosphate, *PPi* inorganic pyrophosphate, *GPP* geranyl diphosphate, *FPP* farnesyl diphosphate, *GGPP* geranylgeranyl diphosphate, *GPPS* geranyl diphosphate synthase, *FPPS* farnesyl diphosphate synthase, *GGPP* geranylgeranyl diphosphate synthase

isoprenoids derive from the C5 units isopentenyl diphosphate (IPP) and its double-bond isomer dimethylallyl diphosphate (DMAPP) (Fig. 1). Condensation of one or several IPP molecules to one DMAPP unit catalyzed by isoprenyl diphosphate synthases (IDSs), also called prenyl transferases (PTs), produces linear isoprenyl diphosphates of different chain length that represent the first intermediates of nearly all isoprenoid groups. IDS enzymes can be classified as *trans*- or *cis*-PTs depending on the stereochemistry of the double bonds formed in the synthesized product, forming two evolutionary distinct groups of proteins [2, 3]. With some exceptions [4–8], most *trans*-PTs generate isoprenyl diphosphates of up to C50, whereas *cis*-PTs synthesize much longer molecules.

Among the *trans*-PTs, short-chain prenyltransferases (SC-PTs) catalyze the sequential head-to-tail addition of up to three molecules of IPP to one molecule of DMAPP producing C10 geranyl diphosphate (GPP), C15 farnesyl diphosphate (FPP) or C20 geranylgeranyl diphosphate (GGPP) [2, 3] (Fig. 1). This enzymatic reaction requires two highly conserved domains, referred to as first aspartate-rich motif (FARM, $DDx_{2-4}D$) and second aspartate-rich motif (SARM, DDx_2D), that are involved in substrate binding [9–14] and catalysis using Mg^{2+} as cofactor [15]. The sequential addition of IPP molecules to the isoprenyl diphosphate product releases a pyrophosphate (PPi) molecule per IPP unit added. This process takes place in the elongation pocket of the SC-PT, a hydrophobic cavity that controls the hydrocarbon tail length by the size of the side chain of some amino acid residues located on the N-terminal side of the FARM. In most cases, the fifth amino acid upstream to this motif is the responsible of the isoprenyl product chain length [16–22]. However, other residues could also play a role in controlling the number of IPP condensations during product elongation [3].

Despite all the studies highlighting the role of the elongation pocket on final product length, predictions are difficult and highly sensitive analytical methods are still required to determine the actual product of uncharacterized SC-PTs. Radioactive enzymatic assays followed by the hydrolysis of the products require specialized equipment for the measurements such as radio-gas chromatography (radio-GC), radio-high-performance liquid chromatography (radio-HPLC), thin layer chromatography (TLC), or liquid scintillation counting (LSC) [22–28]. These methods, however, often fail to conclusively demonstrate the chain length (i.e., identity) of the product. Moreover, they are extremely time-consuming when used to calculate the enzymatic kinetics of the tested SC-PT. Nonradioactive methods based on liquid chromatography coupled with tandem mass spectrometry (LC–MS/MS) have been developed to accurately determine isoprenyl diphosphate metabolites in a much more precise and faster way [29–31]. However, they are still complex and expensive for the enzymatic characterization of SC-PTs (i.e., for the calculation of their optimal pH or kinetic parameters such as V_{max} and K_m).

Here we describe an easy and inexpensive spectrophotometric protocol to enzymatically characterize previously identified SC-PTs using purified enzyme preparations. The described method is an adaptation of the commercial *EnzChek Pyrophosphate Assay Kit (E-6645)*, that allows the detection of the PPi released by a biochemical reaction. The PPi produced in the reaction mix is converted into two molecules of inorganic phosphate (Pi) by an inorganic pyrophosphatase. The Pi then reacts with the substrate 2-amino-6-mercapto-7-methylpurine ribonucleoside (MESG) by a purine nucleoside phosphorylase (PNP) producing ribose

1-phosphate and 2-amino-6-mercapto-7-methyl-purine, which can be measured at 360 nm, hence allowing for the quantification of the PPi released during the enzymatic reaction.

In this article, the method is used to measure the activity of the *Arabidopsis thaliana* GGPPS11 protein (At4g36810), a GGPP synthase (EC 2.5.1.29) that supplies GGPP for the production of carotenoids and other groups of plastidial isoprenoids [32]. This enzyme has been well characterized in previous studies [21, 31, 33], which served as a reference to validate the results obtained with the reported assay.

2 Materials

1. Reaction buffer (20×): 1 M Tris–HCl, pH 7.5, 20 mM $MgCl_2$ (*see* **Note 1**).
2. Solid Tris Base. Tris(hydroxymethyl)aminomethane.
3. Hydrogen chloride (HCl) 37%.
4. Solid magnesium dichloride ($MgCl_2$).
5. 2-Amino-6-mercapto-7-methylpurine ribonucleoside (MESG): 1 mM in milli-Q water (store at −20 °C) (*see* **Note 2**).
6. Purine nucleoside phosphorylase (PNP): 100 U/mL in milli-Q water (store at 4 °C) (*see* **Note 3**).
7. Inorganic pyrophosphatase: 3 U/mL in 1× Reaction buffer (store at 4 °C) (*see* **Note 4**).
8. PPi standard: 500 μM in milli-Q water (*see* **Note 5**).
9. IPP and DMAPP (Echelon Biosciences Inc.): 1 mM in milli-Q water (store at −80 °C) (*see* **Note 6**).
10. Eppendorf Safe-Lock Tubes™ of 1.5 and 2 mL capacity.
11. Sterile polystyrene 96 well-plates (costar®) with low evaporation lid and flat bottom.
12. SpectraMax M3 multi-mode microplate reader (Molecular Devices).
13. GraphPad Prism Software.

3 Methods

3.1 Standard GGPPS Activity Assay

All 200 μL enzyme reactions should be carried out in 96-well plates (Fig. 2). The empty wells can be reused for subsequent experiments. Always perform three technical replicates per condition and at least two independent experiments.

1. Prepare the reaction mixture (Table 1) by adding 10 μL of 20× reaction buffer (*see* **Note 7**), 40 μL of 1 mM MESG substrate,

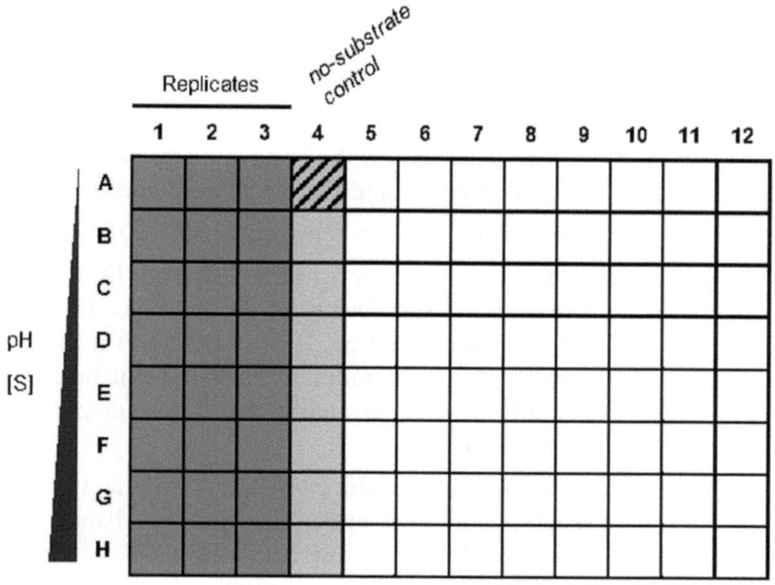

Fig. 2 Experimental design in a 96-well plate format. Black triangle indicates increasing pH or substrate concentration from top to bottom. Dark grey squares (columns 1 to 3) represent standard enzymatic reactions in triplicates per condition. Light grey squares (column 4) indicate no-substrate control. Perform a no-substrate control whenever a reagent is changed among conditions. If only the concentration of a reagent changes but the reagent solution is the same, only one no-substrate control is needed (dashed square)

Table 1
Reagents and volumes to prepare a standard 200 μL GGPPS enzyme reaction

Reagent	Stock concentration	Assay concentration	Volume added
20× reaction buffer[a]	1 M Tris–HCl	50 mM Tris–HCl	10 μL[a]
	20 mM $MgCl_2$	1 mM $MgCl_2$	
	pH 7.5	pH 7.5	
MESG	1 mM	0.2 mM	40 μL
PNP	100 U/mL	1 U/mL	2 μL
Inorganic pyrophosphatase	3 U/mL	0.03 U/mL	2 μL
IPP[b]	1 mM	150 μM	18 μL
DMAPP[b]	1 mM	50 μM	9 μL
Enzyme[c]			3–5 μg[c]
Water			Up to 200 μL

[a]Volume of Reaction buffer used depends on the stock solution concentration (*see* **Note 7**)
[b]Volume of IPP and DMAPP will change when performing the assays for the determination of the kinetic parameters (*see* **Note 12**)
[c]Volume of enzyme will depend on the concentration of the purified enzyme

2 μL of 100 U/mL PNP and 2 μL of the 3 U/mL inorganic pyrophosphatase (*see* **Note 8**). Add enough water to reach a final total volume of 200 μL minus the volumes of the enzyme, IPP and DMAPP that will be added later (Table 1).

2. Add 5 μg of the purified enzyme (*see* **Notes 9** and **10**) to the reaction mixture and mix well by up and down pipetting.
3. Preincubate at room temperature for 10 min (*see* **Note 11**).
4. Add the enzyme substrates IPP and DMAPP as shown in Table 1 for a standard reaction (*see* **Note 12**) and mix well by up and down pipetting. It is very important to always perform a control assay in parallel in which solvent buffer will be added instead of the experimental substrates (*no-substrate control*, *see* **Note 13**, Fig. 2).
5. Immediately introduce the plate in the spectrophotometer, shake it and start measuring the absorbance at 360 nm as a function of time during 1 h at 25 °C. Reading the absorbance every 2 min should be enough to build the activity curve (*see* **Note 14**).
6. When analyzing the data, remember subtracting the *no-substrate control* values from the corresponding experimental samples (*see* **Note 15**).
7. Calculate the enzyme activity obtained for each condition as the increase of PPi production per minute and μg of enzyme (μM PPi min^{-1} μg^{-1}) (*see* **Note 16**). Use them to build a pH curve (*see* **Note 17**) or to obtain the kinetic parameters of the enzyme (V_{max} and K_m) from the Michaelis–Menten curve using the GraphPad Prism software (*see* **Note 18**).

3.2 Standard Curve for PPi Quantification

The linear range for the quantification of PPi using the EnzChek Kit is from 1 μM to 75 μM.

1. Follow the standard reaction (Table 1) adding increasing amounts of the 500 μM PPi standard solution (*see* **Note 19**) and omitting the volumes of the substrates (IPP and DMAPP) and the experimental enzyme.
2. Prepare a *no-PPi control* without PPi (0 μM point, *see* **Note 19**). Also, include a *no-pyrophosphatase control* as a blank with no PPi and no inorganic pyrophosphatase (*see* **Note 11**).
3. After mixing all the reagents, incubate the plate at 25 °C for 60 min (*see* **Note 20**).
4. Measure the absorbance at 360 nm.
5. Subtract the *no-pyrophosphatase control* absorbance value from each experimental reaction and build the PPi standard curve plotting the absorbance at 360 nm as a function of PPi concentration.

4 Notes

1. The 20× Reaction buffer provided in the *EnzChek* Kit contains 2 mM of sodium azide, that acts as a preservative. The Kit allows to perform the activity assay over a pH range of 6.5–8.5, but in this case the buffer must be prepared with the pH of interest (here, sodium azide can be omitted). To generate a pH curve, we recommend to prepare a 20× stock of Reaction buffer at pH 9.5. Then, distribute it in 5 mL aliquots (as many as pH values to be tested). Adjust the pH of every aliquot to the desired value with 37% HCl and then add milli-Q water up to 10 mL to get 10× aliquots of each pH. Remember that the volume of these 10× aliquots added per activity reaction must be doubled (as the recipe in Table 1 contains a 20× Reaction buffer).
2. The Kit includes 6.3 mg (20 μmol) of MESG. Add 20 mL of milli-Q water directly to the container to prepare a 1 mM stock. Immediately, aliquot the homogenized MESG solution and store the aliquots at −20 °C. As each individual reaction requires 40 μL of MESG substrate, and normally triplicates will be performed per condition, we recommend to prepare both 200 μL and 500 μL aliquots. Immediately before use, thaw the required aliquot of MESG at 37 °C (no more than 5 min), vortex strongly and place on ice. If more than one aliquot is needed, we recommend to mix all of them together and use the same solution for all reactions. As described in the manufacturer's instructions, MESG solution is stable on ice at least for 4 h at pH 7.5. Be aware that the half-life of this substrate may change in different conditions. It is not recommended to freeze and reuse MESG leftovers.
3. The Kit provides two vials of 50 U of freeze-dried PNP. To prepare a 100 U/mL stock solution add 0.5 mL of milli-Q water to one of the vials. This solution can be stored at 4 °C for at least 1 month. The nonreconstituted PNP vial may be stored at −20 °C.
4. The Kit contains one vial of 6 U of lyophilized inorganic pyrophosphatase. Add 200 μL of milli-Q water to the vial to prepare a stock of 30 U/mL and store it at 4 °C. Before performing the activity assay, prepare a 3 U/mL aliquot by diluting tenfold the 30 U/mL stock into 1× Reaction buffer (previously diluted from 20× Reaction buffer). Never dilute the inorganic pyrophosphatase into 20× reaction buffer. Leftover 3 U/mL solution can be stored at 4 °C for 1 week.
5. The Kit provides 500 μL of 50 mM $Na_4P_2O_7$ (with 2 mM sodium azide) as a source of PPi. Prepare a 500 μM working solution of PPi standard by diluting 100-fold a portion of the

given stock in milli-Q water. Preparing 500 μL of working solution should be enough for a regular standard curve (*see* **Note 19**).

6. IPP and DMAPP are provided by Echelon Biosciences Inc. as Tris-ammonium salts. Prepare 4 mM stock solutions in milli-Q water, dispense in 50 μL aliquots and store at −80 °C. Prior to use, dilute an aliquot of these substrates in water to prepare 1 mM working solutions. Leftover 1 mM solutions can be frozen again and stored at −80 °C.
7. This volume can change when testing different pH reaction buffers. If 10× buffers are used instead of 20×, 20 μL should be added.
8. Depending on the number of reactions, a mix of the first four reagents can be prepared.
9. The amount of purified enzyme added to the reaction mixture should be determined empirically. It is important that the release of PPi by the enzyme tested does not surpass the activity of the inorganic pyrophosphatase. In our case, 3 μg of the *Arabidopsis* GGPPS11 enzyme also worked well in the assays.
10. *Arabidopsis* GGPPS11 was purified using the pET-G11 construct [31], which harbors a version of the enzyme lacking the plastid targeting peptide and fused to an N-terminal 6x-histidine tag. The recombinant protein was produced in the *E. coli* Rosetta 2 (DE) strain (Novagen, Merck KGaA, Darmstadt, Germany). The cells were grown at 37 °C in 100 mL of 2xYT medium supplemented with the corresponding antibiotics. When they reached an OD_{600} of 0.5–0.8, 1 mM IPTG was added. After the induction, the culture was incubated over night at 18 °C and bacterial cells were then recovered by centrifugation. The pellet was resuspended in 10 mL of buffer A (100 mM Tris–HCl pH 7.5, 100 mM NaCl, 10% glycerol) supplemented with 1 mM DTT, 1 mg/mL lysozyme and one tablet of complete protease inhibitor cocktail (Roche), and incubated in ice for 20 min. After a brief sonication (five pulses of 30 s, 17%), the cell lysate was centrifuged at 12,500 rpm at 4 °C for 20 min. Then 2 μg/mL DNase I and 10 μg/mL RNase A were added directly to the tube and gently mixed. The mixture was incubated 20 min in ice and then centrifuged again at 17,500 rpm at 4 °C for 50 min. The supernatant was filtered using a 0.2 μm filter and incubated for 2 h with 2 mL of nickel-nitrilotriacetic acid (Ni-NTA) beads (Qiagen) previously equilibrated with washing buffer (buffer A + 20 mM imidazole). After the incubation, the mixture was placed into a filter column where the flow through is discarded and the Ni-NTA beads are stacked with the recombinant protein attached. To remove the unspecific proteins retained in the column five

washes of 1 mL of washing buffer were performed, and finally the protein was recovered in one tube by five elutions with 150 μL of elution buffer (buffer A + 150 mM imidazole). The eluted sample was desalted by Thermo Scientific™ Zeba™ Desalting Columns using buffer A as exchange buffer and the purified protein was quantified, aliquoted with glycerol 40% and finally stored at −20 °C.

11. Reagents and enzymes may be contaminated with Pi which will interfere with the measurements. The reagents should be tested in a standard reaction (Table 1) with no inorganic pyrophosphatase (*no-pyrophosphatase control*). If present, contaminating Pi should be subtracted out from the experimental reaction measurements. In addition to the reagents, the purified enzyme may also show Pi contamination that could prevent the obtainment of conclusive data. To reduce the contaminating Pi, the enzyme sample can be preincubated for 10 min before the addition of the substrates (e.g., IPP and DMAPP). In this preincubation step, the reaction catalyzed by the PNP can act as a "Pi mop" lowering the contaminating Pi to submicromolar levels. It is recommended to always carry out this step to avoid any possible Pi interference.

12. The concentration of IPP and DMAPP was empirically calculated to observe an increase of enzymatic activity throughout the time using 3 μg of enzyme. We add threefold more IPP than DMAPP since every GGPP molecule is formed from three molecules of IPP and only one of DMAPP. The concentration and proportion of the substrates may be determined empirically for other SC-PTs.

    For the generation of *pH curves* we used the amount of substrates specified in Table 1.

    For *kinetic parameters determination assays*, the concentration of the substrates will change and so will their required volumes. We calculated the basic kinetic parameters (V_{max} and K_m) for each substrate separately, designing the plate to test different concentrations of one of the substrates fixing the other one in 100 μM (Fig. 2). The range of concentrations used to calculate the kinetics of the enzyme for each substrate was: 5, 10, 20, 50, 75, 100, and 200 μM.

13. The *no-substrate control* must be done whenever a reagent of the *Standard Reaction* changes. For example, when testing the activity of the experimental enzyme in the presence of different pH media, a *no-substrate control* must be included for each pH (column 4 of the Fig. 2, light grey). However, for kinetic parameters determination assays, only one *no-substrate control* is needed (well A4 of the Fig. 2) since only the volume of the substrates changes.

14. The assay may require to set up the measuring time points to establish the linearity region of PPi production versus time.
15. The *no-substrate control* absorbance value should not increase during the one-hour assay, as an indication that there is not too much Pi or PPi contamination that could be transformed after the 10 min "Pi-mop" preincubation. When the absorbance of the *no-substrate control* is stable, the mean of the values obtained in each time point can be calculated and subtracted from the values of the experimental reactions.
16. Use the linear equation from a PPi standard curve (*see* Subheading 3.2) to quantify the PPi produced in each time point and, after plotting it, select the linear range of PPi production to calculate the enzyme activity (μM PPi min^{-1} μg^{-1}) for each condition.
17. To build the pH curve represent the enzyme activities versus the pH. Relative values can be calculated after giving a 100% to the optimal performance.
18. Other softwares may be used to build the Michaelis–Menten curve and calculate the enzyme kinetics. Using Prism, we created an XY table selecting "Enzyme kinetics—Michaelis–Menten" as sample data and added the triplicates of the enzyme activities obtained for each substrate concentration. To obtain the regression curve and the kinetic parameters from the generated graph we performed a "Nonlinear regression" analysis selecting "Enzyme Kinetics—Substrate vs. Velocity" and "Michaelis–Menten equation." As a result of the analysis, the regression curve is superimposed on the graph and a table with the values of V_{max} and K_m together with the statistical parameters is retrieved.
19. A standard curve with nine different PPi concentrations (0, 5, 10, 20, 30, 40, 50, 60, 75 μM) is enough to obtain the linear regression equation with a square of the Pearson correlation coefficient (R^2) > 0.99. Prepare at least three replicates per PPi concentration to build the standard curve.
20. The manufacturer's instructions indicate an incubation of 30–60 min, enough time to transform all the PPi in the reaction mixture into 2-amino-6-mercapto-7-methylpurine. We incubated the *Standard Curve* 60 min adjusting the time to the *Standard Reaction* for GGPPS activity measurement. Around an hour is needed to (1) synthesize the GGPP and release the PPi molecules and then to (2) consume the PPi by the kit enzymes.

References

1. Rodríguez-Concepción M, Boronat A (2015) Breaking new ground in the regulation of the early steps of plant isoprenoid biosynthesis. Curr Opin Plant Biol 25:17–22. https://doi.org/10.1016/j.pbi.2015.04.001
2. Liang PH, Ko TP, Wang AHJ (2002) Structure, mechanism and function of prenyltransferases. Eur J Biochem 269:3339–3354. https://doi.org/10.1046/j.1432-1033.2002.03014.x
3. Vandermoten S, Haubruge É, Cusson M (2009) New insights into short-chain prenyltransferases: structural features, evolutionary history and potential for selective inhibition. Cell Mol Life Sci 66:3685–3695. https://doi.org/10.1007/s00018-009-0100-9
4. Ambo T, Noike M, Kurokawa H, Koyama T (2008) Cloning and functional analysis of novel short-chain cis-prenyltransferases. Biochem Biophys Res Commun 375:536–540. https://doi.org/10.1016/j.bbrc.2008.08.057
5. Sallaud C, Rontein D, Onillon S, Jabes F, Duffe P, Giacalone C, Thoraval S, Escoffier C, Herbette G, Leonhardt N, Causse M, Tissier A (2009) A novel pathway for sesquiterpene biosynthesis from Z,Z-farnesyl pyrophosphate in the wild tomato Solanum habrochaites. Plant Cell 21:301–317. https://doi.org/10.1105/tpc.107.057885
6. Schilmiller AL, Schauvinhold I, Larson M, Xu R, Charbonneau AL, Schmidt A, Wilkerson C, Last RL, Pichersky E (2009) Monoterpenes in the glandular trichomes of tomato are synthesized from a neryl diphosphate precursor rather than geranyl diphosphate. Proc Natl Acad Sci 106:10865–10870. https://doi.org/10.1073/pnas.0904113106
7. Hsieh F-L, Chang T-H, Ko T-P, Wang AH-J (2011) Structure and mechanism of an Arabidopsis medium/long-chain-length prenyl pyrophosphate synthase. Plant Physiol 155:1079–1090. https://doi.org/10.1104/pp.110.168799
8. Akhtar TA, Matsuba Y, Schauvinhold I, Yu G, Lees HA, Klein SE, Pichersky E (2013) The tomato cis-prenyltransferase gene family. Plant J 73:640–652. https://doi.org/10.1111/tpj.12063
9. Marrero PF, Poulter CD, Edwards PA (1992) Effects of site-directed mutagenesis of the highly conserved aspartate residues in domain II of farnesyl diphosphate synthase activity. J Biol Chem 267:21873–21878
10. Joly A, Edwards PA (1993) Effect of site-directed mutagenesis of conserved aspartate and arginine residues upon farnesyl diphosphate synthase activity. J Biol Chem 268:26983–26989
11. Tarshis LC, Yan M, Poulter CD, Sacchettini JC (1994) Crystal structure of recombinant farnesyl diphosphate synthase at 2.6-ANG. resolution. Biochemistry 33:10871–10877. https://doi.org/10.1021/bi00202a004
12. Song L, Poulter CD (1994) Yeast farnesyl-diphosphate synthase: site-directed mutagenesis of residues in highly conserved prenyltransferase domains I and II. Proc Natl Acad Sci U S A 91:3044–3048
13. Koyama T, Tajima M, Sano H, Doi T, Koike-Takeshita A, Obata S, Nishino T, Ogura K (1996) Identification of significant residues in the substrate binding site of Bacillus stearothermophilus farnesyl diphosphate synthase. Biochemistry 35:9533–9538. https://doi.org/10.1021/bi960137v
14. Koyama T, Gotoh Y, Nishino T (2000) Intersubunit location of the active site of farnesyl diphosphate synthase: reconstruction of active enzymes by hybrid-type heteromeric dimers of site-directed mutants. Biochemistry 39:463–469
15. Aaron JA, Christianson DW (2010) Trinuclear metal clusters in catalysis by terpenoid synthases. Pure Appl Chem 82:1585–1597. https://doi.org/10.1351/PAC-CON-09-09-37
16. Ohnuma S, Hirooka K, Hemmi H, Ishida C, Ohto C, Nishino T (1996) Conversion of product specificity of archaebacterial geranylgeranyl-diphosphate synthase. J Biol Chem 271:18831–18837. https://doi.org/10.1074/jbc.271.31.18831
17. Ohnuma SI, Narita K, Nakazawa T, Ishida C, Takeuchi Y, Ohto C, Nishino T (1996) A role of the amino acid residue located on the fifth position before the first aspartate-rich motif of farnesyl diphosphate synthase on determination of the final product. J Biol Chem 271:30748–30754
18. Tarshis LC, Proteau PJ, Kellogg BA, Sacchettini JC, Poulter CD (1996) Regulation of product chain length by isoprenyl diphosphate synthases. Proc Natl Acad Sci 93:15018–15023. https://doi.org/10.1073/pnas.93.26.15018
19. Wang K, Ohnuma S (1999) Chain-length determination mechanism of isoprenyl diphosphate synthases and implications for molecular evolution. Trends Biochem Sci 24:445–451
20. Stanley Fernandez SM, Kellogg BA, Poulter CD (2000) Farnesyl diphosphate synthase. Altering

the catalytic site to select for geranyl diphosphate activity. Biochemistry 39:15316–15321

21. Nagel R, Bernholz C, Vranová E, Košuth J, Bergau N, Ludwig S, Wessjohann L, Gershenzon J, Tissier A, Schmidt A (2015) Arabidopsis thaliana isoprenyl diphosphate synthases produce the C 25 intermediate, geranylfarnesyl diphosphate. Plant J 84:847–859. https://doi.org/10.1111/tpj.13064
22. Wang C, Chen Q, Fan D, Li J, Wang G, Zhang P (2016) Structural analyses of short-chain prenyltransferases identify an evolutionarily conserved GFPPS clade in Brassicaceae plants. Mol Plant 9:195–204. https://doi.org/10.1016/j.molp.2015.10.010
23. Cunillera N, Arró M, Delourme D, Karst F, Boronat A, Ferrer A (1996) Arabidopsis thaliana contains two differentially expressed farnesyl-diphosphate synthase genes. J Biol Chem 271:7774–7780. https://doi.org/10.1074/jbc.271.13.7774
24. Masferrer A, Arró M, Manzano D, Schaller H, Fernández-Busquets X, Moncaleán P, Fernández B, Cunillera N, Boronat A, Ferrer A (2002) Overexpression of Arabidopsis thaliana farnesyl diphosphate synthase (FPS1S) in transgenic Arabidopsis induces a cell death/senescence-like response and reduced cytokinin levels. Plant J 30:123–132
25. Chang T-H, Hsieh F-L, Ko T-P, Teng K-H, Liang P-H, Wang AH-J (2010) Structure of a heterotetrameric geranyl pyrophosphate synthase from mint (Mentha piperita) reveals Intersubunit regulation. Plant Cell 22:454–467. https://doi.org/10.1105/tpc.109.071738
26. Kim OT, Bang KH, Jung SJ, Kim YC, Hyun DY, Kim SH, Cha SW (2010) Molecular characterization of ginseng farnesyl diphosphate synthase gene and its up-regulation by methyl jasmonate. Biol Plant 54:47–53. https://doi.org/10.1007/s10535-010-0007-1
27. Schmidt A, Wächtler B, Temp U, Krekling T, Séguin A, Gershenzon J (2010) A bifunctional geranyl and geranylgeranyl diphosphate synthase is involved in terpene oleoresin formation in Picea abies. Plant Physiol 152:639–655. https://doi.org/10.1104/pp.109.144691
28. Arró M, Manzano D, Ferrer A (2014) Farnesyl diphosphate synthase assay. Methods Mol Biol 1153:41–53. https://doi.org/10.1007/978-1-4939-0606-2_4
29. Henneman L, van Cruchten AG, Denis SW, Amolins MW, Placzek AT, Gibbs RA, Kulik W, Waterham HR (2008) Detection of nonsterol isoprenoids by HPLC-MS/MS. Anal Biochem 383:18–24. https://doi.org/10.1016/j.ab.2008.08.023
30. Nagel R, Gershenzon J, Schmidt A (2012) Nonradioactive assay for detecting isoprenyl diphosphate synthase activity in crude plant extracts using liquid chromatography coupled with tandem mass spectrometry. Anal Biochem 422:433–438. https://doi.org/10.1016/j.ab.2011.12.037
31. Ruiz-Sola MÁ, Barja MV, Manzano D, Llorente B, Schipper B, Beekwilder J, Rodriguez-Concepcion M (2016) A single gene encodes two differentially targeted geranylgeranyl diphosphate synthase isoforms. Plant Physiol 172:1393–1402. https://doi.org/10.1104/pp.16.01392
32. Ruiz-Sola MÁ, Coman D, Beck G, Barja MV, Colinas M, Graf A, Welsch R, Rütimann P, Bühlmann P, Bigler L, Gruissem W, Rodríguez-Concepción M, Vranová E (2016) Arabidopsis GERANYLGERANYL DIPHOSPHATE SYNTHASE 11 is a hub isozyme required for the production of most photosynthesis-related isoprenoids. New Phytol 209:252–264. https://doi.org/10.1111/nph.13580
33. Wang G, Dixon R (2009) A heterodimeric geranyl(geranyl)diphosphate synthase from hop (Humulus lupulus) and the evolution of monoterpene biosynthesis. Proc Natl Acad Sci U S A 106:9914–9919. https://doi.org/10.1073/pnas.0904069106

Chapter 3

Expression, Purification, and Enzyme Activity Assay of Phytoene Synthase In Vitro

Maurizio Camagna and Ralf Welsch

Abstract

Phytoene synthase (PSY) is the rate-limiting step in carotenoid biosynthesis, and accordingly subjected to a number of regulatory mechanisms at various levels, including transcriptional, posttranscriptional, and posttranslational. Several *PSY* genes are present in most taxa and show various degrees of tissue and/or stress-specific responses providing an additional layer of regulating carotenogenesis. Moreover, only a small number of amino acid differences between paralogs or even single nucleotide polymorphisms distinguishing orthologs greatly affect enzyme properties, suggesting that different enzymatic parameters determined by intrinsic properties of PSY protein sequences also determine pathway flux. The characterization of enzyme properties of PSY variants from different origins requires in vitro enzyme assays with recombinant PSY. In this protocol, we present detailed instructions how to purify several milligrams of active PSY enzyme from bacterial lysates, which includes initial recombinant PSY enrichment through inclusion body purification, chaotropic unfolding, refolding in presence of detergents and purification through immobilized metal affinity chromatography. In addition, we provide a protocol to obtain active geranylgeranyl pyrophosphate (GGPP) synthase as active supply of GGPP substrate is a requirement for high in vitro PSY activity. The activity assay requires ^{14}C-labeled substrate and allows to determine its incorporation into phytoene as well as GGPP. The protocol described here was successfully applied to a variety of PSY and GGPP synthase homologs from various plant species.

Key words Phytoene synthase, GGPP synthase, Enzyme assay, Protein purification, Protein refolding, IMAC, Inclusion bodies

1 Introduction

The large diversity of carotenoids results from modifications like desaturation, cyclization, hydroxylation, and epoxidation reactions applied to a basic structure. This basic colorless molecule is phytoene, synthesized by the enzyme phytoene synthase (PSY) through a head-to-head condensation of two molecules geranylgeranyl diphosphate (GGPP). In contrast to the highly lipophilic phytoene, GGPP represents an amphiphilic prenyl phosphate, which is also required for the synthesis of other plant compounds

Manuel Rodríguez-Concepción and Ralf Welsch (eds.), *Plant and Food Carotenoids: Methods and Protocols*, Methods in Molecular Biology, vol. 2083, https://doi.org/10.1007/978-1-4939-9952-1_3,

with crucial functions throughout a plant's life, like chlorophylls, tocopherols, gibberellic acid, and plastoquinones [1, 2].

The activity of PSY was found to be rate-limiting for carotenogenesis [3]. For instance, PSY is light-induced during photomorphogenesis providing enhanced formation of carotenoids for photosynthesis, and is also induced upon abiotic stress to warrant sufficient substrate amounts for the cleavage of xanthophylls into abscisic acid [4–7]. Moreover, increased PSY transcript levels often correlate with increased carotenoid contents in nonphotosynthetic tissues rich in carotenoids, for instance daffodil flowers, tomato fruits, or melons [8–10].

Consequently and corroborating its rate-limiting function, *PSY* overexpression results in large increases in carotenoid levels obtained through biotechnological approaches [11]. In carrot and Arabidopsis roots as well as callus, carotenoid levels are elevated to high levels through *PSY* overexpression which results in crystallization of β-carotene [12, 13].

PSY is subjected to a pronounced regulation at various levels beside transcriptional activation, for example, posttranscriptionally via its 5′UTR and posttranslationally, mediated by the OR proteins [14–16]. In subsequent analysis, OR was found to mediate PSY protein stability and turnover in association with the plastid clp proteinase system [17]. PSY constitutes a small gene family with up to three members in different taxa which sometimes developed specificity for certain physiological requirements. For instance, three *PSY* genes are found in tomato; tomato *PSY2* is expressed in leaves while *PSY1* is fruit-specifically induced and a third homolog is thought to be involved in supporting root strigolactone biosynthesis upon interaction with symbiotic arbuscular mycorrhizal fungi [9, 18, 19]. A functional diversification of the three PSY genes in *Poaceae* is less obvious as two homologs are expressed in leaves while a third version in induced upon abiotic stress [6, 20, 21]. While the endosperm-specific expression of *PSY* in rice resulted only in the accumulation of the colorless phytoene, the generation of Golden Rice required the coexpression of a bacterial desaturase CrtI to proceed to lycopene which is then converted into downstream carotenoids by endogeneous enzymes already present in the endosperm in sufficient quantities [22]. However, the origin of PSY determined the amount of carotenoids accumulating in Golden Rice seeds: for instance, overexpression of daffodil *PSY* and tomato *PSY1* yielded only 1 and 5 μg g^{-1} total carotenoids, respectively, while overexpression of *Poaceae PSY1* from maize and rice yielded up to 30 μg g^{-1} [23]. This suggests differences in intrinsic properties of PSY proteins from various plant taxa. This observation is further supported by largely different carotenoid levels in fruits from different banana varieties while the PSY paralogs differ only slightly in their amino acid sequences [24]. Furthermore, single nucleotide polymorphisms in *PSY* were shown to correlate with changes in carotenoid content of cassava roots and maize seeds [25, 26].

These observations suggest a strong impact of individual amino acids on PSY enzymatic activity. Investigation of the molecular reason for these differences requires characterization of in vitro PSY activity allowing to compare enzymatic properties of different PSY variants in parallel. However, PSY is membrane-associated and its purification is linked with the notorious complications associated with the purification of membrane-proteins. Furthermore, the solubility of the substrate GGPP changes dramatically during the catalysis requiring a biphasic system to accommodate the lipophilic product phytoene (Fig. 1).

In this protocol we describe the purification of large quantities of active PSY and the conditions required to determine its activity. We take advantage of the accumulation of enzymatically largely inactive PSY protein in inclusion bodies which are enriched and cleared from major host impurities after cell lysis. Subsequently, PSY is unfolded by chaotropic agents and refolded in presence of a detergent which warrants its solubility. Solubilized PSY N-terminally equipped with a His-tag is thereafter purified by immobilized metal affinity chromatography (IMAC).

In a recent characterization of the Arabidopsis PSY enzyme activity, we found that freely diffusible GGPP is unsuited to serve as substrate for PSY and that addition of an active GGPP synthase in the assay is required to efficiently convert the precursors IPP and DMAPP into the intermediate GGPP and phytoene [27]. In contrast to the purification of PSY, obtaining active GGPS does not require a complicated protocol and follows standard IMAC procedures which are included in details in this protocol as well.

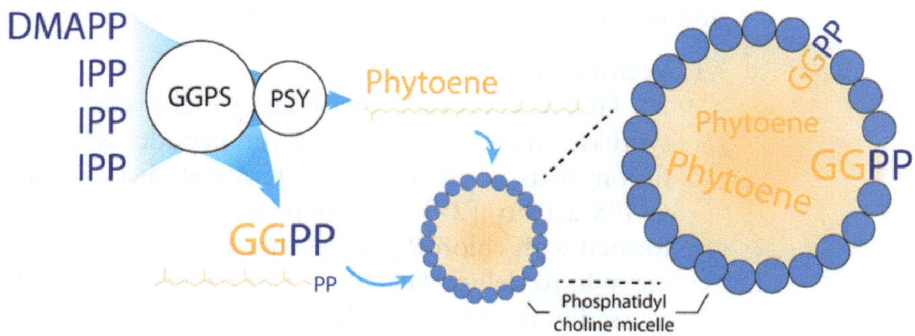

Fig. 1 Reaction scheme for GGPP and phytoene formation in the biphasic system. GGPP synthase catalyses the condensation of the hydrophilic substrates DMAPP and IPP to the highly amphiphilic molecule GGPP. PSY converts two molecules GGPP into phytoene. This reaction relies on the presence of actively synthesizing GGPS suggesting their interaction. In in vitro activity assays, mixed micelles containing phosphatidyl cholin liposomes and the detergent tween 80 function as lipophilic sink both for the final product phytoene as well for the intermediate GGPP. Once formed, GGPP gets sequestered into micelles which makes it inaccessible for phytoene synthesis

In order to determine enzyme parameters, for example, in a reaction kinetic, it is required to individually capture both unmetabolized substrate (IPP) amounts, intermediates (GGPP) as well as the final product phytoene. To achieve this, we have adopted a butanol/heptane-based extraction protocol which is capable to differentially extract these compounds. Moreover, incorporation of ^{14}C-labeled IPP in isotope dilution allows to exactly quantify the amounts of all components through scintillation counting.

The protocol described here applies to the enzyme combination GGPP synthase 11 (GGPS11) and PSY from Arabidopsis. We have recently found that most of the plastid-localized GGPP synthases are capable of providing GGPP for phytoene synthesis by Arabidopsis PSY if coexpressed in yeast [27]. We have almost exclusively performed in vitro assays with GGPS11 and therefore recommend the use of this isoenzyme for the assay. In contrast, we have successfully applied the protocol to purify and determine enzyme activities for PSY variants from different plant sources (e.g., maize and rice PSY1, cassava PSY1 and PSY2, tomato PSY3), and we are therefore confident that this protocol can be used for all plant-derived PSY versions [18, 25].

2 Materials

2.1 Recombinant GGPP Synthase

The requirements to work with radioactive materials in your lab must be met. Safety rules for working with radioactive substances must be followed. Use analytical grade reagents and ultrapure deionized water. All reagents and cellular extracts should be kept on ice during the enzyme extraction and enzyme assay preparation procedure (unless indicated otherwise). Waste disposal regulations including radioactive wastes and rules for the work with bacteria and recombinant DNA must be followed.

1. Vector containing GGPS

   GGPS expression plasmids should contain a GGPP synthase, truncated by its transit peptide as an N-terminal fusion with a 6×His-tag. C-terminal fusions might affect GGPS activity [27]. Transit peptide predictions can be performed with chloroP [28]. Although several GGPP synthases are able to deliver GGPP for phytoene synthesis when coexpressed in yeast [27], we have successfully worked with Arabidopsis GGPP11 (accession number *At4g36810*) and its homolog from *Sinapis* [29] and maize for in vitro assays combined with PSY from various sources.
2. Competent BL21 *E. coli* cells.
3. Ice, heating block at 42 °C, incubator shaker at 37 °C.

4. Super optimal broth with catabolite repression (SOC) medium: 2% (w/v) peptone, 0.5% (w/v) yeast extract, 10 mM NaCl, 2.5 mM KCl, 10 mM $MgCl_2$, 10 mM $MgSO_4$, and 20 mM glucose.
5. Lennox broth (LB) medium: 1% (w/v) Bacto tryptone, 0.5% (w/v) Bacto yeast extract, 0.5% (w/v) NaCl, and LB agar plates: add 1.5 g agar-agar to 100 ml medium. Add selective antibiotic (depending on your GGPS expression vector, e.g., ampicillin 150 μg ml^{-1}).
6. 1 M isopropylthiogalactoside (IPTG).
7. 50 ml plastic centrifugation tubes.
8. Centrifuge, French press.
9. GGPS Buffer A (binding buffer): 20 mM Tris–HCl, pH 8.0; 100 mM NaCl; 10 mM $MgCl_2$; 10% (v/v) glycerol.
10. GGPS Buffer B (elution buffer): GGPS Buffer A supplemented with 100 mM imidazole.
11. TALON® metal affinity resin, empty 5 ml column.

2.2 Recombinant PSY

1. *pCOLD* plasmid containing *PSY* cDNA truncated by its transit peptide.

   Sufficiently high *PSY* expression levels allowing for purification of satisfactory amounts were obtained only when *PSY* ORFs were expressed in the *pCOLD* cold-shock expression system (TaKaRa). We used *pCOLDI* which generates N-terminal 6×His tag fusions and contains a translational enhancer element (TEE). In contrast, high PSY expression levels were not achieved with other expression vectors and tags; for example, *pQE*, *pUC*, *pCR2.1*, *pGEX*, regardless of the origin of the PSY sequence. Sequences encoding transit peptides need to be removed, for example, after prediction by ChloroP [28].
2. Competent Rosetta (DE3) *E. coli* cells.
3. LB-agar and liquid LB medium containing 100 μg ml^{-1} ampicillin and 34 μg ml^{-1} chloramphenicol (LB-A/C).
4. Incubator shaker, adjustable to 37 °C and 15 °C.
5. One 100 ml Erlenmeyer flasks with 50 ml LB-A/C each, three 1 l baffled Erlenmeyer flasks with 350 ml LB-A/C each.
6. 1 M IPTG.
7. Centrifuge and 50 ml plastic centrifugation tubes are required; 500 ml centrifugation bottles and appropriate rotor are recommended.
8. Sonicator, French Press, ice, and overhead shaker.

9. 40 ml Lysis buffer: 50 mM Na_2HPO_4, pH 7.6, 1 mM $MgCl_2$, 5 mM 2-mercapthoethanol.
10. 5 μl benzonase nuclease (10 U μl^{-1}), 5 μl lysozyme (10 mg ml^{-1}).
11. 2 M urea, 1% (v/v) Triton X-100 (10 ml each).
12. 10 ml IB-lysis buffer: 6 M guanidinium–HCl, 0.035% (v/v) 2-mercaptoethanol.
13. Stirrer, 500 ml beaker.
14. 500 ml refolding buffer (RF): 50 mM HEPES, pH 7.6, 1 mM tris(2-carboxyethyl)phosphine (TCEP), 0.2% (w/v) lauryldimethylamine N-oxide (LDAO), 1 mM $MgCl_2$.
15. Fractogel EMD chelate (Merck), coupled with $CoCl_2$ (*see* **Note 1**).
16. 15 ml and 50 ml plastic tubes.
17. 150 ml PSY washing buffer: 50 mM Tris, pH 7.6, 300 mM NaCl, 1 mM TCEP, 1 mM $MgCl_2$, 0.035 (w/v) % LDAO.
18. 50 ml PSY elution buffer: 50 mM Tris, pH 7.6, 150 mM NaCl, 1 mM TCEP, 0.035 (w/v) % LDAO, 100 mM EDTA.
19. 1 L PSY dialysis buffer: 50 mM Tris, pH 7.6, 600 mM NaCl, 1 mM TCEP, 0.035% (w/v) LDAO.
20. Standard regenerated cellulose dialysis membrane with 15 kD cutoff (*see* **Note 2**).
21. Liquid nitrogen.

2.3 PSY Activity Assay

1. 20 mg phosphatidylcholine (e.g., from soybean, Type II-S).
2. 50 ml liposomes incubation buffer: 1 mM DTE, 10 mM $MgCl_2$, 100 mM Tris–HCl, pH 7.2.
3. 2 ml methanol–chloroform (1:2, v/v).
4. 15 ml reaction tubes.
5. French press, sonicator, ice.
6. 10 ml PSY reaction buffer: 100 mM Tris–HCl, pH 7.6, 0.08% (v/v) Tween 80, 20% (v/v) glycerol, 2 mM $MnCl_2$, 10 mM $MgCl_2$, 1 mM TCEP, 600 mM NaCl (*see* **Note 3**).
7. [1-^{14}C]-IPP, triammonium salt in ethanol: 0.15 N NH_4OH (1:1) with 50–60 mCi $mmol^{-1}$.
8. 1 mg ml^{-1} (3.365 mM) DMAPP and IPP (triammonium salt) in ethanol: 0.15 N NH_4OH (1:1).
9. 50 μl 160 mM EDTA.
10. Butanol, 1 M $MgCl_2$ in methanol, heptane.
11. Incubator for reaction tubes at 20 °C, table centrifuge.
12. Scintillation counter, scintillation cocktail, and scintillation vials.

3 Methods

3.1 Expression and Purification of Arabidopsis GGPP Synthase 11

The following protocol yields approximately 1 mg recombinant GGPS which is sufficient for a large number of PSY activity assays.

1. Add about 50 ng of plasmid DNA but maximally 1 μl volume and 10 μl BL21 competent *E. coli* cells and mix gently by gentle snipping. Do not vortex.
2. Incubate on ice for 5 min, then for 30 s at 42 °C, place back on ice for 5 min, add 250 μl SOC medium and incubate at 37 °C for 1 h while shaking at 200 rpm.
3. Plate 50 μl of transformed bacteria on LB-agar plates containing appropriate selective antibiotic. Incubate plate over night at 37 °C.
4. Pick a single colony to inoculate a 37 °C overnight culture of 15 ml of LB medium with appropriate selective antibiotics.
5. The next day, dilute the overnight culture 1:500 in 50 ml LB medium containing 150 μg ml^{-1} ampicillin, and grow at 37 °C to an OD_{600} of 0.5.
6. Add 1 mM IPTG, and incubate for 4 h, then harvest by centrifugation (5000 × *g*, 5 min) and discard the supernatant. The pellet can be stored at −20 °C for several months if the extraction cannot be performed the same day.
7. Resuspend the pellet in 10 ml GGPS buffer A and disrupt via a French Press at 20,000 psi. Centrifuge at 10,000 × *g* in plastic centrifugation tubes for 10 min to pellet the cell debris. Decant the supernatant into a new tube and discard the pellet.
8. Pipette 600 μl TALON resin slurry into a tube, add 10 ml GGPS buffer A, centrifuge at 700 × *g* for 1 min and discard supernatant. Repeat washing one more time with 10 ml GGPS buffer A.
9. Add the equilibrated TALON resin to the lysed bacterial, place the tube horizontal in ice and incubate for 30 min while shaking.
10. Load the suspension onto an empty 5 ml TALON column and let the extract drain by gravity flow, then wash the retained resin with 20 ml GGPS buffer A.
11. Elute the protein with 5 ml GGPS buffer B into 5 tubes with each 1 ml eluate. Number tubes ascending. Figure 2a shows steps of GGPS purification as well as purified GGPS loaded on a Coomassie-stained SDS-PAGE gel.

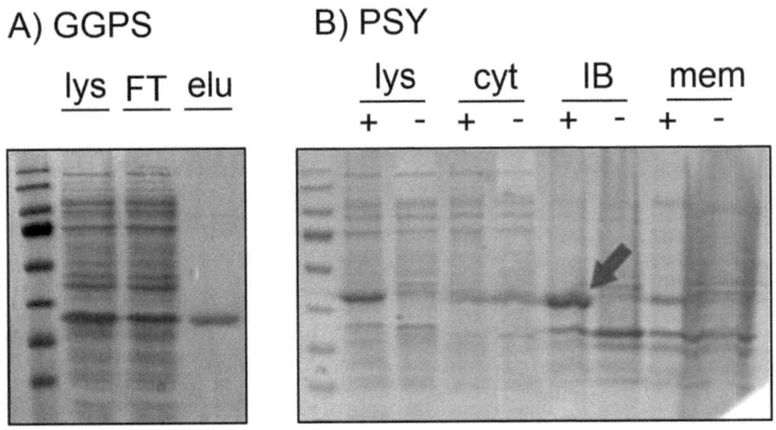

Fig. 2 Expression and purification of GGPS and PSY in *E. coli*. (**a**) SDS-PAGE gel showing the purification of recombinant GGPP synthase from *E. coli* lysate. The protein band eluted from the column (elu) correspond to GGPP synthase 11. The cell lysate (lys), as well as the proteins washed from the column before elution (flow through, FT) are also shown. (**b**) Lysate from *E. coli* cells expressing 6×His-PSY (lys) was subjected to differential centrifugation which revealed a fraction enriched in inclusion bodies (IB), membranes (mem) and cytoplasm (cyt). In both bacterial membranes and the soluble cytoplasma fraction, PSY is present only in minor amounts. Recombinant 6×His-PSY was purified from the inclusion bodies fraction by chaotropic unfolding followed by refolding in presence of detergents. Proteins were separated by SDS-PAGE and Coomassie-stained. +/− indicates induction of protein expression by IPTG

3.2 Expression and Purification of Arabidopsis Phytoene Synthase

1. The following protocol yields up to 5 mg recombinant Arabidopsis PSY which is sufficient for a large number of PSY activity assays. We've successfully applied this protocol to PSYs from other plants as well, but a reduction in yield should be expected.
2. Transform *pCOLD1-PSY* vector into Rosetta (DE3) competent *E. coli* cells as described for GGPS expression vectors above. Spread on LB-A/C agar plate and incubate over night at 37 °C.
3. Pick one single colony and incubate in 2 ml liquid LB-A/C at 37 °C under shaking at 200 rpm for 3–5 h, then transfer into 50 ml LB-A/C and continue incubation at 37 °C overnight. On the next day, use 15 ml each of the overnight culture to inoculate three baffled flasks with a capacity of 1 l each containing 350 ml LB-A/C. Incubate at 37 °C while shaking at 200 rpm.
4. If $OD_{600} = 0.6$ is reached, switch off the shaker and lower the temperature of the incubator to 15 °C. After 30 min, add 1 mM isopropyl-D-1-thiogalactopyranoside (IPTG), switch on the shaker to 200 rpm and incubate for 24 h at 15 °C.

5. On the next day, harvest the cells by centrifugation at 8000 × *g*, 4 °C for 10 min. Pool the bacterial pellets of 3 × 350 ml cultures into one tube (*see* **Note 4**).
6. Resuspend the bacterial pellet of 1 l culture in 20 ml lysis buffer by sonication (Duty Cycle 30/Output Control 2). 1 l bacteria culture corresponds to approximately 2 g of cells after pelleting.
7. After centrifugation (8000 × *g*, 4 °C, 10 min), resuspend the pellet again in 20 ml lysis buffer by sonication. For each gram of *E. coli*, add 2.5 μl Benzonase (10 U μl^{-1}) and 2.5 μl lysozyme (10 mg ml^{-1}). Then incubate the cultures at 4 °C for 30 min on a shaker at 100 rpm.
8. Disintegrate by French press treatment (20,000 psi) and pellet inclusion bodies by centrifugation at 16,000 × *g* for 40 min at 4 °C. Discard the supernatant.
9. Wash pellet containing inclusion bodies (IB) two times with 10 ml Millipore water, one time each with 1% Triton X-100 and 2 M urea. Resuspend pellets by sonication and recover by centrifugation at 20,000 × *g* for 10 min. at 4 °C between each step (*see* **Note 5**).
10. Resuspend pelleted inclusion body in 2 ml water by sonication and add 10 ml IB-lysis buffer. Incubate for 30 min at room temperature.
11. Slowly drop the resuspended inclusion bodies into 450 ml refolding (RF) buffer cooled at 4 °C, approximately one drop every 3 s, while constantly stirring the buffer on ice. The dripping can take up to 20 min. Continue stirring for another 20 min after the inclusion bodies have dripped into the RF buffer (*see* **Note 6**).
12. Transfer RF solution containing 6×His-PSY into a 500 ml centrifugation bottle and pellet undissolved aggregates at 10,000 × *g* for 20 min at 4 °C. Carefully transfer supernatant into empty bottle and discard the pellet.
13. Equilibrate 2 ml $CoCl_2$ Fractogel two times with RF buffer, as described above for TALON equilibration for GGPS purification. Add equilibrated Fractogel to RF supernatant, place on ice and incubate on a shaking platform for 20 min (*see* **Note 7**).
14. Centrifuge the solution at 700 × *g* for 10 min and 4 °C to pellet the beads, carefully discard the supernatant by pipetting and resuspend the beads in 10 ml PSY washing buffer. Transfer the beads to a 15 ml plastic tube, centrifuge for 15 min at 700 × *g* (4 °C) and discard the supernatant by pipetting.
15. Resuspend the pellet in 3 ml PSY elution buffer, incubate for 20 min at 4 °C in an overhead shaker and centrifuge at 700 × *g* for 5 min at 4 °C.

16. Dialyze the eluate against 500 ml PSY dialysis buffer using a standard regenerated cellulose membrane with a 15 kD cutoff, stirring with 100 rpm in a cold room. Replace PSY dialysis buffer with fresh buffer after 1.5 h and continue with dialysis for another 1.5 h (*see* **Note 8**).
17. After dialysis, centrifuge reaction tubes for 20 min at 20,000 × *g* (4 °C), transfer supernatant into new tubes with 100 μl aliquots, immediately freeze in liquid nitrogen and store at −80 °C. Figure 2b shows steps of PSY purification as well as purified PSY loaded on a Coomassie-stained SDS-PAGE gel (*see* **Note 9**).

3.3 Phytoene Synthase Activity Assay

The following protocol includes a method on the preparation of phosphatidylcholine liposomes according to Bozzuto and Molinari [30]. For all subsequent steps regarding PSY activity determination, amounts are given for one measurement in a volume of 200 μl. If kinetics are done, these amounts need to be scaled up according to the number of time points planned. A master mix containing all compounds is then made and samples of 200 μl volume are removed after the desired time.

1. For the preparation of phosphatidylcholine liposomes, dissolve 20 mg phosphatidylcholine in 2 ml methanol–chloroform (1:2, v/v) in a 15 ml reaction tube, evaporate to dryness and add 2 ml liposome buffer. Incubate for 30 min on ice, resuspend by sonication and perform two French Press passages operated at 20,000 psi. Dilute with liposome buffer to 10 μg μl^{-1} phosphatidylcholine and store at −20 °C. Liposomes can be prepared ahead and stored at −20 °C for several months.
2. Prepare the reaction mix which consists of 1 μg GGPS11, an equimolar amount of PSY, 12.5 μg phosphatidylcholine liposomes and add up to 180 μl with PSY reaction buffer. Keep the sample on ice (*see* **Note 10**).
3. Prepare the substrate mix, consisting of 20 μM DMAPP, 16.25 μM IPP and 3.75 μM ^{14}C-IPP (i.e., isotope dilution of 1:5.33) and add up to 20 μl with PSY reaction buffer. Keep the sample on ice.
4. Transfer both enzyme and substrate mix in an incubator at 20 °C and preincubate for 5 min, then start the reaction by adding the substrate to 180 μl reaction mix. Repeatedly pipet up and down to assure that the enzymes and substrate are mixed well.
5. Stop the reaction by adding 50 μl 160 mM EDTA and mix thoroughly.
6. Add one assay volume butanol to the sample, vortex, and centrifuge for 5 min at 15,000 × *g*.

7. Collect the upper butanol phase containing ^{14}C-GGPP and ^{14}C-phytoene. Repeat the extraction, using 100 μl of butanol and join both butanol supernatants in a single tube. The remaining aqueous phase contains ^{14}C-IPP, IPP and DMAPP.
8. Add an equal volume of 1 M $MgCl_2$ in methanol to the butanol sample, mix by vortexing, add another volume heptane, mix again by vortexing. Centrifuge for 5 min at 15,000 × *g*.
9. Recover the upper heptane phase containing ^{14}C-phytoene. Repeat the extraction with 150 μl heptane and pool the heptane supernatants.
10. Transfer extracts into individual vials containing 6 ml of scintillation cocktail and determine the radioactivity in each fraction in a scintillation counter with an exposure time of 1 min (*see* **Note 11**).
11. Calculation of IPP/GGPP/phytoene amounts.
    (a) Empirically determine disintegrations per minute (dpm) of 1 nmol ^{14}C-IPP by scintillation counting 1 μl of ^{14}C-IPP stock in triplicate and determine the average DPM $nmol^{-1}$.
    (b) Convert the quantified DPM of each fraction (IPP, GGPP, phytoene) into nmol ^{14}C-IPP by dividing $DPM_{IPP}/DPM_{GGPP}/DPM_{phytoene}$ by the average DPM $nmol^{-1}$ of ^{14}C-IPP.
    (c) Convert nmol ^{14}C-IPP into total nmol IPP by applying the ^{14}C-IPP:IPP isotope dilution factor, for example, by multiplication with 5.3 given the example above.
    (d) Amounts of GGPP (4 molecules IPP) and phytoene (8 molecules IPP) are then determined by dividing the total nmol IPP in each sample by 4, and 8 respectively.

4 Notes

1. Coupling of Fractogel with Co^{2+} can be performed as follows. Shake Fractogel well to produce a slurry, remove a desired volume for coupling and note down the volume. Remove storage solution by filtering, transfer Fractogel into a beaker and wash two times with Millipore water. Remove water by filtering (avoid Fractogel to become dry). Collect Fractogel in a beaker, add a filtered solution of 1 M $CoCl_2$ to fully cover Fractogel and incubate with slow rotation for 2 h. Remove $CoCl_2$ solution by filtering and wash one time with Millipore water. Resuspend Co^{2+}-coupled Fractogel in the same volume as initially used in storage solution (20% ethanol, 150 mM NaCl). Store at 4 °C.

2. We routinely use a Spectra/Por^R 7 dialysis membrane which is chemically treated to remove any trace levels of heavy metals and sulfide compounds.
3. PSY reaction buffer needs to be prepared always fresh. Stock solutions of all compounds can be made and stored at room temperature (Tris, Tween 80, $MgCl_2$, NaCl). $MnCl_2$ and TCEP stocks needs to be stored in small aliquots at −20 °C to avoid oxidation during multiple freezing–defreezing cycles.
4. It is recommended to use plastic bottles with high capacity (e.g., 500 ml) to reduce centrifugation time. Bacteria pellets in different centrifugation bottles can be pooled after resuspension in 20 ml LB-A/C and recentrifugation.
5. Washed inclusion body pellets can be stored at −20 °C for several months.
6. A constant drip rate can be achieved by using a separatory funnel equipped with a stopcock, mounted on a tripod above the beaker with the stirring refolding buffer. Ideally, place the whole apparatus in the cold room to ensure cooling of resuspended inclusion bodies. Check flow from time to time as inclusion body aggregates might lead to clogging.
7. Highest binding efficiency and yields were achieved with Co^{2+}-coupled Fractogel EMD Chelate if compared with equivalent metal affinity resins from other suppliers. This may be due to the tentacle technology used consisting of long, linear polymer chains which might reduce aggregation of bound 6×His-PSY proteins in close proximity. We observed a yield drop if regular matrices were used.
8. Eluate can be transferred into a 2 ml reaction tube which contains a hole in the lid. Cover the filled tube with one layer of dialysis membrane, carefully close the lid and confirm tightness by reverting the closed tube. Place tube upside down in a floating rack swimming in PSY dialysis buffer.
9. This step is to remove remaining Fractogel and PSY aggregates which might have formed during refolding/elution/dialysis procedure and which compromises efficiency of enzyme activity assays. Refolded PSY can be stored at −80 °C without significant loss in enzyme activity for at least 6 months.
10. Protein concentrations of purified GGPS and PSY should be determined photometrically (OD_{280}, Bradford, etc.) and confirmed on an SDS-PAGE gel with BSA (between 0.5 and 5 μg) and staining with Coomassie. Equimolar amounts of the enzyme pair AtGGPS11 and AtPSY, each truncated by their transit peptide, for one assay of 200 μl volume are 1 μg AtGGPS11 and 1.22 μg AtPSY.

11. Use a universal scintillation cocktail for both hydrophilic and lipophilic samples. A ^{14}C-quenching correction is recommended for scintillation counting.

Acknowledgments

This work was supported in part by the HarvestPlus research consortium (grant 2014H6320.FRE) to R.W. The authors thank the former members of the laboratory of Prof. Peter Beyer for their contributions to establishing this method.

References

1. Beck G, Coman D, Herren E et al (2013) Characterization of the GGPP synthase gene family in Arabidopsis thaliana. Plant Mol Biol 82:393–416
2. Ruiz-Sola MÁ, Coman D, Beck G et al (2016) Arabidopsis GERANYLGERANYL DIPHOSPHATE SYNTHASE 11 is a hub isozyme required for the production of most photosynthesis-related isoprenoids. New Phytol 209:252–264
3. Baranski R, Cazzonelli CI (2016) Carotenoid biosynthesis and regulation in plants. In: Carotenoids. John Wiley & Sons Ltd, Chichester, UK, pp 159–189
4. von Lintig J, Welsch R, Bonk M et al (1997) Light-dependent regulation of carotenoid biosynthesis occurs at the level of phytoene synthase expression and is mediated by phytochrome in Sinapis alba and Arabidopsis thaliana seedlings. Plant J 12:625–634
5. Li F, Vallabhaneni R, Wurtzel ET (2008) PSY3, a new member of the phytoene synthase gene family conserved in the Poaceae and regulator of abiotic stress-induced root carotenogenesis. Plant Physiol 146:1333–1345
6. Welsch R, Wüst F, Bär C et al (2008) A third phytoene synthase is devoted to abiotic stress-induced abscisic acid formation in rice and defines functional diversification of phytoene synthase genes. Plant Physiol 147:367–380
7. Toledo-Ortiz G, Huq E, Rodríguez-Concepción M (2010) Direct regulation of phytoene synthase gene expression and carotenoid biosynthesis by phytochrome-interacting factors. Proc Natl Acad Sci U S A 107:11626–11631
8. Schledz M, Al-Babili S, von Lintig J et al (1996) Phytoene synthase from Narcissus pseudonarcissus: functional expression, galactolipid requirement, topological distribution in chromoplasts and induction during flowering. Plant J 10:781–792
9. Giorio G, Stigliani AL, D'Ambrosio C (2008) Phytoene synthase genes in tomato (Solanum lycopersicum L.) - new data on the structures, the deduced amino acid sequences and the expression patterns. FEBS J 275:527–535
10. Qin X, Coku A, Inoue K, Tian L (2011) Expression, subcellular localization, and cis-regulatory structure of duplicated phytoene synthase genes in melon (Cucumis melo L.). Planta 234:737–748
11. Farré G, Sanahuja G, Naqvi S et al (2010) Travel advice on the road to carotenoids in plants. Plant Sci 179:28–48
12. Maass D, Arango J, Wüst F et al (2009) Carotenoid crystal formation in Arabidopsis and carrot roots caused by increased phytoene synthase protein levels. PLoS One 4:e6373
13. Schaub P, Rodriguez-Franco M, Cazzonelli CI et al (2018) Establishment of an Arabidopsis callus system to study the interrelations of biosynthesis, degradation and accumulation of carotenoids. PLoS One 13:e0192158
14. Zhou X, Welsch R, Yang Y et al (2015) Arabidopsis OR proteins are the major posttranscriptional regulators of phytoene synthase in controlling carotenoid biosynthesis. Proc Natl Acad Sci 112:3558–3563
15. Li L, Yang Y, Xu Q et al (2011) The or gene enhances carotenoid accumulation and stability during post-harvest storage of potato tubers. Mol Plant 5:339–352
16. Álvarez D, Voß B, Maass D et al (2016) Carotenogenesis is regulated by 5'UTR-mediated translation of Phytoene synthase splice variants. Plant Physiol 172:2314–2326
17. Welsch R, Zhou X, Yuan H et al (2017) Clp protease and OR directly control the proteostasis of phytoene synthase, the crucial enzyme

for carotenoid biosynthesis in Arabidopsis. Mol Plant 11:149–162

18. Stauder R, Welsch R, Camagna M et al (2018) Strigolactone levels in dicot roots are determined by an ancestral Symbiosis- regulated clade of the PHYTOENE SYNTHASE gene family. Front Plant Sci 9:255
19. Walter MH, Stauder R, Tissier A (2014) Evolution of root-specific carotenoid precursor pathways for apocarotenoid signal biogenesis. Plant Sci 233:1–10
20. Gallagher CE, Matthews PD, Li F, Wurtzel ET (2004) Gene duplication in the carotenoid biosynthetic pathway preceded evolution of the grasses. Plant Physiol 135(3):1776–1783
21. Li F, Vallabhaneni R, Rocheford TR et al (2008) The maize phytoene synthase gene family: overlapping roles for carotenogenesis in endosperm, photomorphogenesis, and thermal stress tolerance. Plant Physiol 147:1334–1346
22. Beyer P, Al-Babili S, Ye X et al (2002) Golden Rice: introducing the beta-carotene biosynthesis pathway into rice endosperm by genetic engineering to defeat vitamin A deficiency. J Nutr 132:506S–510S
23. Paine JA, Shipton CA, Chaggar S et al (2005) Improving the nutritional value of Golden Rice through increased pro-vitamin A content. Nat Biotechnol 23:482–487
24. Mlalazi B, Welsch R, Namanya P et al (2012) Isolation and functional characterisation of banana phytoene synthase genes as potential cisgenes. Planta 236:1585–1598
25. Welsch R, Arango J, Bar C et al (2010) Provitamin A accumulation in cassava (Manihot esculenta) roots driven by a single nucleotide polymorphism in a phytoene synthase gene. Plant Cell 22:3348–3356
26. Fu Z, Yan J, Zheng Y et al (2010) Nucleotide diversity and molecular evolution of the PSY1 gene in Zea mays compared to some other grass species. Theor Appl Genet 120:709–720
27. Camagna M, Grundmann A, Bär C et al (2019) Enzyme fusion removes competition for geranylgeranyl diphosphate in carotenogenesis. Plant Physiol 179(3):1013–1027
28. Emanuelsson O, Nielsen H, von Heijne G (1999) ChloroP, a neural network-based method for predicting chloroplast transit peptides and their cleavage sites. Protein Sci 8:978–984
29. Kloer DP, Welsch R, Beyer P, Schulz GE (2006) Structure and reaction geometry of geranylgeranyl diphosphate synthase from Sinapis alba. Biochemistry 45:15197–15204
30. Bozzuto G, Molinari A (2015) Liposomes as nanomedical devices. Int J Nanomedicine 10:975–999

Chapter 4

Improved Expression and Purification of the Carotenoid Biosynthetic Enzyme Z-ISO

Eleanore T. Wurtzel and Jesús Beltrán

Abstract

Carotenoids are a large class of pigments that are essential for survival of plants and other species that consume these plant-derived compounds and their bioactive derivatives. The plant biosynthetic pathway is nuclear-encoded and localized in plastids. The pathway enzymes had been known for many years, except for a recently discovered isomerase, 15-*cis*-ζ-carotene isomerase (Z-ISO) which utilizes a novel mechanism to mediate isomerization in response to the redox state of its heme b cofactor. To further study this enzyme, a protocol is described which maximizes purification of a fusion between Maltose Binding Protein and *Zea mays* (maize) Z-ISO (MBP::Z-ISO) expressed in *E. coli* treated with heme biosynthesis precursors which were used to increase heme available for loading into the expressed protein. Further enrichment of the protein was accomplished by improved sonication to release membranes containing Z-ISO, an integral membrane protein, and collection of the membrane fraction which was subjected to Nickel affinity chromatography. The fusion protein bound to the column through a His-tag. The MBP::Z-ISO protein was released using histidine, and not imidazole which binds heme and would interfere with enzyme recovery. Purification of the 75.46 kD MBP::Z-ISO expressed in *E. coli* was accomplished with fivefold improvement of yield and doubled heme content compared to the previously published method Beltrán et al. (Nat Chem Biol 11(8):598–605, 2015). The newer protocol will yield, per liter of culture, 5–6 mg MBP::Z-ISO protein with ~1:1 heme to Z-ISO ratio.

Key words Plant carotenoid biosynthesis, Heme, Isomerase, 15-*cis*-ζ-carotene, Membrane, *E. coli*, Protein purification

1 Introduction

Carotenoids and their bioactive derivatives constitute a large class of pigments that enable plants to capture photosynthetic light energy, protect against high light damage and regulate plant growth and development in a complex environment. Dietary sources of plant carotenoids are important for human health and species survival. With these manifold roles of carotenoids in mind, a multitude of opportunities exist for metabolic engineering and

Manuel Rodríguez-Concepción and Ralf Welsch (eds.), *Plant and Food Carotenoids: Methods and Protocols*, Methods in Molecular Biology, vol. 2083, https://doi.org/10.1007/978-1-4939-9952-1_4,

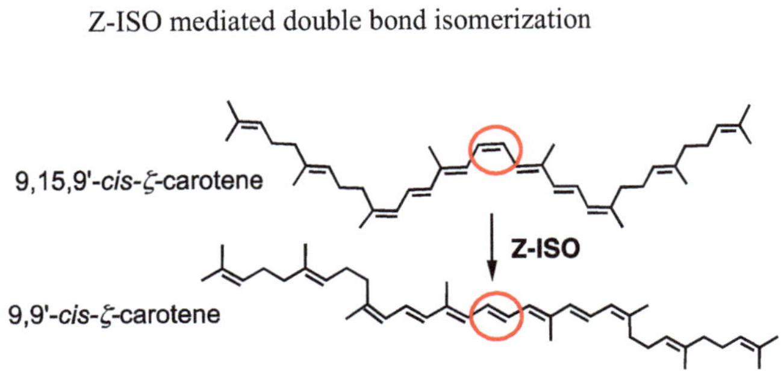

Fig. 1 Double bond isomerization reaction catalyzed by Z-ISO. Red circle denotes the 15–15′ double bond that undergoes *cis* to *trans* isomerization catalyzed by Z-ISO within a membrane environment

synthetic biology applications conceived to meet needs of food security in an ever-changing climate [1, 2].

The biosynthesis of carotenoids in plants is mediated by nuclear-encoded enzymes that are imported into chloroplasts and other plastids [1]. A new enzyme required for the biosynthesis of all plant carotenoids, 15-*cis*-ζ-carotene isomerase (Z-ISO), was recently discovered [3, 4]. Z-ISO is an integral membrane protein having multiple transmembrane domains that mediates carotenoid isomerization by a novel mechanism involving a heme b cofactor. Z-ISO utilizes the heme to catalyze isomerization of the central 15–15′ *cis* double bond of 9,15,9′-tri-*cis*-ζ-carotene to form 9, 9′-di-*cis*-ζ-carotene (*see* Fig. 1) [5]. The heme undergoes ligand switching in response to the redox state of the heme iron, causing a repositioning of the heme within Z-ISO and a conformational change, that alter the heme absorbance and enzyme activity. In the reduced state of the heme (which absorbs maximally at 426 nm), Z-ISO is active, whereas in the oxidized state (which absorbs maximally at 414–415 nm), the enzyme is inactive. The redox state of the Z-ISO heme iron is critical for controlling flux through the pathway. To better understand the Z-ISO catalytic mechanism, a method was developed for the high-level expression of a Maltose Binding Protein fusion with *Zea mays* (maize) Z-ISO (MBP::Z-ISO) so that the protein could be examined by a number of spectroscopic methods [5]. Reported here is a detailed protocol with further improvements. The construct used for expression of MBP::Z-ISO encodes a His-tagged Maltose Binding Protein fused at the N-terminus of Z-ISO. When expressed in *E. coli*, the Z-ISO integrates into the membrane which facilitates MBP::Z-ISO isolation from the membrane fraction. The His tag is used to further purify the protein by Nickel affinity chromatography (Ni-NTA). One additional trick used in the protocol reported here was the use of heme biosynthesis precursors added during protein induction, which improved heme biosynthesis and consequentially increased the amount of *E. coli*-expressed MBP::Z-ISO containing heme. In summary, purification of the 75.46 kD MBP::Z-ISO expressed in *E. coli* was accomplished with fivefold improvement of yield and doubled the heme content as compared to the previously published method [5]. Per liter culture, this newer method describes a yield of purified MBP::Z-ISO of 5–6 mg protein with heme to Z-ISO ratio of ~ 1:1.

2 Materials

1. 1,4-dithiothreitol (DTT).
2. 4-(2-aminoethyl)benzenesulfonyl fluoride hydrochloride (AEBSF).
3. n-Dodecyl β-D-maltoside (DDM) Sol grade.
4. Adenosine triphosphate (ATP).
5. Lysozyme from chicken egg white.
6. Sodium chloride (NaCl).
7. Tris.
8. Yeast extract.
9. Tryptone.
10. Ni-NTA agarose.
11. Slide-A-Lyzer dialysis cassettes G2 20K.
12. Glycerol.
13. Bottle, with cap assembly, polycarbonate, 26.3 mL, 25 × 89 mm, 1 × 3–1/2.
14. Ampicillin (50 mg/mL), filter sterile.
15. Ethylenediaminetetraacetic acid (EDTA).
16. Magnesium chloride ($MgCl_2$).
17. 5-aminolevulinic acid hydrochloride (ALA)
18. Ammonium iron (II) sulfate hexahydrate (ferrous iron).
19. YT medium (2× yeast extract and tryptone): 1% yeast extract, 1.6% tryptone, and 0.5% NaCl, sterilized for 20′.
20. 1 M isopropyl-beta-D-thiogalactopyranoside (IPTG): 2.38 g/ 10 mL. Filter-sterilize. Aliquot 4 mL per tube. Freeze until use.
21. 5-Aminolevulinic Acid Hydrochloride (ALA): 1000× Stock: 235 mM ALA (39.3 mg/mL). Filter-sterilize. Keep on ice until used.
22. Ferrous iron. ammonium iron (II) sulfate hexahydrate: 1000× Stock 35 mM (13.7 mg/mL), weigh ~50 mg into 3.65 mL water. Filter-sterilize. Keep on ice until used.

23. 1 M Tris–HCl, 1 L: Dissolve 121.1 g of Tris base in 800 mL of H_2O. Adjust the pH to 7.6 by adding concentrated HCl (~ 60 mL). Adjust volume to 1 liter with ddH_2O. Sterilize by autoclaving (15 min).
24. 1 M $MgCl_2$, 100 mL: Mix 20.33 g $MgCl_2$ with 70 mL of ddH_2O, add ddH_2O until final volume is 100 mL. Sterilize by autoclaving (15 min).
25. 5 M NaCl, 1 L: Mix 292.2 g of NaCl with 900 mL of ddH_2O by stirring, add ddH_2O until final volume is 1 L. Sterilize by autoclaving (15 min).
26. Buffer A. Modified Resuspension Buffer: 50 mM Tris–HCl pH 7.6, 250 mM NaCl. Autoclave for long-term storage.
27. Buffer B. (For cell resuspension) made on day of use: To 100 mL of Buffer A add 120 mg of AEBSF, 50 μL of 1 M DTT, 15 mg of lysozyme. Store on ice until used.
28. Buffer C. Resuspension Buffer with 5% glycerol (100 mL): 50 mM Tris–HCl pH 7.6, 300 mM NaCl, 5% glycerol.
29. Buffer D. ATP Wash Buffer (6 mL): 40 mM Tris–HCl pH 7.6, 200 mM NaCl, 5 mM $MgCl_2$, 5% Glycerol. Add fresh immediately before use: 16.5 mg of ATP as a powder, 0.6 μL of 1 M DTT, 30 μL of 10% DDM. Adjust to 6 mL with ddH_2O. ATP wash buffer includes ATP and Mg to remove any associated chaperones.
30. Buffer E. Elution Buffer (10 mL): 25 mM Tris–HCl pH 7.6, 200 mM NaCl, 5% Glycerol.

    Add fresh immediately before use: 1 μL of 1 M DTT, 50 μL of 10% DDM, 8 mL of 0.25 M histidine pH 7.1. Adjust to 10 mL with ddH_2O. 0.25 M L-Histidine (3.88 g/100 mL). Make fresh; at room temperature will crystalize and at room temperature will turn yellow. No need to pH.
31. Buffer F. Optional dialysis buffer (2 L, store and use at 4 °C): 20 mM Tris–HCl pH 7.6, 20 mM NaCl, 5% glycerol. Add fresh immediately before use: 0.4 g of DDM, 0.2 mL of 1 M DTT.

3 Methods

DAY 1

1. To 25 mL of 2× YT medium, add 25 μL of ampicillin stock for a final concentration of 50 μg/mL.
2. Inoculate with pMCSG9 Z -ISO E2 in C43 (DE3) cells [5]. Grow overnight with shaking at 37 °C.

DAY 2

1. Prewarm the medium to 37 °C. For 4 × 500 mL each in 3 L Fernbach, add 5 mL overnight culture per 500 mL of 2× YT medium plus antibiotic, 0.5 mL ampicillin stock/500 mL.

   Incubate for 1 h without shaking and then shake at 220 rpm, until OD_{600} equals 0.6 (~3 h) (*see* **Note 1**).
2. IPTG induction with heme biosynthesis reagents (*see* **Note 2**): For each 500 mL, add 0.5 mL each of 1 M IPTG, 235 mM ALA, and 35 mM ferrous iron. Incubate for ~16 h at 28 °C with shaking.

DAY 3

1. Centrifuge cells at 2600 × *g* (3,500 rpm, Eppendorf centrifuge 5810 R) for 15 min at 4 °C and freeze (−80 °C) pellets until use. Use only 2 centrifuge bottles and spin twice to combine cultures (*see* **Note 3**).

DAY 4

1. Resuspend in ~80 mL Buffer B (made with Buffer A and freshly added components (*see* **Note 4**).
2. Sonicate the entire sample on ice in a 250 mL plastic beaker (*10 times*, 30 s each, 60% duty cycle with pulsing (*set at "8"*), with 1 min between sonication) using a Vibra Cell VC600 sonicator equipped with a FLAT tip (Sonics & Materials Inc., Connecticut, USA). The sample should change from milky white to a translucent red (*see* Fig. 2a and **Note 5**).
3. Centrifuge at ~9800 × *g* (11,000 rpm, in a Type 60 Ti rotor) for 15 min, 4 °C (Polycarbonate tubes, 25 × 89 mm, Beckman Coulter). Remove supernatant with pipet and save lysates containing the membrane fraction. The pellet will contain unbroken cells.
4. Centrifuge lysates at ~83,000 × *g* (32,000 rpm, in a Type 60 Ti rotor) for 1 h at 4 °C using Polycarbonate tubes (25 × 89 mm, Beckman Coulter). Invert tubes to drain remaining supernatant. The pellets should appear bright red (*see* Fig. 2b and **Note 6**).
5. Resuspend pellets in 16 mL Resuspension Buffer C t ratio of 8 mL per liter of initial cell culture. Use a glass homogenizer. Do not over homogenize.
6. Increase the volume to a total of 80 mL final at a ratio of 40 mL of Buffer C per liter of starting culture. Divide the sample into two 50 mL Falcon tubes and add to each 0.6 g DDM as a powder (1.5% final concentration). To dissolve the DDM into the buffer, slowly rotate the samples end over end at 4 °C for 15 min using a rotator machine.

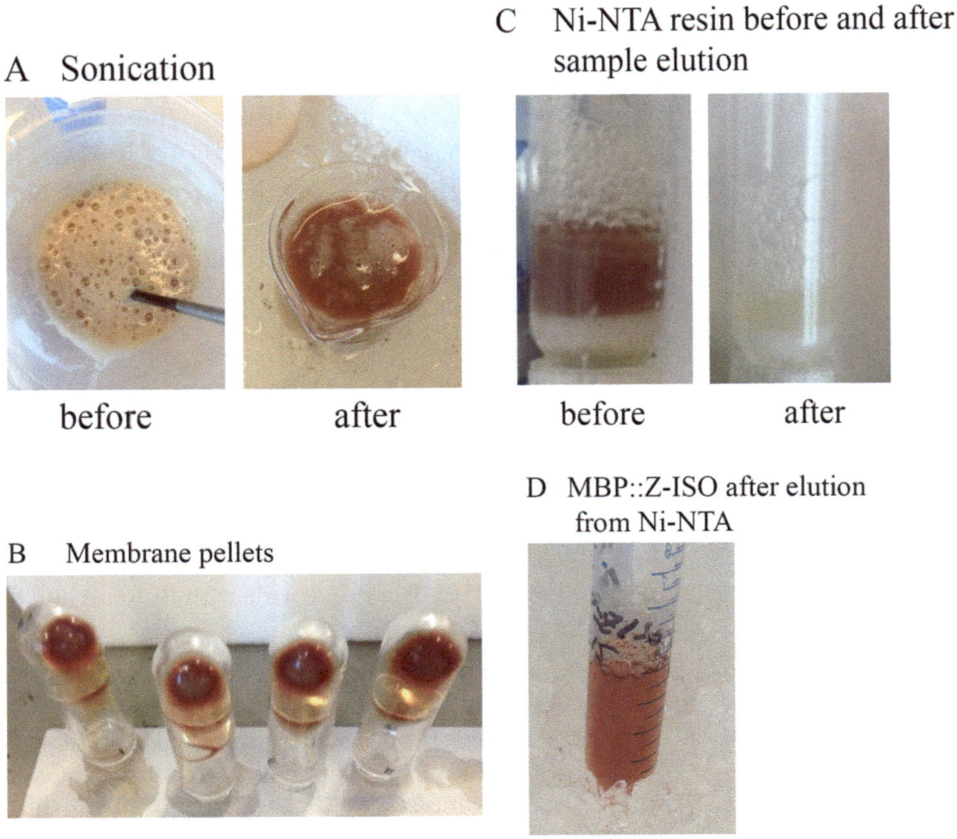

Fig. 2 Stages of MBP::Z-ISO purification. (**a**) Sonication. Upon sonication, the cells should turn from milky white (left) to translucent red (right). (**b**) Membrane pellets recovered after ultracentrifugation. (**c**) Nickel affinity chromatography resin. (**a**) Left to right shows the column resin before and after addition of elution buffer. (**d**) Typical MBP::Z-ISO protein sample eluted from the nickel affinity resin shows a reddish color conferred by the heme

7. Measure ~1.4 mL Ni-NTA containing resin (600 μL resin per 40 mL lysate). Equilibrate the Ni-NTA containing resin twice with Buffer A by centrifugation at 500 × *g*.
8. Add the equilibrated Ni-NTA containing resin to the protein sample at a ratio of ~600 μL resin per 40 mL of lysate in Falcon tubes. Incubate overnight at 4 °C with slow rotation end over end.
9. Centrifuge Falcon tubes for 5 min at 500 × *g*. Careful discard the supernatant and transfer the resin pellet (with a little supernatant) to two 5 mL polypropylene columns (Qiagen) which are capped at the bottom. Let the resin settle and then remove the bottom cap to allow most of the resin to flow through, without letting the resin go dry. Cap the tube at the bottom. The resin should appear red (*see* Fig. 2c).

10. Add 5 resin volumes of ATP Wash Buffer, Buffer D at a ratio of 3 mL per column, carefully mix, and then gently rotate at 4 °C for 30 min. Cap at the top and incubate at 4 °C while rotating the sample gently end over end (*see* **Note 7**). Let the resin settle and then remove the bottom cap to allow most of the resin to flow through, without letting the resin go dry. Cap the tube at the bottom. Save this flow-through fraction to determine the total protein content of unbound protein.
11. Elute MBP::Z-ISO from the resin using Elution Buffer, Buffer E at a ratio of ~1 mL per liter of initial culture. Add 0.5 mL at a time. Mix with entire resin and then elute into the collection tube. Repeat eluting with 0.5 mL Buffer E until the resin is clear (*see* Fig. 2c). The eluate should appear red (*see* Fig. 2d) because the Z-ISO in MBP::Z-ISO carries a heme. After protein and heme quantification below, the best fractions can be pooled (*see* **Note 8**).
12. Quantify the protein by absorbance (280–290 nm) and the heme by absorbance (414–415 nm for oxidized ferric heme or 426 nm for reduced ferrous heme) using a NanoDrop instrument or a spectrophotometer. Blank the instruments with Elution Buffer, Buffer E. For the NanoDrop, use settings of $\varepsilon = 130.75\ \mathrm{L\ mol^{-1}\ cm^{-1}}$ and MW = 75.46 kD. The expected yield is ~5–6 mg $\mathrm{L^{-1}}$ culture with ferric heme (414 nm) to protein (280–290 nm) ratio of ~0.7–0.8 (which is about 1 heme to protein).
13. Optional: The protein can be further dialyzed depending upon further analyses. Using new gloves, prewet a dialysis cassette with Dialysis Buffer, Buffer F, kept at 4 °C. Load the cassette with the protein sample which has been concentrated, if necessary (*see* **Note 9**). Dangle the cassette in 2 L of precooled Dialysis buffer, Buffer F, stirring overnight at 4 °C. A 20 gauge needle can be used to remove the sample, taking care not to damage the membrane. After dialysis, measure the protein and heme concentration as described above using Buffer F to blank the NanoDrop or spectrophotometer. The sample can be further concentrated, depending upon the downstream analysis (*see* **Note 10**).

4 Notes

1. Do not exceed 500 mL culture per 3 L Fernbach flask because yields will be reduced.
2. Addition of heme biosynthesis precursors will improve heme biosynthesis and double the heme loading of the MBP::Z-ISO

fusion protein to increase the amount of Z-ISO containing heme.

3. Repeat multiple rounds of cell cultures to combine for later processing if large quantities of purified protein are desired.
4. From this point on, keep all samples and solutions on ice.
5. Choose a beaker size such that the volume of the sample is about 1/3–1/4 of the beaker and the sonication probe remans immerged in the sample fluid. Sonications should be done in an ice bucket filled with ice and water so that the sample stays cold. To prevent hearing loss, place the ice bucket containing the protein sample in a closed hood and sonicate with ear protectors or ear plugs. After sonication, as the ice melts from the heat, add more ice and remove some water to maintain cold temperature for the sample. The number of sonications depends on the color observed after sonication. The supernatant should go from a milky pink to a deep translucent red (*see* Fig. 2a). If the sample is insufficiently sonicated, the protein yield is reduced. If oversonicated, protein quality is affected. The sample can be divided into multiple batches, if needed.
6. This step is important for purifying the membrane fraction containing MBP::Z-ISO. Carefully remove the supernatant where there is likely an abundance of a truncated MBP product but not the MBP::Z-ISO fusion protein.
7. When adding buffers to the resin, make sure that the entire resin is mixed with buffer.
8. Do not elute the His-tagged Z-ISO with imidazole as imidazole binds to the heme.
9. Protein samples can be concentrated using centrifugal microconcentrators from vendors such as Amicon.
10. Depending on the subsequent analyses, the sample can be further concentrated after dialysis. For electron paramagnetic resonance (EPR) studies, use ~5 mg protein samples at a concentration of ~15.1 mg mL^{-1} (~200 μM); for magnetic circular dichroism (MCD), use 1.6 mg protein samples at 1.57 mg mL^{-1} (~20 μM); for inductively coupled plasma optical emission spectrometry (ICP-OES), use 1 mg at 1 mg mL^{-1} [5].

Acknowledgments

Dr. AImin Liu (The University of Texas at San Antonio) provided helpful advice on optimizing heme biosynthesis for maximal heme-loading of Z-ISO produced in *E. coli*. Research in the Wurtzel laboratory has been funded by the National Institutes of Health (grant GM081160), National Science Foundation, American Cancer Society, Rockefeller Foundation International Rice Biotechnology Program, McKnight Foundation, USDA, DOD, New York State, CUNY, and Lehman College.

References

1. Moise AR, Al-Babili S, Wurtzel ET (2014) Mechanistic aspects of carotenoid biosynthesis. Chem Rev 114(1):164–193
2. Wurtzel ET (2019) Changing form and function through carotenoids and synthetic biology. Plant Physiol 179(3):830–843. https://doi.org/10.1104/pp.18.01122
3. Chen Y, Li F, Wurtzel ET (2010) Isolation and characterization of the Z-ISO gene encoding a missing component of carotenoid biosynthesis in plants. Plant Physiol 153(1):66–79
4. Li F, Murillo C, Wurtzel ET (2007) Maize Y9 encodes a product essential for 15-cis zetacarotene isomerization. Plant Physiol 144(2):1181–1189
5. Beltrán J, Kloss B, Hosler JP, Geng J, Liu A, Modi A, Dawson JH, Sono M, Shumskaya M, Ampomah-Dwamena C, Love JD, Wurtzel ET (2015) Control of carotenoid biosynthesis through a heme-based cis-trans isomerase. Nat Chem Biol 11(8):598–605

Chapter 5

Determination of In Vitro and In Vivo Activities of Plant Carotenoid Cleavage Oxygenases

Lourdes Gómez-Gómez, Gianfranco Diretto, Oussama Ahrazem, and Salim Al-Babili

Abstract

Carotenoid cleavage products, apocarotenoids, are biologically active compounds exerting important functions as chromophore, hormones, signaling molecules, volatiles, and pigments. Apocarotenoids are generally synthesized by the carotenoid cleavage dioxygenases (CCDs) that comprise a ubiquitous family of enzymes. The activity of plant CCDs was unraveled more than 20 years ago, with the characterization of the maize VP14, the first identified CCD. The protocol developed to determine the activity of this enzyme in vitro is still being used, with minor modifications. In addition, in vivo procedures have been developed during these years, mainly based on the exploitation of *Escherichia coli* cells engineered to produce specific carotenoid substrates. Further, technological developments have led to significant improvements, contributing to a more efficient detection of the reaction products. This chapter provides an updated set of detailed protocols suitable for the in vitro and in vivo characterization of the activities of CCDs, starting from well-established methods.

Key words Apocarotenoids, Carotenoids, CCD, Enzymatic activity

1 Introduction

In all clades of life, enzymatic breakage of carotenoids gives rise to biologically active molecules, known as apocarotenoids. In plants, apocarotenoids play important functions, which include their role as hormones, such as abscisic acid (ABA) and strigolactones (SL) [1–3] and as signals regulating different developmental processes [4–6] and involved in thermotolerance and adaptation to environmental stress [7, 8]. In addition, apocarotenoids contribute plant-to-plant communication [9, 10], and to herbivore and pathogen defense [11–14], and attract animals for pollination and seed dispersal [15–18].

Carotenoid cleavage, the initial step in the biosynthesis of apocarotenoids, is generally mediated by nonheme iron

Manuel Rodríguez-Concepción and Ralf Welsch (eds.), *Plant and Food Carotenoids: Methods and Protocols*, Methods in Molecular Biology, vol. 2083, https://doi.org/10.1007/978-1-4939-9952-1_5,

(II) enzymes, the carotenoid cleavage dioxygenases (CCDs) that catalyze the oxidative cleavage of carbon-carbon double bonds in carotenoid backbones, generating products with aldehyde or ketone end groups [19]. Plants CCDs enzymes are classified into two big subfamilies [20]. The nine-*cis*-epoxy-carotenoid-dioxygenases (NCED) subfamily represented by the first characterized CCD, the maize VP14 [1], that catalyzes the synthesis of the ABA precursor xanthoxin by cleaving the C11,C12 double bond in 9-*cis*-violaxanthin and 9′-*cis*-neoxanthin. NCEDs are represented in practically all plant species by a multigene family [21]. The second subfamily is generally referred to as CCD, regardless the type of carotenoid substrates and the cleavage site, and includes all other members of the CCD family, which are not involved in ABA biosynthesis. Indeed, plant CCDs cleave a variety of carotenoid and apocarotenoid substrates at different positions [22–28]. The plant CCD subfamily includes CCD1, CCD2, CCD4, CCD7, and CCD8 enzymes [20]. Large scale genomic sequencing projects on plant species has allowed the identification of many putative CCDs that have expanded this subfamily. However, biochemical studies have only been reported for a limited number of CCDs [20, 29].

Two main strategies, in vitro and/or in vivo studies, have been followed to characterize the activity of CCD enzymes. These studies have been performed with CCDs from many plant species, including Arabidopsis [30], saffron [31], rose [32], petunia [33], tomato [34, 35], potato [22], rice [36], melon [37], maize [38], Vitis [39, 40], citrus [26, 41], and other plant species [42]. The in vivo studies have been done by expressing CCDs in carotenoid accumulating *E. coli* strains containing specific plasmids that allowed the accumulation of different carotenoids substrates (Table 1). The in vitro approach allows testing of much more substrates, including apocarotenoids themselves, and enables comprehensive and quantitative characterization of investigated enzymes. In both cases, products formed have been detected and identified through thin-layer chromatography (TLC), HPLC, LC-MS, or GC-MS.

2 Materials

2.1 In Vivo Assays

1. Bacterial strains. Most common *E. coli* strains, such as JM109, XL1-Blue, and TOPO-10, used for cloning are utilized to accumulate carotenoids.
2. Plasmids. They are listed in (Table 1) and can be purchased from Addgene (https://www.addgene.org).
3. Luria Bertani (LB) medium. Combine 10 g tryptone, 10 g NaCl, and 5 g yeast extract in 900 mL ddH_2O and shake

Table 1
Plasmids for carotenoids production in *E. coli*

Plasmid name	Genes present	Carotenoid produced
pAC-PHYTipi	Contains crtE, crtB, and idi genes of *Erwinia herbicola* (*Pantoea agglomerans*) Eho10	Phytoene
pAC-ZETAipi	Contains crtE, crtB, and idi genes of *Erwinia herbicola* (*Pantoea agglomerans*) Eho10, and crtP gene of *Synechococcus* PCC7942	ζ-Carotene
pAC-NEUR	Contains the *Erwinia herbicola* (*Pantoea agglomerans*) Eho10 genes crtE and crtB, and the *Rhodobacter capsulatus* crtI gene	Neurosporene
pAC-LYCipi	Contains crtE, crtB, crtI, and idi genes of *Erwinia herbicola* (*Pantoea agglomerans*) Eho10	Lycopene
pAC-DELTA	Contains crtE, crtB, and crtI genes of *Erwinia herbicola* (*Pantoea agglomerans*) Eho10, together with the lcyE gene of *Arabidopsis thaliana*	δ-Carotene
pAC-BETAipi	Contains ctrE, crtB, crtI, crtY, and idi genes of *Erwinia herbicola* (*Pantoea agglomerans*) Eho10	β-Carotene
pAC-ZEAXipi	Contains crtE, idi, crtI, crtY, crtB, and crtZ genes of *Erwinia herbicola* (*Pantoea agglomerans*) Eho10	Zeaxanthin
pAC-VIOL	Contains crtE, idi, crtI, crtY, crtB, and crtZ genes of *Erwinia herbicola* (*Pantoea agglomerans*) Eho10 and ZEP from *Arabidopsis thaliana*	Violaxanthin

All the plasmids are low copy bacterial cloning vectors and provide chloramphenicol resistance

until the solutes have dissolved. Adjust the pH to 7.0 with 5 N NaOH. Adjust the final volume of the solution to 1 L with ddH_2O. Sterilize by autoclaving for 20 min at 15 psi. For solid LB medium, add 15 g of agar.

4. Antibiotics stocks. Ampicillin, carbenicillin, and kanamycin are dissolved in ddH_2O at a concentration of 100 mg/mL for ampicillin and 50 mg/mL for carbenicillin and kanamycin. Chloramphenicol is dissolved in EtOH at a concentration of 50 mg/mL.
5. Organic solvents for carotenoid and apocarotenoid extraction and analysis. Acetone, MeOH, $CHCl_2$, $CHCl_3$.
6. Specific oligonucleotides for cloning in an expression vector the CCD to be characterized.
7. Expression vectors. Most frequently used expression vectors are shown in Table 2.
8. 20% arabinose dissolved in ddH_2O and sterilized by filtration.
9. 1 mM isopropyl β-D-thiogalactopyranoside (IPTG) dissolved in ddH_2O and sterilized by filtration.
10. Nitrogen.
11. Vials and columns for HPLC and GC analyses.

Table 2
Expression vectors and conditions used for in vivo characterization of CCDs activities

Expression vector	Resistance	Induction	References
pGEX-2 T	Ampicillin	0.1 mM IPTG	[39, 43]
pGEX 5X-3	Ampicillin	0.1 mM IPTG	[38, 44]
pDEST15	Carbenicillin	0.1% arabinose	[30, 38]
pDEST15	Carbenicillin	0.2% arabinose	[34]
pDEST14	Ampicillin	0.1% arabinose	[45]
pDEST14	Ampicillin	0.2% arabinose	[31, 33]
pBK-CMV	Kanamycin	1 mM IPTG	[37]
pTWIN1	Ampicillin	0.1 mM IPTG	[40]
pBAD/THIO-TOPO®TA	Ampicillin	0.2% arabinose	[22, 46]
pBAD/THIO	Ampicillin	0.2% arabinose	[27, 35]
pBAD/TOPO	Ampicillin	0.08% arabinose	[36]
pTHIO-DAN1	Ampicillin	0.2% arabinose	[25]
pGEX-4T1	Ampicillin	0.1 mM IPTG	[42, 47]
pGEX-4T1	Ampicillin	0.2 mM IPTG	[48]
pET28	Kanamycin	0.5 mM IPTG	[49]
pRSETA	Ampicillin	1% glucose	[50]

2.2 In Vitro Assays

1. Bacterial strains. Most commonly used are *E. coli* strains BL21 DE3pLysS, BL21A1, and XL1-Blue.
2. Plasmids: pGro7, encoding the groES-groEL-chaperone system used to improve protein folding.
3. Luria Bertani (LB) medium. *See* **item 2** in Subheading 2.1.
4. 2YT medium. Dissolve 16 g tryptone, 5 g NaCl and 10 g yeast extract in 900 mL ddH$_2$O, adjust the pH to 7.0 with 5 N NaOH and the final volume of the solution to 1 L with ddH$_2$O. Sterilize by autoclaving for 20 min at 15 psi.
5. Purified carotenoid and apocarotenoid substrates.
6. Carotenoid and apocarotenoid substrates for the assays (*see* **Note 1**). Resuspend the substrates in 200 μL of benzene (or dichloromethane) at a final concentration of 50–100 μM and mixed with 150 μL of an ethanolic detergent mixture consisting of 0.7% (v/v) Triton X-100 and 1.6% (v/v) Triton X-405, for lycopene and prolycopene substrates or with 250 μL of an ethanolic β-octylglucoside solution (4% w/v) for apocarotenoids and other carotenoids. The final mix is evaporated to dryness in a vacuum centrifuge to get a carotenoid gel on the bottom of the tube (*see* **Note 2**).

Table 3
Expression vectors and conditions used for in vitro characterization of CCDs activities

Expression vector	Resistance	Induction	References
pGEX-2T	Ampicillin	0.2 mM IPTG	[43]
pGEX-4T1	Ampicillin	0.2 mM IPTG	[42]
pGEX-4T1	Ampicillin	No induction	[47, 51]
pThio-DAN2	Ampicillin	0.2% arabinose	[22, 24–26, 52]
pBAD/THIO-TOPO®TA	Ampicillin	0.2% arabinose	[46]
pGEX 5X-2	Ampicillin	0.2 mM IPTG	[35, 36]
pGEX 5X-3	Ampicillin	0.2 mM IPTG	[44]
pGEX-6P-1	Ampicillin	0.1 mM IPTG	[41]

7. 2× Modified LEW (Lysis-Equilibration-Wash) buffer: 100 mM NaH_2PO_4, 600 mM NaCl, 1 mg/mL lysozyme, 0.1% Triton X-100 and 1 mM DTT. Adjust pH to 8.0 with NaOH.
8. Detergents: Triton X-100, Triton X-405, and octyl-β-glucoside.
9. Organic solvents for extraction and substrates preparation: acetone, benzene, $CHCl_2$, $CHCl_3$, petroleum, ether petroleum, and diethyl ether.
10. Expression vectors. Most frequently used expression vectors are present in Table 3.
11. 20% Arabinose dissolved in ddH_2O and sterilized by filtration.
12. 1 mM isopropyl β-d-thiogalactopyranoside (IPTG) dissolved in ddH_2O and sterilized by filtration.
13. 2× Assay incubation buffer: 200 mM HEPES–NaOH, pH 7.8, 2 mM Tris(2-carboxyethyl)phosphine hydrochloride (TCEP), 0.4 mM $FeSO_4$, 2 mg mL/L catalase.
14. Vials and columns for HPLC and GC analyses.

2.3 Equipment

1. Table centrifuge.
2. Shaker with controlled temperature.
3. Spectrophotometer.
4. French pressure cell or an ultrasonic homogenizer.
5. Vacuum centrifuge.
6. Nitrogen.
7. Balance.
8. HPLC-PDA.

9. HPLC-PDA-MS.
10. GC-MS.

3 Methods

3.1 In Vivo Assays

1. Create a stock of *E. coli* cells producing different carotenoid substrates to use for your assays. Transform competent cells of the selected *E. coli* strain with each of the pAC vectors, by using standard transformation procedure, such as electroporation [53] or heat-shock [54]. Grow individual colonies to prepare glycerol stocks.
2. Make competent *E. coli* strains engineered to accumulate different carotenoid substrates by a standard procedure (http://cshprotocols.cshlp.org/) and keep them at −80 °C.
3. Transform the competent *E. coli* strains engineered to accumulate different carotenoid substrates with the selected expression vectors harboring the CCD under study (*see* **Notes 3** and **4**) and the empty vector, which is used as a negative control of activity.
4. Take individual colonies to initiate a preculture in a volume of 3 mL LB containing the appropriate antibiotics, and grow overnight at 37 °C.
5. Use 2.5 mL of the overnight cultures to inoculate 50 mL of LB with final concentration of antibiotics reduced to half, in 200 mL flasks. Grow bacteria at 37 °C with shaking (180 rpm) to an OD 600 nm of 0.5, and induced with the appropriate agent (arabinose or IPTG), depending on the expression vector selected. Incubate the cultures with shaking at 28 °C for 6 h, or 20 °C –16 °C overnight in the dark (*see* **Note 5**).
6. Transfer the content of the flask to 50 mL tubes and harvest the *E. coli* cells by centrifugation. The color intensities of the resulting bacterial pellets will be compared by visible inspection to control bacterial cells harboring the empty expression plasmid. Changes in intensities are in many cases an indicator for CCD activity. However, it has been also observed that the expression of the CCD interferes with carotenoid biosynthesis, causing color change without cleavage activity. (You always need a product to be sure that you have a cleavage enzyme and the right substrate.)
7. For pigment extraction for HPLC analyses, add 2.5 mL acetone to the bacterial pellet and vortex until total resuspension. Centrifuge al 9000 × *g* for 10 min and collect the liquid part. If the pellet remain pigmented repeat again the MeOH:$CHCl_3$ extraction. Combine all the extracts and dry under nitrogen. In

addition after the first acetone extraction, the obtained pellet could be extracted by adding to the pellet 2.5 mL of MeOH: $CHCl_3$ (1:1), and vortex until total resuspension. Centrifuge al 9000 × *g* for 10 min and collect the liquid.

8. Resuspend the dry residue in $CHCl_3$, $CHCl_2$ or acetone and stored at −80 °C prior to HPLC-PDA or HPLC-PDA-MS analyses.
9. For volatile analyses, after induction of the cultures (point 5) transfer 5 mL to a tightly closed 50 mL tube. Incubate the tubes at 16 °C for 20 h (125 rpm). An SPME fiber (65 μm polydimethylsiloxane-divinylbenzene) is introduced into the vial through a septum and the headspace volatiles are allowed to be absorbed by the fiber at 45 °C for 30 min. The SPME fiber is keep at −80 °C until the GC analyses.

3.2 In Vitro Assays

1. Amplified the CCD sequence with the corresponding specific oligonucleotides by PCR and cloned it in the selected expression vectors.
2. Transform the competent *E. coli* strain selected with the expression vectors harboring the CCD under study (*see* **Notes 3** and **4**) and the empty vector that is used as a negative control of activity (*see* **Note 6**).
3. Take individual colonies to initiate a preculture in a volume of 3 mL LB containing the appropriate antibiotics, and grow overnight at 37 °C.
4. Use 2.5 mL of the overnight cultures to inoculate 50 mL of 2×YT with the appropriate antibiotics at half strength. Grow bacteria at 28 °C with shaking (150 rpm) to an OD 600 nm of 0.7, and induced with the appropriate agent (arabinose or IPTG), depending on the expression vector selected (Table 3). Incubate the cultures at 28 °C with shaking (150 rpm) for 4-6 h or overnight at 18–20 °C (*see* **Note 5**).
5. Harvest the *E. coli* cells by centrifugation for 6 min at 6000 × *g*. Resuspend the cell pellet in 1 mL 2× LEW buffer, and incubate the tubes for 30 min on ice. Disrupt the cells by sonication, French pressing, or other preferred method [55], and centrifuged at 12,000 × *g* for 30 min at 4 °C. Use the isolated supernatant for the in vitro assays.
6. Prepare the carotenoid and or apocarotenoid substrates for the assays. *See* **item 6** in Subheading 2.2.
7. For the in vitro assay. In a total volume of 200 μL, the carotenoid gel is resuspended by vortex in 100 μL 2× assay incubation buffer (single assay) and sonicated to obtain a clear micellar solution. Add 50 μL of crude protein lysate and ddH_2O to adjust at 200 μL. Transfer the tubes into a shaker (200 rpm) at

28–30 °C in darkness for 2–6 h or overnight. The reaction can be follow by visible inspection of the crude protein containing your CCD to the control extracts. Changes in color indicate activity of the CCD.

8. For HPLC analysis, the assay is stopped by adding 400 μL of acetone, vortex, and add 600 μL petroleum ether: diethyl ether (1:4), and vortex again. Centrifuge the tube at 12,000 × *g* 1 min. After centrifugation, there are two phases visible. Take the epiphase, and transfer it to a new clean tube. Repeat this step with the lower aqueous phase. Combine the two the epiphases and dry in a vacuum centrifuge. Samples can be store at −80 °C until HPLC-PDA or HPLC-PDA-MS analyses.
9. For GC analyses, the assay is performed in screw-top GC glass vials, and volatiles are collected for 30 min after the reaction incubation time, using SPME (solid phase microextraction) fibers (100 μm polydimethylsiloxane).

4 Notes

1. Carotenoid and apocarotenoids can be purchase from commercial providers, but also you can purified by your own the substrates you need. For most of the carotenoid substrates, you can use the bacterial strains that accumulate the different carotenoids and purified them by TLC. For apocarotenoids or most specific carotenoids substrates don't produced by the bacterial strains, you can use the plants known to accumulate those metabolites and proceed with the isolation and purification as referred in the corresponding literature associated to those compounds.
2. Carotenoid and apocarotenoid substrates differ in their solubility in organic substrates, and different detergent. In addition, the detergent can have an effect on the enzyme activity. Different studies have used solvents such as acetone, benzene, hexane, and chloroform and detergents such as Triton X-100, Tween 20, Tween 80, Span 20, Span 80, and β-octylglucoside to form micelles of carotenoids under aqueous conditions. Thus, the formation of detergent micelles of a specific carotenoid or apocarotenoid substrate under aqueous conditions to be used for in vitro conditions will need optimization.
3. Removal of signal peptides may improve the enzymatic activity, as it can interfere with the fold of the CCD and with its activity.
4. At the time of selection of the expression vector, some researches prefer to choose a vector with a fusion partners or tags to improve protein production yields, solubility and

folding, which result in higher activity. CCDs are problematic to express in bacteria cells as they tend to be insoluble.

5. Incubation can be done the first time at 28 °C for 4 h. Although longer incubation times at lower temperature improve the final result. Dark conditions prevent possible photodegradation of carotenoids.

6. Several strategies have been made for efficient production of proteins in *E. coli*, like the coproduction of chaperones to help correct protein folding. The cotransformation of *E. coli* cells with the plasmid pGro7, encoding the groES-groEL-chaperone system, could improve protein folding and further enzyme activity.

References

1. Schwartz SH, Tan BC, Gage DA, Zeevaart JA, McCarty DR (1997) Specific oxidative cleavage of carotenoids by VP14 of maize. Science 276 (5320):1872–1874
2. Akiyama K, Matsuzaki K, Hayashi H (2005) Plant sesquiterpenes induce hyphal branching in arbuscular mycorrhizal fungi. Nature 435 (7043):824–827.. nature03608. https://doi.org/10.1038/nature03608
3. Umehara M, Hanada A, Yoshida S, Akiyama K, Arite T, Takeda-Kamiya N, Magome H, Kamiya Y, Shirasu K, Yoneyama K, Kyozuka J, Yamaguchi S (2008) Inhibition of shoot branching by new terpenoid plant hormones. Nature 455(7210):195–200.. nature07272. https://doi.org/10.1038/nature07272
4. Van Norman JM, Sieburth LE (2007) Dissecting the biosynthetic pathway for the bypass1 root-derived signal. Plant J 49(4):619–628. TPJ2982. https://doi.org/10.1111/j.1365-313X.2006.02982.x
5. Van Norman JM, Zhang J, Cazzonelli CI, Pogson BJ, Harrison PJ, Bugg TD, Chan KX, Thompson AJ, Benfey PN (2014) Periodic root branching in Arabidopsis requires synthesis of an uncharacterized carotenoid derivative. Proc Natl Acad Sci U S A 111(13): E1300–E1309. https://doi.org/10.1073/pnas.14030161111403016111. [pii]
6. Avendano-Vazquez AO, Cordoba E, Llamas E, San Roman C, Nisar N, De la Torre S, Ramos-Vega M, Gutierrez-Nava MD, Cazzonelli CI, Pogson BJ, Leon P (2014) An uncharacterized Apocarotenoid-derived signal generated in zeta-carotene Desaturase mutants regulates leaf development and the expression of chloroplast and nuclear genes in Arabidopsis. Plant Cell 26(6):2524–2537.. tpc.114.123349. https://doi.org/10.1105/tpc.114.123349
7. Havaux M (2014) Carotenoid oxidation products as stress signals in plants. Plant J 79 (4):597–606. https://doi.org/10.1111/tpj.12386
8. Saeed W, Naseem S, Ali Z (2017) Strigolactones biosynthesis and their role in abiotic stress resilience in plants: a critical review. Front Plant Sci 8:1487. https://doi.org/10.3389/fpls.2017.01487
9. Macias FA, Lopez A, Varela RM, Torres A, Molinillo JM (2004) Bioactive apocarotenoids annuionones F and G: structural revision of annuionones a, B and E. Phytochemistry 65 (22):3057–3063.. S0031-9422(04)00429-7. https://doi.org/10.1016/j.phytochem.2004.08.048
10. Yoneyama K, Awad AA, Xie X, Takeuchi Y (2010) Strigolactones as germination stimulants for root parasitic plants. Plant Cell Physiol 51(7):1095–1103. https://doi.org/10.1093/pcp/pcq055pcq055. [pii]
11. Caceres LA, Lakshminarayan S, Yeung KK, McGarvey BD, Hannoufa A, Sumarah MW, Benitez X, Scott IM (2016) Repellent and attractive effects of alpha-, beta-, and Dihydro-beta- ionone to generalist and specialist herbivores. J Chem Ecol 42(2):107–117. https://doi.org/10.1007/s10886-016-0669-z10.1007/s10886-016-0669-z. [pii]
12. Wei S, Hannoufa A, Soroka J, Xu N, Li X, Zebarjadi A, Gruber M (2011) Enhanced beta-ionone emission in Arabidopsis over-expressing AtCCD1 reduces feeding damage in vivo by the crucifer flea beetle. Environ Entomol 40(6):1622–1630. https://doi.org/10.1603/EN11088
13. Park S, Takano Y, Matsuura H, Yoshihara T (2004) Antifungal compounds from the root

and root exudate of Zea mays. Biosci Biotechnol Biochem 68(6):1366–1368

14. Mikhlin ED, Radina VP, Dmitrovskii AA, Blinkova LP, Butova LG (1983) Antifungal and antimicrobial activity of beta-ionone and vitamin a derivatives. Prikl Biokhim Mikrobiol 19 (6):795–803
15. Rodríguez A, Alquézar B, Peña L (2013) Fruit aromas in mature fleshy fruits as signals of readiness for predation and seed dispersal. New Phytol 197(1):36–48. https://doi.org/10.1111/j.1469-8137.2012.04382.x
16. Rivera-Madrid R, Aguilar-Espinosa M, Cárdenas-Conejo Y, Garza-Caligaris LE (2016) Carotenoid Derivates in Achiote (Bixa orellana) seeds: synthesis and health promoting properties. Front Plant Sci 7:1406. https://doi.org/10.3389/fpls.2016.01406
17. Baldermann S, Yamamoto M, Yang Z, Kawahashi T, Kuwano K, Watanabe N (2013) C_{13}-Apocarotenoids: more than Flavor compounds? In: Carotenoid cleavage products, vol 1134. ACS symposium series, vol 1134. American Chemical Society, pp 73–80. doi:https://doi.org/10.1021/bk-2013-1134.ch007
18. Ahrazem O, Rubio-Moraga A, Nebauer SG, Molina RV, Gomez-Gomez L (2015) Saffron: its Phytochemistry, developmental processes, and biotechnological prospects. J Agric Food Chem 63(40):8751–8764. https://doi.org/10.1021/acs.jafc.5b03194
19. Harrison PJ, Bugg TD (2014) Enzymology of the carotenoid cleavage dioxygenases: reaction mechanisms, inhibition and biochemical roles. Arch Biochem Biophys 544:105–111. https://doi.org/10.1016/j.abb.2013.10.005S0003-9861(13)00307-X. [pii]
20. Ahrazem O, Gomez-Gomez L, Rodrigo MJ, Avalos J, Limon MC (2016) Carotenoid cleavage Oxygenases from microbes and photosynthetic organisms: features and functions. Int J Mol Sci 17(11). E1781 [pii]ijms17111781). https://doi.org/10.3390/ijms17111781
21. Seo M, Koshiba T (2002) Complex regulation of ABA biosynthesis in plants. Trends Plant Sci 7(1):41–48. [pii]
22. Bruno M, Beyer P, Al-Babili S (2015) The potato carotenoid cleavage dioxygenase 4 catalyzes a single cleavage of beta-ionone ring-containing carotenes and non-epoxidated xanthophylls. Arch Biochem Biophys 572:126–133. S0003-9861(15)00073-9. https://doi.org/10.1016/j.abb.2015.02.011
23. Bruno M, Vermathen M, Alder A, Wust F, Schaub P, van der Steen R, Beyer P, Ghisla S, Al-Babili S (2017) Insights into the formation of carlactone from in-depth analysis of the CCD8-catalyzed reactions. FEBS Lett 591 (5):792–800. https://doi.org/10.1002/1873-3468.12593
24. Bruno M, Hofmann M, Vermathen M, Alder A, Beyer P, Al-Babili S (2014) On the substrate- and stereospecificity of the plant carotenoid cleavage dioxygenase 7. FEBS Lett 588(9):1802–1807. https://doi.org/10.1016/j.febslet.2014.03.041S0014-5793(14)00260-9. [pii]
25. Frusciante S, Diretto G, Bruno M, Ferrante P, Pietrella M, Prado-Cabrero A, Rubio-Moraga-A, Beyer P, Gomez-Gomez L, Al-Babili S, Giuliano G (2014) Novel carotenoid cleavage dioxygenase catalyzes the first dedicated step in saffron crocin biosynthesis. Proc Natl Acad Sci U S A 111(33):12246–12251. https://doi.org/10.1073/pnas.1404629111 1404629111. [pii]
26. Rodrigo MJ, Alquezar B, Alos E, Medina V, Carmona L, Bruno M, Al-Babili S, Zacarias L (2013) A novel carotenoid cleavage activity involved in the biosynthesis of Citrus fruit-specific apocarotenoid pigments. J Exp Bot 64 (14):4461–4478. ert260 [pii]. https://doi.org/10.1093/jxb/ert260
27. Ahrazem O, Diretto G, Argandona J, Rubio-Moraga A, Julve JM, Orzaez D, Granell A, Gomez-Gomez L (2017) Evolutionarily distinct carotenoid cleavage dioxygenases are responsible for crocetin production in Buddleja davidii. J Exp Bot 68(16):4663–4677. https://doi.org/10.1093/jxb/erx2774068697. [pii]
28. Auldridge ME, McCarty DR, Klee HJ (2006) Plant carotenoid cleavage oxygenases and their apocarotenoid products. Curr Opin Plant Biol 9(3):315–321.. S1369-5266(06)00044-6. https://doi.org/10.1016/j.pbi.2006.03.005
29. Rodriguez-Concepcion M, Avalos J, Bonet ML, Boronat A, Gomez-Gomez L, Hornero-Mendez D, Limon MC, Melendez-Martinez AJ, Olmedilla-Alonso B, Palou A, Ribot J, Rodrigo MJ, Zacarias L, Zhu C (2018) A global perspective on carotenoids: metabolism, biotechnology, and benefits for nutrition and health. Prog Lipid Res 70:62–93. S0163-7827 (17)30039-5. https://doi.org/10.1016/j.plipres.2018.04.004
30. Auldridge ME, Block A, Vogel JT, Dabney-Smith C, Mila I, Bouzayen M, Magallanes-Lundback M, DellaPenna D, McCarty DR, Klee HJ (2006) Characterization of three members of the Arabidopsis carotenoid cleavage dioxygenase family demonstrates the divergent roles of this multifunctional enzyme family. Plant J 45(6):982–993. TPJ2666 [pii]. https://doi.org/10.1111/j.1365-313X.2006.02666.x

31. Rubio A, Rambla JL, Santaella M, Gomez MD, Orzaez D, Granell A, Gomez-Gomez L (2008) Cytosolic and plastoglobule-targeted carotenoid dioxygenases from Crocus sativus are both involved in beta-ionone release. J Biol Chem 283(36):24816–24825.. M804000200. https://doi.org/10.1074/jbc.M804000200
32. Huang FC, Horvath G, Molnar P, Turcsi E, Deli J, Schrader J, Sandmann G, Schmidt H, Schwab W (2009) Substrate promiscuity of RdCCD1, a carotenoid cleavage oxygenase from Rosa damascena. Phytochemistry 70 (4):457–464.. S0031-9422(09)00059-4. https://doi.org/10.1016/j.phytochem.2009.01.020
33. Simkin AJ, Underwood BA, Auldridge M, Loucas HM, Shibuya K, Schmelz E, Clark DG, Klee HJ (2004) Circadian regulation of the PhCCD1 carotenoid cleavage dioxygenase controls emission of beta-ionone, a fragrance volatile of petunia flowers. Plant Physiol 136 (3):3504–3514.. 104.049718. https://doi.org/10.1104/pp.104.049718
34. Simkin AJ, Schwartz SH, Auldridge M, Taylor MG, Klee HJ (2004) The tomato carotenoid cleavage dioxygenase 1 genes contribute to the formation of the flavor volatiles beta-ionone, pseudoionone, and geranylacetone. Plant J 40 (6):882–892. TPJ2263 [pii]. https://doi.org/10.1111/j.1365-313X.2004.02263.x
35. Ilg A, Bruno M, Beyer P, Al-Babili S (2014) Tomato carotenoid cleavage dioxygenases 1A and 1B: relaxed double bond specificity leads to a plenitude of dialdehydes, mono-apocarotenoids and isoprenoid volatiles. FEBS Open Bio 4:584–593. https://doi.org/10.1016/j.fob.2014.06.005S2211-5463(14)00060-6. [pii]
36. Ilg A, Beyer P, Al-Babili S (2009) Characterization of the rice carotenoid cleavage dioxygenase 1 reveals a novel route for geranial biosynthesis. FEBS J 276(3):736–747. EJB6820 [pii]. https://doi.org/10.1111/j.1742-4658.2008.06820.x
37. Ibdah M, Azulay Y, Portnoy V, Wasserman B, Bar E, Meir A, Burger Y, Hirschberg J, Schaffer AA, Katzir N, Tadmor Y, Lewinsohn E (2006) Functional characterization of CmCCD1, a carotenoid cleavage dioxygenase from melon. Phytochemistry 67(15):1579–1589.. S0031-9422(06)00109-9. https://doi.org/10.1016/j.phytochem.2006.02.009
38. Vogel JT, Tan BC, McCarty DR, Klee HJ (2008) The carotenoid cleavage dioxygenase 1 enzyme has broad substrate specificity, cleaving multiple carotenoids at two different bond positions. J Biol Chem 283 (17):11364–11373.. M710106200. https://doi.org/10.1074/jbc.M710106200
39. Mathieu S, Terrier N, Procureur J, Bigey F, Gunata Z (2005) A carotenoid cleavage dioxygenase from Vitis vinifera L.: functional characterization and expression during grape berry development in relation to C13-norisoprenoid accumulation. J Exp Bot 56(420):2721–2731. eri265 [pii]. https://doi.org/10.1093/jxb/eri265
40. Lashbrooke JG, Young PR, Dockrall SJ, Vasanth K, Vivier MA (2013) Functional characterisation of three members of the Vitis vinifera L. carotenoid cleavage dioxygenase gene family. BMC Plant Biol 13:156. https://doi.org/10.1186/1471-2229-13-1561471-2229-13-156. [pii]
41. Ma G, Zhang L, Matsuta A, Matsutani K, Yamawaki K, Yahata M, Wahyudi A, Motohashi R, Kato M (2013) Enzymatic formation of beta-Citraurin from beta-Cryptoxanthin and Zeaxanthin by carotenoid cleavage Dioxygenase4 in the Flavedo of Citrus fruit. Plant Physiol 163(2):682–695. https://doi.org/10.1104/pp.113.223297. pp.113.223297 [pii]
42. Huang FC, Molnar P, Schwab W (2009) Cloning and functional characterization of carotenoid cleavage dioxygenase 4 genes. J Exp Bot 60 (11):3011–3022.. erp137 [pii]. https://doi.org/10.1093/jxb/erp137
43. Schwartz SH, Qin X, Zeevaart JA (2001) Characterization of a novel carotenoid cleavage dioxygenase from plants. J Biol Chem 276 (27):25208–25211. https://doi.org/10.1074/jbc.M102146200. [pii]
44. Schwartz SH, Qin X, Loewen MC (2004) The biochemical characterization of two carotenoid cleavage enzymes from Arabidopsis indicates that a carotenoid-derived compound inhibits lateral branching. J Biol Chem 279 (45):46940–46945. https://doi.org/10.1074/jbc.M409004200. [pii]
45. Booker J, Auldridge M, Wills S, McCarty D, Klee H, Leyser C (2004) MAX3/CCD7 is a carotenoid cleavage dioxygenase required for the synthesis of a novel plant signaling molecule. Curr Biol 14:7
46. Alder A, Holdermann I, Beyer P, Al-Babili S (2008) Carotenoid oxygenases involved in plant branching catalyse a highly specific conserved apocarotenoid cleavage reaction. Biochem J 416(2):289–296.. BJ20080568. https://doi.org/10.1042/BJ20080568
47. Sergeant MJ, Li JJ, Fox C, Brookbank N, Rea D, Bugg TD, Thompson AJ (2009) Selective inhibition of carotenoid cleavage dioxygenases: phenotypic effects on shoot

branching. J Biol Chem 284(8):5257–5264.. M805453200. https://doi.org/10.1074/jbc.M805453200

48. Garcia-Limones C, Schnabele K, Blanco-Portales R, Luz Bellido M, Caballero JL, Schwab W, Munoz-Blanco J (2008) Functional characterization of FaCCD1: a carotenoid cleavage dioxygenase from strawberry involved in lutein degradation during fruit ripening. J Agric Food Chem 56(19):9277–9285. https://doi.org/10.1021/jf801096t
49. Ahrazem O, Rubio-Moraga A, Berman J, Capell T, Christou P, Zhu C, Gomez-Gomez L (2016) The carotenoid cleavage dioxygenase CCD2 catalysing the synthesis of crocetin in spring crocuses and saffron is a plastidial enzyme. New Phytol 209(2):650–663. https://doi.org/10.1111/nph.13609
50. Sun Z, Hans J, Walter MH, Matusova R, Beekwilder J, Verstappen FW, Ming Z, van Echtelt E, Strack D, Bisseling T, Bouwmeester HJ (2008) Cloning and characterisation of a maize carotenoid cleavage dioxygenase (ZmCCD1) and its involvement in the biosynthesis of apocarotenoids with various roles in mutualistic and parasitic interactions. Planta 228(5):789–801. https://doi.org/10.1007/s00425-008-0781-6
51. Schmidt H, Kurtzer R, Eisenreich W, Schwab W (2006) The carotenase AtCCD1 from Arabidopsis thaliana is a dioxygenase. J Biol Chem 281(15):9845–9851.. M511668200. https://doi.org/10.1074/jbc.M511668200
52. Bruno M, Koschmieder J, Wuest F, Schaub P, Fehling-Kaschek M, Timmer J, Beyer P, Al-Babili S (2016) Enzymatic study on AtCCD4 and AtCCD7 and their potential to form acyclic regulatory metabolites. J Exp Bot 67(21):5993–6005.. erw356. https://doi.org/10.1093/jxb/erw356
53. Gonzales MF, Brooks T, Pukatzki SU, Provenzano D (2013) Rapid protocol for preparation of Electrocompetent Escherichia coli and Vibrio cholerae. J Vis Exp 80:50684. https://doi.org/10.3791/50684
54. Froger A, Hall JE (2007) Transformation of plasmid DNA into E. coli using the heat shock method. J Vis Exp 6:253. https://doi.org/10.3791/253
55. Benov L, Al-Ibraheem J (2002) Disrupting Escherichia coli: a comparison of methods. J Biochem Mol Biol 35(4):428–431

Chapter 6

Expression and Characterization of Mammalian Carotenoid Cleavage Dioxygenases

Linda Dora Thomas, Srinivasagan Ramkumar, and Johannes von Lintig

Abstract

Carotenoid cleavage dioxygenases (CCDs) are nonheme iron enzymes that catalyze double bond processing of carotenoids and their apocarotenoid metabolites. Mammalian genomes encode three members of this protein family, namely BCO1, BCO2, and RPE65. Mutations and genetic polymorphism in the corresponding genes are associated with inherited blinding diseases, vitamin A deficiency, and high carotenoid plasma levels. Here we describe a method for the heterologous expression of mammalian BCO1 and BCO2 in *E. coli* and the biochemical characterization of these recombinant enzymes. Dissecting the enzymatic properties of CCDs will advance our knowledge of the biochemical processes that are govern by these disease-associated enzymes and may assist the design of interventions directed against these disease states.

Key words Carotenoids, Retinoids, Carotenoid cleavage dioxygenases, Protein expression, Enzyme assays

1 Introduction

In mammals, carotenoids can be chemically converted to apocarotenoid metabolites [1]. These compounds include vitamin A aldehyde, from which all biologically active retinoids can be synthesized, including visual chromophore and retinoic acid [2]. Retinoic acid regulates gene expression throughout the mammalian life cycle and results in the effects of vitamin A in development, immunity, and cellular differentiation [3–5].

Enzymatic oxidative cleavage of carotenoids at a specific position of the polyene chain has been proposed as the method for apocarotenoid production. This conversion is catalyzed by an ancient family of carotenoid cleavage dioxygenases (CCDs) [6]. CCDs are nonheme iron oxygenases with a characteristic tertiary structure comprised of a rigid seven-bladed β-propeller covered by a half-dome [7]. The ferrous iron in the active center is

Manuel Rodríguez-Concepción and Ralf Welsch (eds.), *Plant and Food Carotenoids: Methods and Protocols*, Methods in Molecular Biology, vol. 2083, https://doi.org/10.1007/978-1-4939-9952-1_6,

accessible through a long, nonpolar tunnel lined with hydrophobic residues that likely act as conduits for passage of the lipophilic substrates. The role of ferrous iron in double bond cleaving CCDs is to activate diatomic oxygen for cleavage of carotenoids. The subsequent reaction follows a dioxygenase mechanism in which both oxygen atoms are incorporated into the apocarotenoid products [8].

The mammalian genome encodes three different types of CCDs. The β-carotene-15,15′-dioxygenase (BCO1) and the β-carotene-9′,10′-dioxygenase (BCO2) enzymes catalyze oxidative cleavage across double bonds at the C15, C15′ and C9′, C10′ positions on the carbon backbone of carotenoids, respectively [9]. A third family member, the retinal pigment epithelium-specific 65 kDa protein (RPE65) does not possess the double bond cleavage activity of BCO1 and BCO2. It catalyzes both the ester cleavage and double bond isomerization reactions at position C10, C11 of retinyl esters (RE) into 11-*cis*-retinol [10].

For the biochemical characterization of mammalian BCO1 and BCO2, *E. coli* expression systems have played a critical role. Initially, expression of these enzymes in carotenoid producing *E. coli* strains provided a robust and reliable platform to biochemically characterize these enzymes [9, 11]. The expression of BCO1 resulted in the conversion of the yellow β-carotene into colorless retinoids. The color shift of the bacteria can be observed with the naked eye and spectroscopically quantified. Additionally, methods for the biochemical characterization of mammalian CCDs in cell free systems were established. The challenging biochemistry of the reactions catalyzed by CCDs initially led to conflicting results. For instance, RPE65 was proposed to be a retinoid binding protein [12, 13] and the existence of a vitamin A forming enzyme was even disputed based on such assays [14]. Later, inconsistent results have been published with regards to the biochemical activity of recombinant primate BCO2 [15–17]. Some of these discrepancies can be explained by the demanding in vitro assay conditions for these enzymes. CCDs catalyze reactions at the lipid aqueous interface which poses a challenge with regards to substrate delivery to the soluble enzymes. Also, chemical purity of the substrate is critical to catalysis and to prevent misinterpretations due to nonenzymatic oxidation of the substrates [14]. Moreover, heterologous expression of CCDs in *E. coli* expression system faces known obstacles such as protein solubility and inclusion body formation [16]. This chapter summarizes the advanced state of methodological knowledge about the biochemical characterization of recombinant mammalian BCO1 and BCO2. It provides a simple and efficient protocol for the heterologous expression and biochemical characterization for this class of enzymes (Fig. 1) and discusses possible pitfalls and problems.

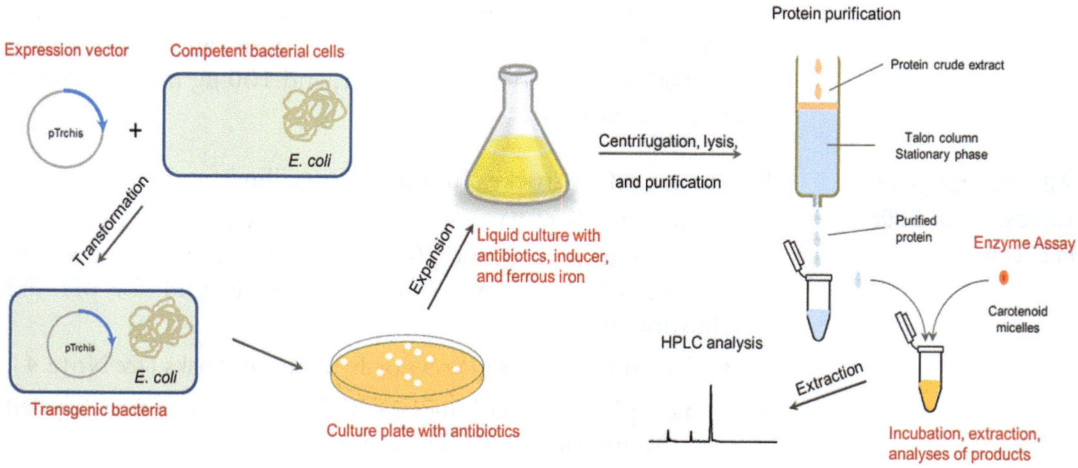

Fig. 1 Schematic workflow of the method for the biochemical characterization of recombinant mammalian CCDs. The experimental details are described in the main text

2 Materials

Use analytical grade reagents and ultrapure deionized water. All solvents used for liquid chromatography should be HPLC grade. Extra care should be taken for carotenoid substrates. Most commercially available carotenoids contain significant impurities, which affect the enzyme activity. Additionally, carotenoids and apocarotenoids are light sensitive and as a result can be oxidized and can display limited storage times. Therefore, carotenoid and apocarotenoid substrates need to be prepared fresh and purified by chromatographic methods when needed. All reagents and cellular extracts should be kept on ice during the enzyme extraction and enzyme assay preparation procedure (unless indicated otherwise). Waste disposal regulations and rules for the work with bacteria and recombinant DNA must be meticulously followed.

2.1 CCD Expression Plasmid Transformation into E. coli

1. BL21 competent *E. coli* cells (*Stratagene*) (*see* **Note 1**).
2. CCD expression plasmid (*see* **Note 2**).
3. General purpose water bath.
4. **S**uper **O**ptimal broth with **C**atabolite repression (**SOC**) Outgrowth medium: 2% (w/v) vegetable peptone, 0.5% (w/v) yeast extract, 10 mM NaCl, 2.5 mM KCl, 10 mM $MgCl_2$, 10 mM $MgSO_4$, and 20 mM glucose. SOC outgrowth medium can be made to any volume depending on the volume that is needed.
5. Console incubator shaker (*New Brunswick E25/E25R or equivalent*).

6. Lennox Broth (LB) agar culture plates: 1% (w/v) Bacto tryptone, 0.5% (w/v) Bacto yeast extract, 0.5% (w/v) NaCl, 1.5 g per 100 mL agar, deionized water, and 100 μg mL^{-1} of ampicillin (*see* **Note 3**).

2.2 Heterologous Expression of CCDs in E. coli

1. 100 mg mL^{-1} stock solution of ampicillin.
2. Lennox Broth (LB) media: 1% (w/v) Bacto tryptone, 0.5% (w/v) Bacto yeast extract, 0.5% (w/v) NaCl, deionized water, 100 ug mL^{-1} of ampicillin, and 2 M NaOH for pH 7.2 adjustment.
3. Baffled culture flasks (*Pyrex*) or Erlenmeyer flasks (*see* **Note 4**).
4. An agar plate with colonies of BL21 *E. coli* cells transformed with the appropriate CCD expression plasmid.
5. Ferrous iron/ascorbic acid solution: 60 mg $FeSO_4$ and 10 mg L-ascorbate in 10 mL H_2O (*see* **Note 5**).
6. 1 M stock solution of isopropyl β-D-thiogalactopyranoside (IPTG).
7. Console incubator shaker (*New Brunswick E25/E25R or equivalent*).
8. Cell density meter (*Biochrom WPA CO8000 Cell Density Meter or equivalent*).
9. Cuvettes.
10. Centrifuge tubes/bottles (*see* **Note 6**).
11. Centrifuge (*Eppendorf 5804 R or comparable model*).

2.3 Preparation of CCD Protein Extracts

1. BL21 competent *E. coli* pellet.
2. Extraction buffer: 20 mM Tricine/KOH (pH 7.4) and 150 mM NaCl (*see* **Note 7**).
3. 10 mg mL^{-1} stock solution of lysozyme.
4. 100 mM stock solution of Phenylmethylsulfonyl fluoride (PMSF).
5. 100 mM stock solution of Tris (2-carboxyethyl) phosphine hydrochloride (TCEP), pH 7.5.
6. 2 mL Eppendorf tubes.
7. Ultrasonic liquid processor (e.g., *Q125 Sonicator from QSonica*) (*see* **Note 8**).
8. Refrigerated centrifuge.

2.4 Small Scale Purification of Recombinant 6× His-Tag-CCDs

1. Talon® superflow metal affinity resin.
2. Binding buffer: 50 mM sodium phosphate, 300 mM NaCl, 1 mM TCEP, pH 7.0.
3. Wash buffer: 50 mM sodium phosphate, 300 mM NaCl, 10–20 mM imidazole, 1 mM TCEP, pH 7.0.

4. Elution buffer: 50 mM sodium phosphate, 300 mM NaCl, 250 mM imidazole, 1 mM TCEP, pH 7.0.
5. Amicon® Ultra—0.5 mL 10 K centrifugal filters or similar.

2.5 Carotenoid Substrate Preparation for CCDs Enzymatic Assay

1. Thin Layer Liquid Chromatography (TLC) developing tank.
2. TLC solution for carotenoid purification: 30 mL of acetone, 50 mL of Petroleum ether (PE), and 20 mL of Diethyl ether (DEE) (*see* **Note 9**).
3. Thin Layer Chromatography plates (*Millipore Sigma TLC Silica gel 60 F254*).
4. Ruler and pencil.
5. 2 mL solvent tight reaction tubes.
6. Chloroform.
7. Carotenoids (*see* **Note 10**).
8. Glass cylinder tube.
9. Flat edge spatula.
10. Dark-colored glass storage bottle.
11. 10 × 10 cm Aluminum foil.
12. UV spectrophotometer (*Shimadzu UV-1280 or equivalent*).
13. Concentrator (*Vacufuge plus from Eppendorf or similar*).

2.6 Enzymatic Assay for Recombinant CCDs

1. Decyl maltose neopentyl glycol (DMN): 3% (w/v) of DMN in 100% ethanol (*see* **Note 11**).
2. Carotenoid substrate.
3. Eppendorf tubes.
4. Bacterial protein extract and/or purified CCD enzyme.
5. Diethyl ether (DEE).
6. Petroleum ether (PE).
7. Acetone.
8. Deionized water.
9. Nonrefrigerated microcentrifuge.
10. Thermomixer (*Eppendorf*).
11. Concentrator (*Vacufuge plus from Eppendorf or similar*).

3 Methods

3.1 CCD Expression Plasmid Transformation into E. coli

1. Mix 1 μL of CCD expression plasmid (approx. 50 ng of DNA) and 10 μL of BL21 competent *E. coli* cells (*Stratagene*) together by gently stirring. Do not mix by pipetting.
2. Incubate on ice for 5 min.

3. Heat-shock cells at 42 °C for 30 s.
4. Immediately place cells back on ice.
5. Add 250 μL of SOC media.
6. Allow cells to incubate in the console incubator shaker at 37 °C for 1 h while shaking at 200 rpm.
7. Afterward, plate 50 μL of transformed cells on LB-agar plates containing 0.1 mg mL^{-1} of ampicillin or another plasmid selective antibiotic.
8. Incubate the plates upside down for 16 h at 37 °C.
9. After incubation, check to observe the presence of colonies on the plates.

3.2 Heterologous Expression of CCDs in E. coli

1. For the generation of a bacterial preculture, mix 5 μL of 100 mg mL^{-1} stock solution of ampicillin into 5 mL of LB media.
2. Select a single bacteria colony from the agar plate using a sterile pipette tip and place it into the 5 mL LB media.
3. Allow bacteria culture to grow for 16 h at 37 °C in an incubator shaker set to 200 rpm.
4. The generation of the expression culture involves adding 50 μL of stock ampicillin into a 50 mL LB media and 100 μL of stock iron/ascorbic acid solution.
5. Add 500 μL of fresh overnight preculture to each 50 mL LB expression culture.
6. Incubate the expression culture at 37 °C until the OD_{600} equals 0.6 in an incubator shaker set to 200 rpm. The optical density (OD) of the culture can be determined using a cell densitometer.
7. Once an OD_{600} of 0.6 is reached, expression can be induced by adding IPTG (0.1–0.5 mM final concentration) from the stock solution and incubated at 16 °C for 24 h in an incubator shaker set to 200 rpm.
8. Harvest cells by centrifugation at 6000 × *g* for 15 min at 4 °C.
9. Discard supernatant carefully and store the bacterial pellet at −80 °C until further use.

3.3 Preparation of CCD Protein Extracts

1. Thaw the bacterial pellet on ice.
2. Add 20 μL of 100 mM PMSF and 20 μL of 100 mM TCEP to 2 mL of the extraction buffer.
3. Add the solution to the bacterial pellet and mix homogeneously by vortexing.
4. Add 4 μL of lysozyme solution (10 mg mL^{-1}), vortex again, and allow to stand on ice for 30 min.

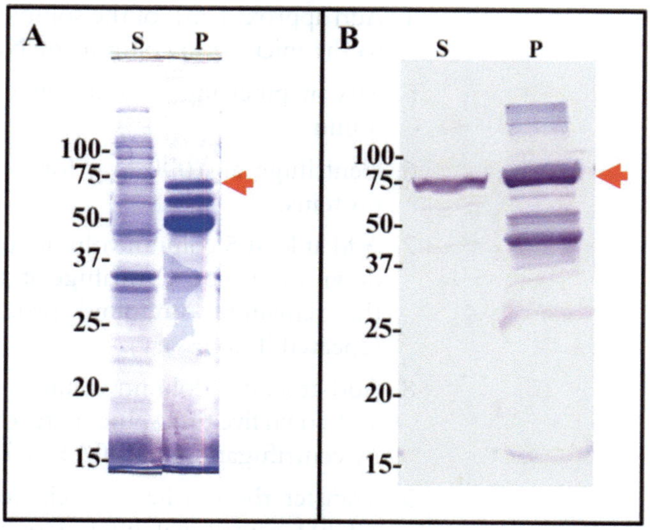

Fig. 2 Analysis of protein extracts of *E. coli* expressing recombinant mouse BCO2. (**a**) Soluble (S for supernatant) and insoluble (P for pellet) fraction of a protein lysate was separated on SDS-PAGE gel (40 μg per lane). Coomassie Brilliant Blue was used to stain total protein. (**b**) Western blot using anti-his-tag antibody to detect the recombinant mouse BCO2. *Red arrows* indicate BCO2 protein band

5. At the end of the incubation period, vortex vigorously again for 10 s.
6. Set the amplitude and the time on the sonicator to 35% intensity and 1 min, respectively. Be sure that the sonicator probe has been cleaned first with water and then 70% ethanol twice through and wiped clean with paper cloth.
7. While the sample is kept on ice, sonicate each sample for 1 min and place it on ice for 1 min. Repeat the procedure four times.
8. Centrifuge the homogenate to separate soluble and insoluble fractions (Fig. 2) for 20 min at 20,000 × *g* at 4 °C (*see* **Note 12**). Pipet the soluble supernatant fraction through a 0.2 μm Millipore PVDF filter to remove putative debris and store at 4 °C until further use.

3.4 Small Scale Purification of Recombinant 6× His-Tagged CCDs

1. Prepare the column by pipetting 0.2 mL Talon resin slurry into a micro spin column (this amount is equivalent to approx. 100 μL of packed resin).
2. Wash resin with at least 5 column volumes (500 μL) of distilled water. Centrifuge at 1000 × *g* for 1 min to pack resin.
3. Equilibrate the column with at least 5 column volumes of binding buffer. Centrifuge at 1000 × *g* for 1 min to remove the buffer.

4. Add approx. 1 mL of the soluble fraction of the bacterial lysate to the micro spin column (*see* **Note 13**).
5. Mix by pipetting up and down five times. Incubate for up to 5 min.
6. Centrifuge at 1000 × *g* for 1 min to remove the unbound proteins.
7. Add at least 5 column volumes of wash buffer and pipet up and down five times. Centrifuge at 1000 × *g* for 1 min to remove the remaining unbound proteins (the wash step can be repeated if necessary).
8. Add at least 5 column volumes of elution buffer and pipet up and down five times. Incubate for 5 min and harvest the protein by centrifugation at 1000 × *g* for 1 min.
9. Transfer the purified protein to an Amicon® Ultra—0.5 mL 10 kD centrifugal filter. Concentrate the protein solution. Dilute the concentrated protein solution with extraction buffer containing 1mM final concentration of TCEP if necessary.
10. The purity of the recombinant protein can be analyzed by SDS-PAGE and Coomassie brilliant blue staining using standard protocols.

3.5 Carotenoid Substrate Preparation for Enzymatic Assay

1. In the TLC developing tank, place the following: 40 mL of PE, 10 mL of acetone, and 10 mL of DEE. Mix the contents that are in the TLC developing tank and cover with glass cover to prevent solvent vapors from escaping.
2. Take one TLC silica plate and use a ruler and a pencil to draw a straight line that is 2.54 cm (1 inch) from the bottom of the TLC silica plate.
3. Place about 1 mg of carotenoid powder into an Eppendorf tube and immediately add 200–300 μL of chloroform and mix to dissolve.
4. Using a pipette tip, distribute the dissolved carotenoid chloroform solution on the pencil line and let the chloroform evaporate.
5. Place the TLC silica plate into the TLC developing tank for separation of the carotenoid. Separation of pure carotenoid substrate from impurities is complete when the solvent reaches the upper top of the TLC silica gel.
6. Add 4 mL of acetone into a glass tube.
7. Remove TLC silica plate from the TLC developing tank and use a flat edge spatula to scrape the silica with carotenoid (large yellow or red line) onto the aluminum foil.

8. Transfer the silica powder containing the carotenoid from the aluminum foil into the glass tube containing the 4 mL of acetone.
9. Allow the solution to sit for 10 min to let the white silica powder settle to the bottom of the glass tube, leaving a colored carotenoid solution at the top.
10. Carefully transfer the carotenoid acetone solution into a collection tube.
11. Using the concentrator, dry vacuum the content of the Eppendorf tubes for about 20 min or until liquid has evaporated.
12. Redissolve the carotenoid substrate in an appropriate solvent and determine the optical density of the carotenoid in a spectrometer (*see* **Note 14**). Use the molar extinction coefficient to calculate the molar concentration of the carotenoid.

3.6 Enzymatic Assay for Recombinant CCDs

1. Set thermomixer to 35 °C and 600 rpm.
2. Mix 33 μL of 3% (w/v) of DMN with the appropriate amount of carotenoid substrate volume (*see* **Note 14**).
3. Using the concentrator, dry vacuum the samples for 10–20 min, until liquid has completely evaporated.
4. Pipet 100 μL of either the soluble protein extract or purified enzyme solution into the Eppendorf tube containing the dried carotenoid/detergent residue.
5. Vortex vigorously for 20–30 s to form micelles.
6. Place samples on the thermomixer and incubate for various time points. The progress of the enzymatic reaction is indicated by a color shift of the solution from yellow to pale (Fig. 3).
7. Stop the reaction by adding in the following order 100 μL distilled water, 400 μL acetone, 400 μL DEE, and 100 μL PE.
8. Using a table top centrifuge, centrifuge for 30 s at 6800 × *g* at room temperature to separate the phases.
9. Transfer the solvent layer (top layer), which should be ~600 μL to a new labeled tube.
10. Using a concentrator, dry vacuum samples for ~10–20 min or until liquid has completely evaporated.
11. Redissolve the dried pellet in running agents and analyze product formation by high-performance liquid chromatography (HPLC) analysis (Fig. 4) (*see* **Note 15**).

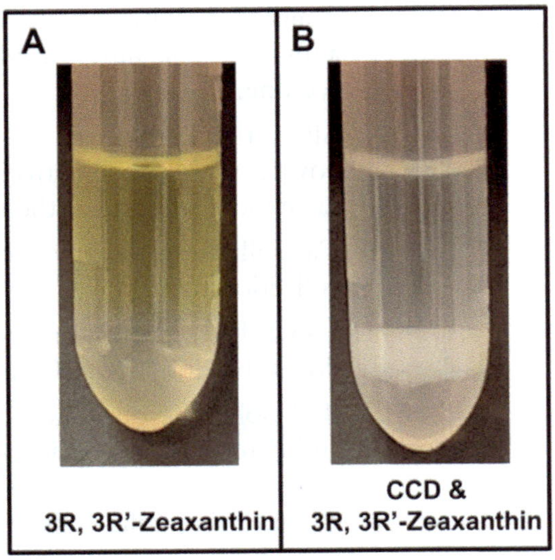

Fig. 3 Color shift as an indicator of CCD enzymatic activity. (**a**) Lipid extract of 3R,3R′-zeaxanthin DNM micelles incubated in enzyme assay buffer. (**b**) Lipid extract of 3R,3R′-zeaxanthin DNM micelles incubated in presence of an protein extract containing mouse BCO2. Note the visible color shift from yellow to pale white between the control and enzyme sample, which is indicative for the conversion of the carotenoid substrate to apocarotenoid products

4 Notes

1. BL21 *E. coli* cells are best used with the expression vectors with modified *lac* and *T7* promoters.
2. CCD expression plasmid should contain the CCD gene insert of interest in the correct orientation. CCDs can be expressed as N-terminal and C-terminal fusion proteins. We routinely express CCDs with polyhistidine tags in *E coli.* Verify the plasmid constricts by DNA sequencing.
3. Agar plates are made by first combining all solid components together into an Erlenmeyer flask. Secondly, add the appropriate amount of deionized water and mix until the solid components have dissolved. Thirdly, autoclave agar solution to sterilize all components. After autoclaving, allow flask to cool. Once the agar solution has cooled, add the desired type and amount of antibiotics and mix well by swirling. Pour 20 mL of agar solution into sterile culture plates and allow plates to solidify.
4. Baffled flasks enhance bacterial growth due to the increased amount of oxygen in the media from agitation created by the ridges in the baffled flasks.

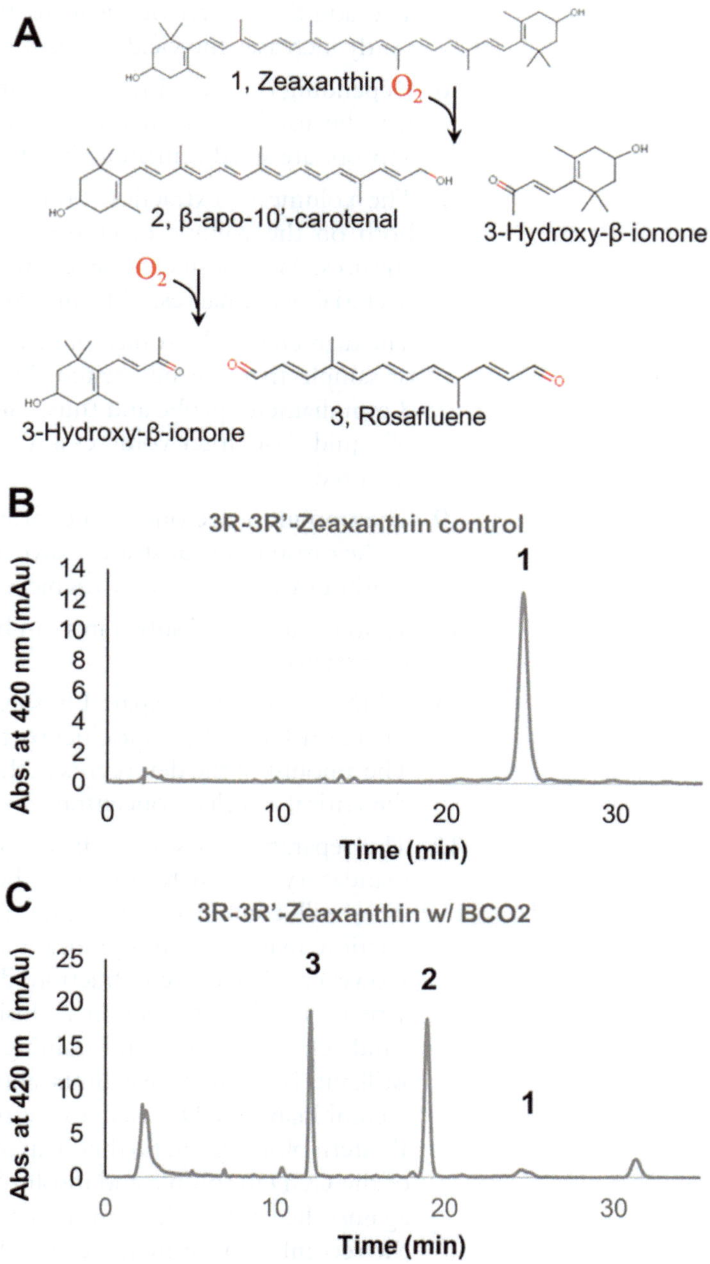

Fig. 4 HPLC analysis of carotenoid substrates and apocarotenoid products. (**a**) Schematic illustration of the enzymatic conversion of zeaxanthin by BCO2. (**b**) HPLC trace at 420 nm of a lipid extract of a control incubation of zeaxanthin in the absence of BCO2 enzyme. (**c**) HPLC trace at 420 nm of a lipid extract of an incubation of zeaxanthin in the presence of mouse BCO2

5. The addition of ferrous iron to the culture medium significantly increases the yield of enzymatically active protein.
6. Depending on the expression volume, 50 mL centrifuge tubes may be used for a 50 mL culture. For larger volumes, the appropriate sized centrifuge bottles are required.
7. The volume of extraction buffer that is needed will depend both on the volume of cultures and the bacterial density of cultures. We routinely use 2 mL of extraction buffer for a bacterial pellet harvested from a 50 mL of bacterial culture.
8. The diameter of the sonicator probe will determine the volume of sample that can be lyzated. For example, the Q125 has a 3 mm diameter probe and thus, can sonicate 500 μL to 15 mL of liquid. For larger volumes a sonicator with a larger probe is required.
9. The amount of acetone can be varied depending on the polarity of the carotenoid substrate. Carotenes need less acetone, while xanthophylls need more to achieve proper separation.
10. Various carotenoid substrates can be purified using Thin Layer Chromatography.
11. DMN needs to be prepared fresh each time. Other detergents such as n-Octyl β-D-thioglucopyranoside can be used as well. The amount of the detergents in the enzyme assay must exceed the critical micellar concentration of the respective substance.
12. The separation of soluble from insoluble protein fractions is mandatory. In our hands, only the soluble fraction (supernatant) will contain enzymatically active CCD. The insoluble fraction may contain significant amounts of CCD protein. However, this protein fraction does not display enzymatic activity and likely exists in inclusion bodies. SDS polyacrylamide electrophoresis and staining of the gels with Coomassie Brilliant Blue may give hints about the distribution of the recombinant CCD when expressed in significant amounts. Western blotting can further help to determine the distribution of the CCD in soluble and insoluble fractions when antibodies against the CCD and/or fusion parts (polyhistidine-tag etc.) of the recombinant protein are available.
13. The amount of lysate depends on the expression levels and solubility of the recombinant CCD. Lower amounts might be used when the protein exist in high levels in the soluble fraction, whereas higher amounts are needed when the CCD is expressed at low levels.
14. We routinely use up to 2000 pmol carotenoid substrate per 100 μL enzyme assay. This amount of substrate can be solubilized in DMN detergent micelles. The concentration of the carotenoid solution can be determined spectrophotometrically

and calculated by Beer's law. The molar extinction coefficient ε at 450 nm of β-carotene in hexane is 125,000 L × mol^{-1} × cm^{-1}, zeaxanthin in ethanol 133,000 L × mol^{-1} × cm^{-1}, and for lutein in ethanol 145,000 L × mol^{-1} × cm^{-1}.

15. We routinely carry out HPLC analyses with an Agilent 1260 Infinity Quaternary HPLC system (*Santa Clara, CA, USA*) equipped with a pump (*G1312C*) with an integrated degasser (*G1322A*), a thermostated column compartment (*G1316A*), an autosampler (*G1329B*), a diode-array detector (*G1315D*), and online analysis software (*Chemstation*). The analyses are carried out at 25 °C using a normal-phase Zorbax Sil (5 μm, 4.6 × 150 mm) column (*Agilent Technologies, Santa Clara, CA*) protected with a guard column with the same stationary phase. As mobile phase we use hexanes ethyl acetate mixes at a flow rate of 1.4 mL × min^{-1}. The composition of the mobile phase depends on the polarity of the carotenoid and apocarotenoids.

Acknowledgments

The research was supported in part by U.S. National Institute of Health grants EY020551. The authors thank present and former members of the von Lintig laboratory for their manifold contributions to establishing this method.

References

1. Harrison EH, Quadro L (2018) Apocarotenoids: emerging roles in mammals. Annu Rev Nutr 38:153–172
2. von Lintig J (2012) Metabolism of carotenoids and retinoids related to vision. J Biol Chem 287:1627–1634
3. Rhinn M, Dollé P (2012) Retinoic acid signalling during development. Development (Cambridge England) 139:843–858
4. Ross AC (2012) Vitamin a and retinoic acid in T cell-related immunity. Am J Clin Nutr 96:1166S–1172S
5. Tang XH, Gudas LJ (2011) Retinoids, retinoic acid receptors, and cancer. Annu Rev Pathol 6:345–364
6. Giuliano G, Al-Babili S, von Lintig J (2003) Carotenoid oxygenases: cleave it or leave it. Trends Plant Sci 8:145–149
7. von Lintig J, Kiser PD, Golczak M, Palczewski K (2010) The biochemical and structural basis for trans-to-cis isomerization of retinoids in the chemistry of vision. Trends Biochem Sci 35:400–410
8. Sui X, Golczak M, Zhang J, Kleinberg KA, von Lintig J, Palczewski K, Kiser PD (2015) Utilization of Dioxygen by carotenoid cleavage Oxygenases. J Biol Chem 290:30212–30223
9. Kiefer C, Hessel S, Lampert JM, Vogt K, Lederer MO, Breithaupt DE, von Lintig J (2001) Identification and characterization of a mammalian enzyme catalyzing the asymmetric oxidative cleavage of provitamin a. J Biol Chem 276:14110–14116
10. Redmond TM, Poliakov E, Yu S, Tsai JY, Lu Z, Gentleman S (2005) Mutation of key residues of RPE65 abolishes its enzymatic role as isomerohydrolase in the visual cycle. Proc Natl Acad Sci 102:13658–13663
11. Redmond TM, Gentleman S, Duncan T, Yu S, Wiggert B, Gantt E, Cunningham FX Jr (2001) Identification, expression, and substrate specificity of a mammalian beta-carotene 15,15′-dioxygenase. J Biol Chem 276:6560–6565
12. Gollapalli DR, Maiti P, Rando RR (2003) RPE65 operates in the vertebrate visual cycle

by stereospecifically binding all-trans-retinyl esters. Biochemistry 42:11824–11830

13. Xue L, Gollapalli DR, Maiti P, Jahng WJ, Rando RR (2004) A palmitoylation switch mechanism in the regulation of the visual cycle. Cell 117:761–771

14. Hansen S, Maret W (1988) Retinal is not formed in vitro by enzymatic central cleavage of beta-carotene. Biochemistry 27:200–206

15. Li B, Vachali PP, Gorusupudi A, Shen Z, Sharifzadeh H, Besch BM, Nelson K, Horvath MM, Frederick JM, Baehr W, Bernstein PS (2014) Inactivity of human beta, beta-carotene-9′,10′-dioxygenase (BCO2) underlies retinal accumulation of the human macular carotenoid pigment. Proc Natl Acad Sci U S A 111:10173–10178

16. Babino D, Palczewski G, Widjaja-Adhi MA, Kiser PD, Golczak M, von Lintig J (2015) Characterization of the role of beta-carotene 9,10-Dioxygenase in macular pigment metabolism. J Biol Chem 290:24844–24857

17. Palczewski G, Amengual J, Hoppel CL, von Lintig J (2014) Evidence for compartmentalization of mammalian carotenoid metabolism. FASEB J 28:4457–4469

Chapter 7

Transportomics for the Characterization of Plant Apocarotenoid Transmembrane Transporters

Olivia Costantina Demurtas, Rita de Brito Francisco, Enrico Martinoia, and Giovanni Giuliano

Abstract

Apocarotenoids are carotenoid derivatives produced by the nonenzymatic or enzymatic cleavage of carotenoids, followed by different enzymatic modifications. In plants, apocarotenoids play different roles, such as attraction of pollinators and seeds dispersal, defense against pathogens and herbivores, protection against photo-oxidative stresses, stimulation and inhibition of plant growth and regulation of biological processes in the case of phytohormones abscisic acid and strigolactones. While carotenoids are in general plastid-localized metabolites, apocarotenoids can reach different final destinations inside or outside the cell. The mechanisms of apocarotenoid transport through biological membranes have been poorly studied. This chapter describes a method to characterize transmembrane transporters involved in the transport of polar and amphipathic apocarotenoids. This protocol was successfully used to in vitro characterize the transport activity of ATP-binding cassette (ABC) and multidrug and toxic extrusion (MATE) in microsomes isolated from *Saccharomyces cerevisiae* expressing these plant transporters.

Key words Carotenoids, Apocarotenoids, ATP-binding cassette (ABC) transporters, Multidrug and toxic compound extrusion (MATE) transporters, Yeast microsomes, LC-MS

1 Introduction

Apocarotenoids are produced from carotenoids through both enzymatic and nonenzymatic cleavage. In plants, they play important biological roles as pigments, volatiles and signaling molecules [1–3]. In mammals, apocarotenoids play roles in signaling and regulation of metabolism [4].

While the key enzymatic steps that lead to the formation of apocarotenoids in different plant species have been studied, the proteins mediating their intra- and intercellular transport are still poorly investigated. Many apocarotenoids, and their glycosylated

Olivia Costantina Demurtas and Rita de Brito Francisco contributed equally to this work.

Manuel Rodríguez-Concepción and Ralf Welsch (eds.), *Plant and Food Carotenoids: Methods and Protocols*, Methods in Molecular Biology, vol. 2083, https://doi.org/10.1007/978-1-4939-9952-1_7,

derivatives, are polar molecules that do not readily diffuse across membranes, and thus require an active transport mechanism to cross them.

Two main classes of plant transmembrane transporters have been associated with secondary metabolite transport: ATP-binding cassette (ABC) and multidrug and toxic compound extrusion (MATE) transporters. However, both these classes have also been shown to be involved in plant hormone transport. ABC transporters are localized at the plasma membrane, the tonoplast, chloroplasts, mitochondria, and peroxisomes and mediate active transport driven by ATP hydrolysis [5]. MATE transporters, on the other hand, have been reported to be localized to the tonoplast, the plasma membrane or the chloroplast and are driven by a transmembrane proton gradient [6]. ABC transporters participate in the translocation of the apocarotenoid hormones abscisic acid (ABA) and strigolactone. In *Arabidopsis thaliana*, two plasma membrane-localized ABC transporters, AtABCG40 and AtABCG30, are responsible of the import of ABA from the apoplast into the cytosol [7, 8], and two, ABCG25 and ABCG31 of the export from xylem parenchyma cells or from the seed coat [8, 9]. A third ABC transporter, AtABCC2 is involved in the vacuolar sequestration of ABA glucosyl ester (ABA-GE) [10]. In *Petunia hybrida* the ABC transporter PDR1 is involved in strigolactone exudation from roots [11]. To date, no transport of apocarotenoids by a MATE transporter has been demonstrated.

Two different experimental approaches are available for the characterization of transmembrane transporters: reverse genetics, by knocking out or overexpressing the corresponding gene and observing the phenotype [11, 12]; this approach can be complemented by isolating organelles such as chloroplasts, mitochondria or vacuoles from wild-type plants and the corresponding mutants and/or overexpressors and performing transport experiments, usually with the silicon oil technique [10, 13]. A second approach is heterologous expression in animal, plant or yeast cells, followed by observation of transport in intact cells or by isolation of membrane vesicles [12, 14, 15], that can be used either directly or can be further fractionated to enrich a specific membrane-type [13]. Depending on the transport direction, transport activity will be determined using either outside-out or inside-out vesicles.

While transport by plant ABC and MATE transporters has been characterized for a series of compounds for which purified substrates are commercially available (for example nicotine, flavonoids, folates, etc., [5, 16–18]), only few apocarotenoids are commercially available as standards. To overcome this problem, a complex crude plant extract can be used to perform transport assays and the uptake of several metabolites can be evaluated simultaneously by liquid chromatography-photodiode array–high resolution mass spectrometry (LC-PDA-HRMS) (Fig. 1). This approach has been

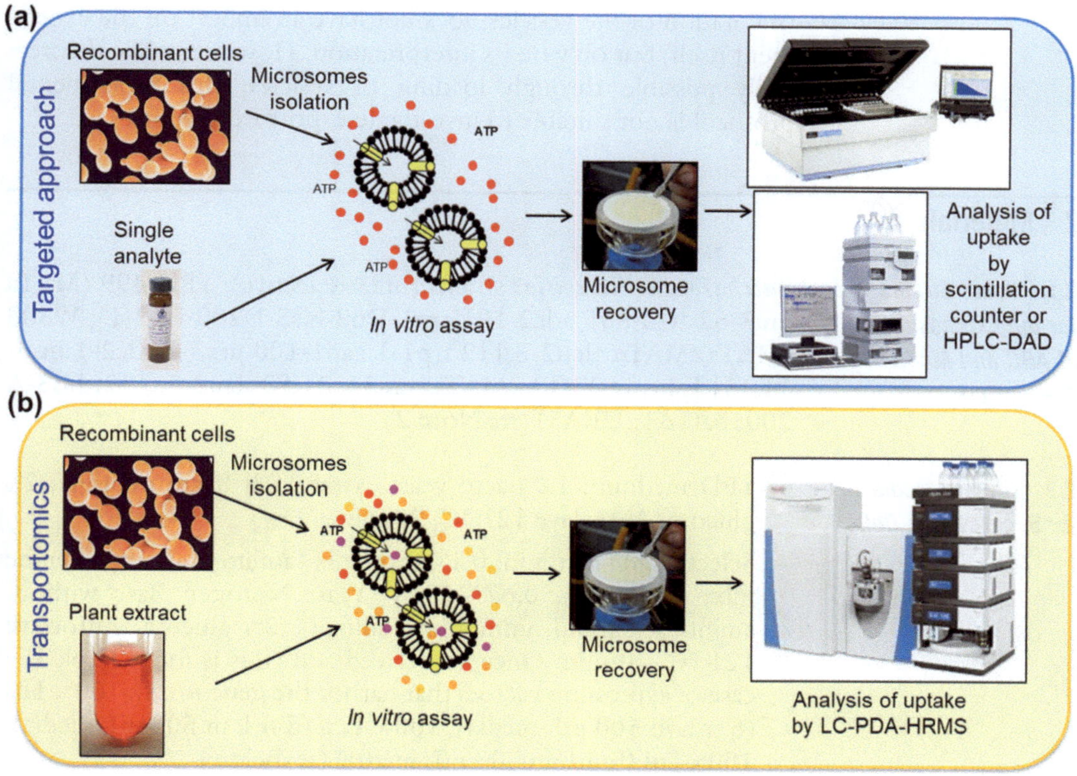

Fig. 1 Schematic representation of the targeted (**a**) or transportomics (**b**) approaches used to characterize transmembrane transporters. Microsomes obtained from recombinant yeast cells expressing the transporter are incubated with a single substrate (**a**) or with a total plant extract (**b**) in the presence of MgATP. The transport assay is performed by the rapid filtration technique [20] and the uptake by vesicles analyzed by scintillation counting or HPLC-PDA (in the case of a single substrate, **a**), or by LC-PDA-HRMS (**b**)

named "transportomics" [19]. It can be applied for the characterization of any plant transporter and potentially allows the characterization of the full complement of metabolites transported within a given cell type (the transportome).

The method described below can be applied to transport experiments using membrane vesicles. It is not suitable for experiments with whole organelles, since in most cases the metabolite of interest already resides in the organelle and it is often difficult to figure out the difference between endogenous and transported metabolite. Depending on the transport mechanism involved, there may be further limitations. The small size of the vesicles requires that a substrate has to accumulate within the vesicles (internal concentration higher than outside) in order to be sure that the substrate observed is not just adsorbed by the lipid bilayer. Hence only energized transport processes can give unequivocal results. Both inside-out and outside-out vesicles are suitable for transport experiments as described below. Since most of the vesicle isolation protocols yield vesicles in both orientations, the orientation of the vesicles does not have an impact on the experiment itself, but only on its interpretation. However, while theoretically possible through loading of vesicles, this experimental protocol is not suitable to investigate export from vesicles.

2 Materials

2.1 Yeast Strains for the Expression of ABC and MATE Tranporters

Saccharomyces cerevisiae strains (ura3-defective): YPH 499 (MATa ura3-52 lys2-801 ade2-101 trpl-D63 his3-D200 leu2-1); W303 (MATa/MATα {leu2-3,112 trp1-1 can1-100 ura3-1 ade2-1 his3-11,15} [phi+]); YBT1 (MATa; ura3Δ::HIS3; leu2-3, 112; his3-Δ 200; bat1Δ1::URA3) (*see* **Note 2**).

2.2 Growth Media for S. cerevisiae *Cells*

- YPD medium: 1% Bacto yeast extract, 2% Bacto peptone, 2% glucose. Autoclave 121 °C, 20 min.
- Selective SD medium: 0.1% drop mix (amino acids and adenine mix, *see* below), 0.67% YNB (Yeast Nitrogen Base without amino acids and ammonium sulfate), 2% glucose. Autoclave 121 °C, 20 min. Once autoclaved add (this is for example the case of expressing vectors that harbor the gene *ura3*): 100× His (5 mL in 500 mL media), 100× Leu (5 mL in 500 mL media), 100×Trp (5 mL in 500 mL media) (*see* below).
- Drop mix:

	Final concentration (mg/L)	g per 14.5 g mix
Adenine	40	0.4
L-arginine	20	0.2
L-tyrosine	30	0.3
L-isoleucine	30	0.3
L-phenylalanine	50	0.5
L-glutamic acid	100	0.1
L-aspartic acid	100	1.0
L-threonine	200	0.2
L-serine	400	4.0
L-valine	150	1.5
L-methionine	150	1.5
L-lysine	180	1.8

- 100× Ura: in 100 mL bidistilled water (bd H_2O) dissolve 0.4 g of L-uracil (0.8 g/200 mL).

- 100× His: in 100 mL bd H_2O, dissolve 0.6 g of L-histidine (1.2 g/200 mL).
- 100× Leu: in 100 mL bd H_2O, dissolve 2.6 g of L-leucine (0.8 g/200 mL).

  These three stocks can be autoclaved at 121 °C for 20 min and then store at 4 °C.
- 100× Trp: in 100 mL bd H_2O dissolve 0.8 g of L-tryptophan (4 g/500 mL). Do not autoclave. Filter-sterilize (0.22 μm), store at 4 °C in the dark.

2.3 Solutions for Microsome Preparation

- Digestion medium: 1.1 M sorbitol, 20 mM Tris. Adjust pH to 7.6 with 6 M HCl. Just before starting the digestion step add 1 mM DTT and 57 U Lyticase/mL digestion buffer (Lyticase from *Arthrobacter luteus*) (*see* **Note 1**).
- Homogenization buffer: 1.1 M Glycerol anhydrous, 50 mM Tris, 5 mM EDTA. Adjust pH to 7.4 with ascorbic acid (powder). Just before starting the homogenization step freshly add 1 mM DTT, 1.5% (w/v) PVP [Polyvinylpyrrolidone (K30), 1000 (water max 5%)], 2 mg/mL BSA (Albumin, from bovine serum), 1 mM PMSF (100 or 200 mM stock solution in 70% ethanol) (*see* **Note 1**).
- Storage buffer: 1.1 M glycerol anhydrous, 50 mM Tris, 1 M EDTA. Adjust pH to 7.4 with MES (powder). Just before starting the storage step freshly add: 1 mM DTT, 2 mg/mL BSA, 1 mM PMSF (*see* **Note 1**).

2.4 Solutions for In Vitro Transport Assays

- Transport buffer: 0.4 M Glycerol anhydrous, 100 mM KCl, 20 mM Tris. Adjust pH to 7.4 with MES (powder) (*see* **Note 1**).
- Transport buffer + DTT: transport buffer freshly supplemented with 1 mM DTT (*see* **Note 1**).

2.5 Equipment

1. Incubator.
2. Spectrophotometer.
3. Electroporator.
4. Ultracentrifuge.
5. Glass Potter homogenizer.
6. Vacuum centrifuge.
7. Liquid nitrogen.
8. Vacuum manifold.
9. Shaker with controlled temperature.
10. Biological hood.
11. Vortex.

12. Homogenizer (use for example the Mixer Mill MM 400, Retsch).
13. Vacuum concentrator.
14. HPLC-PDA-HRMS apparatus.

3 Methods

3.1 Preparation of Electrocompetent Yeast Cells

1. Inoculate the yeast strain in 5–10 mL YPD medium. Grow overnight (O/N) in a rotary shaker at 30 °C, 150 rpm.
2. Inoculate 50 mL YPD with saturated culture (measure OD; starting OD_{600} should be about 0.2) (*see* **Note 3**).
3. Grow cells until $OD_{600} = 0.6–0.8$.
4. Collect the cells by centrifugation (1200 × *g*, 5 min, 4 °C).
5. Wash cells with 20 mL of sterile bd H_2O and centrifuge (1200 × *g*, 5 min, 4 °C).
6. Add 5 mL 1 M sorbitol and 0.1 mL 0.5 M DTT.
7. Incubate at 30 °C, 10 min, 150 rpm.
8. Centrifuge (1200 × *g*, 5 min, 4 °C). From now on work on ice.
9. Wash cells with 5 mL of cold 1 M sorbitol (no DTT) and centrifuge (1200 × *g*, 5 min, 4 °C). Repeat this step twice.
10. Resuspend cells in 0.4 mL cold 1 M sorbitol.
11. Aliquot 40 μL cells per reaction tube.

3.2 Yeast Transformation by Electroporation

1. Mix 40 μL cells with DNA (100 ng vector/300 ng insert) [the volume of DNA should be ≤5 μL]. Use chilled/sterile cuvettes.
2. Pulse at 1.5 kV. 25 μF, 200 Ω (5 ms).
3. Add immediately 0.2 mL 1 M sorbitol.
4. Plate on selective SD medium.

3.3 Isolation of Microsomes from Recombinant Yeast Cells

1. Inoculate 5–10 mL liquid media (adapted to your yeast strain) with one yeast colony.
2. Grow on selective SD medium O/N at 30 °C, 150 rpm.
3. Inoculate 900 mL liquid culture (selective SD medium) with an aliquot of the O/N culture (OD_{600} = at least 0.2) and let it grow O/N.
4. Sediment the yeast cells by centrifugation 10 min at 1200 × *g*. Discard supernatant.
5. Resuspend the pellet in 900 mL YPD media and let them grow for 1 h at 30 °C, continuous shaking (This step helps to soften the cell walls).

6. Repeat **step 4**.
7. Dissolve the pellet in sterile bd water (600 mL).
8. Repeat **steps 4** and 7.
9. Repeat **step 4** and dissolve the pellet in digestion medium; the volume of digestion medium is established as such that OD_{600} is 10.
10. Incubate the cells in digestion medium at 30 °C, continuous shaking.
11. After 40 min establish the ratio OD_{600} in digestion medium/ OD_{600} in water. If the ratio is approximately 3, stop the digestion; if not, continue the digestion reaction and measure again the ratio OD_{600} in digestion medium:OD_{600} in water after 30 min; stop digestion after 2 h at the latest.
12. Centrifuge the digestion mix for 10 min at 1200 × *g* and start working on ice.
13. Resuspend the pellet in 25 mL homogenization buffer.
14. Disrupt cells via homogenization (35–50 × up/down in a glass potter homogenizer).
15. Pellet the disrupted cells by centrifugation at 6000 × *g*, 10 min, 4 °C.
16. Collect supernatant in ultracentrifuge tubes and resuspend pellet in 10 mL homogenization buffer.
17. Repeat disruption (35–50 × up/down in the glass potter).
18. Repeat **step 15**.
19. Recover the supernatant and transfer to the ultracentrifuge tubes containing the supernatant of the previous centrifugation.
20. Pellet down total microsomes in an ultracentrifuge (100,000 × *g*, 45 min, 4 °C).
21. Resuspend pellet in storage buffer to a final OD_{600} of 4.
22. Aliquot microsomes into reaction tubes (250 μL/tube).
23. Snap-freeze tubes in liquid nitrogen, and store them at −80 °C.

3.4 Total Plant Extract for Transportomics Assay (Quantity for 25–50 Assays)

1. Lyophilize plant material in a freeze-dryer.
2. Reduce the dried material in powder by agitation in the presence of tungsten beads in a homogenizer or by mortar and pestle.
3. Weight 3–30 mg powder according to the plant matrix.
4. For extraction of polar and semipolar metabolites add 750 μL of cold 50–75% (v/v) LC-MS grade methanol (*see* **Notes 5** and **8**).

5. Vortex 10 s and put the vials in a homogenizer.
6. Shake with the homogenizer for 30 s at 15–20 Hz.
7. Keep at room temperature for 5 min.
8. Repeat **steps 5–7**.
9. Centrifuge at 20,000 × *g* for 20 min.
10. Transfer 600 μL of supernatant to HPLC filter tubes (e.g., Mini-Uniprep PTFE filter).
11. Take 50 μL, add formic acid (0.1% v/v) and internal standard and characterize apocarotenoid content by HPLC-PDA-HRMS (*see* **Notes** 7 and **8**).
12. Use a freshly prepared plant extract to perform transport assays. Depending on the stability of the analytes, the extract can be stored at 4 °C for 1–2 days.

3.5 Transportomics Assay Using the Rapid Filtration Technique [20]

1. Thaw microsomes on ice.
2. Calculate the total protein content of microsome preparations by Bradford assay [21]. Generally, "good" microsome preparations contain ≥400 μg of total proteins/100 μL of microsomes with an $OD_{600} = 4$). If necessary, dilute microsomes to appropriate concentration.
3. Prepare two reaction mixtures on ice, one minus and one plus ATP. The minus ATP mixture contains 1 mM DTT, 1 mM $MgSO_4$, 100 μg/mL creatine kinase, 10 mM creatine phosphate, and 10–20 μL of total plant extract in transport buffer +DTT (*see* **Note 6**). The plus ATP mixture contains all of the above, plus 5 mM $MgSO_4$ and 4 mM MgATP. Aliquot 550 μL per reaction and incubate 5–10 min at 24 °C.
4. Start the reactions by adding 100 μL of microsomes (final volume of reaction = 650 μL). Incubate the assays at 24 °C.
5. Stop the reaction at different time points (at least one after 30 s and one after 15 min) by loading 100 μL of the reaction mixture on a prewetted nitrocellulose filter (0.45 μm pore size) placed under vacuum (around 25 inches Hg) on a vacuum manifold. Perform three technical replicates for each time point. As soon as the mixture has been filtered, immediately wash the filter three times with 2.5 mL of ice-cold transport buffer (*see* **Note 9**).
6. Recover the filters and place on a paper towel. Let the filters dry out completely at room temperature.
7. Place the filters inside a 2 mL amber glass vial.

3.6 Analysis of Transport by HPLC-PDA-HRMS

1. For the analysis of polar and semipolar substrate dissolve the filter-bound microsomes by adding 1 mL of cold 75% (v/v)

methanol (LC-MS grade) containing the internal standard (e.g., 0.5 μg/mL formononetin) to the glass vial.

2. Transfer the glass vial into a shaker with controlled temperature. For heat-stable compounds perform the incubation through continuous agitation for 30 min at 30–37 °C. For heat-labile compounds perform the incubation at 20–25 °C.
3. To remove microsome lipids, transfer 800 μL of eluted samples in a safe-lock microcentrifuge tube and add 400 μL of chloroform. Vortex for 30 s and shake in a homogenizer, for example use a Mixer Mill MM 400 (Retsch) or similar for 5 min at 20 Hz frequency. Add 200 μL ultrapure water to separate phases, vortex and centrifuge at 20,000 × *g* for 20 min.
4. Recover 800 μL of the upper phase and transfer into a new tube.
5. Dry the sample in a vacuum concentrator and resuspend it in 80 μL of 50% (v/v) methanol; centrifuge at 20,000 × *g* for 10 min to remove precipitates.
6. Analyze an aliquot of the supernatant by LC-PDA-HRMS (*see* **Note 8**).

4 Notes

1. Do not autoclave solutions described in Subheadings 2.3 and 2.4; store at 4 °C for few weeks.
2. Select an adequate yeast strain to perform the transport assays. For example, *S. cerevisiae* contains five vacuolar localized ABCC members (Ycf1p, Bpt1p, Ybt1p/Bat1p, Nft1p, and Vmr1p) [22]. So, in case of studying a plant ABCC transporter it is advisable to test yeast strains mutated in one or multiple ABCCs present in *S. cerevisiae* and perform transport assays using microsomes purified from yeasts that heterologous express the transporter of interest.
3. A critical point to obtain intact and functional yeast microsomes is the yeast growth. Using our experimental conditions, we obtained good results when OD600 is 3–3.5 after O/N growth. Furthermore it is suggested to test the physiological intactness of the isolated microsomes before transport assays. This can be performed by the vesicles lumen acidification method as described in [23].
4. Choose as Internal standard (IS) a compound that has similar chemical properties to your analytes, is not present in your plant extract, is chemically stable, ionizes easily, and its retention time does not coincide with that of any analytes (to avoid ion suppression). For example formononetin or reserpine are

secondary metabolites that can be considered suitable IS for the characterization of polar and semipolar extract from many plant species. Furthermore formononetin is restricted to a limited number of plants.

5. When plant extracts are prepared for the transportomics assay (Subheading 3.4, **step 4**), add the internal standard (IS) to characterize the extract by HPLC-PDA-HRMS and not when performing transport assays.
6. The volume of plant extract suggested to perform the transport assays is 10–20 μL in 650 μL of reaction mixture. The final metabolite concentration in the reaction mixture should be similar to their estimated cellular concentration (or, if not detectable in LC-PDA-HRMS, up to tenfold higher). This step needs to be carefully optimized by the operator.
7. If the apocarotenoids present in your extract are glycosylated, don't add formic acid. This avoids deglycosylation of the respective apocarotenoids.
8. Use LC-MS grade solvents for the sample extraction and chromatography separation.
9. Some apocarotenoids can stick to the nitrocellulose filter. Make a first trial by loading the reaction mixture in the absence of ATP on nitrocellulose filter and proceed with washes. If the substrate sticks to the filter try with other types, such as acetate cellulose filters.

Acknowledgments

Work in GG's lab was partially supported by the European Union's Horizon 2020 programme, project "Newcotiana", Grant Agreement 760331, and by a grant from the Lazio Region, project "ProBioZaff".

References

1. Havaux M (2014) Carotenoid oxidation products as stress signals in plants. Plant J 79 (4):597–606. https://doi.org/10.1111/tpj.12386
2. McQuinn RP, Giovannoni JJ, Pogson BJ (2015) More than meets the eye: from carotenoid biosynthesis, to new insights into apocarotenoid signaling. Curr Opin Plant Biol 27:172–179. https://doi.org/10.1016/j.pbi.2015.06.020
3. Hou X, Rivers J, Leon P, McQuinn RP, Pogson BJ (2016) Synthesis and function of Apocarotenoid signals in plants. Trends Plant Sci 21 (9):792–803. https://doi.org/10.1016/j.tplants.2016.06.001
4. Harrison EH, Quadro L (2018) Apocarotenoids: emerging roles in mammals. Annu Rev Nutr 38:153–172. https://doi.org/10.1146/annurev-nutr-082117-051841
5. Kang J, Park J, Choi H, Burla B, Kretzschmar T, Lee Y, Martinoia E (2011) Plant ABC transporters. The Arabidopsis Book:e0153
6. Martinoia E, Meyer S, De Angeli A, Nagy R (2012) Vacuolar transporters in their physiological context. Annu Rev Plant Biol

63:183–213. https://doi.org/10.1146/annurev-arplant-042811-105608

7. Kang J, Hwang JU, Lee M, Kim YY, Assmann SM, Martinoia E, Lee Y (2010) PDR-type ABC transporter mediates cellular uptake of the phytohormone abscisic acid. Proc Natl Acad Sci U S A 107(5):2355–2360. https://doi.org/10.1073/pnas.0909222107
8. Kang J, Yim S, Choi H, Kim A, Lee KP, Lopez-Molina L, Martinoia E, Lee Y (2015) Abscisic acid transporters cooperate to control seed germination. Nat Commun 6:8113. https://doi.org/10.1038/ncomms9113
9. Kuromori T, Miyaji T, Yabuuchi H, Shimizu H, Sugimoto E, Kamiya A, Moriyama Y, Shinozaki K (2010) ABC transporter AtABCG25 is involved in abscisic acid transport and responses. Proc Natl Acad Sci U S A 107(5):2361–2366. https://doi.org/10.1073/pnas.0912516107
10. Burla B, Pfrunder S, Nagy R, Francisco RM, Lee Y, Martinoia E (2013) Vacuolar transport of abscisic acid glucosyl ester is mediated by ATP-binding cassette and proton-antiport mechanisms in Arabidopsis. Plant Physiol 163(3):1446–1458. https://doi.org/10.1104/pp.113.222547
11. Kretzschmar T, Kohlen W, Sasse J, Borghi L, Schlegel M, Bachelier JB, Reinhardt D, Bours R, Bouwmeester HJ, Martinoia E (2012) A petunia ABC protein controls strigolactone-dependent symbiotic signalling and branching. Nature 483(7389):341–344. https://doi.org/10.1038/nature10873
12. Tejada-Jimenez M, Galvan A, Fernandez E (2011) Algae and humans share a molybdate transporter. Proc Natl Acad Sci U S A 108(16):6420–6425. https://doi.org/10.1073/pnas.1100700108
13. Larsen B, Xu D, Halkier BA, Nour-Eldin HH (2017) Advances in methods for identification and characterization of plant transporter function. J Exp Bot 68(15):4045–4056. https://doi.org/10.1093/jxb/erx140
14. Miller AJ, Zhou JJ (2000) Xenopus oocytes as an expression system for plant transporters. Biochim Biophys Acta 1465(1–2):343–358
15. Ozvegy C, Litman T, Szakacs G, Nagy Z, Bates S, Varadi A, Sarkadi B (2001) Functional characterization of the human multidrug transporter, ABCG2, expressed in insect cells. Biochem Biophys Res Commun 285(1):111–117. https://doi.org/10.1006/bbrc.2001.5130
16. Shoji T (2014) ATP-binding cassette and multidrug and toxic compound extrusion transporters in plants: a common theme among diverse detoxification mechanisms. Int Rev Cell Mol Biol 309:303–346. https://doi.org/10.1016/b978-0-12-800255-1.00006-5
17. Remy E, Duque P (2014) Beyond cellular detoxification: a plethora of physiological roles for MDR transporter homologs in plants. Front Physiol 5:201. https://doi.org/10.3389/fphys.2014.00201
18. Yazaki K (2006) ABC transporters involved in the transport of plant secondary metabolites. FEBS Lett 580(4):1183–1191. https://doi.org/10.1016/j.febslet.2005.12.009
19. Krumpochova P, Sapthu S, Brouwers JF, de Haas M, de Vos R, Borst P, van de Wetering K (2012) Transportomics: screening for substrates of ABC transporters in body fluids using vesicular transport assays. FASEB J 26(2):738–747. https://doi.org/10.1096/fj.11-195743
20. Tommasini R, Evers R, Vogt E, Mornet C, Zaman GJ, Schinkel AH, Borst P, Martinoia E (1996) The human multidrug resistance-associated protein functionally complements the yeast cadmium resistance factor 1. Proc Natl Acad Sci U S A 93(13):6743–6748
21. Bradford MM (1976) A rapid and sensitive method for the quantitation of microgram quantities of protein utilizing the principle of protein-dye binding. Anal Biochem 72:248–254
22. Paumi CM, Chuk M, Snider J, Stagljar I, Michaelis S (2009) ABC transporters in Saccharomyces cerevisiae and their interactors: new technology advances the biology of the ABCC (MRP) subfamily. Microbiol Mol Biol Rev 73(4):577–593. https://doi.org/10.1128/mmbr.00020-09
23. Gomez C, Terrier N, Torregrosa L, Vialet S, Fournier-Level A, Verries C, Souquet JM, Mazauric JP, Klein M, Cheynier V, Ageorges A (2009) Grapevine MATE-type proteins act as vacuolar H+-dependent acylated anthocyanin transporters. Plant Physiol 150(1):402–415. https://doi.org/10.1104/pp.109.135624

Part III

Analytics

Chapter 8

Applications of Visible Spectroscopy and Color Measurements in the Assessments of Carotenoid Levels in Foods

Paula Mapelli-Brahm, Francisco J. Rodríguez-Pulido, Carla M. Stinco, Francisco J. Heredia, and Antonio J. Meléndez-Martínez

Abstract

The color of the food can be evaluated visually or instrumentally. The instrumental measurement allows to measure the color objectively and can be carried out by different techniques, most of them fast, cheap, and simple. Carotenoids are responsible for the color of many plants and foods. The measurement of color can be an effective tool to determine indirectly the concentration of carotenoids in plants and foods. In this chapter, we describe both traditional and recent methodologies for color measurement.

Key words Spectrophotometry, Spectroradiometry, Digital image analysis (DIA), Tristimulus colorimetry, Transmission, Reflection

1 Introduction

1.1 Color Definition

According to Wyszecki and Stiles [1] color could be defined as a mental response to the stimulus that a visible radiation produces on the retina, which is transmitted to the brain by the optical nerve. In other words, color is that aspect of visual perception by which an observer may detect differences between two structure-free fields of views of the same size and shape, such as may be caused by differences in the spectral composition of the radiant energy concerned in the observation. The perception of color is a very complex psychophysical phenomenon, related to aspects of different nature, namely the physiology of vision, the psychology of the observer, and the spectral radiant energy of a source of visible light.

In the human retina there are different kinds of photoreceptors. On the one hand, the rods are sensitive to white and black. On the other hand, there are cones of three different types, each of which is selectively sensitive to light in an interval of wavelengths. The

Manuel Rodríguez-Concepción and Ralf Welsch (eds.), *Plant and Food Carotenoids: Methods and Protocols*, Methods in Molecular Biology, vol. 2083, https://doi.org/10.1007/978-1-4939-9952-1_8,

absorption maxima of these cones are: red ($\lambda \sim 650$ nm), green ($\lambda \sim 530$ nm), and blue ($\lambda \sim 430$ nm). When visible light (coming either directly from a light source or from an object that was illuminated by a light source) contact these photoreceptor cells of the retina, they simultaneously send information via the optic nerve to the brain, which eventually interprets it in terms of color. This trichromatic or three-dimensional color perception is the foundation of the Trichromatic Theory of color, which allows each color stimulus to be numerically defined and represented in a color space.

Nowadays, the most used color spaces for the color specification include CIEXYZ, CIELUV and CIELAB, being the latter the one recommended by the International Commission of Illumination (CIE) as the most appropriate for the color specification in the food industry, due to its uniformity, precision and accuracy. The CIELAB color space is a Cartesian system defined by three rectangular colorimetric coordinates $L*$, $a*$, $b*$ dimensionless magnitudes. The parameters of this space are the following [2]:

- $L*$ is defined as lightness and can take values between 0 (corresponding to black) and 100 (corresponding to white).
- $a*$ takes positive values for reddish colors and negative values for the greenish ones.
- $b*$ takes positive values for yellowish colors and negative values for the bluish ones.

From the colorimetric parameters $a*$ and $b*$, the psychometric parameters chroma ($C*_{\text{ab}}$) and hue-angle (h_{ab}) are defined [2]:

- $C*_{\text{ab}}$ allows to determine for each hue its degree of difference in comparison to a gray color with the same lightness and is related to the quantitative attribute of colorfulness.
- Hue-angle (h_{ab}) is the qualitative attribute according to which colors have been traditionally defined as reddish, greenish, and so on. Its values, therefore, vary between 0° and 360°.
- Mathematically, these psychometric parameters are defined as follows [2]:

$$C^*_{\text{ab}} = \sqrt{a^{*2} + b^{*2}} \tag{1}$$

$$h_{\text{ab}} = \text{arctg}\left(b^*/a^*\right) \tag{2}$$

1.2 Carotenoids

Carotenoids are natural pigments that give color to a wide variety of fruits and vegetables. These compounds possess a characteristic system of conjugated double bonds (c.d.b.) which is their chromophore, that is, the part of the molecule responsible for the absorption of visible light color and therefore of their color (Fig. 1) [3].

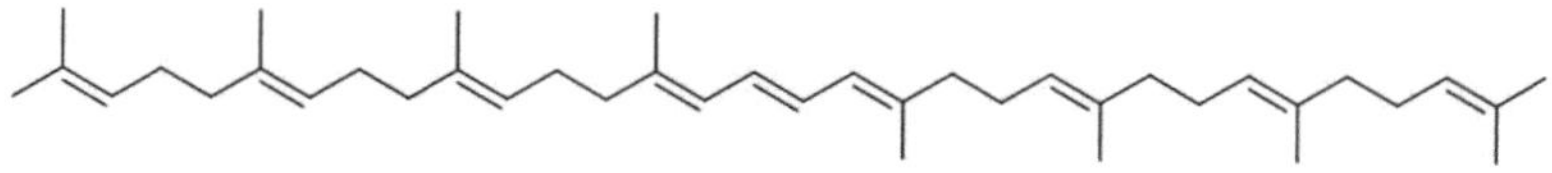

Phytoene (3 c.d.b.)

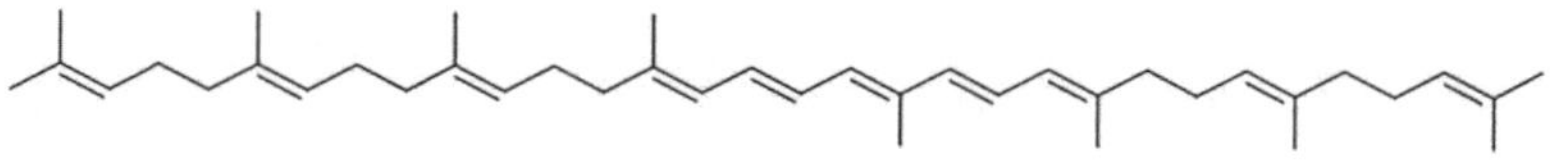

Phytofluene (5 c.d.b.)

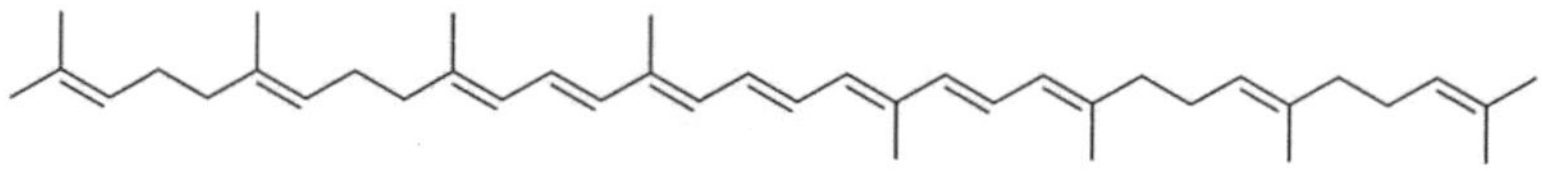

ζ-Carotene (7 c.d.b.)

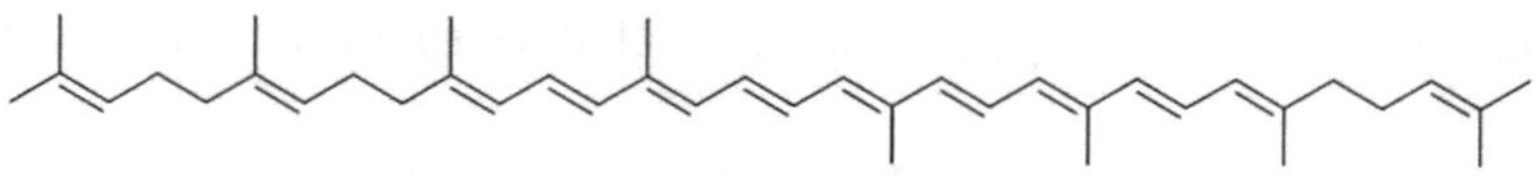

Neurosporene (9 c.d.b.)

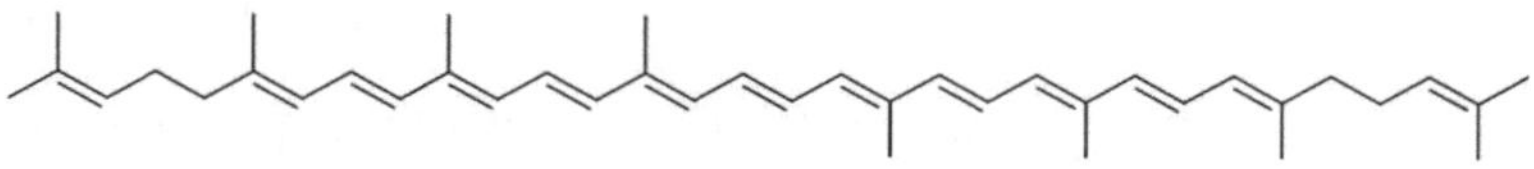

Lycopene (11 c.d.b.)

Fig. 1 Chemical structures of some linear carotenoids with different chromophores

Table 1
Number of conjugated double bonds (c.d.b.) and absorption maxima (λ_{max}) in hexane of some carotenoids [5]

Carotenoid	c.d.b.	λ_{max} (nm)
Phytoene	3	286
Phytofluene	5	348
ζ-Carotene	7	400
Neurosporene	9	440
Lycopene	11	471

When the electrons of atoms or molecules absorb a quantum of electromagnetic radiation there are transitions from one energy state to another of higher energy. For this radiation absorption to

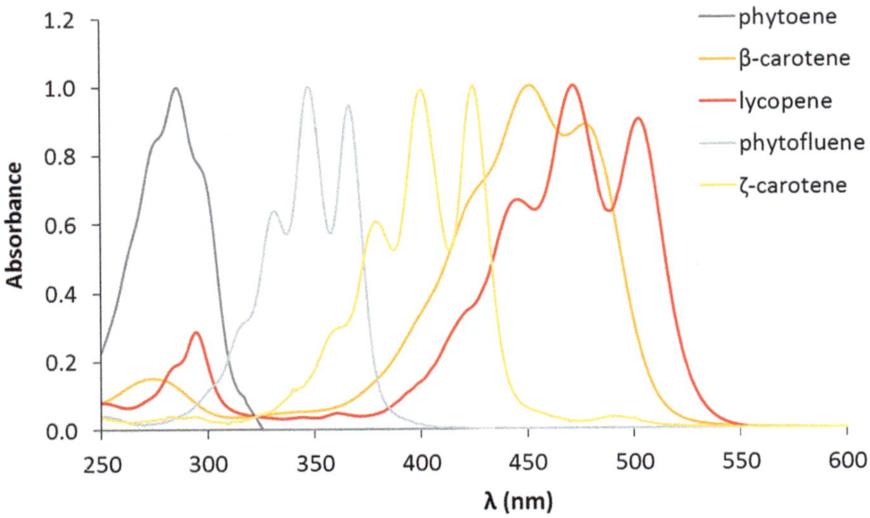

Fig. 2 Visible absorption spectra in methanol of some representative carotenoids

occur, its energy (E) must coincide with the energy difference between the two states. The energy of the radiation is inversely related to its wavelength according to the formula:

$$E = h\nu = \mathrm{hc}/\lambda \tag{3}$$

where E is the energy of the quantum, h is Planck's constant, ν the frequency of the radiation, c the velocity of light, and λ the wavelength.

The typical absorption spectra of atoms are characterized by the presence of single absorption lines. The reason is that in the case of atoms, the electron in its ground state (that is, the lowest energy level) absorbs radiation of a definite wavelength and is eventually promoted to an excited state of higher energy. In the case of molecules, the situation is more complex as an electron can exist in different vibrational energy levels and each of these in a number of rotational energy levels. Thus, in this case, the electronic transition can be due to a range of energy quanta matching the energy differences between electronic basal or excited states in the various vibrational and rotational energy levels. As a result, the absorption spectra of molecules are characterized by the presence of absorption bands [4, 5].

In the case of carotenoids, the delocalization of the electrons in their characteristic polyene chain (system of c.d.b.) makes the energy necessary for the electronic transitions to take place decrease as compared to more saturated compounds so that, in most cases, the energy of the visible light (ca. 400–500 nm) is sufficient to cause these transitions and, therefore, is responsible for the color of these compounds. The greater the number of c.d.b., the greater the wavelength of the light absorbed (Table 1) [4].

The relationships between the chemical structure of carotenoids and the features of their UV–visible absorption spectra have been long-known and have been treated in great detail in the bibliography [5]. The translation of this information into color parameters is more recent though [6, 7]. The carotenoids absorb mainly violet and bluish light (ca. 400–500 nm), so they typically exhibit yellow, orange, or reddish tones (Fig. 2) [3]. In particular, it is considered that at least 7 c.d.b. are necessary for a carotenoid to exhibit color [6, 7]. Thus, phytoene and phytofluene, with 3 and 5 c.d.b. respectively, are colorless carotenoids. In addition to the chemical structure of carotenoids, its colors may depend on other factors such as concentration, aggregation of molecules or interaction with other molecules [3].

1.3 Importance of Color in Foods

The color of food is one of the first sensory characteristics that we perceive and one of the most important attributes for the consumer. The visual perception of color is a factor of great importance because it can have an impact on the emotions of the consumers and therefore can be a decisive factor in the food purchasing decisions, being, in many cases, even more important than the perception of taste and smell [8]. In fact, it is well known that consumers can relate color to other sensory characteristics, like sweetness and flavor [9]. As an example, in olive oil, whose color is mainly due to carotenoids and chlorophylls, green color is associated with bitter and spicy tastes while yellow with fruity flavors [10]. The importance of colour in foods owing their color mainly to carotenoids like citrus and olive oils as well as visual and instrumental color measurement methodologies is reviewed elsewhere [11, 12]. Therefore, one of the major interests of the food industry for carotenoids lies in the use of them as colorings in food and feed. Both carotenoids (like β-apo-8′-carotenal, canthaxanthin, lutein, lycopene, or β-carotene,) and carotenoid containing products (paprika, annatto, diverse extracts) are approved for this purpose by diverse regulatory bodies [3, 13]. Among the foods carotenoid can be used as colorant additives are sauces, dairy, soft drinks, fruit juices, butter, margarines, and confectionery, among others [14].

Carotenoids are also used in feed additives to confer appealing colours to foods of animal origin like egg yolk or fish flesh. Both, carotenoids obtained by chemical synthesis (such as astaxanthin or canthaxanthin, among others) or products containing these carotenoids (such as petals of *Tagetes erecta* as source of lutein, red pepper as source of capsanthin or microbial or crustacean sources of astaxanthin) have been used [15–22]. The livestock industry is also making important efforts to increase the consumption of carotenoids in livestock to obtain meat with more attractive colors for the consumer [23]. Given the unarguable significance of the color of foods for the consumers, the use of affordable and reliable methods for their objective color specification is very important for the food industry.

1.4 Color Measurement

The color of the food can be evaluated visually or instrumentally. The visual analysis of the color consists in the evaluation of its characteristics by means of the sense of sight/vision. Information concerning the training and selection of the panelists, sample presentation, illumination requirements, observation geometry, etc. for the visual assessments can be found in classical reference texts [24, 25]. The visual color assessment can also be carried out by comparing the color of a test sample with those of standards present in color scales or atlases (DIN system, Munsell system, OSA-UCS system, and Ostwald system, among others). The main advantages of the visual evaluation of color is that is a cheap and simple methodology which does not usually require sophisticated instrumentation, although the color specification achieved is subjective [4].

Among the main advantages of the instrumental evaluation of color in relation to visual methods is that the inevitable subjective component of the latter can be eliminated. Other advantages include rapidity, minimal or no manipulation of the samples, the possibility of automation, versatility, the possibility of the acquisition of a large volume of information (colorimetric parameters, spectroscopic data), among others [4]. In these types of analysis, the color is expressed by means of the color coordinates. Certain characteristics related to the light, the object and the observer must be adjusted beforehand to ensure a correct measurement of the color (*see* **Note 1**). There are two main types of instruments that measure the color of an object: those that directly measure the tristimulus values, called colorimeters, and those that measure the radiometric or photometric magnitude that characterizes the source or object from which the tristimulus values are calculated (spectrophotometer and spectroradiometer) [26].

1.4.1 Colorimeters

Colorimeters directly measure the color of primary radiation sources (which emit light) and secondary radiation sources (which reflect or transmit external light), obtaining the tristimulus values X, Y, Z optically, not mathematically. The colorimeter reproduces the response of only a standard observer and a preestablished standard illuminant, so the values obtained may differ from one instrument to the other [27].

1.4.2 Spectrophotometers

Spectrophotometers measure the spectral distribution of transmittance or reflectance of an object from which the color can be calculated under different theoretical conditions. In the spectrophotometer the *transmittance* measure is the ratio between the amount of transmitted light that reaches the detector once it has passed through the sample and the amount of light incident on it. *Reflectance* is the ratio between the response of the instrument to the sample and the response to a known standard, usually a highly reflective sample [28]. Both transmittance and reflectance are intrinsic properties of the object, which do not depend on the illumination received or the observer, in contrast to what occurs with color. The efficiency, resolution, sensitivity and spectral range of the spectrophotometers will depend on the design variables and the selection of the optical components that comprise it [29].

1.4.3 Spectroradiometers

Spectroradiometers typically obtain the radiometric measurement of the spectral distribution of a primary or secondary radiation source. As in the case of spectrophotometers, tristimulus values are calculated through mathematical integration. The spectroradiometer has the same components as the spectrophotometer with the exception of the light source which, in this case, is external to the instrument and, therefore, variable. These instruments can also be used to measure the transmittance or reflectance of any object [26].

All these classic instruments (colorimeters, spectrophotometers and spectroradiometers) require a certain chromatic uniformity in the samples so that the measure of the color is representative. This requirement is sometimes complicated for small samples. In these cases, the use of *digital image analysis* (DIA) for the color measurement is highly recommended. An artificial vision device consists of a lighting system, a device that captures the images and converts the analogue signal into digital (digital camera), and a computer with the appropriate software for processing images and interpreting results [30]. In this process, the digital camera receives the images in its CCD device, which register the color in gradations of three basic colors: red, green, and blue (RGB), in accordance with the three-dimensional character of color. Thus, this methodology is based on the acquisition of an image for further analysis, from which the total color of the object can be obtained. One of the advantages of this technique is that it can be used to measure not only the average color, but also the distribution of the color within the sample. The use of digital cameras in relation to color measurement has been developed and spread widely due to the fidelity in the measurement and the quality of the image. One of the most used systems is the DigiEye system [31] consisting of a Nikon D-80 digital camera connected to a controlled lighting camera (VeriVide DigiEye®, Leicester, UK) and a computer with the DigiPix software.

1.5 Color Measurement and Carotenoid Content

Given the advantages of the instrumental measurements of color, it is not surprising that it has been long used to rapidly estimate carotenoid levels in products like citrus fruits carrots or tomatoes, as typical examples. In this sense, it can be harnessed for different purposes, including [4]:

- The quantitative/qualitative analysis of carotenoids and other pigments.
- The monitoring of quantitative/qualitative changes in pigments as a result of diverse processes (postharvest treatments, industrial processing, storage).
- The evaluation of stages of maturity on the basis of the changes in the pigments contents,
- The grading of foods according to their quality, since color is an important quality attribute and some carotenoids have nutritional relevance.

More information about the diverse types of studies harnessing UV-Vis spectroscopic information to obtain information about carotenoid contents can be found in dedicated revisions of the literature [4, 7, 12].

2 Materials

2.1 Spectrophotometry (Transmission Measurement)

1. Glass cuvettes (2, 5, and 10 mm path length).
2. Sample filtration system: Syringe (13 mm diameter), filter holder and filters (2.0 μm pore).
3. Centrifuge.
4. Plastic tubes.
5. UV–visible scanning spectrophotometer (e.g., HP8452 Hewlett-Packard).
6. CROMALAB Software [32].

2.2 Spectroradiometry (Reflection Measurement)

1. Glass cuvettes (10 mm path length) for liquid samples.
2. Spectroradiometer (e.g., CAS 140 B Instrument Systems).
3. External lamp.
4. Top 100 telescope optical probe.
5. Tamron zoom.
6. Cabin with grey walls ($L* = 50$) (surround).
7. Pressed barium sulfate plate as white reference (background).
8. Software IS-SPECWIN.

2.3 Digital Image Analysis (Reflection Measurement)

1. "DigiEye" imaging system, comprised by a cabin, a digital camera and a CPU.
2. "DigiEye" software.
3. "Digitizer" certified colour chart.

3 Methods

3.1 Spectrophotometry (Transmission Measurement)

This approach is normally used to measure the absorbance of a carotenoid extract and/or characterize its colour. Thus, this approach was used to establish relationships colour-structure in common dietary carotenoids [7]. Examples of classical spectrophotometric determination of carotenoids are provided elsewhere [4].

1. Centrifuge the samples at 3220 × *g* for 10 min at 20 °C (*see* **Note 2**).
2. Filter the samples through the filters before the colour determination.
3. Turn on the spectrophotometer to stabilize it about 30 min before the measurements.
4. Click on the SPT HP8452 program icon to run the software.
5. Open the method (previously created, *see* **Note 3**).
6. Fill the cuvette with an appropriate blank (usually the solvent the carotenoids are dissolved in) and place it in the cuvette holder.
7. Click on the icon *Blank* to perform the measurement of the blank.
8. Fill the cuvette with the filtered sample and place it in the cuvette holder (*see* **Notes 4** and **5**).
9. Click on the icon *Sample* to perform the measurement of the sample.
10. Save the measure: File/Save (Select the place to save the spectra of the samples)/OK.
11. To obtain the colour parameters, it is necessary to export the spectra in DX format.
12. Open the sample file and select the spectrum.
13. Export the selected spectrum as DX format.
14. Open the DX files with the CromaLab program to obtain the colorimetric parameters.
15. From the carotenoid content determined by analytical techniques and the colorimetric parameters measured by spectrophotometry, look for statistical models that allow to estimate the carotenoid concentration in other samples from the colour measurements.

3.2 Spectroradiometry (Reflection Measurement)

This methodology has been applied to characterize the colour of citrus and tomato products and, in some cases, to relate it with their carotenoid levels or even theoretical vitamin A or in vitro antioxidant capacity of carotenoid-containing extracts [33–38].

1. Turn on the external lamp and wait about 30 min to stabilize it before the measurements.
2. Turn on the CAS140B spectroradiometer.
3. Click on the IS-SPECWIN program icon to run the software.
4. Click on the Reflection icon.
5. Open the method (previously created, *see* **Note 6**).
6. Perform the measurement of the dark current:
7. The dark current can be measured manually by pressing the corresponding button. Thus, the detector physically closes the light beam and measures the electrical signal produced without external stimulus. This measurement is stable and does not have to be repeated very often. In any case, it is recommended to repeat the black measurement every time the operating conditions are changed (e.g., exposure time).
8. Perform the measurement of the blank (*see* **Note 7**):

   For liquid samples: Place a cuvette filled with distilled water in the cuvette holder against a reference pressed barium sulfate plate inside the cabin.

   For solid samples: Place the pressed barium sulfate plate as white reference material inside the cabin.
9. Focus the cuvette or the plate in the "view" position of the telescopic probe. Once focused, switch to the "measure" position.
10. Click the *R* icon to perform the blank measurement.
11. Perform the measurement of the sample:

    For liquid samples: Place a cuvette filled with the sample in the cuvette holder against a reference pressed barium sulfate plate inside the cabin.

    For solid samples: Place the sample against a reference pressed barium sulfate plate inside the cabin.
12. Click on the icon ▶ (or press F8) to perform the color measurement of the sample.
13. Save the measure: File/Save (Select the place to save the spectra of the samples)/OK.
14. From the spectrum, the IS-SPECWIN program calculates the colorimetric parameters.
15. From the carotenoid content determined by analytical techniques and the colorimetric parameters measured by spectrophotometry, look for statistical models that allow to estimate the carotenoid concentration in other samples from the colour measurements.

3.3 Digital Image Analysis (Reflection Measurement)

This methodology has been applied to characterize the color of tomato products and to estimate lycopene levels, as compared to spectroradiometry [39] (*see* **Note 8**).

1. Connect plug power adaptor to the cabin.
2. Switch the lamps on by pressing twice the "Enter" button.
3. Wait for 10 min for light stabilization.
4. Select the diffuse or direct light by means of the action of the two handles located at the ends of the lower part of the cabin.
5. Turn on the computer and run the "DigiEye" software.
6. The grey table must then be removed from the tray and the calibration chart is placed instead on the protrusions on the surface of the tray.
7. An image of the chart is made in the menu "Camera" → "Release" → "Release".
8. Then select the menu "Calibration" → "Camera".
9. A pop-up window will ask if the marks are on the colored frames. After making sure and accepting, a graph and values will appear. These values show in terms of color differences (CIEDE 2000) the difference between the measured units and what is stored by the software. The average must be less than 1.5 and the median less than 1.2. If the lamps and the calibration chart are in good conditions this process does not have to give error.
10. From this moment on, the camera is calibrated, and it is possible to make the desired images of the samples. To do this, click again on the menu "Camera" → "Release" → "Release" (*see* **Note 9**).
11. In the top menu there are selection tools like an image editing program. Select one of the different tools and select the desired area by dragging the mouse (*see* **Notes 10** and **11**).
12. Once the area has been selected, press the "DigiPix" menu, select below the sample from where you want to acquire the data (S1, S2, S3, ...) and press "Reflectance" (*see* **Note 12**).
13. In a pop-up window a graph will appear with the reflectance spectrum of the marked area and below, the colorimetric coordinates according to the type of illuminant.
14. The images can be also saved as TIF or JPG formats in the menu "File" → "Save".
15. From the carotenoid content determined by analytical techniques and the colorimetric parameters measured by Digital Image Analysis, look for statistical models that allow to estimate the carotenoid concentration in other samples from the color measurements.

4 Notes

1. The illuminant D65 and the 10° Standard Observer are recommended for food color measurements.
2. For spectrophotometric measurements it is necessary to centrifuge and filter the samples for removing suspended particles.
3. For spectrophotometric measurements the method for the measurement should be previously created to cover the entire visible spectrum (380–770 nm) at constant intervals ($\Delta\lambda = 2$ nm).
4. If the samples are very dark, it is necessary to use cuvettes with smaller path length for spectrophotometric measurements.
5. For spectrophotometric measurements and for the spectroradiometric measurements of liquid samples the cuvettes used to make the blank measurements should be the same as those used to make the measurements of the samples.
6. For the spectroradiometric measurements the method for the measurement should be previously created. The following parameters should be defined: Integration time; signal level; averages (the spectroradiometer should be set to take at least three consecutive measurements of each sample); resampled from 380 to 780 nm (to cover the entire visible spectrum); Top100 aperture (If 10 mm path length-cuvettes are used, the opening of the top 100 should be adjusted to blende 3). Anyway, this depends on the distance between sample and Top100 probe.
7. System geometry of the spectroradiometer:

   The zoom, to which the probe was attached, was held at a fixed distance of 50 cm in a straight line from the sample or the blank. Incident illumination: 45°.
8. The potential of the digital image analysis system is very high. However, small deviations in color measurements have been observed in the blue zone of the CIELAB color space. Notwithstanding, these hues are not common in foods, and even less so in foods where carotenoids are to be evaluated.
9. In the digital image analysis, in rare occasions and when pressing the button that acquires an image, the equipment gives error in the focus. This is because there is no sharp-edged object just below the camera. This is usually solved by slightly moving the sample in the tray inside the booth and recapturing the image.
10. You may want to process the image with software other than the one DigiEye incorporates. In fact, it is common to use automatic tools programmed under Matlab or Python. In

these cases, the use of a sample surface with a clearly different color than the sample is recommended. This will improve the process of identifying the region of interest. The use of white, grey or black paperboard is recommended depending on the nature of our samples.

11. It is not necessary a very powerful computer for the digital image analysis, but it is necessary to bear in mind that depending on our RAM memory, the equipment can be blocked when there are many images open. It is also recommended to close images that have already been measured or stored.
12. When using the DigiPix color measurement tool, a reflectance spectrum appears. This spectrum is simulated, since the camera is not a spectrophotometer and only measures three bands R, G, and B. Therefore, the spectrum can be used for educational or for color interpretation but not for analytical purposes.

References

1. Wyszecki G, Stiles WS (1982) Color science. Concepts and methods. Quantitative data and formulae. John Wiley & Sons, New York
2. CIE (2004) Colorimetry, vol 142. CIE Central Bureau,Publ, Vienna, Austria, p 2001
3. Rodríguez-Concepcion M, Avalos J, Bonet ML et al (2018) A global perspective on carotenoids: metabolism, biotechnology, and benefits for nutrition and health. Prog Lipid Res 70:62–93
4. Meléndez-Martínez AJ, Vicario IM, Heredia FJ (2011) Applicability of the measurement of visible spectroscopic data and colour in the analysis of food carotenoid. In: Carotenoids: properties, effects, and diseases. Nova Science Publishers, Hauppauge, N.Y
5. Britton G, Liaaen-Jensen S, Pfander H (1995) Carotenoids. Volume 1B: Spectroscopy. Birkhäuser, Basel, Switzerland
6. Meléndez-Martínez AJ, Britton G, Vicario IM et al (2006) Relationship between the colour and the chemical structure of carotenoid pigments. Food Chem 101(3):1145–1150
7. Meléndez-Martínez AJ, Vicario IM, Heredia FJ (2017) Aplicación de las medidas de espectroscopía visible y de color al análisis de carotenoides. In: Carotenoides en agroalimentación y salud. Editorial Terracota SA, México, pp 118–140
8. Simon JE, Decker EA, Ferruzzi MG et al (2017) Establishing standards on colors from natural sources. J Food Sci 82(11):2539–2553
9. Clydesdale FM (1993) Colour as a factor in food choice. Crit Rev Food Sci Nutr 33:83–101
10. Ryan D, Robards K, Lavee S (1998) Assessment of quality in olive oil. Olivae 72:23–41
11. Meléndez-Martínez AJ, Vicario IM, Heredia FJ (2005) Instrumental measurement of orange juice colour: a review. J Sci Food Agric 85(6):894–901. https://doi.org/10.1002/jsfa.2115
12. Moyano MJ, Heredia FJ, Meléndez-Martínez AJ (2010) The color of olive oils: the pigments and their likely health benefits and visual and instrumental methods of analysis. Compr Rev Food Sci Food Saf 9(3):278–291
13. Lehto S, Buchweitz M, Klimm A et al (2017) Comparison of food colour regulations in the EU and the US: a review of current provisions. Food Addit Contam Part A 34(3):335–355. https://www.tandfonline.com/. https://doi.org/10.1080/19440049.2016.1274431
14. Martins N, Roriz CL, Morales P et al (2015) Food colorants: challenges, opportunities and current desires of agro-industries to ensure consumer expectations and regulatory practices. Trends Food Sci Technol 52:1–15
15. Breithaupt DE (2007) Modern application of xanthophylls in animal feeding - a review. Trends Food Sci Technol 18(10):501–506. http://www.sciencedirect.com/science/article/B6VHY-4NKJ0D6-4/2/3c2f8dae3a7adbb884fb5d3ad9069a9f
16. Breithaupt DE, Bamedi A, Wirt U (2002) Carotenol fatty acid esters: easy substrates for

digestive enzymes? Comp Biochem Physiol Part B 132:721–728
17. Breithaupt DE, Weller P, Grashorn MA (2003) Quantification of carotenoids in chicken plasma after feeding free or esterified lutein and capsanthin using high-performance liquid chromatography and liquid chromatography-mass spectrometry analysis. Poult Sci 82 (3):395–401. http://www.scopus.com/inward/record.url?eid=2-s2.0-0038548600&partnerID=40
18. Maoka T (2011) Carotenoids in marine animals. Mar Drugs 9(2):278–293
19. Shahidi F, Metusalach BJA et al (1998) Carotenoid pigments in seafoods and aquaculture. Crit Rev Food Sci Nutr 38(1):1–67
20. Wade NM, Gabaudan J, Glencross BD (2017) A review of carotenoid utilisation and function in crustacean aquaculture. Rev Aquac 9 (2):141–156
21. de Carvalho CCR, Caramujo MJ (2017) Carotenoids in aquatic ecosystems and aquaculture: a colorful business with implications for human health. Front Mar Sci 4:93):1–93)14
22. Spada FP, Selani MM, Coelho AAD et al (2016) Influence of natural and synthetic carotenoids on the color of egg yolk. Sci Agric 73 (3):234–242
23. Álvarez R, Meléndez-Martínez AJ, Vicario IM et al (2014) Effect of pasture and concentrate diets on concentrations of carotenoids, vitamin a and vitamin E in plasma and adipose tissue of lambs. J Food Compos Anal 36(1–2):59–65
24. MacDougall D (2002) Colour in food, improving quality, 1st edn. Woodhead Publishing Ltd, Cambridge
25. Hutchings JB (1994) Food colour and appearance. Blackie Academic and Professional Publication, Glasgow
26. Wyszecki G, Stiles WS (2000) Color science. Concepts and methods. Quantitative data and formulae, vol 2. John Wiley & Sons, Inc, New York
27. Artigas JM (2002) Medida del color. In: Capilla P, Artigas JM, Pujol JE (eds) Fundamentos de colorimetría. Valencia, España
28. Workman J, Jerome J (2016) Ultraviolet spectroscopy. In: Workman J, Jerome J (eds) Concise handbook of analytical spectroscopy, the: theory, applications, and reference materials. World Scientific, Singapore
29. Skoog D, Holler FJ, Crouch SR et al (2008) Principios del Análisis Instrumental. Editorial Cengage Learning, México
30. Wang HH, Sun DW (2002) Correlation between cheese meltability determined with a computer vision method and with Arnott and Schreiber tests. J Food Sci 67:745–749
31. Luo MR, Cui GH, Li C (2001) British patent entitled apparatus and method for measuring colour (DigiEye System). Application No 0124683.4. Derby University Enterprises Limited, 4 October
32. Heredia FJ, Álvarez C, González-Miret ML et al (2004) CromaLab, Analisis de color. Registro General de la Propiedad Intelectual:2004
33. Stinco CM, Escudero-Gilete ML, Heredia FJ et al (2016) Multivariate analyses of a wide selection of orange varieties based on carotenoid contents, color and in vitro antioxidant capacity. Food Res Int 90:194–204
34. Stinco CM, Heredia FJ, Vicario IM et al (2016) In vitro antioxidant capacity of tomato products: relationships with their lycopene, phytoene, phytofluene and alpha-tocopherol contents, evaluation of interactions and correlation with reflectance measurements. LWT-Food Sci Technol 65:718–724
35. Meléndez-Martínez AJ, Gómez-Robledo L, Melgosa M et al (2011) Color of orange juices in relation to their carotenoid contents as assessed from different spectroscopic data. J Food Compos Anal 24(6):837–844
36. Meléndez-Martínez AJ, Ayala F, Echávarri JF et al (2011) A novel and enhanced approach for the assessment of the total carotenoid content of foods based on multipoint spectroscopic measurements. Food Chem 126 (4):1862–1869
37. Meléndez-Martínez AJ, Vicario IM, Heredia FJ (2007) Rapid assessment of vitamin a activity through objective color measurements for the quality control of orange juices with diverse carotenoid profiles. J Agric Food Chem 55 (8):2808–2815
38. Meléndez-Martínez AJ, Vicario IM, Heredia FJ (2003) Application of tristimulus colorimetry to estimate the carotenoids content in ultra-frozen orange juices. J Agric Food Chem 51 (25):7266–7270
39. Stinco CM, Rodríguez-Pulido FJ, Escudero-Gilete ML et al (2013) Lycopene isomers in fresh and processed tomato products: correlations with instrumental color measurements by digital image analysis and spectroradiometry. Food Res Int 50(1):111–120. https://doi.org/10.1016/j.foodres.2012.10.011

Chapter 9

A Routine Method for the Extraction and HPLC-DAD Profiling of Major Plant and Food Carotenoids

Dámaso Hornero-Méndez

Abstract

Carotenoids are ubiquitously present in Nature, and especially in plants and derived foods. The carotenoid profiling is necessary to understand relevant aspects in relation to their biochemistry and genetics, as well as their important roles on human health and animal ecophysiology. Here we propose a simple methodology for the routine extraction and efficient HPLC separation of the most common plant and food carotenoids from plant and animal origins.

Key words HPLC, UV/Visible, DAD, Carotenoids, Carotenes, Xanthophylls

1 Introduction

Among natural pigments, carotenoids are those more ubiquitously found in living organisms [1]. Carotenoids are responsible for the yellow, orange, and red colors of most fruits and vegetables, including fruits, flowers, roots, leaves, and seeds; and also of some animal tissues and derived foods (i.e., egg-yolk, skin, etc.) [2]. The most common carotenoids are C40 isoprenoids characterized by a polyene skeleton with a long conjugated double-bond system, which constitutes the chromophore. Longer (C50 bacterial carotenoids) and shorter (for instance apocarotenoids) chain carotenoids are also found. About 750 naturally occurring carotenoids have been isolated and described [1]. Carotenoids can only be synthesized de novo by plants, and some bacteria and fungi. In contrast, animals must necessarily uptake carotenoids from diet since they are unable to synthesize them [3]. Carotenoids play essential roles in the light-harvesting process and the photoprotection mechanisms of plants. Moreover, after ingestion by animals, carotenoids exert a wide range of health-promoting functions and actions such as provitamin A (only some of them), antioxidant, reduction in the risk of developing cardiovascular and other degenerative diseases (such as

Manuel Rodríguez-Concepción and Ralf Welsch (eds.), *Plant and Food Carotenoids: Methods and Protocols*, Methods in Molecular Biology, vol. 2083, https://doi.org/10.1007/978-1-4939-9952-1_9,

age-related macular degeneration (AMD) and cataracts), enhancement of the immune response, and cell protection against free radicals [3, 4].

Thus, the qualitative and quantitative analysis of carotenoids is of paramount importance for the investigation of all these associated aspects. Chromatography, and especially HPLC, has always been the choice methodology for attaining this task. In fact, the invention and development of the initial chromatography technique by Mikhail Tswett in the early 1900s was closely linked to the study of chloroplast pigments [5, 6]. The first HPLC application for the separation of carotenoids was reported by Stewart and Wheaton in 1971 for the investigation of the carotenoid composition of *Citrus* extracts [7]. Since then, important advances and innovations have been made in the HPLC technique, including the development of new stationary phases well as sensible, selective and robust detectors (mainly DAD and MS detectors), which have been applied to the carotenoid research field [8]. Consequently, a vast diversity of HPLC methods can be found in the carotenoid literature. The protocol here presented has been used routinely in the author's lab during the last 25 years for analysing carotenoids in great variety of plant and food matrices [9]. This method was firstly developed for the separation and quantification of the complex carotenoid profile in red pepper fruits (*Capsicum annuum*) and derived products such as paprika powder and oleoresin [9]. The HPLC method uses a C18 reverse-phase and a binary gradient composed by acetone and deionized water, providing a rapid (β-carotene elutes at 21 min) and very reproducible separation. The optimised conditions have given excellent results when used for the major carotenoid analysis in a wide variety of samples such as green vegetables, apple fruits, potato tubers, cereal grains, dates, strawberry fruits, rose hips, sarsaparilla berries, fungi extracts, and orange juice [9–20]. Additionally, the present HPLC conditions have been applied to the analysis of carotenoids in fungi and bacteria [21, 22], and in animal samples such as plasma, egg-yolk, and skin [23–25].

2 Materials

Reagents should be at least of analytical grade, unless stated otherwise.

2.1 Sample Pretreatment

Plant material and food-related samples can be analyzed either in fresh (solid or liquid) or dehydrated state (*see* **Note 1**). Solid and semisolid materials, such as fruits and vegetables, should be chopped and mixed with a household blender right before extraction. In the case of harder material (i.e., cereals and other seeds) a lab mill can be used (*see* **Note 2**).

1. Household blender.
2. Lab mill.
3. Lyophilizer.

2.2 Extraction of Carotenoids

1. Homogenizer (Ultraturrax model T-25 or similar).
2. Solvent for extraction: Acetone containing 0.1% butylated hydroxytoluene (BHT).
3. Diethyl ether containing 6 ppm BHT (*see* **Note 3**).
4. NaCl solution (10% w/v).
5. Anhydrous Na_2SO_4 solution (2% w/v).
6. Refrigerated centrifuge with rotor for 50 mL and 15 mL tubes.
7. Refrigerated centrifuge with rotor for microcentrifuge tubes (1.5–2 mL) with a minimum centrifuge force of 12,000 × *g*.
8. 15 mL and 50 mL centrifuge tubes (Falcon type or similar).
9. 1.5 mL microcentrifuge tubes (Eppendorf type or similar).
10. Separation funnel (1 L).
11. Vortex.
12. Anhydrous Na_2SO_4 (solid).
13. Rotary evaporator.
14. Nitrogen (gas).
15. Acetone (HPLC grade).
16. Sonication bath.
17. Sample vials for HPLC (*see* **Note 4**).

2.3 Saponification of Carotenoid Extract

1. 15 mL centrifuge tube (Falcon type or similar).
2. Diethyl ether (containing 6 ppm BHT).
3. KOH–methanol solution (10% w/v).
4. Vortex.
5. pH test strips.
6. Anhydrous Na_2SO_4 (solid).
7. Nitrogen (gas).
8. Acetone (HPLC-grade).

2.4 High-Performance Liquid Chromatography (HPLC)

1. HPLC equipment with a diode array detector (*see* **Note 5**). In our case the HPLC system consisted of a Waters 2695 Alliance chromatograph fitted with a Waters 2998 photodiode array detector (DAD), and controlled with Empower2 software (Waters Corporation, Milford Massachusetts, USA).
2. Solvent A: Acetone (HPLC-grade).

3. Solvent B: Deionized water (HPLC-grade deionized ultrapure water produced with a Milli-Q Advantage A10 system, Merck Millipore, Madrid, Spain).
4. HPLC column. A reversed-phase C18 column (200 mm × 4.6 mm i.d., 3 μm, Mediterranea SEA18; Teknokroma, Barcelona, Spain), fitted with a guard column of the same packing material, should be used (*see* **Note 6**).

2.5 Preparation of Calibration Curves

1. UV/visible (UV/Vis) spectrophotometer. In our case the system consisted of a Hewlett-Packard UV/Vis diode array spectrophotometer model 8452A.
2. Quartz cuvette (0.5 mL).
3. 1 mL volumetric flasks.
4. Carotenoid standards (*see* **Note 7**).
5. Solvents (analytical-grade) for individual stock solutions of carotenoids: petroleum ether, ethanol, *n*-hexane.
6. Acetone (HPLC-grade).

3 Methods

3.1 General Precautions

Carotenoids are labile compounds which are prone to degradation upon exposure to oxygen and light, thus precautions should be taken to minimize the adverse effects of these factors. It is highly recommended that all operations should be carried out under dimmed light to prevent isomerization and photodegradation of carotenoids. Follow "Good Laboratory Practice" procedures so that all operations involving the manipulation of solvents are performed in a fume hood.

3.2 Extraction of Carotenoids (Standard Procedure)

1. Sample size. One to three grams taken from a homogeneous and representative sample is generally sufficient to provide accurate and reproducible data (*see* **Note 8**). In the case of dehydrated samples the sample size can be scaled down by a factor of 5–10 (*see* **Note 9**).
2. Place sample in a 50 mL centrifuge tube (Falcon tube or similar) and extract it with 20 mL of acetone (containing 0.1% BHT) by using a homogenizer during 1 min (*see* **Note 10**). If an internal standard (IS) is used, this should be added at this step (the IS amount should be adjusted to the expected average carotenoid content) (*see* **Note 11**).
3. Separate extract from sample debris by centrifugation at 4500 × *g* at 4 °C during 5 min.
4. Repeat the extraction procedure 3–4 times until complete extraction of color (*see* **Note 12**).

5. Collect and pool extracts into a separation funnel (1 L volume).
6. Add 50 mL of diethyl ether (containing 6 ppm BHT) and 100 mL of NaCl solution (10% w/v), and shake vigorously during 1 min to facilitate carotenoid transfer to ether phase. Allow to stand for the separation of two phases, and discard the aqueous phase (lower phase).
7. The upper phase, containing the carotenoid pigments is washed three additional times with 10% NaCl and finally with 2% (w/v) Na_2SO_4 solution for removing the water contained in the organic fraction (*see* **Note 13**). Aqueous phases are discarded after each washing.
8. Filter the extract through an anhydrous Na_2SO_4 bed into a round-bottom flask. Wash the filtering material with additional diethyl ether until no color is observed.
9. Evaporate the extract to dryness under vacuum in a rotary evaporator at temperature not higher than 30 °C (*see* **Note 14**).
10. The dry extract is dissolved in 3–5 mL of acetone (*see* **Note 15**) and an aliquot (0.5 mL) is stored at −30 °C until HPLC analysis (*see* **Note 16**).
11. Extraction should be performed in at least triplicate ($n = 3$).

3.3 Extraction of Carotenoids (Microscale Procedure)

1. Sample size. Use 0.2–0.5 g depending on the sample nature (*see* **Note 17**).
2. Place sample in a 15 mL centrifuge tube (Falcon tube or similar) and extract carotenoids with 3 mL of acetone (containing 0.1% BHT) by using a homogenizer during 1 min (*see* **Notes 9** and **10**). Add IS in this step (if used).
3. Separate extract by centrifugation at 4500 × g at 4 °C during 5 min.
4. Repeat the extraction procedure 3–4 times until complete extraction of color.
5. Collect and pool extracts into a clean tube.
6. Reduce solvent volume down to 2–3 mL under a gentle nitrogen stream (*see* **Note 18**).
7. Add 3 mL of diethyl ether and 4 mL of NaCl solution (10% w/v), and shake vigorously during 30 s to facilitate carotenoid transfer to the ether phase.
8. Separate phases by centrifugation at 4500 × g at 4 °C during 5 min.
9. Transfer the upper phase (diethyl ether phase), with a glass Pasteur pipette, into a clean tube, and repeat (2–3 times) the extraction of the aqueous phase with 2 mL of diethyl ether. Discard the lower phase (aqueous phase).

10. Collect the ether phases and evaporate to dryness under vacuum in a rotary evaporator at temperature not higher than 30 °C. Alternatively, solvent can be evaporated under N_2 stream.
11. Dissolve the dry extract in 0.5–1.0 mL of acetone (*see* **Note 15**) and stored at −30 °C until HPLC analysis.
12. Extraction should be performed in at least triplicate.

3.4 Saponification

If saponification is required for the hydrolysis of lipids and carotenoid esters, the following micro-scale procedure can be used.

1. 1 mL of the pigment extract obtained in Subheading 3.2 (**step 10**) is placed in a 15 mL centrifuge tube (Falcon type or similar) and dried under a nitrogen stream (*see* **Note 19**).
2. Add 2 mL of diethyl ether and 1 mL of 10% (w/v) KOH–methanol.
3. Shake the mixture in a vortex for 30 s and left to react for 20 min at room temperature under nitrogen atmosphere and with periodic shaking (*see* **Note 20**).
4. Add 8 mL of distilled water for washing, and separate phases by centrifugation (at 4500 × *g* at 4 °C during 5 min).
5. Discard the lower aqueous phase and repeat the washing operation (*see* **Note 13**) until the pH of the aqueous phase is neutral (use pH test strips).
6. Discard the lower phase (last washing).
7. Filter the ether phase through an anhydrous Na_2SO_4, and rinse the filtering material with more diethyl ether. Collect the filtrate in a clean test tube.
8. Evaporate the ether to dryness under a nitrogen stream.
9. Dissolve the extract in 1 mL of acetone and store at −30 °C until subsequent chromatographic analysis.

3.5 High Performance Liquid Chromatography (HPLC)

1. Equilibrate the HPLC equipment to the initial solvent conditions (75% solvent A and 25% solvent B) and 25 °C for column temperature.
2. Set flow rate at 1 mL/min.
3. Prior to injection, samples are centrifuged at >13,000 × *g* for 5 min at 4 °C for particle removal (*see* **Notes 21** and **22**).
4. Inject 10 μL of sample into the HPLC system (*see* **Notes 23** and **24**).
5. Run the binary-gradient elution profile: Initial conditions (75% A; 25% B) increases linearly to 95% A in 10 min, then hold 95% A for 7 min and raises to 100% A in 3 min, and is maintained constant (100% A) for 3 min (*see* **Note 25**). Initial conditions are restored in 5 min.

6. Detection is performed at 450 nm, and the online spectra are acquired in the 330–700 nm wavelength range with a resolution of 1.2 nm.
7. Integrate chromatogram at 450 nm (*see* **Note 26**), and record peak areas for quantification according to calibration curves prepared with pure carotenoid standards (*see* Subheading 3.5). As examples, Figs. 1–5 show the HPLC chromatograms and the online UV/Vis spectra obtained for carotenoid extracts prepared from spinach leaves, tomato fruits, red pepper fruits, sarsaparilla berries and egg-yolk, respectively.
8. Calculate concentration from calibration curves (*see* **Note 27**) and give quantitative result in adequate units (i.e., μg/g, mg/kg, mg/100 g, mg/L, in either fresh or dry weight basis).

3.6 Preparation of Calibration Curves

1. Prepare a stock solution for each carotenoid standard by dissolving a small amount (300–500 μg) in 5 mL of acetone or petroleum ether for xanthophylls or carotenes, respectively.
2. Use UV/Vis spectrophotometry for measuring the concentration of the stock solutions by using Beer–Lambert law and the specific absorption coefficient ($A_{1cm}^{1\%}$) for each carotenoid in a particular solvent at a characteristic wavelength (λ(nm)). An appropriate dilution of the stock solution with the selected solvent is needed. Table 1 presents a list of common carotenoids including specific absorption coefficients, wavelength for measurement and solvent. Check the literature for a more detailed list [1, 26, 27].
3. Calculate the concentration (μg/mL) of the stock solution from the following expression

$$C = \frac{A \times DF \times 10^4}{A_{1cm}^{1\%}}$$

where C is the concentration (μg/mL), A is the absorbance at the selected λ(nm), $A_{1cm}^{1\%}$ is the specific absorption coefficient, and DF is the dilution factor.

4. For each carotenoid prepare a set of eight working solutions in the concentration range 0.05–40 μg/mL by diluting appropriated volumes of the stock solution with acetone.
5. Centrifuge an aliquot of each working solution at >13,000 × g for 5 min at 4 °C.
6. Inject 10 μL of each working solution into the HPLC system. Do this by triplicate.
7. Integrate chromatograms at 450 nm (or other selected wavelength for particular carotenoids) and annotate peak areas.

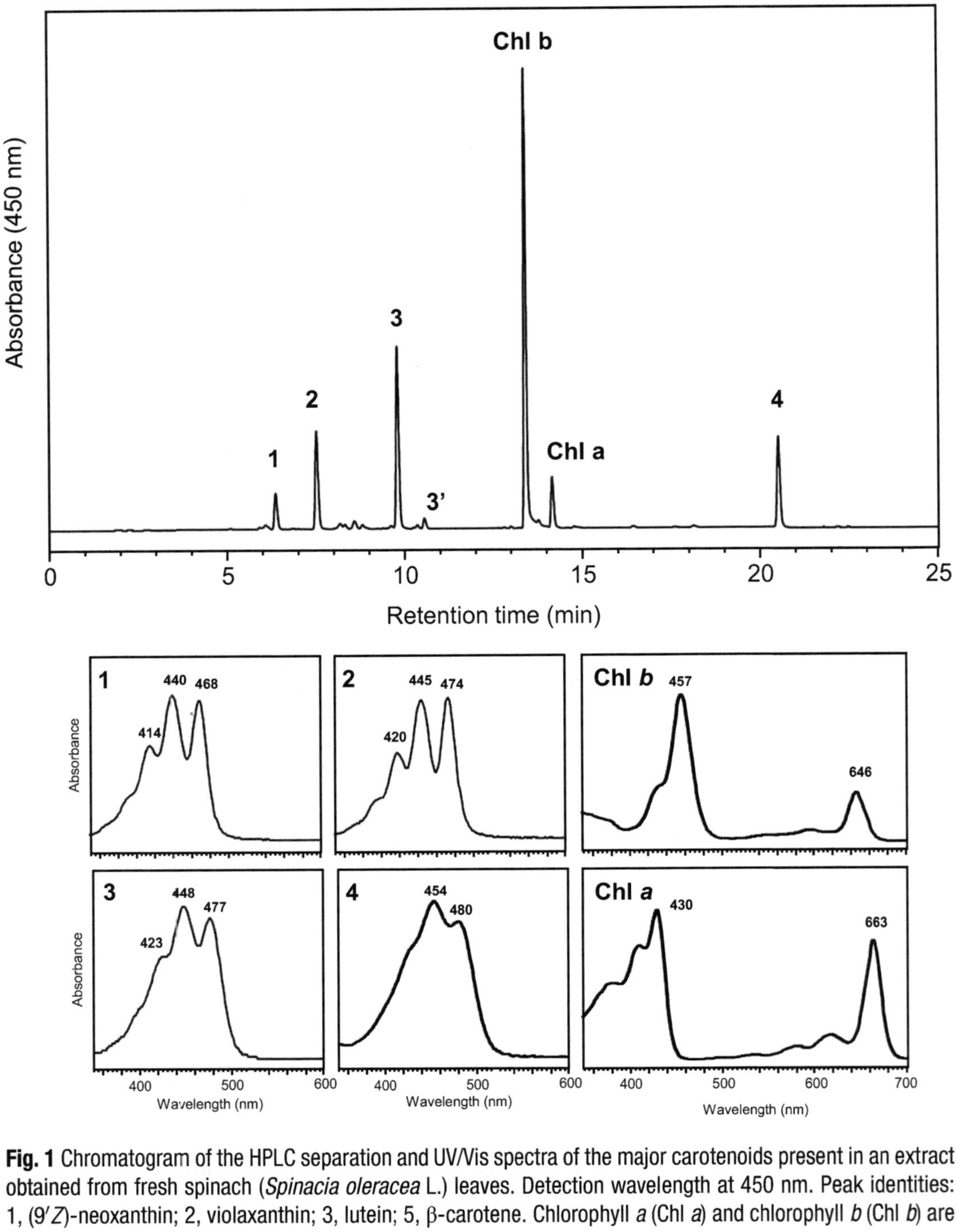

Fig. 1 Chromatogram of the HPLC separation and UV/Vis spectra of the major carotenoids present in an extract obtained from fresh spinach (*Spinacia oleracea* L.) leaves. Detection wavelength at 450 nm. Peak identities: 1, (9′*Z*)-neoxanthin; 2, violaxanthin; 3, lutein; 5, β-carotene. Chlorophyll *a* (Chl *a*) and chlorophyll *b* (Chl *b*) are also present in the chromatogram

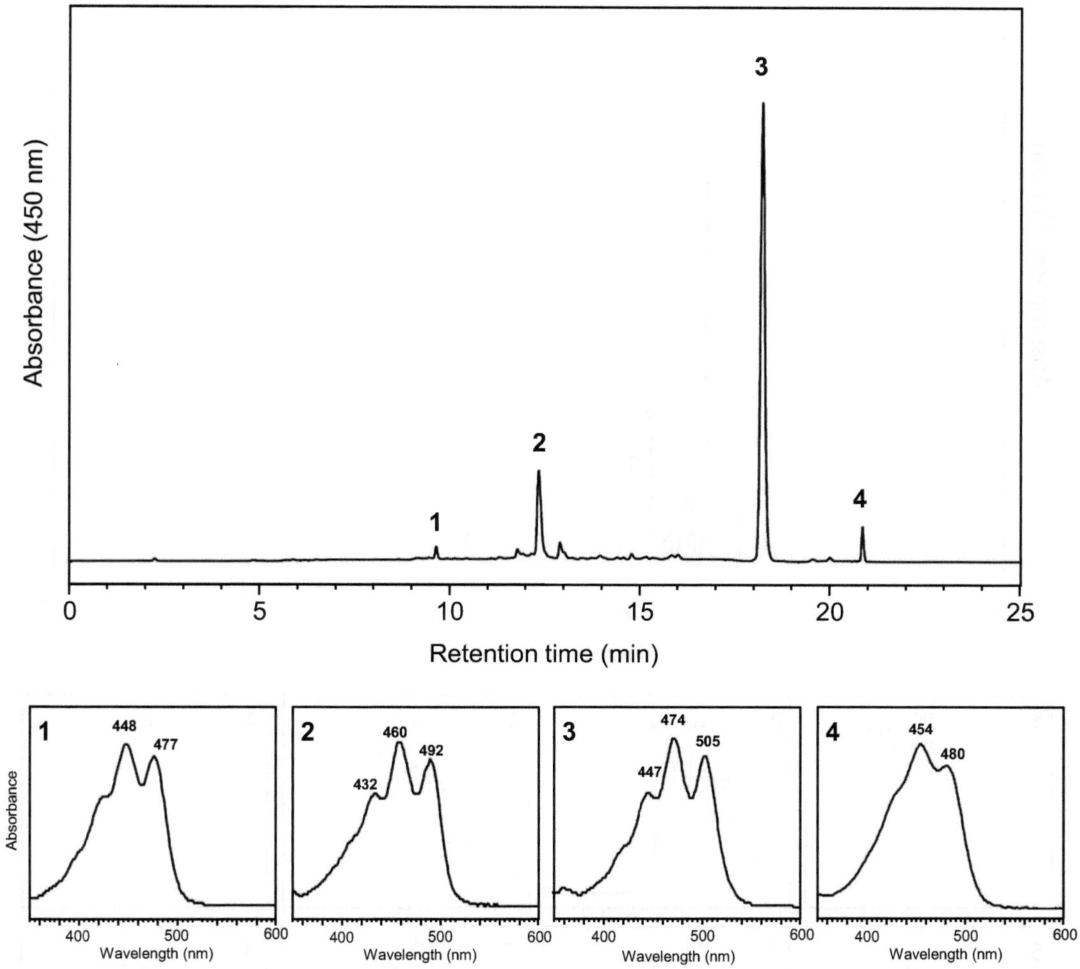

Fig. 2 Chromatogram of the HPLC separation and UV/Vis spectra of the major carotenoids present in an extract obtained from tomato (*Solanum lycopersicum* L.) fruits. Detection wavelength at 450 nm. Peak identities: 1, lutein; 2, lycophyll (dihydroxy-lycopene); 3, lycopene; 4, β-carotene

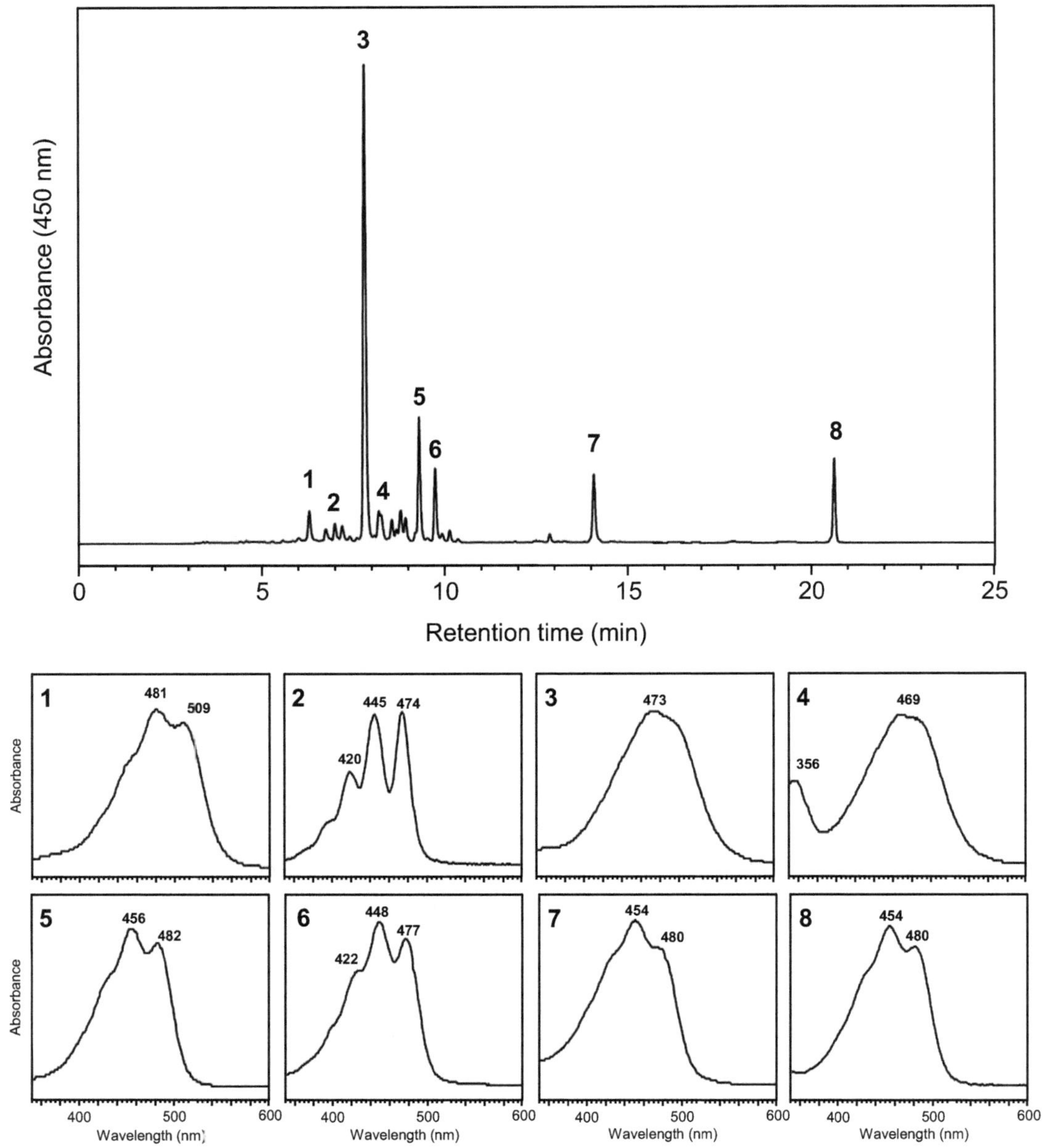

Fig. 3 Chromatogram of the HPLC separation and UV/Vis spectra of the major carotenoids present in an extract (saponified) obtained from red pepper (*Capsicum annuum* L.) fruits. Detection wavelength at 450 nm. Peak identities: 1, capsorubin; 2, violaxanthin; 3, capsanthin; 4, *cis*-capsanthin; 5, zeaxanthin; 6, cucurbitaxanthin A; 7, β-cryptoxanthin; 8, β-carotene

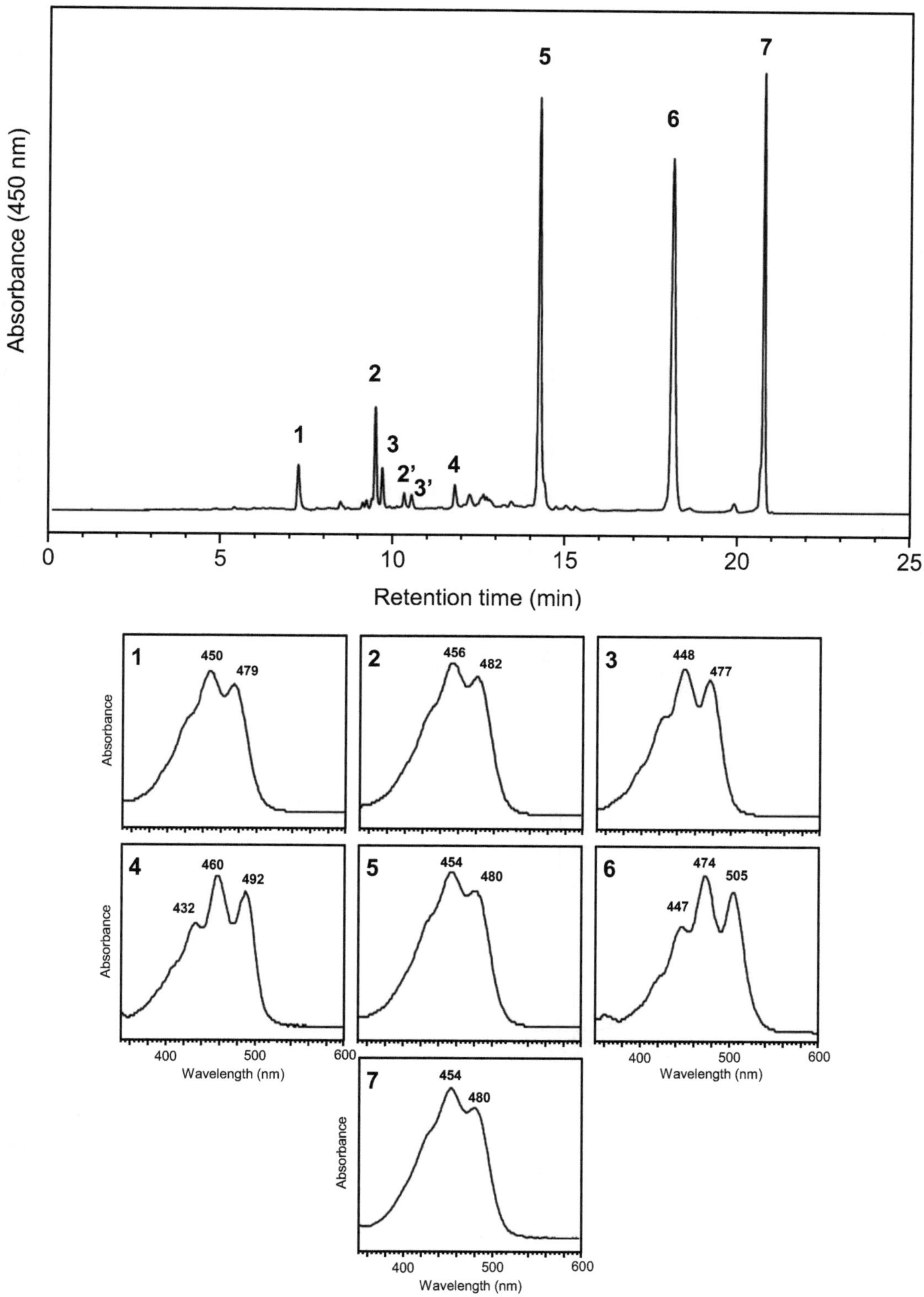

Fig. 4 Chromatogram of the HPLC separation and UV/Vis spectra of the major carotenoids present in an extract (saponified) obtained from sarsaparilla berries (*Smilax aspera* L.). Detection wavelength at 450 nm. Peak identities: 1, antheraxanthin; 2, zeaxanthin; 2′, *cis*-zeaxanthin; 3, lutein; 3′, *cis*-lutein; 4, lycophyll; 5, β-cryptoxanthin; 6, lycopene; 7, β-carotene

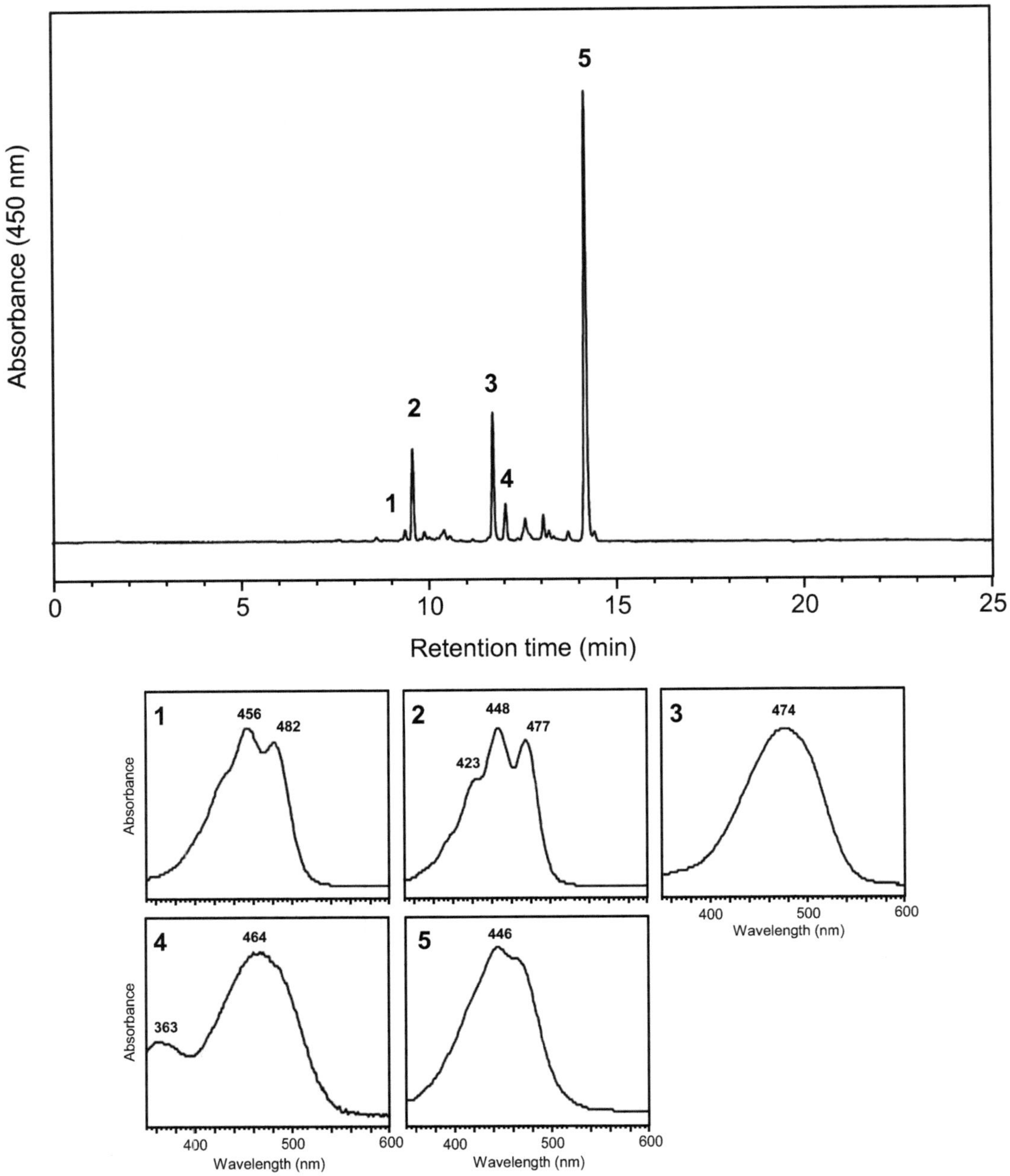

Fig. 5 Chromatogram of the HPLC separation and UV/Vis spectra of the major carotenoids present in an extract obtained from egg-yolks from farmed hens. Detection wavelength at 450 nm. Peak identities: 1, zeaxanthin; 2, lutein; 3, canthaxanthin; 4, *cis*-canthaxanthin; 5, β-apo-8′-carotenoic acid ethyl ester

Table 1
Specific absorption coefficients ($A^{1\%}_{1cm}$) used for quantitative spectrophotometric determination of carotenoids

Carotenoid	$A^{1\%}_{1cm}$	λ(nm) for measurement	Solvent
Antheraxanthin	2350	446	Ethanol
Canthaxanthin	2200	466	Light petroleum
Capsanthin	2072	483	Benzene
Capsorubin	2200	489	Benzene
α-Carotene	2800	444	Light petroleum
β-Carotene	2592	449	Light petroleum
β-Cryptoxanthin	2386	449	Light petroleum
Lutein	2550	445	Ethanol
Lycopene	3450	470	Light petroleum
Neoxanthin	2243	439	Ethanol
Phytoene	1250	286	Light petroleum
Phytofluene	1350	348	Light petroleum
Violaxanthin	2250	440	Ethanol
Zeaxanthin	2348	449	Light petroleum

8. Plot peak area versus concentration (μg/mL) to calculate the calibration curve expression by the "linear least-squares" method (*see* **Note 27**). Calibration curve is given as follows:

$$\text{Peak area} = \text{Response factor} \times C\ (\mu g/mL) + \text{Intercept}$$

4 Notes

1. Lyophilization, also known freeze-drying, is the preferred method for dehydrating or drying biological materials as it preserves intact most of chemical constituents.
2. Avoid the increase of temperature by using short milling times.
3. Commercial diethyl ether is usually supplied with added butylated hydroxytoluene (BHT), an antioxidant which reduces the formation of peroxides.
4. If possible, amber glass vials should be used in order to prevent the photodegradation of pigments.
5. As minimum requirements, the HPLC equipment should deliver binary solvent gradients, control constant column

temperature and should have an UV/Vis spectrophotometric detector (if possible a diode array detector (DAD)).

6. Similar reversed-phase C18 columns from other suppliers can be used, however selectivity may change in some cases and elution order can be affected. For example, in our column zeaxanthin elutes earlier than lutein, but in most C18 columns zeaxanthin elutes right after lutein. Therefore, elution order should be checked during method implementation by using either carotenoid standards or natural extracts with well-known carotenoid profile.
7. Traditionally, few carotenoids have been commercially available in pure form. Consequently, authentic reference carotenoid compounds have been isolated from natural sources with well-known carotenoid profile and where the aimed compound, if possible, is a major one. A combination of solvent partition and chromatographic techniques (TLC, open column chromatography, analytical and semipreparative HPLC, etc.) are used for this purpose. Guidance for the selection of appropriate natural sources, and recommendations for analytical procedures for isolation, purification and identification are found in the literature [8, 9, 27–31]. In the last decade, reliable commercial suppliers have become available (e.g., CaroteNature GmbH, Extrasynthese SAS, and Sigma-Aldrich). Highly experienced carotenoid labs still prefer to isolate their own carotenoid standards since in most cases the purity is higher than the commercial ones.
8. Follow adequate sampling procedures for guaranty representativeness as well as to minimize errors during quantification. For carotenoids this has been recently reviewed by Rodríguez-Amaya [30].
9. In order to improve the extraction of carotenoids, dehydrated and lyophilized samples should be reconstituted with water up to the level present in the raw material. As a general rule five parts of water are added for one part of dry sample.
10. Select a homogenization speed so that the formation of foam is minimized and consequently reduce the degradation pigments by decreasing contact with air.
11. β-Apo-8′-carotenal is the most common carotenoid used as IS. The carotenoid used as IS needs to conform to some important requirements, the first one being its absence in the sample to be studied. The full of requirements for carotenoid IS, as well as guidance in their use can be found in Khachik [8].
12. The dispersing tool used for homogenization is cleaned during every successive extraction.

13. The organic phase frequently reduces its volume due to the high volatility of diethyl ether. If this happens the original diethyl ether volume should be approximately restored with more solvent before proceeding to next step.
14. If some water residues are still present after solvent evaporation, add a few milliliters (3–5 mL) of absolute ethanol and evaporate again in the rotary evaporator. The addition of ethanol will facilitate the evaporation of water by forming an azeotrope.
15. Introduce the sample into a sonication bath during 10 s for ensuring solubilization of carotenoids.
16. If carotenoids are not analyzed immediately after the extraction, samples should be stored at −30 °C (or lower) until chromatographic analysis (HPLC) in order to prevent degradation of carotenoids. If possible, the air from the sample vial should be displaced with a stream of N_2.
17. A preliminary test sample should be analyzed for an adequate selection of sample size.
18. A heating thermo-block for tubes at temperature <30 °C can be used to speed up the process.
19. In the case of the extract prepared by the "Microscale procedure" (Subheading 3.4), the entire extract can be submitted to saponification.
20. A single phase should be obtained. In the event that two phases are separated, add a few drops of methanol until a single phase is formed, and mix thoroughly.
21. Alternatively, samples can be filtered through 0.45 μm nylon filters. However, this practice is not recommended when using volatile solvents as evaporation during extract manipulation will cause overestimation of carotenoid content due to sample concentration.
22. Samples which have been kept at low temperature in the freezer should be left at room temperature for 5–10 min to ensure solubilization of pigments. For the same reason these samples cannot be centrifuged right out of the freezer.
23. If available, the injection should be achieved with an automatic injector for better reproducibility and automatization. In addition the sample compartment of the automatic injector should be fitted with a cooling system for sample vials in order to prevent thermo-degradation as well as sample concentration due to solvent evaporation. The recommended temperature range is 10–15 °C. Lower temperature should not be used as carotenoid can precipitate in the vial thus leading to errors in the quantification.

24. Avoid using injection volumes higher than 20 μL for preventing peak distortion during chromatography.
25. If xanthophyll esters are present in the sample and saponification has not been performed then the last step of the gradient elution at 100% A should be increased to 10 min to allow separation and detection of the very nonpolar diesters.
26. In general, 450 nm is a detection wavelength that can be used for most carotenoids. However, for some particular carotenoids the detection wavelength needs to be adequate to their chromophore properties. For instance, this is the case of phytoene and phytofluene with absorption maxima at 286 and 348 nm, respectively.
27. Calibration curves can be used for the estimation of other carotenoids such as the case of *cis* isomers (or *Z*- isomers), which are usually quantified using the calibration curve of the *trans* (or all-*E* isomers) counterpart.

Acknowledgments

The author thanks to the Spanish Government (projects AGL2017-85368P and CaRed Network BIO2015-71703-REDT and BIO2017-90877-REDT) for financial support.

References

1. Britton G, Liaaen-Jensen S, Pfander H (2008) Carotenoids handbook. Birkhäuser Verlag, Basel
2. Britton G, Hornero-Méndez D (1997) Carotenoids and colour in fruit and vegetables. In: Tomás-Barberán FA, Robins RJ (eds) Phytochemistry of fruit and vegetables. Clarendon Press, Oxford, pp 11–27
3. Rodriguez-Concepcion M, Avalos J, Bonet ML, Boronat A, Gomez-Gomez L, Hornero-Mendez D, Limon MC, Meléndez-Martínez AJ, Olmedilla-Alonso B, Palou A, Ribot J, Rodrigo MJ, Zacarias L, Zhu C (2018) A global perspective on carotenoids: metabolism, biotechnology, and benefits for nutrition and health. Prog Lipid Res 70:62–93. https://doi.org/10.1016/j.plipres.2018.04.004
4. Britton G, Liaaen–Jensen S, Pfander H (2009) Carotenoids, Nutrition and health, vol 5. Birkhäuser Verlag, Basel, Switzerland
5. Ettre LS, Sakodynskii KI (1993) M. S. Tswett and the discovery of chromatography I: early work (1899–1903). Chromatographia 35 (3):223–231. https://doi.org/10.1007/BF02269707
6. Ettre LS, Sakodynskii KI (1993) M. S. Tswett and the discovery of chromatography II: completion of the development of chromatography (1903–1910). Chromatographia 35 (5):329–338. https://doi.org/10.1007/BF02277520
7. Stewart I, Wheaton TA (1971) Continuous flow separation of carotenoids by liquid chromatography. J Chromatogr 55:325–336
8. Khachik F (2009) Analysis of carotenoids in nutritional studies. In: Britton G, Liaaen–Jensen S, Pfander H (eds) Carotenoids, vol 5: Nutrition and health, vol 5. Birkhäuser Verlag, Basel, Switzerland, pp 7-44
9. Mínguez-Mosquera MI, Hornero-Méndez D (1993) Separation and quantification of the carotenoid pigments in red peppers (*Capsicun annuum* L.), paprika and oleoresin by reversed-phase HPLC. J Agric Food Chem 41:1616–1620
10. Atienza SG, Ballesteros J, Martin A, Hornero-Méndez D (2007) Genetic variability of carotenoid concentration and degree of esterification among tritordeum (x*Tritordeum* Ascherson et Graebner) and durum wheat

accessions. J Agric Food Chem 55 (10):4244–4251. https://doi.org/10.1021/jf070342p

11. Delgado-Pelayo R, Gallardo-Guerrero L, Hornero-Méndez D (2014) Chlorophyll and carotenoid pigments in the peel and flesh of commercial apple fruit varieties. Food Res Int 65:272–281. https://doi.org/10.1016/j.foodres.2014.03.025
12. Delgado-Pelayo R, Gallardo-Guerrero L, Hornero-Méndez D (2016) Carotenoid composition of strawberry tree (*Arbutus unedo* L.) fruits. Food Chem 199:165–175. https://doi.org/10.1016/j.foodchem.2015.11.135
13. Delgado-Pelayo R, Hornero-Méndez D (2012) Identification and quantitative analysis of carotenoids and their esters from sarsaparilla (*Smilax aspera* L.) berries. J Agric Food Chem 60(33):8225–8232. https://doi.org/10.1021/jf302719g
14. Fernandez-Orozco R, Gallardo-Guerrero L, Hornero-Méndez D (2013) Carotenoid profiling in tubers of different potato (*Solanum* sp) cultivars: accumulation of carotenoids mediated by xanthophyll esterification. Food Chem 141(3):2864–2872. https://doi.org/10.1016/j.foodchem.2013.05.016
15. Hornero-Méndez D, Mínguez-Mosquera MI (2000) Xanthophyll esterification accompanying carotenoid overaccumulation in chromoplast of *Capsicum annuum* ripening fruits is a constitutive process and useful for ripeness index. J Agric Food Chem 48(5):1617–1622. https://doi.org/10.1021/jf9912046
16. Hornero-Méndez D, Mínguez-Mosquera MI (2000) Carotenoid pigments in rosa mosqueta hips, an alternative carotenoid source for foods. J Agric Food Chem 48(3):825–828. https://doi.org/10.1021/jf991136n
17. Mellado-Ortega E, Hornero-Méndez D (2015) Carotenoid profiling of *Hordeum chilense* grains: the parental proof for the origin of the high carotenoid content and esterification pattern of tritordeum. J Cereal Sci 62:15–21. https://doi.org/10.1016/j.jcs.2014.12.005
18. Mellado-Ortega E, Hornero-Méndez D (2012) Isolation and identification of lutein esters, including their regioisomers, in tritordeum (x*Tritordeum* Ascherson et Graebner) grains: evidence for a preferential xanthophyll acyltransferase activity. Food Chem 135 (3):1344–1352. https://doi.org/10.1016/j.foodchem.2012.05.046
19. Mattera MG, Hornero-Méndez D, Atienza SG (2017) Lutein ester profile in wheat and tritordeum can be modulated by temperature: evidences for regioselectivity and fatty acid preferential of enzymes encoded by genes on chromosomes 7D and 7H(ch). Food Chem 219:199–206. https://doi.org/10.1016/j.foodchem.2016.09.133
20. Boudries H, Kefalas P, Hornero-Méndez D (2007) Carotenoid composition of Algerian date varieties (Phoenix dactylifera) at different edible maturation stages. Food Chem 101 (4):1372–1377. https://doi.org/10.1016/j.foodchem.2006.03.043
21. Avalos J, Pardo-Medina J, Parra-Rivero O, Ruger-Herreros M, Rodriguez-Ortiz R, Hornero-Mendez D, Limon MC (2017) Carotenoid biosynthesis in fusarium. J Fungi (Basel) 3(3):39. https://doi.org/10.3390/jof3030039
22. Garrido-Fernandez J, Maldonado-Barragan A, Caballero-Guerrero B, Hornero-Mendez D, Ruiz-Barba JL (2010) Carotenoid production in lactobacillus plantarum. Int J Food Microbiol 140(1):34–39. https://doi.org/10.1016/j.ijfoodmicro.2010.02.015
23. Hornero-Mendez D, Cerrillo I, Ortega A, Rodriguez-Grinolo MR, Escudero-Lopez B, Martin F, Fernandez-Pachon MS (2018) Beta-Cryptoxanthin is more bioavailable in humans from fermented orange juice than from orange juice. Food Chem 262:215–220. https://doi.org/10.1016/j.foodchem.2018.04.083
24. Negro JJ, Grande JM, Tella JL, Garrido J, Hornero D, Donázar JA, Sanchez-Zapata JA, BenÍtez JR, Barcell M (2002) An unusual source of essential carotenoids. Nature 416:807. https://doi.org/10.1038/416807a
25. Blanco G, Hornero-Mendez D, Lambertucci SA, Bautista LM, Wiemeyer G, Sanchez-Zapata JA, Garrido-Fernandez J, Hiraldo F, Donazar JA (2013) Need and seek for dietary micronutrients: endogenous regulation, external signalling and food sources of carotenoids in new world vultures. PLoS One 8(6):e65562. https://doi.org/10.1371/journal.pone.0065562
26. Britton G (1995) UV/visible spectroscopy. In: Britton G, Liaaen–Jensen S, Pfander H (eds) Carotenoids, Spectroscopy, vol 1B. Birkhäuser Verlag, Basel, Switzerland, pp 13–62
27. Mínguez-Mosquera MI, Hornero-Méndez D, Pérez-Gálvez A (2002) Carotenoids and provitamin a in functional foods. In: Hurst WJ (ed) Methods of analysis for functional foods and nutraceuticals. CRC Press, Boca Raton, pp 101–157
28. Kimura M, Rodriguez-Amaya DB (2002) A scheme for obtaining standards and HPLC quantification of leafy vegetable carotenoids. Food Chem 78(3):389–398. https://doi.org/10.1016/S0308-8146(02)00203-0

29. Britton G, Young AJ (1993) Methods for the isolation and analysis of carotenoids. In: Young A, Britton G (eds) Carotenoids in photosynthesis. Chapman & Hall, London, pp 409–457
30. Rodriguez-Amaya DB (2016) Qualitative and quantitative analyses. In: Rodriguez-Amaya DB (ed) Food carotenoids: chemistry, biology, and technology. John Wiley & Sons, Ltd., London, pp 47–81
31. Schiedt K, Liaaen-Jensen S (1995) Isolation and analysis. In: Britton G, Liaaen–Jensen S, Pfander H (eds) Carotenoids, Isolation and analysis, vol 1A. Birkhäuser Verlag, Basel, Switzerland, pp 81–108

Chapter 10

Acquisition of Mass Spectrometry Data of Carotenoids: A Focus on Big Data Management

Antonio Pérez-Gálvez, Isabel Viera, and María Roca

Abstract

Accurate determination of carotenoid profile in plant tissues and food samples requires the application of hyphenated analytical resources including HPLC with high-resolution hybrid mass spectrometers. The high analytical power of modern MS equipment means the generation of Big Data resulting from the independent and complementary physicochemical properties of the target compounds that requires a complex processing to unravel the results. The present protocol describes a complete pipeline methodology for high-throughput analysis of carotenoids based on mass spectrometry (MS). After an exhaustive extraction, the workflow includes the acquisition of HPLC-hr-MS and MS^2 spectra assisted step by step by specific post-processing software.

Key words Carotenoids, Mass spectrometry, Product ions, Exact mass, Isotopic pattern, HPLC-MS, qQ-TOF, APCI, Tandem mass spectra, Software

1 Introduction

The carotenoids are a family of natural pigments with a steadily increasing number of compounds and expanding research interest. Probably due to the even improved analytical techniques, new carotenoid structures are continuously characterized in addition to the exciting identification of derivative products resulting from their metabolism. Several public databases have compiled part of the chemical knowledge of carotenoids, relieving the work devoted to the survey of certain structures. For example, the Carotenoids Database (http://carotenoiddb.jp/), includes information of almost 1200 different carotenoids in 700 organisms [1], while the Prokaryotic Carotenoid Database (http://bioinfo.imtech.res.in/servers/procardb/) is specific for prokaryote organisms, and contains more than 1800 entries [2]. To face this high diversity, it is necessary to apply current accurate metabolomics tools to unravel the carotenoid profile in a specific tissue. Existing methodologies

Manuel Rodríguez-Concepción and Ralf Welsch (eds.), *Plant and Food Carotenoids: Methods and Protocols*, Methods in Molecular Biology, vol. 2083, https://doi.org/10.1007/978-1-4939-9952-1_10,

are based on pioneer studies on mass spectrometry of carotenoids [3–8], which have been recently reviewed [9]. Hence, fast atom bombardment (FAB) ionization has been replaced by atmospheric pressure ionization (API) methods for the ionization of most natural products, and such improvement reached the carotenoids. Although electrospray ionization (ESI) and atmospheric pressure chemical ionization (APCI) show both advantages and drawbacks for ionization of carotenoids, the use of APCI source has probably become the standard choice. Similarly, ion trap (IT) and time of flight (TOF) are the conventional hardware designs for mass analysis at the ionization downstream, IT for structural elucidation through ion accumulation of selected protonated molecular ions which are subjected to MS^n analysis (tandem-in-time), and TOF for acquisition of accurate mass and isotopic pattern values by filtering protonated molecular ions (tandem-in-space). Anyhow, with the modern soft ionization techniques, tandem mass spectrometry has emerged as an essential strategy, as in addition to increase the accuracy of mass measurements, it allows to figure out certain reaction patterns during the fragmentation process of the protonated molecular ions. Accordingly, the final aim is to allocate qualifier ions that simplify the identification process. Furthermore, it is possible to reach high-resolution mass spectrometry (*hr*-MS) with the modern equipment, so that the possibility of inaccurate identifications is considerably reduced [10]. These analytical advancements empower the performance of metabolomics analysis, and two approaches can be pursued: targeted or untargeted. The former applies when the study is focused in ***known*** carotenoids, while the latter is performed if the analysis comprised the study of all carotenoids present, either known or unraveled structures. In any case, the complexity of metabolomics studies demands the existence of specific workflows to consistently analyze carotenoids from different sources. The present protocol describes a specific workflow for mass spectrometry analysis of carotenoids, with special emphasis in the assistance of post-processing software. Although the protocol has been focused on licensed software, free applications with comparable capabilities can be downloaded on the internet.

2 Materials

2.1 Equipment

1. Homogenizer, filter paper, separating funnel.
2. Rotary evaporator.
3. LC-MS system and software for instrument control.
4. Analytical column.
5. Computer and software for MS data analysis.

2.2 Reagents

1. For carotenoid extractions, the quality of the solvents should reach HPLC-grade, while for MS analysis the solvents must be LC-MS-grade (*see* **Note 1**).
2. Diethyl ether, sodium chloride 10% (w/v), anhydrous Na_2SO_4, acetone.
3. 0.22 μm PTFE syringe filter.
4. Calibration mix of synthetic polymer poly(ethylene glycol) (PEG 400, 10 μg/mL).

3 Methods

3.1 Carotenoid Extraction

The user should note that the extraction is the sum of simple steps, but slight modifications can be introduced at any of them that result in an improved success of the entire extraction procedure. Hence, each sample material requires almost a tailored protocol, so that the present description should be considered as a sequence of criteria and description of critical points to diminish common drawbacks affecting the subsequent HPLC-MS analysis. Carotenoids are usually stable in diminished light and avoiding excessive working temperature and continuous contact with air (*see* **Note 2**). Extracts are stored at −20 °C for 1 month.

1. Mix 10 g of fresh tissue with 30 mL of extracting solvent and break up the tissue with a homogenizer, filter the solution and pour the solvent in a separating funnel. Repeat the operation until no more color is extracted from the remaining tissue (*see* **Note 3**).
2. Mix all the combined extracts with 100 mL of diethyl ether and 100 mL of 10% (w/v) sodium chloride. After gentle shaking let the phases to separate (*see* **Note 4**).
3. Discard the water layer and filter the remaining organic phase through a solid bed of anhydrous Na_2SO_4 (25 g).
4. Evaporate the diethyl ether in a rotary evaporator and dissolve the residue in 100 μL acetone.
5. Filter the extract through a 0.22 μm PTFE syringe filter before chromatographic analysis.

3.2 LC–MS Data Acquisition

1. Set up the LC-MS system with the isocratic or gradient conditions of the selected HPLC method. Once the system is equilibrated, analyze the sequence of samples in the autosample tray. The following description is the practice routinely used in our laboratory [11]. A binary solvent system consisting of methyl-*tert*-butyl ether–methanol–water 85:15:4 (solvent A) and methyl-*tert*-butyl methyl–methanol–water 7:90:3 (solvent B) is used at a flow rate of 1 mL min^{-1}. Carotenoid separation is

performed on a C30 analytical column (3 μm particle size, 25 × 0.46 cm i.d.). Injection volume is 20 μL per sample. LC separation is performed using the following gradient setting: hold at 100% A for 10 min, then gradient from 100 to 50% A in 30 min, 50 to 100% B in 10 min, 100% A in 5 min, and finally isocratic 100% A for 5 min for reequilibrating column.

2. Set the photodiode array detector to acquire the common spectrum range for carotenoids (300–700 nm), although monitoring at 400–480 nm is commonly made during the run.
3. A split post-column of 0.4 mL min^{-1} is directly introduced on the APCI source, which is operated in positive mode.
4. A calibration routine of the MS instrument is made on each HPLC run (start-and-end mode) with the calibration mix that yields product ions in the entire mass range of the instrument. Hence, it is not necessary to create different calibration files, and any modification of both the ion source and the optics is individually collected for each sample.
5. The MS scan modes are tailored depending on the mass analyzer available at the lab of the user (Q, IT, Q-IT, TOF, Q-TOF). The strategy to follow at this point is whether the acquisition of "only" protonated molecular ions is required (full-scan mode, FS) or the choice of the user is to acquire both protonated molecular ions and product ions (broad band collision-induced dissociation mode, bbCID). Additionally, a specific scan mode could be selected, the precursor ion scan mode (PI) where some protonated molecular ions are designated for transmitting their corresponding product ions. Depending on the MS scan mode, additional screening tasks are subsequently required (*see* **Note 5**). The *m/z* scan in the UHR-TOF is in the 50–1200 Da range. In our laboratory, MS data are acquired in bbCID mode to obtain MS and tandem MS spectra simultaneously. The conditions of the APCI source are optimized for each LC-MS system. However, the user should consider that the subsequent settings of the mass analyzer affect the precondition of the APCI source (*see* **Note 6**).

3.3 Peak Prediction and Annotation

To get the effective information that finally drives the user to build a scenario for the extract within a biological context, it is necessary to apply a systematic workflow for peak annotation and identification. Figure 1 depicts the MS analysis of an extract from human milk with the automatically annotated protonated molecular ions to indicate the challenge of the study of MS data means. The pipeline methodology is represented in Fig. 2, allowing the user to take advantage of the complementary action of the inclusive analytical approach, and the application of bioinformatics tools for a comprehensive study of data [10].

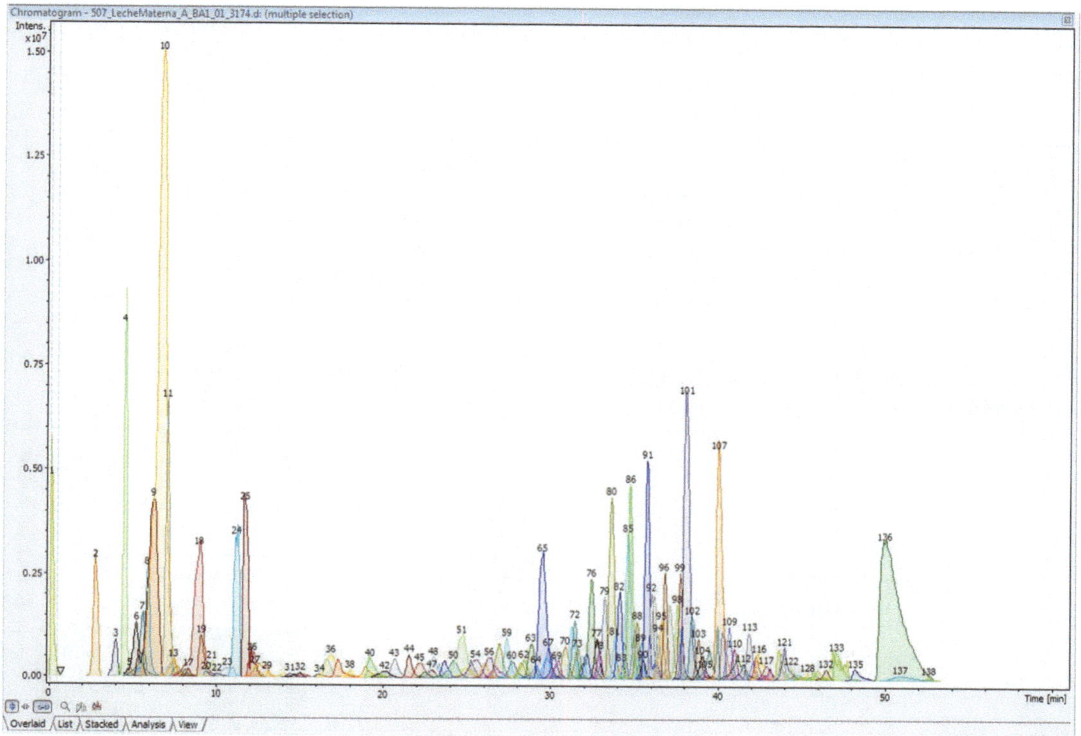

Fig. 1 Full scan MS chromatogram of an extract from human milk with the automatically annotated protonated molecular ions

1. The user should collect in an *in-house* database the calculated *m/z* values and the elemental composition of the predicted carotenoids in the analyzed extract (targeted strategy) or to compile the comprehensive information regarding the current knowledge of carotenoids and the anticipated structures the user proposes (untargeted strategy). Additionally, the calculated *m/z* values and the elemental composition of the product ions should be included (*see* Fig. 3 and **Note 7**).
2. Perform the mass screening with the software to all the raw data (FS) in a batch mode. To increase the accuracy of the screening, the software allows the user to include filtering rules (*see* **Note 8**). Commonly, the mass filtering is the first rule applied for the protonated molecular ions, setting a mass error value for MS peak match between the experimental and the theoretical values. The lower the mass error value (5 ppm, 3 ppm, 1 ppm…) the higher the accuracy of the predicted match. Additionally, modern MS software includes a filtering rule for the isotopic pattern value of the calculated and experimental MS data. This filtering rule is managed by the software with the application of an algorithm that yields a numerical value. The lower the deviation between the calculated and

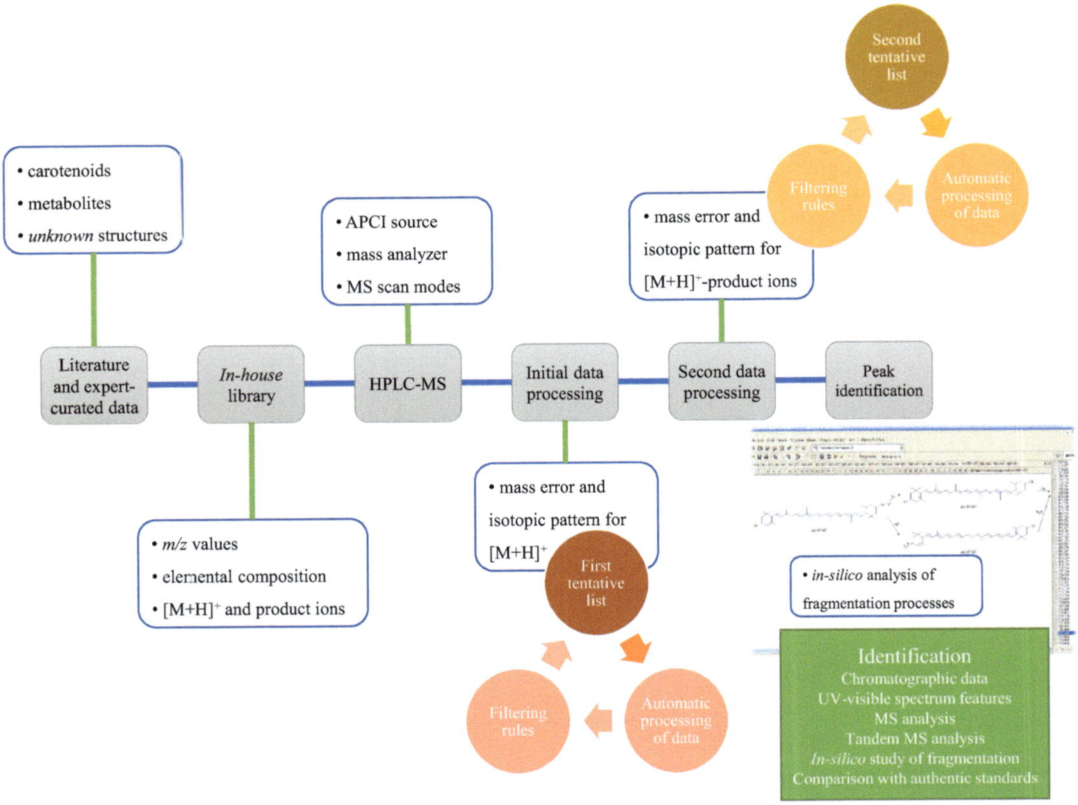

Fig. 2 Complete pipeline methodology for *hr*-MS analysis of carotenoids

Carotenoid	Elemental composition	$[M+H]^+$	RT (min)	UV-Vis max	$[M+H-18]^+$	$[M+H-H_2O]^+$	$[M+H-80]^+$	$[M+H-C_6H_8]^+$	[M+H-92]+	$[M+H-C_7H_8]^+$	$[M+H-106]^+$	$[M+H-C_8H_{10}]^+$	$[M+H-124]^+$	$[M+H-C_9H_{16}]^+$
Lutein	$C_{40}H_{56}O_2$	569,4370	14,50	420, 444, 472	551,4247	$C_{40}H_{54}O$			477,3727	$C_{33}H_{48}O_2$	463,357	$C_{32}H_{46}O_2$		
Zeaxanthin	$C_{40}H_{56}O_2$	569,4370	17,20	425, 450, 476	551,4247	$C_{40}H_{54}O$	489,3727	$C_{34}H_{48}O_2$	477,3727	$C_{33}H_{48}O_2$	463,357	$C_{32}H_{46}O_2$		
β- carotene	$C_{40}H_{56}$	537,4434	36,60	424, 451, 477			457,3829	$C_{34}H_{48}$	445,3829	$C_{33}H_{48}$			413,3203	$C_{31}H_{40}$

Fig. 3 Example of an in-house database including *m/z* and elemental composition of parent compound and product ions, retention times, and UV-Vis maxima

experimental isotopic patterns, the higher the accuracy of the predicted match (Fig. 4).

3. A first list is obtained including the tentatively identified carotenoids in the extract. The report is commonly exported to a Microsoft Excel based document where the user can obtain additional information regarding the peak area, retention time, UV–visible maxima as columns.

4. A second screening is performed with the data obtained with the bbCID or PI mode. This screening is made with specific applications built in the software for data managing (*see* **Note 8**).

3.4 Peak Identification

Peak identification is subjected to the best practices and expertise of the user, including the tentative lists generated in Subheading 3.3

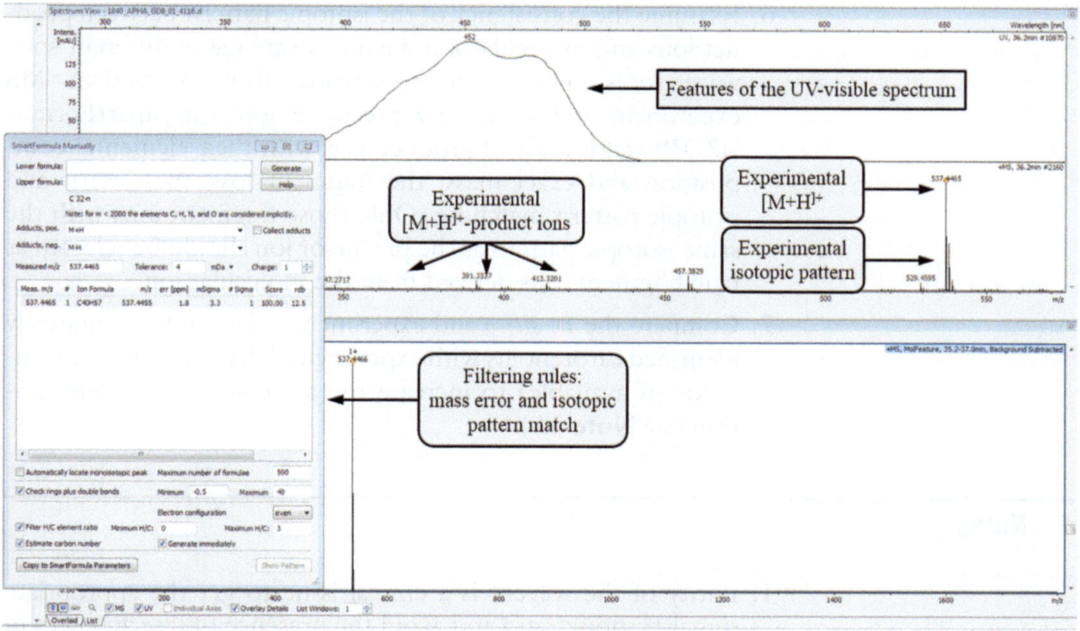

Fig. 4 Carotenoid MS screening according to UV-Vis spectrum, mass accuracy, and isotopic pattern for $[M+H]^+$ at 537.4465 Da

and the final compilation of chromatographic data, UV-visible spectra and MS information. The aim at this point of the data processing is to confirm that specific product ions arise from the protonated molecular ions, so that the experimental fingerprint is correlated with the common fragmentation reactions observed for carotenoids [12]. This step is critical for the identification of unanticipated carotenoid structures.

1. Draw the structure of the carotenoid (*see* **Note 9**).
2. Apply the software for the *in silico* generation of product ions (*see* **Note 10**).
3. Compare the list of the *in silico* product ions with the list of experimental product ions for the selected carotenoid structure.
4. Review the structures proposed for each product ion and select the correct one. The common fragmentation rules of carotenoids should be followed.
5. Review the proposed mechanisms for each product ion. The software also suggests different mechanisms to generate the same fragment ion structure. Revise and select the most appropriate mechanism proposed for each fragment ion. The mechanism should be consistent with the chemical behavior of carotenoids. In principle, the simplest should be the most probable [10].

6. Examine the consistency of the isotopic pattern between product ions and molecular ion, taking advantage of different software tools. For example, starting from a tandem MS experiment and selecting a precursor ion, the SmartFormula3D™ yields a list of product ions with their elemental composition and exact mass, the fragment loss, mass error and isotopic pattern matching. Only those fragment ions with the same isotopic pattern as the precursor ion should be selected as true fragment ions derived from the molecular ion.
7. Compare the *in silico* and experimental data of the tentatively identified carotenoids with experimental data of authentic standards (if available) to increase the accuracy of the identification (*see* **Note 11**).

4 Notes

1. Purity of the solvent is a critical issue to get the appropriate signal-to-noise ratio and avoid the presence of interferences in the MS analysis. Our experience has shown that even those LC-MS-grade solvents need to be checked in the applied experimental conditions. If none of them yields the required standard, then the use of pesticide residue analysis (PRA)-grade solvents is advised.
2. Avoid the extensive use of plastic lab supplies as contaminants dissolve with the extract and appear as artifacts in the MS analysis.
3. The extracting solvent and the ratio of tissue and solvent depend on the pigment polarity range and concentration in the tissue, as well as on the kind of matrix for extraction. It is convenient to perform some preliminary tests to establish amount of solvent and number of extractions. Methanol and acetone are frequently applied in this step, although solvent mixtures with water are also used. In any case, the user should consider safety and ease of use in the solvent selection.
4. If the sample lacks esterified carotenoids, chemical hydrolysis step could be applied at this point to remove interference with chlorophylls and triacyl glycerides [5].
5. The filtering rules are independent and complementary physicochemical properties available at the experimental and theoretical level, including chromatographic behavior, UV-visible spectrum, *m/z* values (for both $[M+H]^+$ and the product ions) and isotopic patterns (for both $[M+H]^+$ and the product ions) that once organized and compared (experimental vs. theoretical values) allow the characterization within a database of targeted compounds.

6. In our laboratory, the APCI conditions commonly applied are: ion spray voltage set −4500 eV and the turbo gas temperature at 700 °C. The nebulizing gas is set at 70 psi, the heating gas at 30 psi, the curtain gas at 30 psi and the collision gas at 10 psi.
7. The library is a Microsoft Excel document that contains information from literature and data generated in the lab of the user. This library should be continuously updated and enriched. Hence, for a specific carotenoid the user can include new proposed fragmentations in addition to the already known product ions.
8. Producers of MS equipment offer up-to-date software for instrument control and tools for data analysis. These tools may be embedded as a module or option in the main software distributed with the MS instrument. In our laboratory we apply the DataAnalysis™ software with the TargetAnalysis™ and SmartFormula3D™ utilities from Bruker, but other alternatives exist (Xcalibur, Analyst, MassHunter, MSFact, XCMS2).
9. Several chemical drawing software tools are freely available on the internet (ChemDraw Prime, ChemSketch, BKChem, Avogadro).
10. Predictive software tools commonly applied in this step are MassFrontier™, HighChem, Mass Fragmenter, Advanced Chemistry Development, and EPIC. Alternatively, open versions can be found in internet, as OpenMS-Simulator, MetFragCL, CFM-ID, MAGMa+, or MS-FINDER.
11. Particularly useful could be to visit the webpage ms-utils.org. It is a repository and updated list of free software for analysis of mass spectrometry data.

Acknowledgments

This work was supported by the Ministerio de Ciencia, Investigación y Universidades, Agencia Estatal de Investigación y Fondo Europeo de Desarrollo Regional (FEDER), project numbers RTI2018-095415-B-I00 and AGL2017-87884-R. Antonio Pérez-Gálvez is a tenured scientist at the Spanish National Research Council (CSIC).

References

1. Yabuzaki J (2017) Carotenoids database: structures, chemical fingerprints and distribution among organisms. Database 2017:bax004
2. Nupur LNU, Vats A, Dhanda SK, Raghava GPS, Pinnaka AK, Kumar A (2016) ProCarDB: a database of bacterial carotenoids. BMC Microbiol 16:96
3. van Breemen RB (1997) Liquid chromatography/mass spectrometry of carotenoids. Pure Appl Chem 69:2061–2066

4. Schweiggert U, Kammerer DR, Carle R, Schieber A (2005) Characterization of carotenoids and carotenoid esters in red pepper pods (*Capsicum annuum* L.) by high-performance liquid chromatography/atmospheric pressure chemical ionization mass spectrometry. Rapid Commun Mass Spectrom 19:2617–2628
5. de Rosso VV, Mercadante AZ (2007) Identification and quantification of carotenoids, by HPLC-PDA-MS/MS, from Amazonian fruits. J Agric Food Chem 55:5062–5072
6. Dugo P, Herrero M, Kumm T, Giuffrida D, Dugo G, Mondello L (2008) Comprehensive normal-phase x reversed-phase liquid chromatography coupled to photodiode array and mass spectrometry detection for the analysis of free carotenoids and carotenoid esters from mandarin. J Chromatogr A 1189:196–206
7. Ornelas-Paz JJ, Yahia EM, Gardea AA (2008) Changes in external and internal color during postharvest ripening of 'Manila' and 'Ataulfo' mango fruit and relationship with carotenoid content determined by liquid chromatography-APcI+−time-of-flight mass spectrometry. Postharvest Biol Technol 50:145–152
8. van Breemen RB, Dong L, Pajkovic ND (2012) Atmospheric pressure chemical ionization tandem mass spectrometry of carotenoids. Int J Mass Spectrom 312:163–172
9. Rivera SM, Christou P, Canela-Garayoa R (2014) Identification of carotenoids using mass spectrometry. Mass Spectrom Rev 33:353–372
10. Pérez-Gálvez A, Roca M (2019) Recent developments in the analysis of carotenoids by mass spectrometry. In: Zepka L, Jacob-Lopes J, De Rosso VV (eds) Progress in carotenoid research. IntechOpen, London, pp 17–44
11. Ríos JJ, Xavier AAO, Díaz-Salido E, Arenilla-Vélez I, Jarén-Galán M, Garrido-Fernández J, Aguayo-Maldonado J, Pérez-Gálvez A (2017) Xanthophyll esters are found in human colostrum. Mol Nutr Food Res 61:1700296
12. Pérez-Gálvez A, Sánchez-García A, Garrido-Fernández J, Ríos JJ (2018) MS tools for a systematic approach in survey for carotenoids and their common metabolites. Arch Biochem Biophys 650:85–92

Chapter 11

cis/trans Carotenoid Extraction, Purification, Detection, Quantification, and Profiling in Plant Tissues

Yagiz Alagoz, Namraj Dhami, Chris Mitchell, and Christopher I. Cazzonelli

Abstract

Reverse phase high-performance liquid chromatography (HPLC) is the method of choice used in biological, health, and food research to identify, quantify, and profile carotenoid species. The identification and quantification of *cis*- and/or *trans*-carotene and xanthophyll isomers in plant tissues can be affected by the method of sample preparation and extraction, as well as the HPLC column chemistry and the solvent gradient. There is a high degree of heterogeneity in existing methods in terms of their ease, efficiency, and accuracy. We describe a simple carotenoid extraction method and two different optimised HPLC methods utilizing C18 or C30 reverse-phase columns. We outline applications, advantages, and disadvantages for using these reverse phase columns to detect xanthophylls and *cis*-carotenes in wild-type photosynthetic leaves and mutant dark-grown etiolated seedlings, respectively. Resources are provided to profile individual species based upon their spectral properties and retention time, as well as quantify carotenoids by their composition and absolute levels in different plant tissues.

Key words Carotenoid, *cis*-carotene, Molar coefficient, Composition, Plants, HPLC, C30, C18

1 Introduction

Carotenoids are secondary metabolites comprised of 40 carbon atoms synthesized from isoprene subunits connected head-to-tail. They are hydrophobic compounds, stored within the lipophilic environment of organelle membranes from plastid organelles. Carotenoids are mainly biosynthesized in plants, algae as well as some photosynthetic bacteria and fungi. In plants, carotenoids are required for photosynthesis and photoprotection [1]. They can be cleaved into smaller apocarotenoids, that function as phytohormones and/or signaling metabolites to mediate developmental processes (e.g., root and shoot branching), plant-microorganism associations, and intracellular as well as organelle communications [2–7]. In humans, carotenoids serve as antioxidants to protect against some cancer types [8], cardiovascular, neurodegenerative, and eye diseases [9–11]. There is a high nutritional value associated

Manuel Rodríguez-Concepción and Ralf Welsch (eds.), *Plant and Food Carotenoids: Methods and Protocols*, Methods in Molecular Biology, vol. 2083, https://doi.org/10.1007/978-1-4939-9952-1_11,

with carotenoids and research has been primed towards enhancing the abundance of carotenoids and specific isoforms in agriculture and protected cropping [12, 13], aquaculture industries [14], as well as in skin medicines and micronutrient supplements [15]. Since enzymatic and nonenzymatic oxidation as well as *cis/trans* isomerization (triggered by light and temperature) affect cleavage and carotenoid configuration, their stability, bioavailability, and biological functions can be difficult to elucidate [16, 17]. Therefore, the identification, characterization, and absolute quantification of different carotenoid isomers can be crucial to resolve complex metabolic phenomena in plants and animals.

In plants, carotenoid biosynthesis and storage can vary in different organs [18, 19] during development (young vs. old leaves and stage of fruit ripening) [20, 21], and in response to changes in temperature [22], light intensity and/or spectra [23], oxidative stress [24], drought [25, 26], as well as the presence of biotic factors like pathogens [27] or mutualistic organism like mycorrhizal fungi [28]. Depending on the plant tissue and plastid sink, carotenoids may not be present in their free forms and some additional processes can be required to enable quantification. For instance, carotenoids present in callus and seed tissues are highly esterified, which hinders their extraction and identification using chromatographic methods like high-performance liquid chromatography (HPLC). A process known as saponification is necessary to remove fatty acids bound to free carotenoids. Additional details about this process can be found elsewhere [29].

Carotenoid extraction can be performed using organic solvents like hexane, acetone, tetrahydrofuran (THF), or mixed solvents like acetone–ethyl acetate (3:2, v/v) [30], hexane–isopropanol (1:1, v/v) [31], THF–methanol (1:1, v/v) [32], and ethanol–hexane (4:3, v/v) [33]. Here we refer the reader to a detailed report describing frequently used carotenoid extraction methods [34]. Individual carotenoids have different chemical properties in terms of their molecular weights, structure, ionic charge, and hydrophobicity. When carotenoid extracts are passed through a reverse-phase column using an HPLC solvent gradient, these chemical properties enable carotenoid separation by retention time. The efficiency of HPLC separation methods depend on the chemistry of both stationary and mobile phases which are comprised of a column and organic solvents, respectively. Due to the similarity of some geometric and positional carotenoid isomers, chromatographic separation can be a demanding task. There is a range of carotenoid extraction methods, columns, and reverse phase solvent gradients reported [35]. Therefore, selecting the correct combination of organic solvents for extraction, reverse-phase column chemistry, and the solvent gradient improve the chromatographic technique leading to a higher resolving power in carotenoid binding and separation. This can improve detection of individual carotenoids based upon their individual light absorption properties and unique spectral conformations, which in some cases isoforms can be difficult to distinguish within a crude carotenoid extract mixture without clear separation by retention time.

Here we describe a rapid and simple carotenoid extraction protocol that is suitable for both photosynthetic and nonphotosynthetic plant tissues, and compare two solvent gradient-based HPLC methods using C18 and C30 reverse-phase column chemistry to separate *cis/trans* carotenoids. We also describe how to identify carotenoids and their isoforms as well as develop standards for absolute quantification.

2 Materials

Use of HPLC-grade (or analytical) reagents and ultrapure deionized water is recommended. For health and safety concerns, it is advised to comply with associated risk assessments, standard operating procedures, and waste disposal protocols.

2.1 Crude Extraction of Carotenoids from Plants

1. Extraction buffer: Acetone–ethyl acetate (EtOAc) (6:4, v/v). For 1 L extraction buffer, mix 600 mL of 100% acetone with 400 mL of 100% EtOAc then add and dissolve 1 g butylated hydroxytoluene (BHT) in the mixture (*see* **Note 1**).
2. 100% ultrapure H_2O (to be used as a phase separation buffer).
3. 1.5 and 2 mL microcentrifuge tubes and micropipettes with suitable tips.
4. Liquid nitrogen and ice to be used while/after grinding and to store ground plant tissue samples.
5. Stainless steel beads or pestle and mortar for mechanical grinding of plant tissues (*see* **Note 2**).
6. A tissue grinder or cell lysis system with appropriate adapters (*see* **Note 3**).
7. Refrigerated centrifuge, vortexer, and/or mixers.

2.2 Reverse-Phase HPLC-Based Quantification of Carotenoids

1. Dark amber colored HPLC vials (2 mL), glass inserts (0.2 mL) and screw cap 8 mm open with silicone/polytetrafluoroethylene (PTFE) septa.
2. C18 HPLC column: C18 (250 × 4.6, 4-μm).
3. C30 HPLC column: C30 (250 × 4.6 mm, S-5 μm).
4. HPLC mobile phases for C18 column:
   (a) Mobile phase A—acetonitrile–H_2O–triethylamine (TEA) (9:1:0.01, v/v/v). For 1 L mobile phase A, mix 900 mL of 100% acetonitrile, 100 mL of ultrapure H_2O and 1 mL TEA in a 1 L glass bottle. Gently invert the bottle 2–3

times for mixing. Filter and purify to remove contaminating particles and/or air bubbles if necessary.

(b) Mobile phase B—100% EtOAc.

(c) Mobile phase C—100% acetonitrile.

5. HPLC mobile phases for C30 column:

(a) Mobile phase A—100% methanol (MeOH).

(b) Mobile phase B—100% methyl tert-butyl ether (MTBE).

6. An HPLC instrument equipped with a Diode Array Detector (DAD).

2.3 Fraction Collection of Individual Carotenoid Peaks

1. Dark amber-colored HPLC vials, glass inserts, and screw-cap covers.
2. Sample concentrator with nitrogen gas supply.
3. 100% organic solvent and 100% EtOAc (*see* **Note 4**).
4. UV-vis spectrophotometer.
5. Plastic disposable or quartz cuvettes (preferentially) with 10 mm path-length for spectrophotometric measurements.
6. Agilent 1260 Infinity fraction collector or equivalent fraction collector.

3 Methods

Keeping samples on ice (constant low temperature) and away from direct high energy light during extraction as well as downstream analytical steps as a routine practice to avoid carotenoid oxidation and isomerization. Working with a high number of samples (>50) is not encouraged for routine carotenoid extraction or HPLC in a single run due to an increased risk of solvent evaporation and carotenoid oxidation. When attempting to quantify *cis*-carotenes in extracts, 10–20 samples/run are ideal as some carotenes are highly sensitive to isomerization.

3.1 Plant Tissue Collection

The collection and handling of plant tissue (sample) is a key step underpinning the efficiency and accuracy of carotenoid quantification. Carotenoids can undergo isomeric transformation during tissue maturation [20], and/or in response to changing environmental conditions. Therefore, newly collected plant tissues should be rapidly snap-frozen after determining the fresh weight of the sample. Alternatively, the water content can be removed from a tissue sample using a cold vacuum dryer (4 °C or below) and dry weight measured using a microscale.

For routine analysis of Arabidopsis tissues, we recommend 20–50 mg of green leaves, 75–100 mg of etiolated seedlings, up to 200 mg of roots, and 10–15 mg of any carotenoid-rich tissues (e.g., dried leaf tissues and tomato fruits) that can be extractable in a single step with 0.5–1 mL extraction buffer (*see* **Note 2**). While collecting fresh tissues, it is recommended the excessive moisture to be removed from tissues by blotting onto a clean lint-free paper, thereby enhancing accuracy in determination of the fresh weight between independent samples. Immediately transfer samples into liquid nitrogen avoiding exposure to high temperatures and/or light (*see* **Note 5**). Samples should be stored at −80 °C until pigment extraction. In this work, we performed metabolic analysis using Arabidopsis cotyledon tissues collected from 7-day-old dark-grown wild-type (WT) and mutant lines *ZETA-CAROTENE ISOMERASE* (*z-iso*; *zic1-3*, Salk_136385) and *CAROTENOID ISOMERASE* (*crtiso*; *ccr2.1*) [36]. Additionally, 14-day-old dark-grown tomato *CAROTENOID ISOMERASE* mutant (*crtiso*; *tangerine*^{mic}) [37] is used to confirm peak retention times of identified carotenoids.

3.2 Carotenoid Extraction

The complete extraction of carotenoids from tissues enhances the accuracy of quantification. A single-step extraction method works effectively for most tissue types (e.g., leaves, roots, etiolated tissues, and fruits). Extraction under an ambient lighting condition is preferable to limit carotenoid photoisomerisation. The carotenoid extraction standard operating procedure is as follows;

1. Weigh plant tissues and place in a 1.5 or 2 mL microcentrifuge tube (flat-bottom tubes are recommended) with 2 stainless steel beads (3 mm diameter). Seal the tubes and snap freeze in liquid N_2. Prepare 3–5 biological replicates per treatment (*see* **Note 2**).
2. Prepare the cell lysis system for mechanical grinding of samples by adjusting to 20 Hertz frequency for 2 min of shaking (this ensures complete grinding) (*see* **Note 2**).
3. Place frozen tubes containing plant material into prechilled (−80 °C) cell lysis adaptors, balance as necessary, and begin tissue grinding (*see* **Note 3**).
4. After grinding, return tubes containing the ground tissue back to liquid N_2 (*see* **Note 6**).
5. For pigment extraction, transfer tubes from liquid N_2 to ice and add 500 μL extraction buffer and vortex vigorously for 1 min.
6. Add an equal volume (500 μL) of water and invert tubes gently 2–3 times. Avoid vigorous shaking, which can emulsify the extract and hinder phase separation.
7. Centrifuge the tubes for 5 min at 15,000 rpm (~21,100 × *g*) at 4 °C.

8. Collect the upper ethyl acetate phase (~250 μL) enriched in carotenoid and chlorophyll pigments and transfer to a new 1.5 mL microcentrifuge tube (*see* **Note 7**).
9. Centrifuge the upper phase for 5 min, as described in **step 7**.
10. Transfer approximately 150 μL of the ethyl acetate phase into a new 1.5 mL microcentrifuge tube. Avoid pipetting near the bottom of the tube that might have contaminating cellular debris which can hinder HPLC separation.
11. Transfer 50 μL of the crude extract into an HPLC insert within a 2 mL amber colored glass vial and secure a sealing cap. Immediately transfer vials to an HPLC sample collector chamber or temporarily store at −20 °C in the dark. Higher temperature and light exposure may cause isomerization. Extracted samples can be stored at −80 °C for several days.

3.3 Carotenoid Separation: Comparison of C18 vs. C30 Columns and Suitable Solvent Gradients

1. Filter mobile phase solvents to remove oxygen and contaminating particles before filling bottles (1–2 L).
2. Turn on HPLC instrument units and install the appropriate column. Here, we make reference to the Agilent 1260 Infinity HPLC system equipped with a Diode Array Detector (DAD) (*see* **Note 8**).
3. Open HPLC instrument software, turn off autosampler compartment illumination and set to 4 °C (*see* **Note 9**) while maintaining the column temperature at room temperature (RT).
4. Load sample vials in the autosampler tray including a blank (100% EtOAc) in the first position, that help to equilibrate the column and test for residual pigmentation that could occur from a previous run if the column was not washed sufficiently.
5. Program the reverse phase gradient method using instrumentation software. We compared the performance of the C18 and C30 column using reverse-phase solvent gradient as previously described in [30] and [38], respectively, with only minor modifications. Separating carotenoid isomers can be difficult and might require slight modification to gradient programming depending upon the HPLC instrument. We recommend C18 column chemistry for the analysis of major carotenoids like Antheraxanthin and Zeaxanthin. However, the C30 column provides a better resolution for the identification of *cis*-carotene isomers. The selection of the method also depends on the carotenoid composition of the sample to be analysed. Figures 1 and 2 compare carotenoid profiles from plants grown under light and continuous dark, respectively. The composition and reverse-phase solvent gradient used for both columns are as follows:

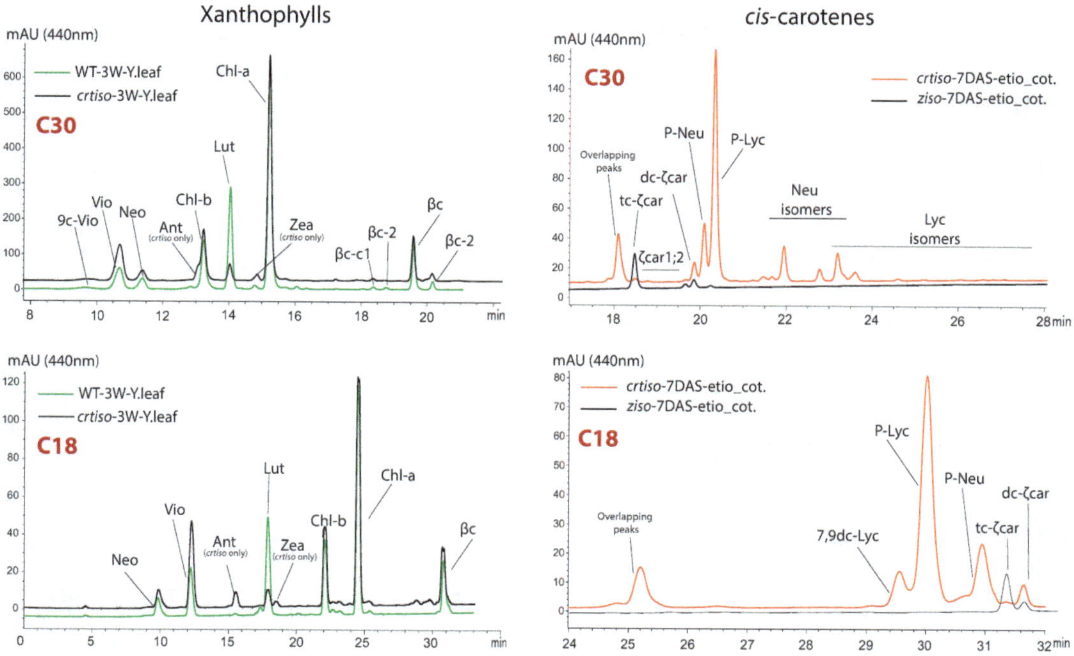

Fig. 1 Distribution of xanthophylls and carotenes on C18 vs. C30 columns. Tissue samples for HPLC were collected from 3-week-old young Arabidopsis leaves and 7-day-old dark-grown cotyledons of Arabidopsis *ziso* (*zic1-3*) and *crtiso* (*ccr2.1*) mutants. Abbreviations: 3W, 3-week-old; cot, cotyledon; etio, etiolated; mAU, milli-Absorbance Units; Y.leaf, young leaf. For the full nomenclature of carotenoid abbreviations; *see* Table 1

(a) C18 column: Separate carotenoids using C18-Mobile phase A and C18-Mobile phase B with the following gradient: 0 min 100% A (1 mL/min); 3 min 100% A (1 mL/min); 31 min 33.3% A and 66.7% B (1 mL/min); 31.2 min 100% B (1.5 mL/min); 32 min 100% B (1.5 mL/min), 35 min 100% A (1 mL/min).

(b) C30 column: Separate carotenoids using C30-Mobile phase A and C30-Mobile phase B with the following gradient: 0 min 100% A; 5 min 100% A; 25 min 100% B; 30 min 100% A, 35 min (or until column equilibration is achieved depending upon isomers detected) 100% A with the flow rate of 1 mL/min.

6. Adjust the DAD detection range to record signals at 286 nm, 348 nm, 400 nm, 440 nm, 454 nm, and 470 nm (*see* **Note 13**).
7. Start HPLC instrument software to inject 20–50 μL of pigment extract through the HPLC column and facilitate carotenoid separation as per programmed gradient (*see* **Note 10**).

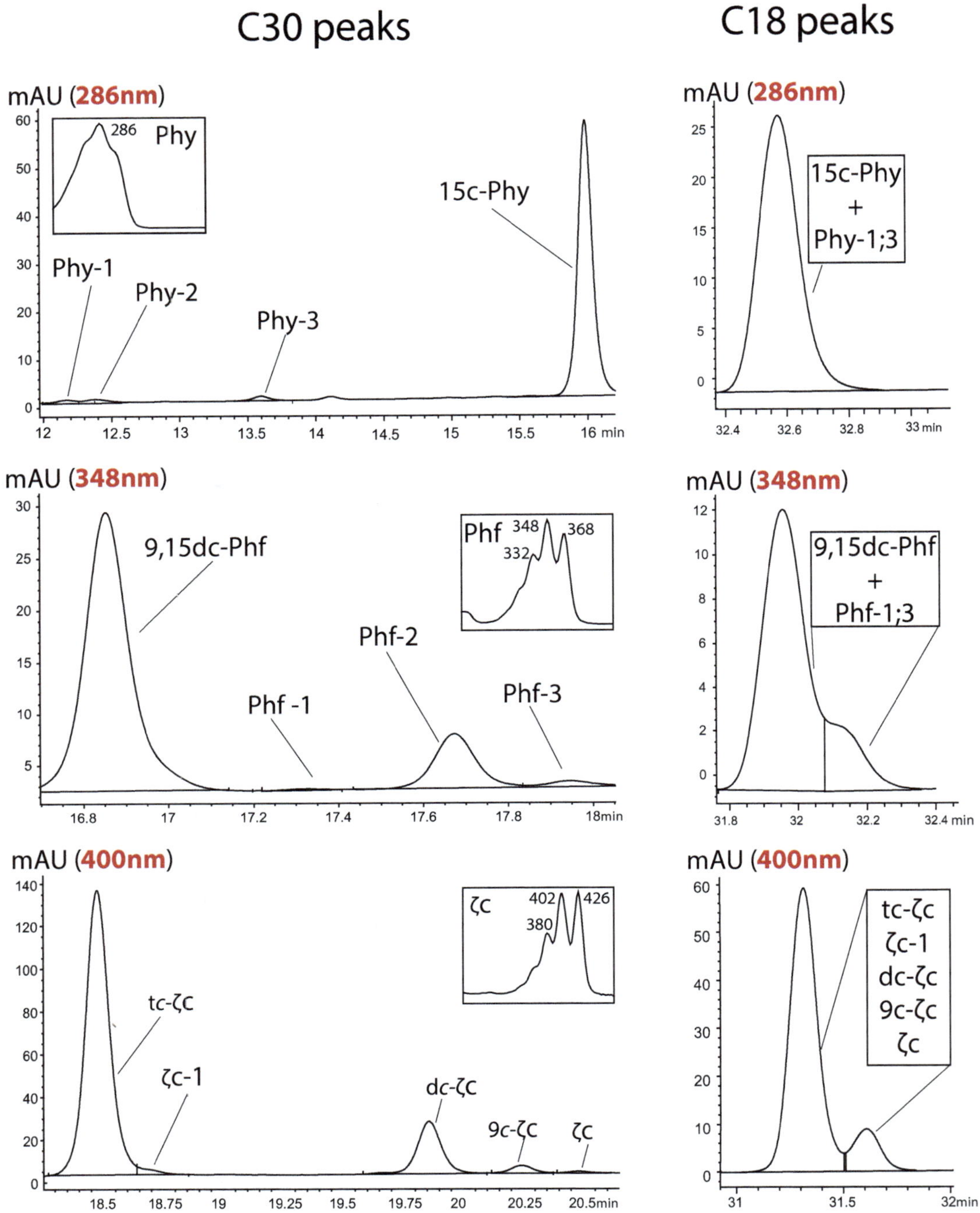

Fig. 2 Comparison of C18 and C30 columns for *cis/trans* carotenoid identification. Separation of *cis*-carotene isomers was performed more effectively and with a higher resolution on a C30 column rather than C18. HPLC peaks were collected from 7 day-old dark-grown cotyledons of Arabidopsis *ziso* (*zic1-3*) mutant. Abbreviations: mAU, milli-Absorbance Units. For the full nomenclature of carotenoid abbreviations; *see* Table 1

3.4 Quantification by Absolute Levels and/or Determining the Percent Composition

1. Open data files in the HPLC instrument software (e.g., Agilent ChemStation off-line software).
2. Identify carotenoid species according to their retention time, spectral features, and ratios of maximum absorption peaks (λ) (*see* **Notes 11** and **12**). A detailed DAD spectra analysis (Fig. 3) and retention times of some of the identified carotenoids on C30 and C18 columns have been provided in Tables 1 and 2.
3. Perform peak integration and calculate individual peak area using HPLC instrument software (*see* **Note 13**).
4. If calibration curves and slope were predetermined, use the formula below to calculate carotenoid abundance in grams per gram fresh weight (g/gfw). Otherwise, perform fraction collection (**step 4**) and/or establish standard calibration curves using commercially available carotenoids (**step 15**) (*see* **Note 14**).

   Formula:

$$\text{Carotenoid weight(g/gfw)} = \text{Peak area (AU)} \times 1/\text{Slope} \times \text{Dilution factor (DF)} \times \text{g (grams)}$$

5. Program HPLC instrument software and fraction collector to collect individual carotenoid fractions according to their retention times or spectral identification.
6. Collect each fraction in a new dark amber colored glass vial.
7. Validate the purity of carotenoid fractions using HPLC according to the instructions given at **step 2**.
8. Repeat **steps 5** and **6** until pure carotenoid fractions have been prepared.
9. Evaporate the mobile phase within the collected fractions under a steam of nitrogen gas.
10. Redissolve carotenoid precipitate in 1 mL of organic solvent (*see* **Notes 15** and **16**).
11. Prepare the spectrophotometer at RT and zero the spectrometer with the same solvent used to dissolve the precipitated carotenoid (*see* **Note 17**).
12. Transfer 1 mL of redissolved carotenoid sample (from **step 10**) into a glass or disposable plastic cuvette and measure absorbance at specific wavelengths (λ) (*see* Table 3) for each carotenoid using a UV-vis spectrophotometer (*see* **Note 18**).
13. Calculate the carotenoid concentration using this formula:

    Formula:

$$\frac{\text{Absorbance read(AU)} \times \text{Dilution factor(DF)} \times \text{Concentration(mg/mL)}}{A^{1\%}{}_{1\text{cm}}}$$

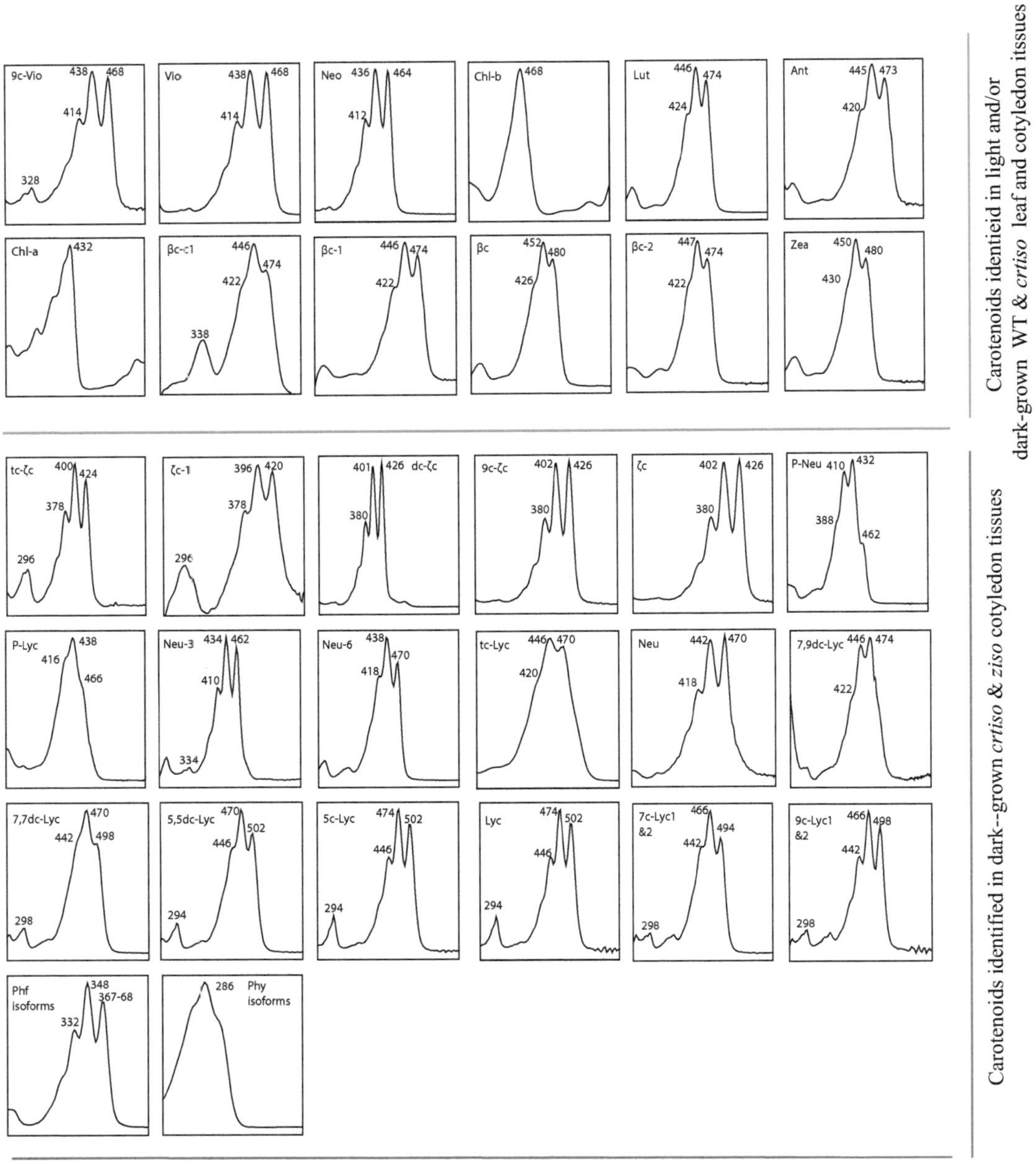

Fig. 3 HPLC diode array detector (DAD) peaks of carotenoids identified in dark and/or light-grown Arabidopsis seedlings with the C30 column. Retention time (RT) and maximum wavelength (λ_{max}) values were used to validate each carotenoid peak. To discriminate each isomer with high accuracy and precision, crude carotenoid extracts of dark-grown cotyledon tissues from *ziso* (*zic1-3*) and *crtiso* (*ccr2.1*) mutants of Arabidopsis and *crtiso* mutant of tomato (*tangerine*^{mic}) were used. For the full nomenclature of carotenoid abbreviations; *see* Table 1

14. Evaporate the organic solvent using nitrogen gas.
15. Prepare a set of standards by dissolving the pure carotenoid standards in 100% EtOAc and serially diluting (usually 1/2 or 1/5 by volume) to cover an appropriate concentration range relevant to the tissue samples being analyzed.

Table 1
Identification of carotenoid isomers with C30 column in light-grown Arabidopsis leaves and dark-grown cotyledons of Arabidopsis *ziso* and *crtiso* mutants

Carotenoid (C30)	RT (min)	λ (nm)	%III/II	Chemical formula
9-*cis*-violaxanthin (9c-Vio)	9.62	(328) 414, 438, 468	87.36	$C_{40}H_{56}O_4$
All-*trans*-violaxanthin (Vio)	10.66	414, 438, 468	97.13	$C_{40}H_{56}O_4$
All-*trans*-neoxanthin (Neo)	11.38	412, 436, 464	91.69	$C_{40}H_{56}O_4$
Chlorophyll-B (Chl-b)	13.26	_, 468, _	N/A	$C_{55}H_{70}MgN_4O_6$
All-*trans*-lutein (Lut)	14.02	424, 446, 474	66.07	$C_{40}H_{56}O_2$
All-*trans*-zeaxanthin (Zea)	14.94	430, 450, 480	30.39	$C_{40}H_{56}O_2$
Chlorophyll-A (Chl-a)	15.25	_, 432, _	N/A	$C_{55}H_{72}MgN_4O_5$
β-carotene cis-isomer-1 (βc-c1)	18.36	(338) 422, 446, 474	N/A	$C_{40}H_{56}$
β-carotene isomer-1 (βc1)	18.77	422, 446, 474	62.91	$C_{40}H_{56}$
β-carotene (βc)	19.62	426, 452, 480	29.11	$C_{40}H_{56}$
β-carotene isomer-2 (βc2)	20.13	422, 447, 474	35.52	$C_{40}H_{56}$
All-*trans*-antheraxanthin (Ant)	13.16	420, 445, 473	59.57	$C_{40}H_{56}O_3$
9,15,9′-tri-*cis*-ζ-carotene (tc-ζc)	18.37	(296) 378, 400, 424	75.45	$C_{40}H_{60}$
ζ-carotene cis-isomer-1 (ζc-1)	18.62	(296) 376, 396, 420	94.03	$C_{40}H_{60}$
9,9′-di-*cis*-ζ-carotene (dc-ζc)	19.85	380, 401, 426	103.01	$C_{40}H_{60}$
9-*cis*-ζ-carotene (9c-ζc)	20.15	380, 402, 426	106.84	$C_{40}H_{60}$
All-*trans*-ζ-carotene (ζc)	20.37	380, 402, 426	105.71	$C_{40}H_{60}$
7,9,9′-tri-*cis*-neurosporene (P-Neu)	20.07	388, 410, 432, 462	N/A	$C_{40}H_{58}$
7,9,9′,7′-tetra-*cis*-lycopene (P-Lyc)	20.3	416, 438, 466	N/A	$C_{40}H_{56}$
Neurosporene isomer-3 (Neu-3)	21.67	410, 434, 462	84.66	$C_{40}H_{58}$
Neurosporene isomer-6 (Neu-6)	23.05	418, 438, 470	49.77	$C_{40}H_{58}$
Tri-*cis*-lycopene (tc-Lyc)	23.47	420, 446, 470	N/A	$C_{40}H_{58}$
All-*trans*-neurosporene (Neu)	23.65	418, 442, 470	93.61	$C_{40}H_{58}$
7,9-di-*cis*-lycopene (7,9 dc-Lyc)[a]	23.91	422, 446, 474	N/A	$C_{40}H_{56}$
7-*cis*-lycopene isomer-1 (7c-Lyc1)[a]	24.59	(298) 442, 466, 494	36.65	$C_{40}H_{56}$
7-*cis*-lycopene isomer-2 (7c-Lyc2)[a]	24.95	(298) 442, 466, 494	37.01	$C_{40}H_{56}$
9-*cis*-lycopene isomer-1 (9c-Lyc1)[a]	25.32	(298) 442, 466, 498	75.00	$C_{40}H_{56}$
9-*cis*-lycopene isomer-2 (9c-Lyc2)[a]	25.55	(298) 442, 466, 498	76.08	$C_{40}H_{56}$
7,7′-di-*cis*-lycopene (7,7 dc-Lyc)[a]	26.45	(298) 442, 470, 498	N/A	$C_{40}H_{56}$
5,5′-di-*cis*-lycopene (5,5 dc-Lyc)[a]	27.02	(294) 446, 470, 502	40.39	$C_{40}H_{56}$
5-*cis*-lycopene isomer (5c-Lyc)[a]	27.24	(294) 446, 474, 502	73.84	$C_{40}H_{56}$

(continued)

Table 1
(continued)

Carotenoid (C30)	RT (min)	λ (nm)	%III/II	Chemical formula
All-*trans*-lycopene (Lyc)[a]	27.47	(294) 446, 474, 502	75.07	$C_{40}H_{56}$
9,15-di-*cis*-phytofluene (9,15 dc-Phf)	16.74	332, 348, 367	66.24	$C_{40}H_{62}$
Phytofluene isomer-1 (Phf-1)	17.21	332, 348, 368	80.95	$C_{40}H_{62}$
Phytofluene isomer-2 (Phf-2)	17.57	332, 348, 368	88.44	$C_{40}H_{62}$
Phytofluene isomer-3 (Phf-3)	17.82	332, 348, 368	102.38	$C_{40}H_{62}$
Phytoene isomer-1 (Phy-1)	12.04	286	N/A	$C_{40}H_{64}$
Phytoene isomer-2 (Phy-2)	12.31	286	N/A	$C_{40}H_{64}$
Phytoene isomer-3 (Phy-3)	13.45	286	N/A	$C_{40}H_{64}$
15-*cis*-phytoene (15c-Phy)	15.87	286	N/A	$C_{40}H_{64}$

N/A (Not available) In cases where there are only two maximum peaks present
Carotenoids were identified either based on their spectral features and change to their retention times or using pure standards. The retention times (RT), absorption maxima (AM), chemical formulas, and absorbance wavelengths (λ) are given below. The majority of lycopene isomers ([a]) are present in trace amounts in etiolated Arabidopsis. Dark-grown cotyledon tissue from tomato *tangerine*^{Mic} was used to validate measured absorbance values
[a]*% III/II* = (mAU PeakIII-mAU Baseline)/(mAU PeakII- mAU Baseline) × 100 (Baseline value represents the lowest absorbance value between peak III and II)

16. Perform HPLC by injecting 20–50 μL of diluted samples through the column following the same HPLC operating procedure (as described previously) to develop a standard curve for absolute carotenoid quantification.
17. Follow the instruction given in **steps 2** and **3** to calculate peak area and carotenoid abundance.
18. Calculate the slope of the dilution series and use this coefficient to calculate the absolute carotenoid abundance as g/gfw (*see* **Note 19**). Below is an example of a calibration curve for β-carotene prepared using C30 column chemistry (*see* Fig. 4).

4 Notes

1. The extraction buffer can be prepared in large volume as stocks without BHT and stored according to the risk assessments. BHT should be added fresh prior to performing carotenoid extraction.
2. It's recommended to use 2 steel beads (3 mm diameter) for each 2 mL microcentrifuge tube. This is optimized to grind up soft plant tissues with the tissue grinder system. It is better to use a 2 mL rather than a 1.5 mL microtube as it gives more space for the steel beads to freely rotate and crush the plant

Table 2
Identification of carotenoid isomers with C18 column in light-grown Arabidopsis leaves and dark-grown cotyledons of Arabidopsis *ziso* and *crtiso* mutants

Carotenoid (C18)	RT (min)	λ (nm)	Chemical formula
Neoxanthin (Neo)	9.84	414, 438, 467	$C_{40}H_{56}O_4$
Violaxanthin (Vio)	12.22	418, 442, 471	$C_{40}H_{56}O_4$
Antheraxanthin (Ant)	15.48	422, 447, 477	$C_{40}H_{56}O_3$
Lutein (Lut)	17.92	422, 448, 476	$C_{40}H_{56}O_2$
All-*trans*-zeaxanthin (Zea)	18.6	430, 450, 480	$C_{40}H_{56}O_2$
Chlorophyll-B (major peak) (Chl-b)	22.04	_, 458, _	$C_{55}H_{70}MgN_4O_6$
Chlorophyll-A (major peak) (Chl-a)	24.53	_, 430, _	$C_{55}H_{72}MgN_4O_5$
β-carotene (βc)	30.82	428, 456, 482	$C_{40}H_{56}$
7,9-di-*cis*-lycopene (dc-Lyc)	29.53	442, 468	$C_{40}H_{56}$
Tetra-*cis*-lycopene (P-Lyc)	30.03	420, 442, 468	$C_{40}H_{56}$
Tri-cis-neurosporene (P-Neu)	30.95	388, 410, 435, 464	$C_{40}H_{58}$
Tri-*cis*-ζ-carotene (tc-ζc)	31.37	380, 400, 425	$C_{40}H_{60}$
di-*cis*-ζ-carotene (dc-ζc)	31.63	382, 402, 428	$C_{40}H_{60}$
Phytofluene isomer-1 (Phf-1)	31.98	332, 349, 368	$C_{40}H_{62}$
Phytofluene isomer-2 (Phf-2)	32.2	332, 350, 368	$C_{40}H_{62}$
Phytoene (Phy)	32.64	_, 286, _	$C_{40}H_{64}$

material without getting stuck at the bottom of the tube. For samples like roots, fruits, and large seeds, we recommend using a pestle and mortar. The 2 mL tubes are ideal in case 1 mL extraction buffer is particularly needed where the required tissue weight is more than 50 mg for extraction.

3. The tissue grinder adapters should be prechilled in a −80 °C freezer for at least 1 h prior to tissue grinding. Do not snap-freeze in liquid nitrogen as it causes damage.
4. There are multiple options available to be used as organic solvents like ethanol, benzene, acetone, petroleum, and chloroform. Keep in mind that a specific absorption coefficient ($A^{1\%}{}_{1\text{cm}}$) is unique to that particular organic solvent used to calculate the absolute quantities after spectrophotometric measurements. An expanded list of carotenoid absorption coefficients for various solvents can be found elsewhere [39–41].
5. The tissue samples from mutant lines that hyper-accumulate *cis*-carotenes (like dark-grown etiolated cotyledon tissues)

Table 3
UV absorption coefficients of some identified carotenoids

Carotenoid	λ (nm)	$*\varepsilon_{mol}$	$A^{1\%}_{1cm}$	MW (g/mol)	Solvent	Chemical formula
Violaxanthin (Vio)	419, 440	133,400	2550	600.884	E	$C_{40}H_{56}O_4$
Zeaxanthin	428, 450	133,400	2480	568.886	E	$C_{40}H_{56}O_2$
Neoxanthin (Neo)	418, 442	117,000	2380	600.884	E	$C_{40}H_{56}O_4$
Lutein (Lut)	422, 445	127,000	2550	568.886	E	$C_{40}H_{56}O_2$
β-Carotene (βc)	450, 476	140,400	2620	536.888	E	$C_{40}H_{56}$
Antheraxanthin (Ant)	444,472	137,200	2350	584.885	E	$C_{40}H_{56}O_3$
Tetra-*cis*-(pro)lycopene (P-Lyc)	414, 436	102,900	1920	536.888	H	$C_{40}H_{56}$
All-*trans*-neurosporene (Neu)	415, 440	157,000	2920	538.904	H, P	$C_{40}H_{58}$
All-*trans*-lycopene (Lyc)	444, 470	184,900	3450	536.888	P	$C_{40}H_{56}$
Phytoene (Phy)	276, 286	68,000	915	542.936	H, P	$C_{40}H_{64}$
All-*trans*-ζ-carotene (ζc)	378, 400	138,000	2555	540.92	H	$C_{40}H_{60}$
All-*trans*-phytofluene (Phf)	331, 347	73,200	1580	542.936	H, P	$C_{40}H_{62}$

ε_{mol} molar extinction coefficient ($\varepsilon_{mol} = (A\ 1\%\ \text{x molecular mass})/10$), $A^{1\%}_{1cm}$ UV-vis absorbance of 1 cm layer of the specific solution, *MW* molecular weight

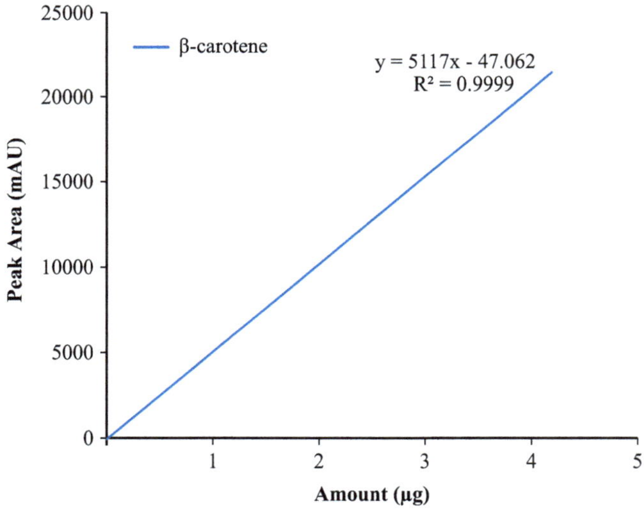

Fig. 4 Establishing a standard calibration curve for individual carotenoids. The calibration curve analysis for the β-carotene standards where the slope is "*m*", and the *y*-intercept is "*n*" on a given slope formula ($y = mx + n$). R^2 value represents the accuracy in preparation of dilution series of a standard

should be collected under nonphotosynthetic dim green LED light to prevent carotenoid photo-oxidation and photoisomerisation.

6. Ground tissue powder can stick to the edges of the microcentrifuge lid following grinding. Gently tap the tubes several times before opening to transfer tissue to the bottom of the vial. Ensure no liquid N_2 enters the microcentrifuge tube as this can result in an explosion after sealing the tube.
7. Carotenoids are positioned within the upper ethyl acetate phase, so the acetone–EtOAc ratio can be changed (e.g., 4:6, v/v) if a higher volume of extract is needed. Consider any changes to this dilution when calculating the carotenoid abundances in samples.
8. Before performing HPLC analysis, consider using a guard column integrated between the injector and the analytical column (C18 & C30). Guard columns protect the analytical column against blockage that can arise due to the presence of impurities inside the crude extract and or solvents if not filtered in advance.
9. A longer HPLC run (~10 h or longer) when the auto-sampler compartment is set to 4 °C can cause condensation of water in the compartment. It is preferable to maintain the auto-sampler compartment at 6–7 °C when analysing large numbers of samples in a single run.
10. We recommend running HPLC with 100% mobile phase B prior to sample injection until the back pressure stabilizes.
11. Pure carotenoid standards can be readily purchased and used to validate retention times and carotenoid spectra.
12. Calculation of % III/II ratio can facilitate the characterization of spectral fine structures in identified peaks. The %III/II value is dependent to the organic solvents being used and can be calculated by dividing the maximum absorption unit (λ_{max}) of peak III by λ_{max} of peak II by taking the minimum absorption value in-between as a baseline ($\lambda_{baseline}$), multiplied by 100 ($[\lambda_{maxIII}-\lambda_{baseline}]/[\lambda_{maxII}-\lambda_{baseline}] \times 100$). The λ_{maxIII}, λ_{maxII} and $\lambda_{baseline}$ can be calculated as shown in Fig. 5.
13. Due to their different spectral characteristics, carotenoid abundance should be quantified by taking specific measurements at the maximum appropriate wavelength. It is recommended to use absorbance values taken at 440 nm for the majority of carotenoids except for phytoene (at 286 nm), phytofluene (at 348 nm), ζ-carotene (at 400 nm), β-carotene (at 454 nm), and lycopene isomers (at 470 nm) excluding 7,9,9′,7′-tetra-*cis*-lycopene (prolycopene) (at 440 nm).

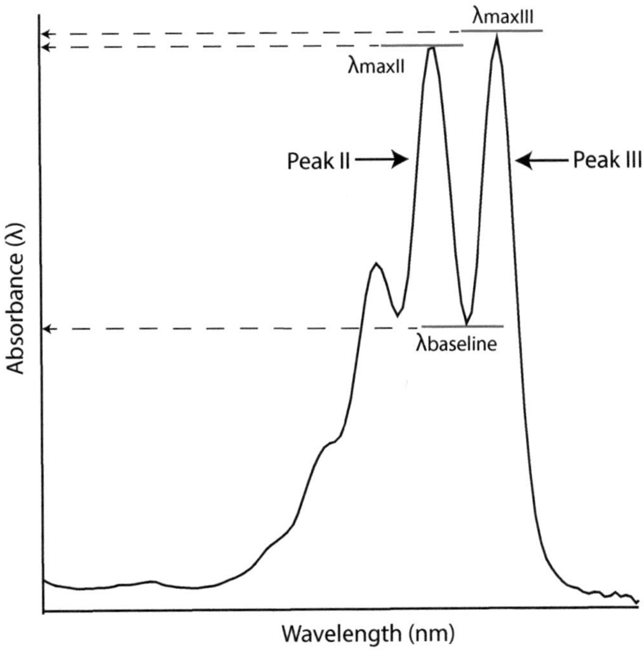

Fig. 5 Carotenoid peak ratios help determine isomers. A basic demonstration of how to calculate %III/II ratios

14. If absolute quantification of carotenoids is necessary, perform calibration for each specific carotenoid that you need to quantify with available standards according to steps between 15 and 18. Alternatively, collect carotenoid fractions by performing the steps between 4 and 14, then perform calibration analysis again. It is recommended to perform a calibration analysis every time when you start using a new column and/or instrument or any other HPLC protocol as coefficients for the absolute quantification might be slightly different depending on these variables.
15. It might be necessary to use higher volume of solvents as necessary to completely dissolve carotenoids before spectral measurements.
16. Once prepared, stock solutions can be stored at −20 °C freezer for a short period of time. Bring standards to RT before taking UV-Vis measurements as temperature impacts absorption measurements.
17. Make sure that there is no substance (dust from gloves, finger prints, etc.) to block the passage of light through the cuvette surfaces during spectrophotometric measurements. Avoid touching the sides of the cuvettes. Lint-free wipes are recommended to clean the cuvette surface.

18. Make sure that the absorbance values (AU) for your UV-VIS measurements are in a linear absorption bandwidth for accurate quantification (between 0.2 and 0.8). Higher or lower concentrations are not reliable for absorbance measurements. If necessary, stock solutions can be diluted with the same organic solvent before taking the measurements. Multiple measurements for the same sample may ensure the accuracy of your calculations. If the $A^{1\%}{}_{1cm}$ values for *cis*-isomers are not available, values belong to all-*trans* carotenoids can be used as an alternative if necessary, although not recommended.
19. Calculate the slope where *X*-axis is defined as μg carotenoid/ 20–50 μL volume injection and where *Y*-axis is peak area (mAU).

Acknowledgments

This work is supported by Australian Research Council Discovery Grant DP130102593 (to C.I.C). We acknowledge the Western Sydney University and Hawkesbury Institute for the Environment for providing a Ph.D. scholarship award to Y.A. and N.D. and financial support to purchase chemicals required for this research.

Author contributions: Y.A. wrote the methods chapter, prepared figures and tables. Y.A., N.D. and C.M. optimised methods for carotenoid extraction and HPLC separation. C.I.C outlined and edited the methods chapter. C.I.C supervised Y.A. and N.D. All authors contributed to editing this chapter.

References

1. Baranski R, Cazzonelli CI (2016) Carotenoid biosynthesis and regulation in plants. In: Carotenoids: nutrition, analysis and technology. Wiley Blackwell, London, pp 161–190
2. Cuttriss AJ, Chubb AC, Alawady A, Grimm B, Pogson BJ (2007) Regulation of lutein biosynthesis and prolamellar body formation in Arabidopsis. Funct Plant Biol 34(8):663–672
3. Van Norman JM, Zhang J, Cazzonelli CI, Pogson BJ, Harrison PJ, Bugg TD, Chan KX, Thompson AJ, Benfey PN (2014) Periodic root branching in Arabidopsis requires synthesis of an uncharacterized carotenoid derivative. Proc Natl Acad Sci 111(13):E1300–E1309
4. Dickinson AJ, Lehner K, Mi J, Jia K-P, Mijar M, Dinneny J, Al-Babili S, Benfey PN (2019) β-Cyclocitral is a conserved root growth regulator. Proc Natl Acad Sci 116 (21):10563–10567
5. Al-Babili S, Bouwmeester HJ (2015) Strigolactones, a novel carotenoid-derived plant hormone. Annu Rev Plant Biol 66:161–186
6. Cazzonelli CI, Hou X, Alagoz Y, Rivers J, Dhami N, Lee J, Marri S, Pogson BJ (2019) A *cis*–carotene derived apocarotenoid regulates etioplast and chloroplast development. bioRxiv 528331. https://doi.org/10.1101/528331
7. Wang JY, Haider I, Jamil M, Fiorilli V, Saito Y, Mi J, Baz L, Kountche BA, Jia K-P, Guo X (2019) The apocarotenoid metabolite zaxinone regulates growth and strigolactone biosynthesis in rice. Nat Commun 10(1):810
8. Iskandar AR, Miao B, Li X, Hu K-Q, Liu C, Wang X-D (2016) β-Cryptoxanthin reduced lung tumor multiplicity and inhibited lung cancer cells motility by down-regulating nicotinic acetylcholine receptor α7 expression. Cancer Prev Res 9(11):875–886

9. Beydoun MA, Chen X, Jha K, Beydoun HA, Zonderman AB, Canas JA (2018) Carotenoids, vitamin a, and their association with the metabolic syndrome: a systematic review and meta-analysis. Nutr Rev 77(1):32–45
10. Cho KS, Shin M, Kim S, Lee SB (2018) Recent advances in studies on the therapeutic potential of dietary carotenoids in neurodegenerative diseases. Oxidative Med Cell Longev 2018:4120458
11. Von Lintig J (2012) Metabolism of carotenoids and retinoids related to vision. J Biol Chem 287(3):1627–1634
12. Che P, Zhao Z-Y, Glassman K, Dolde D, Hu TX, Jones TJ, Gruis DF, Obukosia S, Wambugu F, Albertsen MC (2016) Elevated vitamin E content improves all-trans β-carotene accumulation and stability in biofortified sorghum. Proc Natl Acad Sci 113 (39):11040–11045
13. Beyer P, Al-Babili S, Ye X, Lucca P, Schaub P, Welsch R, Potrykus I (2002) Golden rice: introducing the β-carotene biosynthesis pathway into rice endosperm by genetic engineering to defeat vitamin a deficiency. J Nutr 132 (3):506S–510S
14. Nogueira M, Enfissi EM, Valenzuela MEM, Menard GN, Driller RL, Eastmond PJ, Schuch W, Sandmann G, Fraser PD (2017) Engineering of tomato for the sustainable production of ketocarotenoids and its evaluation in aquaculture feed. Proc Natl Acad Sci 114 (41):10876–10881
15. Boelsma E, Hendriks HF, Roza L (2001) Nutritional skin care: health effects of micronutrients and fatty acids. Am J Clin Nutr 73 (5):853–864
16. Hou X, Rivers J, León P, McQuinn RP, Pogson BJ (2016) Synthesis and function of apocarotenoid signals in plants. Trends Plant Sci 21 (9):792–803
17. Alagoz Y, Nayak P, Dhami N, Cazzonelli CI (2018) Cis-carotene biosynthesis, evolution and regulation in plants: the emergence of novel signaling metabolites. Arch Biochem Biophys 654:172–184
18. McQuinn RP, Wong B, Giovannoni JJ (2018) AtPDS overexpression in tomato: exposing unique patterns of carotenoid self-regulation and an alternative strategy for the enhancement of fruit carotenoid content. Plant Biotechnol J 16(2):482–494
19. Enfissi EM, Nogueira M, Bramley PM, Fraser PD (2017) The regulation of carotenoid formation in tomato fruit. Plant J 89(4):774–788
20. Dhami N, Tissue DT, Cazzonelli CI (2018) Leaf-age dependent response of carotenoid accumulation to elevated CO_2 in Arabidopsis. Arch Biochem Biophys 647:67–75
21. D'andrea L, Simon-Moya M, Llorente B, Llamas E, Marro M, Loza-Alvarez P, Li L, Rodriguez-Concepcion M (2018) Interference with Clp protease impairs carotenoid accumulation during tomato fruit ripening. J Exp Bot 69(7):1557–1568
22. Li F, Vallabhaneni R, Yu J, Rocheford T, Wurtzel ET (2008) The maize phytoene synthase gene family: overlapping roles for carotenogenesis in endosperm, photomorphogenesis, and thermal stress tolerance. Plant Physiol 147 (3):1334–1346
23. Llorente B, D'andrea L, Ruiz-Sola MA, Botterweg E, Pulido P, Andilla J, Loza-Alvarez P, Rodriguez-Concepcion M (2016) Tomato fruit carotenoid biosynthesis is adjusted to actual ripening progression by a light-dependent mechanism. Plant J 85 (1):107–119
24. Han S-H, Sakuraba Y, Koh H-J, Paek N-C (2012) Leaf variegation in the rice zebra2 mutant is caused by photoperiodic accumulation of tetra-Cis-lycopene and singlet oxygen. Mol Cells 33(1):87–97
25. Chak RK, Thomas TL, Quatrano RS, Rock CD (2000) The genes ABI1 and ABI2 are involved in abscisic acid-and drought-inducible expression of the Daucus carota L. Dc3 promoter in guard cells of transgenic Arabidopsis thaliana (L.) Heynh. Planta 210(6):875–883
26. Saez A, Robert N, Maktabi MH, Schroeder JI, Serrano R, Rodriguez PL (2006) Enhancement of abscisic acid sensitivity and reduction of water consumption in Arabidopsis by combined inactivation of the protein phosphatases type 2C ABI1 and HAB1. Plant Physiol 141 (4):1389–1399
27. Zhou J, Zeng L, Liu J, Xing D (2015) Manipulation of the xanthophyll cycle increases plant susceptibility to Sclerotinia sclerotiorum. PLoS Pathog 11(5):e1004878
28. Liu G, Pfeifer J, Brito Francisco R, Emonet A, Stirnemann M, Gübeli C, Hutter O, Sasse J, Mattheyer C, Stelzer E (2018) Changes in the allocation of endogenous strigolactone improve plant biomass production on phosphate-poor soils. New Phytol 217 (2):784–798
29. Schaub P, Rodriguez-Franco M, Cazzonelli CI, Álvarez D, Wüst F, Welsch R (2018) Establishment of an Arabidopsis callus system to study the interrelations of biosynthesis, degradation and accumulation of carotenoids. PLoS One 13(2):e0192158

30. Pogson B, McDonald KA, Truong M, Britton G, DellaPenna D (1996) Arabidopsis carotenoid mutants demonstrate that lutein is not essential for photosynthesis in higher plants. Plant Cell 8(9):1627–1639
31. Mezzomo N, Maestri B, dos Santos RL, Maraschin M, Ferreira SR (2011) Pink shrimp (P. brasiliensis and P. paulensis) residue: influence of extraction method on carotenoid concentration. Talanta 85(3):1383–1391
32. Zaghdoudi K, Framboisier X, Frochot C, Vanderesse R, Barth D, Kalthoum-Cherif J, Blanchard F, Guiavarc'h Y (2016) Response surface methodology applied to supercritical fluid extraction (SFE) of carotenoids from persimmon (Diospyros kaki L.). Food Chem 208:209–219
33. Strati IF, Gogou E, Oreopoulou V (2015) Enzyme and high pressure assisted extraction of carotenoids from tomato waste. Food Bioprod Process 94:668–674
34. Saini RK, Keum Y-S (2018) Carotenoid extraction methods: a review of recent developments. Food Chem 240:90–103
35. Amorim-Carrilho K, Cepeda A, Fente C, Regal P (2014) Review of methods for analysis of carotenoids. TrAC Trends Anal Chem 56:49–73
36. Park H, Kreunen SS, Cuttriss AJ, DellaPenna D, Pogson BJ (2002) Identification of the carotenoid isomerase provides insight into carotenoid biosynthesis, prolamellar body formation, and photomorphogenesis. Plant Cell 14(2):321–332
37. Isaacson T, Ronen G, Zamir D, Hirschberg J (2002) Cloning of tangerine from tomato reveals a carotenoid isomerase essential for the production of β-carotene and xanthophylls in plants. Plant Cell 14(2):333–342
38. Alba R, Payton P, Fei Z, McQuinn R, Debbie P, Martin GB, Tanksley SD, Giovannoni JJ (2005) Transcriptome and selected metabolite analyses reveal multiple points of ethylene control during tomato fruit development. Plant Cell 17(11):2954–2965
39. Britton G, Liaaen-Jensen S, Pfander H (2003) Handbook of carotenoids. Birkhäuser, Basel
40. Britton G, Liaaen-Jensen S, Pfander H (1995) Carotenoids: spectroscopy, vol 1. Springer, Basel
41. Hengartner U, Bernhard K, Meyer K, Englert G, Glinz E (1992) Synthesis, isolation, and NMR-spectroscopic characterization of fourteen (Z)-isomers of lycopene and of some acetylenic didehydro-and tetradehydrolycopenes. Helv Chim Acta 75(6):1848–1865

Chapter 12

Determination of Plant Volatile Apocarotenoids

José L. Rambla and Antonio Granell

Abstract

Volatile compounds derived from carotenoid cleavage are biologically relevant molecules in plants, as they are involved in different processes of plant communication and also have a significant effect on our perception of food flavor.

Here we describe two methods for the determination of volatile apocarotenoids in fruit based on extraction by means of headspace solid-phase microextraction (HS-SPME) and separation and detection by means of gas chromatography coupled to mass spectrometry (GC-MS). The first method is optimised for the detection of the complete volatile profile in *Citrus* fruit, including those compounds derived from carotenoids. The second is a shorter method focused to the detection of volatile apocarotenoids in the tomato fruit.

Key words Volatile compounds, Aroma, Apocarotenoid, SPME, Gas chromatography, Mass spectrometry

1 Introduction

Volatile apocarotenoids are short-chain nonpolar molecules derived from the cleavage of carotenoids which can be released from the plant and detected by the same or another plant or even a different organism.

Volatile apocarotenoids can be produced enzymatically by the action of carotenoid cleavage dioxygenases (CCD) [1, 2] but also nonenzymatically by chemical degradation of carotenoids [3]. Apocarotenoid volatiles have been detected in a wide range of plant species and in almost any plant organ/tissue, including leaves [4], flowers [5] or fruit [6–8].

In some species such as tomato, volatile apocarotenoid production increases in the ripe fruit, participating in fruit flavor and aroma. In the case of the ripe tomato fruit, three volatile apocarotenoids have been described to impact the human perception of tomato flavor [9]. Nevertheless, production of most of the apocarotenoid volatiles peaks at earlier stages of fruit development in

Manuel Rodríguez-Concepción and Ralf Welsch (eds.), *Plant and Food Carotenoids: Methods and Protocols*, Methods in Molecular Biology, vol. 2083, https://doi.org/10.1007/978-1-4939-9952-1_12,

species such as strawberry or grape [6], thus suggesting their participation in different roles along fruit development. Given the role of plant apocarotenoids in inter- and intraorganism communication, determining their levels is a prerequisite to understand their functions.

Here we describe first a method for the analysis of apocarotenoid volatile compounds which is based on a general method for volatiles which includes extraction by means of HS-SPME and separation and detection by GC-MS. A typical chromatogram obtained in orange juice by means of this method is shown in Fig. 1. In this chromatogram, over one hundred volatile compounds have been identified [10], including ten apocarotenoids (Table 1).

In the second part we describe another method, also based on HS-SPME coupled to GC-MS, which is particularly focused in the apocarotenoids. It is well known that the particular conditions in sample preparation and extraction have a key effect on the compounds that can be detected in the analysis. Many of the products after carotenoid cleavage are semivolatile C_{10}–C_{13} compounds. Sensitivity in the detection of such compounds increases by increasing the temperature during volatile capturing [11]. Therefore, the

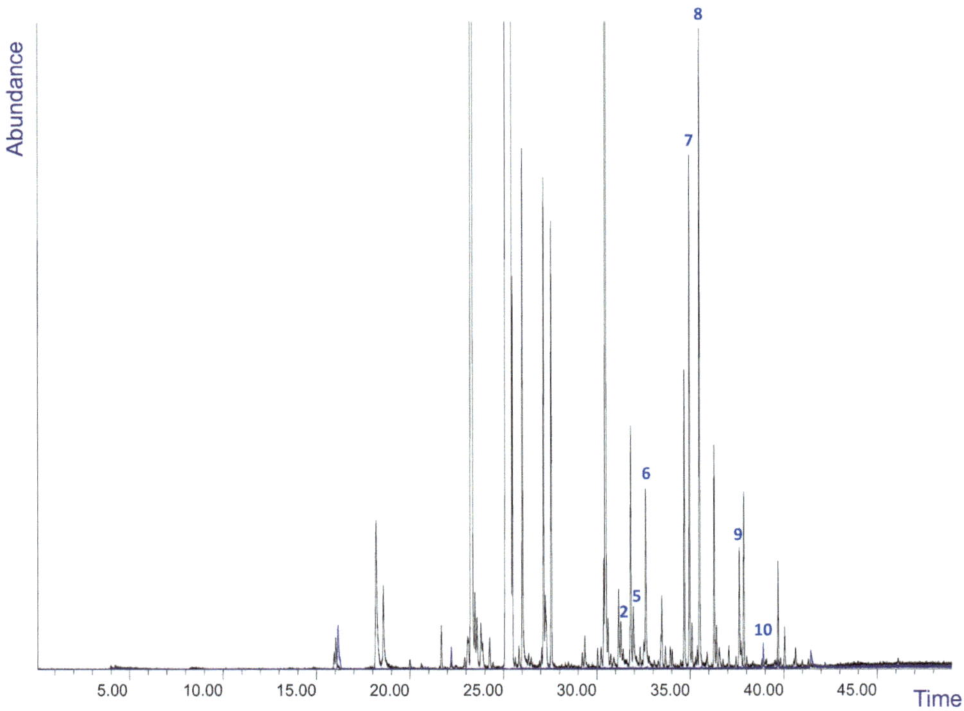

Fig. 1 GC-MS extracted ion chromatogram (EIC) of orange (*Citrus sinensis*) juice showing ions 69 and 177 (characteristic for several volatile apocarotenoids). Over 100 volatile compounds have been identified in this chromatogram [10]. The most prominent peaks corresponding to apocarotenoids have been indicated (numbers according to those in Table 1). Time scale is expressed in minutes

Table 1
List of unequivocally identified apocarotenoid volatile compounds usually detected in *Citrus* juice, with the m/z of the specific ion used for the integration of the peak for each compound, and the retention time in the chromatographic conditions described

ID	Compound	Q Ion	RT	RI
1	6-methyl-5-hepten-2-one	108	23.03	985.2
2	Nerol	93	31.28	1233.0
3	β-Cyclocitral	137	31.63	1248.4
4	Neral	84	31.81	1252.8
5	Geraniol	69	31.95	1256.4
6	Geranial	69	32.61	1276.9
7	Neryl acetate	69	34.97	1360.4
8	Geranyl acetate	69	35.50	1377.8
9	Geranylacetone	43	37.62	1456.2
10	β-Ionone	177	38.90	1503.1

Q Ion m/z of the ion used for integration, *RT* Retention Time (min), *RI* Kovats Retention Index

second method here described combines a higher incubation temperature with a shorter chromatographic separation (but still adequate to separate C_8–C_{13} compounds, which include most of the apocarotenoids). In the cases in which apocarotenoids are the only volatiles of interest, this provides a higher sensitivity and productivity method for their analysis. A typical chromatogram obtained of the tomato fruit volatile fraction by means of this method is shown in Fig. 2. Table 2 shows the set of volatile apocarotenoids typically identified in tomato fruit with this method. The chemical structure of the compounds identified with both methods is represented in Fig. 3.

2 Materials

2.1 Volatile Apocarotenoid Method

1. A 50/30 μm DVB/CAR/PDMS coating SPME fiber (*see* **Note 1**).
2. Manual orange juice extractor.

2.2 Specific Apocarotenoid Method

3. Liquid nitrogen.
4. 20 mL screw cap containers.
5. 15 mL clip cap glass vials.
6. Finely powdered solid $CaCl_2{\cdot}2H_2O$ (*see* **Note 2**).

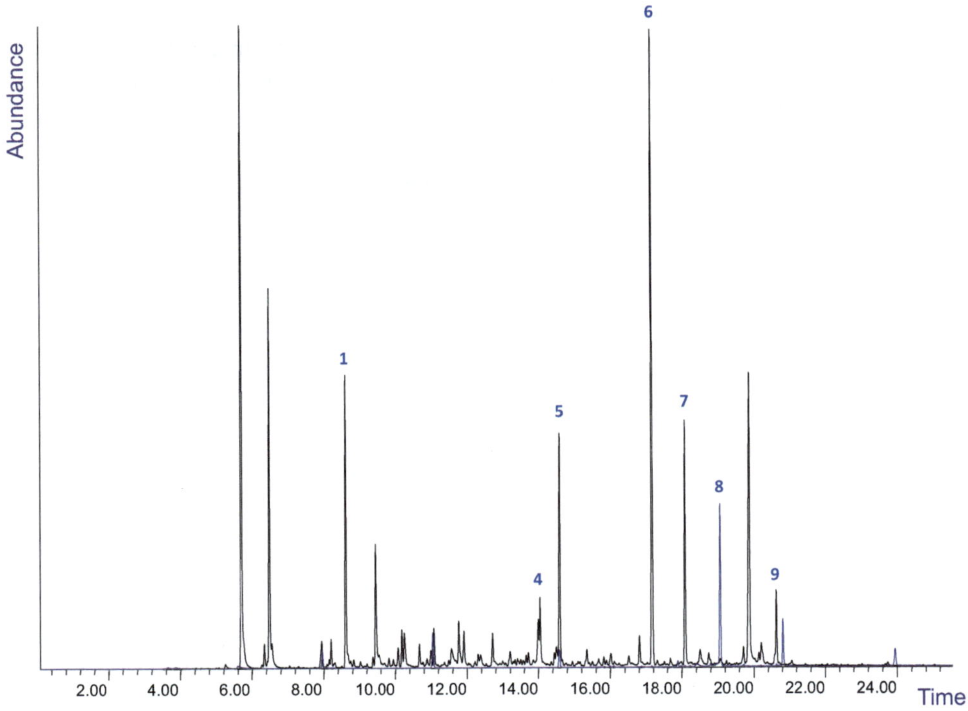

Fig. 2 GC-MS extracted ion chromatogram (EIC) of red ripe tomato fruit showing ions 69 and 177 (characteristic for several volatile apocarotenoids). Major peaks corresponding to apocarotenoids have been indicated (numbers according to those in Table 2). Time scale is expressed in minutes

Table 2
List of unequivocally identified apocarotenoid volatile compounds usually detected in red ripe tomato fruit, with the *m/z* of the specific ion used for the integration of the peak for each compound, and the retention time in the chromatographic conditions described

ID	Compound	Q Ion	RT	RI
1	6-methyl-5-hepten-2-one	108	8.62	984.5
2	6-methyl-5-hepten-2-ol	89	8.75	991.8
3	β-Cyclocitral	137	14.03	1251.0
4	Neral	69	14.05	1252.0
5	Geranial	41	14.60	1278.7
6	β-Damascenone	69	17.16	1407.2
7	Geranylacetone	69	18.09	1456.5
8	β-Ionone	177	19.07	1508.8
9	Pseudoionone	109	20.63	1595.9

Q Ion m/z of the ion used for integration, *RT* Retention Time (min), *RI* Kovats Retention Index

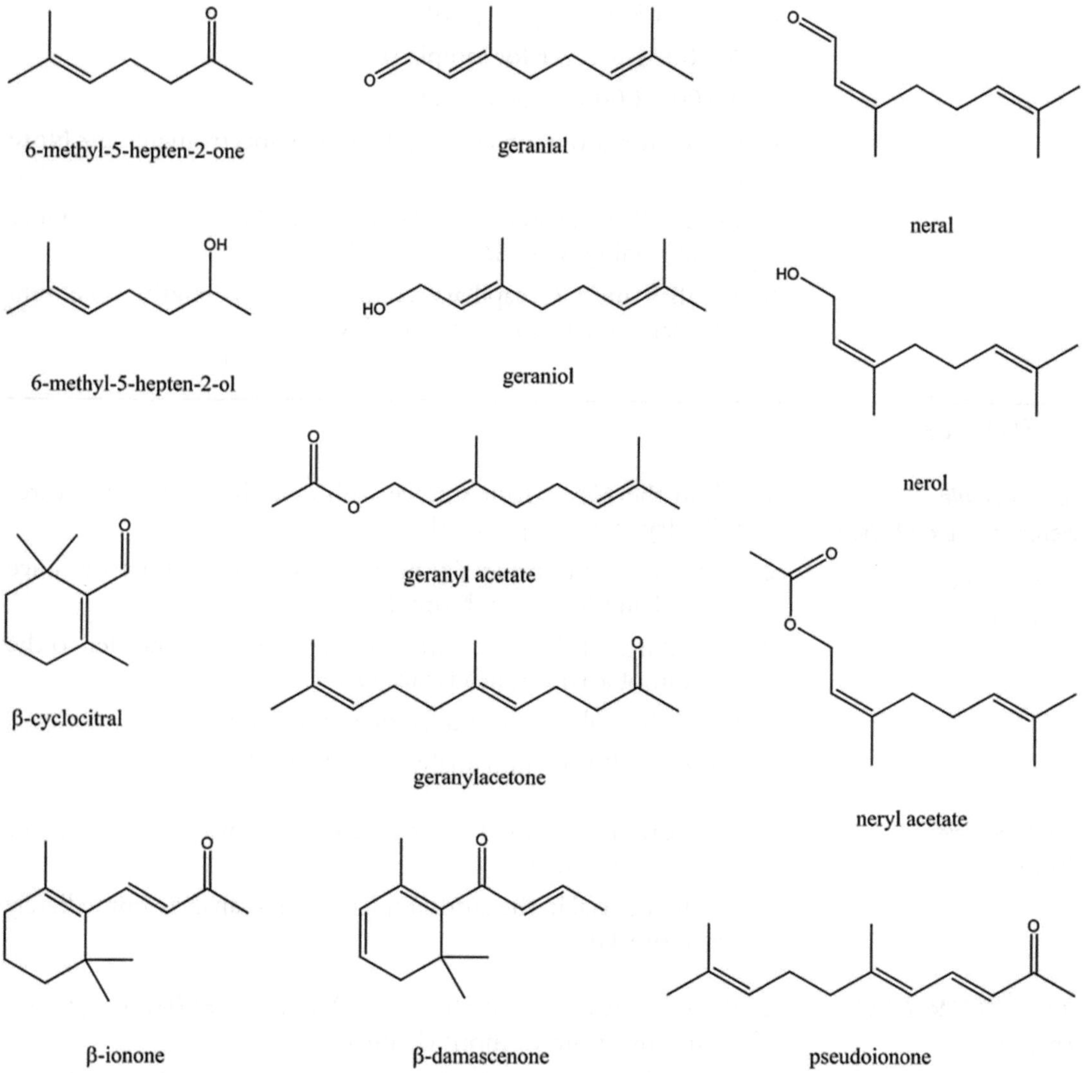

Fig. 3 Chemical structure of volatile apocarotenoids shown in Tables 1 and 2

7. A 100 mM EDTA solution, pH 7.5 adjusted with NaOH (*see* **Note 3**).
8. A 65 μm DVB/PDMS coating SPME fiber (*see* **Note 1**).
9. Cryogenic mill (*see* **Note 4**).
10. Water bath.
11. Ultrasonic bath.
12. Precision scale (±1 mg).

2.3 Common to Both Methods

1. 10 mL headspace screw cap vials and 18 mm magnetic caps with silicone-PTFE septa for automated SPME (*see* **Note 5**).
2. Pure standards of all the compounds identified (Table 1).

3. Helium 5.0 or higher purity.
4. A −80 °C freezer for sample storage.
5. A 100–1000 μL micropipette.
6. Gas Chromatograph coupled to a mass spectrometer (*see* **Note 6**).
7. Combi-PAL autosampler adapted for SPME, including a Fiber Conditioning Module (*see* **Note 7**).
8. A DB-5 ms GC capillary column (60 m length, 0.25 mm internal diameter, 1.0 μm film thick).

3 Methods

3.1 Volatile Apocarotenoid Method

3.1.1 Sampling Procedure

1. Clean the *Citrus* fruit surface with deionized water and carefully dry with a paper cloth.
2. Cut the fruit transversally in two pieces and extract the juice with a hand juicer (*see* **Note 8**).
3. Take 2 mL of the juice with a micropipette and transfer to the bottom of a 10 mL headspace vial.
4. Close the vial tightly and freeze immediately.
5. Store at −80 °C until analysis (*see* **Note 9**).

3.1.2 Sample Preparation

1. Incubate the vial for 10 min at 20 °C in a water bath to allow the sample to thaw.
2. Locate the vial in the autosampler tray and analyze immediately (*see* **Note 10**).

3.1.3 HS-SPME GC-MS Analysis

1. Preincubate de vials at 50 °C for 10 min with 500 rpm agitation by means of the CombiPAL autosampler.
2. Insert the SPME holder in the vial, and expose the 50/30 μm DVB/CAR/PDMS fiber to the headspace for volatile extraction at 50 °C for 20 min with 500 rpm agitation.
3. Desorb the volatiles by introducing the fiber in the injection port at 250 °C for 1 min in splitless mode.
4. Clean the fiber in the Fiber Conditioning Module with Helium flow at 250 °C for 5 min to prevent cross-contamination.
5. Gas chromatography conditions:
   - Constant Helium flow of 1.2 mL/min.
   - Oven temperature program: 40 °C for 2 min, 5 °C/min ramp until 250 °C, held at 250 °C for 5 min.

     Chromatographic time is 49 min. Total run time, including oven cooling, is about 1 h.
6. Mass spectrometry conditions:

- Transfer line 260 °C, ionization source 230 °C, quadrupole 150 °C.
- Ionization energy 70 eV (EI).
- Scan mode acquisition in the m/z range 35–250 (6.2 scans per second) (*see* **Note 11**).

3.1.4 Compound Identification and Quantitation

1. Unequivocal compound identification is performed by both mass spectrum similarity and retention time coelution. Check the coincidence of both with those of pure standards injected under the same chromatographic conditions.
2. Integrate the area of each peak from the extracted ion chromatogram with the compound specific m/z ion indicated in Table 1 (*see* **Note 12**).
3. An external standard should be introduced in the analytical sequence (i.e., each 8–10 samples) and integrated as any other chromatogram for data normalization. An admixture of all the samples analyzed in the experiment is preferable for this purpose (*see* **Note 13**).
4. For data normalization, in each chromatogram divide the area of each peak to that of the average area of the same peak in the closest reference chromatograms before and after that particular chromatogram. The result is a relative quantification for each compound (the ratio of its abundance in the sample compared to that in the external standard).

3.2 Specific Apocarotenoid Method

3.2.1 Sampling Procedure

1. Take the ripe tomato fruit with great care not to produce any physical damage (*see* **Note 14**).
2. Clean the fruit surface with deionized water and carefully dry with a paper cloth.
3. Excise the pericarp, discarding all the other fruit tissues. Dry smoothly the inner part of the pericarp with a paper cloth to prevent contamination with seeds or locular tissue (*see* **Note 15**).
4. Cut the pericarp into pieces (depending on fruit size) and flash freeze in liquid nitrogen (*see* **Note 16**).
5. Homogenize the samples with a cryogenic mill to produce a fine powder. Take care to keep the sample frozen all the time.
6. Store at −80 °C in either plastic or glass tubes until analysis (*see* **Note 9**).

3.2.2 Sample Preparation

1. Weigh 500 g of frozen pericarp powder in a 15 mL glass vial (*see* **Note 17**).
2. Close the vial and incubate for 10 min at 37 °C in a water bath.

3. Add 1.1 g of solid $CaCl_2 \cdot 2H_2O$ and 500 μL of a 100 mM EDTA solution (pH 7.5), close the vial and mix gently.
4. Ultrasonicate for 5 min.
5. Transfer 1 mL of the homogenized mixture to the bottom of a 10 mL screw-cap headspace vial and close hermetically (*see* **Notes 18** and **19**).

3.2.3 HS-SPME GC-MS Analysis

1. Preincubate de vials at 80 °C for 10 min with 500 rpm agitation by means of the CombiPAL autosampler (*see* **Note 20**).
2. Insert the SPME holder in the vial, and expose the 65 μm DVB/PDMS fiber to the headspace for volatile extraction at 80 °C for 20 min with 500 rpm agitation.
3. Introduce the fiber for 1 min in the gas chromatograph injection port at 250 °C for desorption of volatile compounds. Sample desorption is performed in splitless mode.
4. Clean the fiber in the Fiber Conditioning Module with Helium flow at 250 °C for 5 min to prevent cross-contamination.
5. Gas chromatography conditions:
   - Constant Helium flow of 1.3 mL/min.
   - Oven temperature program: 120 °C for 1 min, 7 °C/min ramp until 250 °C, then held at 250 °C for 5 min.

     Chromatographic time is 25.5 min. Total run time, including oven cooling, is around 30 min.
6. Mass spectrometry conditions:
   - Transfer line 260 °C, ionization source 230 °C, quadrupole 150 °C.
   - Electron impact ionization (EI), electron energy 70 eV.
   - Scan mode acquisition in the m/z range 35–250 (6.2 scans per second) (*see* **Note 11**).

3.2.4 Compound Identification and Quantitation

Do exactly as indicated in Subheading 3.1.4, except that retention time and m/z for integration of compounds are those in Table 2.

4 Notes

1. A variety of fiber coatings exists, and the most adequate should be selected depending on the chemical nature of the most important target compounds in the experiment. In the case of volatile apocarotenoids, a PDMS coating would also be a good option due to the chemical nature of these compounds.
2. Solid $CaCl_2 \cdot 2H_2O$ is available in different presentations. It must be as a fine powder so that can be rapidly dissolved.

3. Sometimes dissolving EDTA can be a slow process. Addition of some NaOH for pH adjustment while dissolving EDTA will help it dissolve completely.
4. In case only a small amount of material is available (i.e., below 3–4 g) sample homogenization will be better performed with a homogenizer for low volume samples such as a Retsch Mixer Mill MM400 with grinding jars and balls, taking care to pre-cool them with liquid nitrogen.
5. When automated SPME injection is performed, vials should be sealed with silicone septa. Harder materials in the septum can produce fiber damage. Additionally, a thin PTFE film in the inner side of the septum is recommended to prevent sample contamination.
6. The gas GC injection port should have a specific liner for SPME (i.e., SPME liner 0.75 mm × 6.35 × 78.5 for Agilent GCs IP Deact., RESTEK). Standard injection liners have a higher internal volume and will produce wider peaks and poor chromatographic resolution, particularly in the region at the beginning of the chromatogram.
7. In case a Fiber Conditioning Module is not available, the fiber can be cleaned by remaining an extended time (*i.e.* 5 min) in the GC injection port.
8. A manual juicer rather than automated or press type extractors is highly recommended to prevent contamination with essential oils from the flavedo.
9. Samples should not be stored at higher temperature such as −20 °C, because the volatile profile, including the apocarotenoids, would show a quick evolution along the time of storage.
10. Volatile compounds in the juice are not stable. Therefore, samples must be thawed immediately before analysis.
11. Volatile compounds are low molecular weight molecules, most often under 220 amu. When using a quadrupole detector, a narrow m/z scan range (as the one indicated in this method) provides either better peak shape (more points per peak) or higher sensitivity.
12. Criteria for the selection of the *m/z* of the ion used for integration are adequate abundance of the signal and specificity with the coeluting peaks. Depending on the volatile profile of each set of samples, some of the peaks might be integrated better with a different *m/z* ion than those proposed in Tables 1 and 2. Therefore, the ions indicated in the table should be taken as a first choice and further optimized for the type of matrix under study if necessary.
13. The use of an external standard such as the sample admixture is essential to correct for variations in detector sensitivity and

particularly fiber aging. This is a critical issue for data normalization when extracting volatile compounds with SPME.

14. Make sure that all the samples are at the same ripening stage. The production of volatile compounds, including those derived from carotenoids, changes dramatically during the process of ripening. Additionally, biological material should be manipulated with care. Any stress, such as physical damage, will have a remarkable effect on the volatile profile.
15. When comparing different types of samples, analysis of pericarp produces better comparable metabolomic profiles than using the whole fruit, which also includes other tissues with a quite different chemical composition, such as seeds or locular tissue. These other tissues tend to be in different proportions, depending on many reasons such as fruit size, locule number or seed content, among others. Nevertheless, the same method can be used with other tissues or the whole fruit, if preferred.
16. Fruit should be cut into pieces small enough to be introduced in the container, but preferably not smaller.
17. Freeze-drying is an adequate technique for the storage of plant material for some other metabolomic analysis, but not for volatile analysis. The freeze-drying process produces depletion in the levels of most volatile compounds.
18. When the frozen tomato powder is thawed, it tends to form aggregates. Therefore, the end of the tip should be cut to ease pipetting.
19. When prepared as indicated, the composition in most of the volatile compounds in the samples is rather stable for about 10–12 h [12].
20. High incubation temperature favours the detection of semivolatile apocarotenoids. Nevertheless, it needs to be verified that they are not produced by thermal degradation from carotenoids present in the matrix.

References

1. Rubio A, Rambla JL, Santaella M, Gomez MD, Orzaez D, Granell A, Gomez-Gomez L (2008) Cytosolic and plastoglobule-targeted carotenoid dioxygenases from Crocus sativus are both involved in beta-ionone release. J Biol Chem 283:24816–24825
2. Rubio-Moraga A, Rambla JL, Fernández-del-Carmen A, Trapero-Mozos A, Ahrazem O, Orzáez D, Granell A, Gómez-Gómez L (2014) New target carotenoids for CCD4 enzymes are revealed with the characterization of a novel stress-induced carotenoid cleavage dioxygenase gene from Crocus sativus. Plant Mol Biol 86:555–569
3. Boon CS, Mc Clements DJ, Weiss J, Decker EA (2010) Factors influencing the chemical stability of carotenoids in foods. Crit Rev Food Sci Nutr 50:515–532
4. López-Gresa MP, Lisón P, Campos L, Rodrigo I, Rambla JL, Granell A, Conejero V, Bellés JM (2017) A non-targeted metabolomics approach unravels the VOCs associated with the tomato immune response against pseudomonas syringae. Front Plant Sci

8:1188. https://doi.org/10.3389/fpls.2017.01188

5. Moraga AR, Rambla JL, Ahrazem O, Granell A, Gomez-Gomez L (2009) Metabolite and target transcript analyses during Crocus sativus stigma development. Phytochemistry 70:1009–1016
6. Rambla JL, Trapero-Mozos A, Diretto G, Rubio-Moraga A, Granell A, Gomez-Gomez L, Ahrazem O (2016) Gene-metabolite networks of volatile metabolism in airen and tempranillo grape cultivars revealed a distinct mechanism of aroma bouquet production. Front Plant Sci 7:1619. https://doi.org/10.3389/fpls.2016.01619
7. Rambla JL, Medina A, Fernandez-del-Carmen A, Barrantes W, Grandillo S, Cammareri M, Lopez-Casado G, Rodrigo G, Alonso A, Garcia-Martinez S, Primo J, Ruiz JJ, Fernandez-Munoz R, Monforte AJ, Granell A (2017) Identification, introgression, and validation of fruit volatile QTLs from a red-fruited wild tomato species. J Exp Bot 68:429–442
8. Esteras C, Rambla JL, Sánchez G, López-Gresa MP, González-Mas MC, Fernández-Trujillo JP, Bellés JM, Granell A, Picó MB (2018) Fruit flesh volatiles and carotenoid profile analysis within the Cucumis melo L. species reveals unexploited variability for future genetic breeding. J Sci Food Agric 98:3915–3925. https://doi.org/10.1002/jsfa.8909
9. Tieman D, Zhu G, Resende MFR, Lin T, Nguyen C, Bies D, Rambla JL, Ortiz-Beltran KS, Taylor M, Zhang B, Ikeda H, Liu Z, Fisher J, Zemach I, Monforte AJ, Zamir D, Granell A, Kirst M, Huang S, Klee HJ (2017) A chemical genetic roadmap to improved tomato flavor. Science 355:391–394
10. González-Mas MC, Rambla JL, Alamar MC, Gutierrez A, Granell A (2011) Comparative analysis of the volatile fraction of fruit juice from different citrus species. PLoS One 6: e22016. https://doi.org/10.1371/journal.pone.0022016
11. Rambla JL, Alfaro C, Medina A, Zarzo M, Primo J, Granell A (2015) Tomato fruit volatile profiles are highly dependent on sample processing and capturing methods. Metabolomics 11:1708–1720. https://doi.org/10.1007/s11306-015-0824-5
12. Tikunov Y, Lommen A, de Vos CHR et al (2005) A novel approach for nontargeted data analysis for metabolomics. Large-scale profiling of tomato fruit volatiles. Plant Physiol 139:1125–1137

Chapter 13

A Method for Extraction and LC-MS-Based Identification of Carotenoid-Derived Dialdehydes in Plants

Jianing Mi, Kun-Peng Jia, Aparna Balakrishna, and Salim Al-Babili

Abstract

We developed a chemical derivatization based ultra-high performance liquid chromatography-hybrid quadrupole-Orbitrap mass spectrometer (UHPLC-Q-Orbitrap MS) analytical method to identify low-abundant and instable carotenoid-derived dialdehydes (DIALs, diapocarotenoids) from plants. Application of this method enhances the MS response signal of DIALs, enabling the detection of diapocarotenoids, which is crucial for understanding the function of these compounds and for elucidating the carotenoid oxidative metabolic pathway in plants.

Key words Carotenoid-derived dialdehydes, Apocarotenoids, Diapocarotenoids, Chemical derivatization, Arabidopsis, UHPLC-MS

1 Introduction

Carotenoids are a large class of mostly C_{40} terpenoids characterized by an extended conjugated double bond system that enables them to absorb light in the blue-green (450–550 nm) range of the visible spectrum and to contribute to plant light-harvesting via extending the range of light absorption of the photosynthetic apparatus beyond that of chlorophyll [1–9]. In addition, carotenoids are indispensable photoprotective pigments that prevent destructive oxidation processes by quenching both singlet and triplet state chlorophylls [10–15]. Furthermore, carotenoids are important determinants of the fluidity of plastid membranes and are accumulated in flowers and fruits to attract animals for pollination and seed dispersal [9, 16, 17]. Carotenoid-derived metabolites also exert important biologic functions. Thus, the class of carotenoid-derivatives includes hormones, such as abscisic acid and strigolactone, [18–25], retinoids, that is, retinal, retinol (vitamin A), and retinoic acid [26–28], the fungal pheromone trisporic acid [29, 30], the stress signaling molecule β-cyclocitral [31–33], the

Manuel Rodríguez-Concepción and Ralf Welsch (eds.), *Plant and Food Carotenoids: Methods and Protocols*, Methods in Molecular Biology, vol. 2083, https://doi.org/10.1007/978-1-4939-9952-1_13,

recently discovered growth regulators anchorene and zaxinone [34, 35], pigments, such as crocin in saffron [36], citraurin in citrus fruits [37], and the fungal neurosporaxanthin [38, 39], and volatiles, such as safranal [40, 41], geranial [42], and β-ionone [43, 44]. All of these bioactive metabolites (named apocarotenoids) are generated by oxidative cleavage of carotenoids, which is generally catalyzed by carotenoid cleavage dioxygenases (CCDs) or by nonenzymatic oxidation caused by reactive oxygen species (ROSs) [25, 32, 45, 46]. In Arabidopsis, there are five different types of CCDs. The nine-*cis*-epoxycarotenoid cleavage dioxygenases (NCED represented by 5 enzymes; NCED2, 3, 5, 6, and 9) are involved in the biosynthesis of abscisic acid, while the four other CCD types, designated as CCD1, CCD4, CCD7, and CCD8 exert different biological functions and have, accordingly, different substrates and regiospecificities [23, 25, 44, 47]. For example, CCD1 enzymes catalyze the conversion of carotenoids into a plenitude of volatiles, such as geranial, pseudoionone, and β-ionone, and are likely involved in scavenging of destructed carotenoids/apocarotenoids; the Arabidopsis CCD4 converts all-*trans*-β-carotene into β-ionone and all-*trans*-β-apo-10′-carotenal and determines carotenoid content; The strigolactone biosynthesis enzyme CCD7 (MAX3) cleaves 9-*cis*-β-carotene to form β-ionone and 9-*cis*-β-apo-10′-carotenal, the substrate of CCD8 (MAX4) that forms the central strigolactone biosynthesis intermediate carlactone.

Carotenoid cleavage yields dialdehyde products (DIALs), besides monocarbonyl apocarotenoids that have been in the focus of carotenoid research. Carotenoid-derived DIALs are defined as a class of carotenoid oxidative products incorporating two aldehyde functional groups, which are produced from repeated oxidative cleavage of double bonds within carotenoids or apocarotenoids (Fig. 1). Recent in vitro studies showed that several plant and cyanobacterial CCDs can cleave carotenoids or apocarotenoids to produce DIALs. Adrian et al expected that all-*trans*-β-apo-10-′-carotenal can be cleaved by CCD8 to form all-*trans*-4-methyl-2,4,6-octatrienedial [48], which has been confirmed in Wang et al.'s work. Scherzinger et al. demonstrated that cyanobacterial retinal-forming enzymes can catalyze the conversion of β-apo-8′-carotenoids to all-*trans*-2,6-dimethyl-2,4,6-octatrienedial [49]. Ilg et al. reported tomato CCDs 1A and 1B convert many apocarotenoid substrates including apo-8′-lycopenal, apo-10′-lycopenal, apo-12′-lycopenal, and apo-15′-lycopenal in in vitro assays, leading to a series DIALs with different carbon chains (e.g., all-*trans*-2,6-dimethyl-2,4,6,8-decatetraenedial, all--*trans*-3,7-dimethyl-2,4,6,8-decatetraenedial, all-*trans*-2,6,11-trimethyl-2,4,6,8,10-dodecapentaenedial, all-*trans*-2,6,11-trimethyl-2,4,6,8,10,12-quattuordecahexaenedial, all-*trans*-2,7,11-trimethyl-2,4,6,8,10,12-quattuordecahexaenedial,

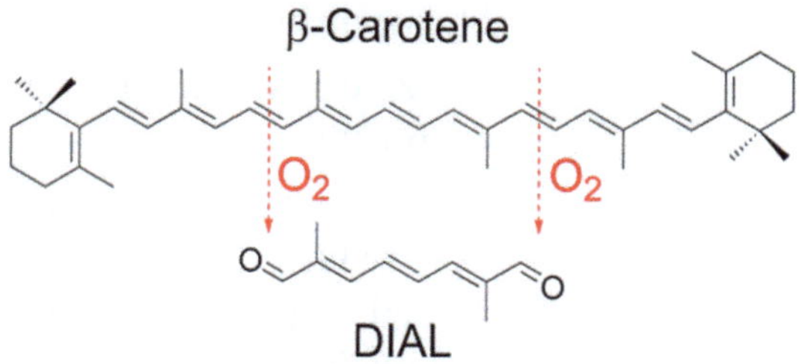

Fig. 1 Structures of β-carotene and the generation of a representative carotenoid-derived DIAL

all-*trans*-4,9,13-trimethyl-2,4,6,8,10,12,14-sedecaheptaenedial, all-*trans*-2,6,11,15-tetramethyl-2,4,6,8,10,12,14-sedecaheptaenedial, all-*trans*-2,6,11,15-tetramethyl-2,4,6,8,10,12,14,16-octodecaoctaenedial) [50]. Moreover, studies on the biosynthesis of bixin (also known as annatto, one of the oldest pigment widely used in foods and cosmetics) [51, 52] and crocin (the major saffron pigment) [53], demonstrate the role of DIALs as pigments [41, 54]. To our knowledge, all carotenoid-derived DIALs reported so far were identified from enzymatic in vitro assays that usually produce them in relatively high amounts, which allow their detection by HPLC or LC-MS. However, endogenous DIALs occur at very low levels and are unstable in plants, making their direct detection using present methods very difficult.

Understanding the biological function of DIALs and determining their role in plant carotenoid metabolism require an analytical method that enables sensitive and reliable determination of these compounds. In this work, we developed a chemical derivatization based UHPLC-Q-Orbitrap MS method for the analysis of carotenoid-derived DIALs from Arabidopsis. A crude plant DIALs extract is prepared, which is added to a chemical derivatization regent consisting of 10 mg/mL N^2,N^2,N^4,N^4-tetraethyl-6-hydrazineyl-1,3,5-triazine-2,4-diamine (T3) methanol solution with 1% formic acid. Following derivatization, DIALs are detected using UHPLC-Q-Orbitrap MS.

2 Materials

Use HPLC grade reagents for the extraction. All reagents and water used for UHPLC-Q-Orbitrap MS analyses should be LC-MS grade. All LC-MS solvent additives used must be LC-MS grade. All reagents should be kept on ice during extraction. Waste disposal regulations must be meticulously followed.

2.1 Plant Material Preparation

1. Arabidopsis seedlings.
2. Liquid nitrogen.

3. Microcentrifuge tubes of 2 mL capacity.
4. Freeze dry system (Labconco FreeZone 6 Liter Console, or equivalent).
5. Bead beater (Biospec Products, INC. Mini bead beater, or equivalent).
6. Analytical precision balance (Secura® analytical balance, or equivalent).

2.2 Crude DIALs Extraction

1. Stock extraction solution: 10 mL acetonitrile with 0.1% butylated hydroxytoluene (BHT) (*see* **Note 1**) [55]. The stock extraction solution is made to a volume of 10 mL and can be stored for 1 month at 4 °C.
2. Microcentrifuge tubes of 2 mL capacity.
3. Ultrasonic bath (Branson 5510EDTH, or equivalent).
4. Refrigerated microcentrifuge (Eppendorf centrifuge 5424R, or equivalent).
5. Concentrator (Eppendorf concentrator plus, or equivalent).

2.3 DIAL Derivatization

1. 1% Formic acid methanol stock solution (*see* **Note 2**). The stock solution is made to a volume of 10 mL and can be stored for 1 month at 4 °C.
2. Stock derivatization solution: 1 mL of 10 mg/mL N^2,N^2,N^4,N^4-tetraethyl-6-hydrazineyl-1,3,5-triazine-2,4-diamine (T3) methanol solution with 1% formic acid (*see* **Note 3**) [56]. The stock derivatization solution is made to a volume of 1 mL and can be stored for 1 month at 4 °C.
3. Ultrasonic bath (Branson 5510 EDTH, or equivalent).
4. ThermoMixer (HLC Heating-ThermoMixer MHR 23, or equivalent).

2.4 UHPLC-Q-Orbitrap MS Detection

1. DIAL standard stock solution: add 1.6 mg all-*trans*-2,7-Dimethyl-2,4,6-octatrienedial (DIAL10) to a volumetric flask and make up to 10 mL with methanol (*see* **Note 4**). The stock DIAL solution is made to a volume of 10 mL and can be stored for 3 months at −20 °C.
2. Ultra-high performance liquid chromatography (UHPLC) solvents: 450 mL water, 50 mL acetonitrile, 1 mL formic acid (solvent A); 450 mL acetonitrile, 50 mL isopropanol, 1 mL formic acid (solvent B) (*see* **Note 5**).
3. 1 mL Syringe (BD 1 mL Syringe, or equivalent).
4. 0.22 μm Syringe filter (Thermo Scientific 4 mm PTFE syringe filter, or equivalent).
5. 1.5 mL Amber glass vial (VWR, or equivalent).

6. Micro glass insert (VWR 28 × 5.6 mm, or equivalent).
7. Acquity UPLC BEH C_{18} column (100 × 2.1 mm; 1.7 μm; Waters) with an Acquity UPLC BEH C_{18} guard column (5 × 2.1 mm; 1.7 μm; Waters), or equivalent.
8. Dionex UltiMate 3000 UHPLC system (Thermo Scientific) or equivalent.
9. Heated-electrospray Ionization-Hybrid Quadrupole-Orbitrap Mass Spectrometer (HESI-Q-Orbitrap MS, Thermo Scientific Q Exactive Plus) or equivalent.

3 Methods

All steps used for extraction and derivatization of DIALs should be carried out under dim light, unless otherwise specified. This method has been optimized for detecting DIALs in *Arabidopsis thaliana*. Detecting DIALs in other plant species or tissues might require additional optimization.

3.1 Plant Material Preparation

1. Collect fresh Arabidopsis seedlings (10-day) in 2 mL microcentrifuge tubes and keep them frozen in liquid nitrogen for 5 min (*see* **Note 6**).
2. Transfer tubes with plant material to the freeze-dry system and dry the material for 12 h (*see* **Note 7**).
3. Add 3 beads in each tube after drying. Then transfer tubes in liquid nitrogen for 1 min (*see* **Note 8**).
4. Transfer tubes to the bead beater and homogenize plant material for 1 min.
5. Weigh between 25 and 30 mg dry weight plant material in a 2 mL microcentrifuge tube. Work quickly and cover the weighed material with aluminum foil. Note the exact weight.

3.2 Crude DIALs Extraction

1. Prepare the extraction solution by adding 10 mg of BHT to a volumetric flask and make up to 10 mL with acetonitrile. Keep the extraction solution on ice.
2. Add 0.5 mL of extraction solution to each tube and mixture gently. Transfer tubes to an ultrasonic bath for the extraction for 15 min at 25 °C. After that, transfer tubes to a precooled microcentrifuge at 18,000 × *g* for 5 min at 4 °C. Collect the supernatant and reextract the pellet with 0.5 mL of extraction solution. Combine the two supernatants and dry it using a concentrator (*see* **Note 9**).

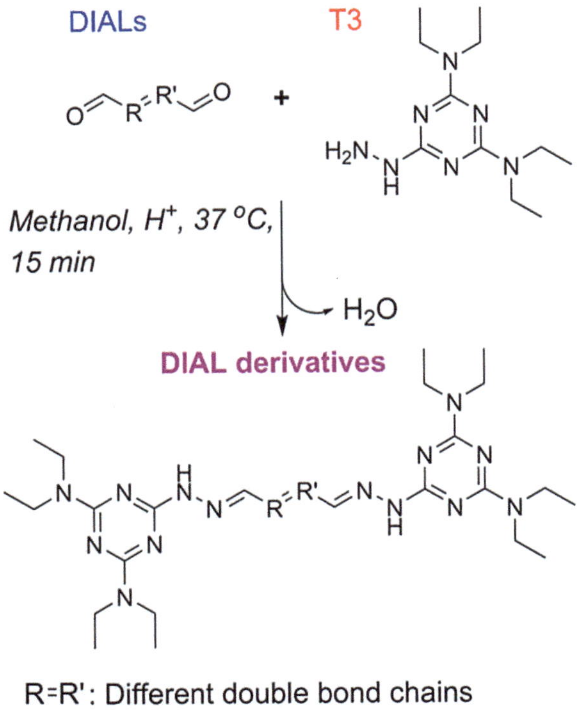

Fig. 2 Scheme of DIAL derivatization

3.3 DIALs Derivatization

1. Prepare 1% formic acid methanol stock solution by adding 100 μL LC-MS grade formic acid to a volumetric flask and make up to 10 mL with methanol.
2. Prepare the derivatization regent by adding 10 mg of T3 to a volumetric flask and make up to 1 mL with 1% formic acid methanol stock solution. Keep the derivatization regent on ice.
3. Add 50 μL of the derivatization regent to the DIAL extract and transfer tubes to ultrasonic bath for the sonication of 30 s (*see* **Note 10**). Than transfer tubes to a thermomixer for the incubation at 37 °C for 15 min (Fig. 2).
4. Dilute the DIAL derivatization solution with 50 μL of 1% Formic acid methanol stock solution.

3.4 UHPLC-Q-Orbitrap MS Detection

1. Prepare the DIAL standard solution by adding 10.25 μL of DIAL10 standard stock solution to a volumetric flask and make up to 1 mL with methanol (*see* **Note 11**). Keep the DIAL10 standard in dark.
2. Prepare the DIAL10 standard derivatization solution: Add 2 μL of DIAL10 standard solution and 50 μL of the derivatization regent in microcentrifuge tube (*see* **Note 12**). Transfer the tube to a thermomixer at 37 °C for 15 min. Then dilute the solution with 50 μL of 1% formic acid methanol stock solution.

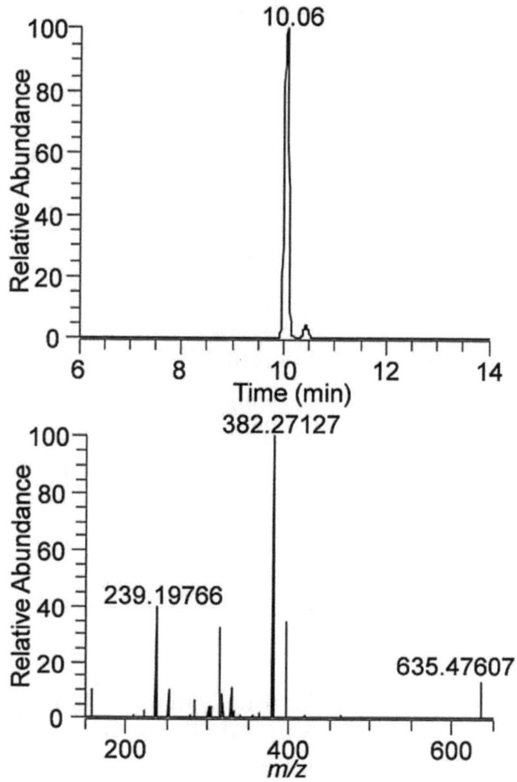

Fig. 3 Identification of derivatized DIAL10 by using UHPLC-MS/MS. Extracted ion chromatogram of (upper) and MS/MS spectrum (down) of derivatized DIAL10

3. Prepare LC-MS sample solution: The DIAL derivatization solution is filtered to 1.5 mL amber glass vial with micro glass insert using 1 mL syringe with 0.22 μm syringe filter.
4. Optimize the instrument parameters for detecting DIALs by infusing a DIAL10 standard derivatization solution into the mass spectrometer using the optimization procedure under positive ionization mode. The optimized parameters will be different depending on the used LC-MS instrument, the optimized parameters of an HESI-Q-Orbitrap MS were full MS scan, scan range of 150–900 *m/z*, resolution of 280,000, AGC target of $3e^6$, and maximum inject time of 150 ms.
5. Liquid chromatographic separation of DIAL derivatives is achieved with a BEH C_{18} column (Acquity UPLC BEH C_{18} column) using the following LC parameters: column temperature = 35 °C, injection volume = 15 μL, flow rate = 0.2 mL min^{-1}, total run time = 22 min. Use a solvent gradient program starting with a linear gradient from 20 to 100% solvent B over 15 min, isocratic separation at 100% B

until 18 min, return to initial conditions of 20% B at 19 min and equilibrate at 20% B until 22 min (*see* **Note 13**).

6. Ionization was achieved using heated-electrospray ionization with an atmospheric pressure ionization source operating under positive ionization mode. The spray voltage was 4000 V, the sheath gas flow rate was 40 arbitrary units, the auxiliary gas flow rate was 10 arbitrary units, the capillary temperature was 350 °C, the auxiliary gas heater temperature was 400 °C, and the S-lens RF was 50. For MS/MS, NCE was 20 eV. The ionization parameters also need to be optimized for different LC-MS systems.
7. Xcalibur 2.2 software (Thermo Scientific) was used for data acquisition and processing. Identify the DIAL derivatives produced by the DIAL derivatization reaction using high-resolution MS and MS/MS data, confirmed by synthetic DIAL10 standard (Fig. 3).

4 Notes

1. We usually make a 10 mL solution by dissolving 10 mg of BHT in 10 mL acetonitrile in a volumetric flask. Then transfer solution to a new glass bottle and keep it at 4 °C.
2. Prepare 1% formic acid methanol stock solution by adding 100 μL LC-MS grade formic acid to a volumetric flask and make up to 10 mL with methanol. To decrease evaporation of formic acid, we usually add 2 mL methanol to the volumetric flask firstly, then add 100 μL LC-MS grade formic acid to the volumetric flask with methanol using pipette. Lastly, make up to 10 mL with methanol in the volumetric flask. 1% Formic acid methanol stock solution should be kept at 4 °C. It is helpful for storage. Use protective clothing, gloves and protection glasses when using formic acid.
3. Prepare the derivatization regent by adding 10 mg of T3 to a volumetric flask and make up to 1 mL with 1% formic acid methanol stock solution. The stock derivatization solution is stored for 1 month at 4 °C. To avoid degradation, do not suggest keeping the stock derivatization solution for more than 1 month.
4. Prepare DIAL standard stock solution by adding 1.6 mg all-*trans*-2,7-dimethyl-2,4,6-octatrienedial (Tokyo Chemical Industry Co., Ltd.) to a brown volumetric flask and make up to 10 mL with methanol. DIAL is unstable under light or oxygen conditions. In the DIAL stock solution preparation, methanol should be precooled on ice to reduce the amount of oxygen. Work quickly and keep the DIAL standard stock

solution in a brown bottle for 3 month at −20 °C as soon as possible to minimize degradation.

5. UHPLC solvents were prepared with LC-MS grade regents and water. Mix 450 mL water and 50 mL acetonitrile in mobile phase A glass bottle, and 450 mL acetonitrile and 50 mL isopropanol in mobile phase B glass bottle, then degas them in ultrasonic bath for 20 min. After that, add 1 mL formic acid in mobile phase A glass bottle and mobile phase B glass bottle, respectively. The expiry period of mobile phase A and mobile phase B are 5 days and 2 weeks, respectively.
6. Collect fresh Arabidopsis seedlings in 2 mL microcentrifuge tubes and keep them frozen in liquid nitrogen for 5 min. Work quickly and quench the metabolism in plant materials as soon as possible.
7. Make sure that the freeze-dry system is ready, and plant materials is in a frozen state. It is difficult to dry the plant material if it is in a defrozen state.
8. Transfer tubes with dry plant materials to liquid nitrogen for 1 min. It makes it easier to homogenize plant material.
9. The drying time of the supernatant in a concentrator is not fixed. It depends on the samples number and the liquid volume. It is better to test the drying time, as too long drying time leads to degradation of DIALs.
10. The sonication of 30 s in ultrasonic bath is used to dissolve DIALs and to mix DIALs with the derivatization regent.
11. Prepare the 10 pmol μL^{-1} of DIAL standard solution by adding 10.25 μL of DIAL standard stock solution to a volumetric flask and make up to 1 mL with methanol. It makes it easier to take 20 pmol DIAL for the derivatization with a minimum amount of solvent.
12. Based on its MS response signal in UHPLC-Q-Orbitrap MS, we use 20 pmol DIAL standard (2 μL of DIAL standard solution) to prepare the DIAL standard derivatization solution. The amount of DIAL standard also need to be optimized for different LC-MS systems with different sensitivity.
13. To minimize the effect of free derivatization regent on MS, LC system was switched to waste channel—disconnected from MS system—in the first 6 min of the solvent gradient program. After 6 min, LC system was switched to MS system for detecting DIAL derivatives.

Acknowledgments

This work was supported by base line funding and a Competitive Research Grant (CRG4) given to Salim Al-Babili from the King Abdullah University of Science and Technology (KAUST).

References

1. Cogdell BJ, Gillbro T, Andersson PO, Liu RSH, Asato AE (1994) Carotenoids as accessory light-harvesting pigments. Pure Appl Chem 66:1041–1046
2. Horton P, Ruban AV, Walters RG (1996) Regulation of light harvesting in green plants. Annu Rev Plant Physiol Plant Mol Biol 47:655–684
3. Ruban AV, Pascal AA, Robert B (2000) Xanthophylls of the major photosynthetic light-harvesting complex of plants: identification, conformation and dynamics. FEBS Lett 477:181–185
4. Polívka T, Sundström V (2004) Ultrafast dynamics of carotenoid excited states–from solution to natural and artificial systems. Chem Rev 104:2021–2072
5. Fraser PD, Bramley PM (2004) The biosynthesis and nutritional uses of carotenoids. Prog Lipid Res 43:228–265
6. Polívka T, Frank HA (2010) Molecular factors controlling photosynthetic light harvesting by carotenoids. Acc Chem Res 43:1125–1134
7. Mendes-Pinto MM, Galzerano D, Telfer A, Pascal AA, Robert B, Ilioaia C (2013) Mechanisms underlying carotenoid absorption in oxygenic photosynthetic proteins. J Biol Chem 288:18758–18765
8. Moise AR, Al-Babili S, Wurtzel ET (2014) Mechanistic aspects of carotenoid biosynthesis. Chem Rev 114:164–193
9. Nisar N, Li L, Lu S, Khin NC, Pogson BJ (2015) Carotenoid metabolism in plants. Mol Plant 8:68–82
10. Demmig-Adams B, Gilmore AM, Adams WW (1996) Carotenoids 3: in vivo function of carotenoids in higher plants. 3rd. FASEB J 10:403–412
11. Niyogi KK, Grossman AR, Bjorkman O (1998) Arabidopsis mutants define a central role for the xanthophyll cycle in the regulation of photosynthetic energy conversion. Plant Cell 10:1121–1134
12. Pogson BJ, Niyogi KK, Bjorkman O, DellaPenna D (1998) Altered xanthophyll compositions adversely affect chlorophyll accumulation and nonphotochemical quenching in Arabidopsis mutants. Proc Natl Acad Sci U S A 95:13324–13329
13. Barber J, Anderson JM, Telfer A (2002) What is β-carotene doing in the photosystem II reaction Centre? Phil Trans R Soc Lond B 357:1431–1440
14. Jahns P, Holzwarth AR (2012) The role of the xanthophyll cycle and of lutein in photoprotection of photosystem II. BBA-Bioenergetics 1817:182–193
15. Ballottari M, Mozzo M, Girardon J, Hienerwadel R, Bassi R (2013) Chlorophyll triplet quenching and photoprotection in the higher plant monomeric antenna protein Lhcb5. J Phys Chem B 117:11337–11348
16. Hirschberg J (2001) Carotenoid biosynthesis in flowering plants. Curr Opin Plant Biol 4:210–218
17. Maresca JA, Graham JE, Bryant DA (2008) The biochemical basis for structural diversity in the carotenoids of chlorophototrophic bacteria. Photosynth Res 97:121–140
18. Parry AD, Horgan R (1991) Carotenoids and abscisic acid (ABA) biosynthesis in higher plants. Physiol Plant 82:320–326
19. Schwartz SH, Leon-Kloosterziel KM, Koornneef M, Zeevaart JA (1997) Biochemical characterization of the aba2 and aba3 mutants in Arabidopsis thaliana. Plant Physiol 114:161–166
20. Nambara E, Marion-Poll A (2005) Abscisic acid biosynthesis and catabolism. Annu Rev Plant Biol 56:165–185
21. Abuauf H, Haider I, Jia KP, Ablazov A, Mi J, Blilou I, Al-Babili S (2018) The Arabidopsis DWARF27 gene encodes an all-*trans*–/9-*cis*-β-carotene isomerase and is induced by auxin, abscisic acid and phosphate deficiency. Plant Sci 277:33–44
22. Xie X, Yoneyama K, Yoneyama K (2010) The strigolactone story. Annu Rev Phytopathol 48:93–117
23. Alder A, Jamil M, Marzorati M, Bruno M, Vermathen M, Bigler P et al (2012) The path from β-carotene to carlactone, a strigolactone-like plant hormone. Science 335:1348–1351

24. Al-Babili S, Bouwmeester HJ (2015) Strigolactones, a novel carotenoid-derived plant hormone. Annu Rev Plant Biol 66:161–186
25. Jia KP, Baz L, Al-Babili S (2018) From carotenoids to strigolactones. J Exp Bot 69:2189–2204
26. Moise AR, Von Lintig J, Palczewski K (2005) Related enzymes solve evolutionarily recurrent problems in the metabolism of carotenoids. Trends Plant Sci 10:178–186
27. Eroglu A, Harrison EH (2013) Carotenoid metabolism in mammals, including man: formation, occurrence, and function of apocarotenoids. J Lipid Res 54:1719–1730
28. Grune T, Lietz G, Palou A, Ross AC et al (2010) β-Carotene is an important vitamin a source for humans. J Nutr 140:2268S–2285S
29. Austin DJ, Bu'lock JD, Drake D (1970) The biosynthesis of trisporic acids from b-carotene via retinal and trisporol. Experientia 26:348–349
30. Medina HR, Cerdá-Olmedo E, Al-Babili S (2011) Cleavage oxygenases for the biosynthesis of trisporoids and other apocarotenoids in Phycomyces. Mol Microbiol 82:199–208
31. Ramel F, Birtic S, Ginies C, Soubigou-Taconnat L, Triantaphylidès C, Havaux M (2012) Carotenoid oxidation products are stress signals that mediate gene responses to singlet oxygen in plants. Proc Natl Acad Sci U S A 109:5535–5540
32. Havaux M (2014) Carotenoid oxidation products as stress signals in plants. Plant J 79:597–606
33. Lv F, Zhou J, Zeng L, Xing D (2015) β-Cyclocitral upregulates salicylic acid signalling to enhance excess light acclimation in Arabidopsis. J Exp Bot 66:4719–4732
34. Jia KP, Dickinson AJ, Mi J, Cui G et al (2018) Anchorene is an endogenous diapocarotenoid required for anchor root formation in Arabidopsis. bioRxiv. https://doi.org/10.1101/496737
35. Wang JY, Haider I, Jamil M, Fiorilli V et al (2018) The apocarotenoid metabolite zaxinone regulates growth and strigolactone biosynthesis in rice. Nat Commun 10:810
36. Demurtas OC, Frusciante S, Ferrante P et al (2018) Candidate enzymes for saffron crocin biosynthesis are localized in multiple cellular compartments. Plant Physiol 177:990–1006
37. Rodrigo MJ, Alquézar B, Alós E, Medina V, Carmona L, Bruno M et al (2013) A novel carotenoid cleavage activity involved in the biosynthesis of citrus fruit-specific apocarotenoid pigments. J Exp Bot 64:4461–4478
38. Estrada AF, Youssar L, Scherzinger D, Al-Babili S, Avalos J (2008) The ylo-1 gene encodes an aldehyde dehydrogenase responsible for the last reaction in the Neurospora carotenoid pathway. Mol Microbiol 69:1207–1220
39. Díaz-Sánchez V, Estrada AF, Trautmann D, Al-Babili S, Avalos J (2011) The gene carD encodes the aldehyde dehydrogenase responsible for neurosporaxanthin biosynthesis in Fusarium fujikuroi. FEBS J 278:3164–3176
40. Beltran JC, Stange C (2016) Apocarotenoids: a new carotenoid-derived pathway. Subcell Biochem 79:239–272
41. Frusciante S, Diretto G, Bruno M, Ferrante P et al (2014) Novel carotenoid cleavage dioxygenase catalyzes the first dedicated step in saffron crocin biosynthesis. Proc Natl Acad Sci U S A 111:12246–12251
42. Ilg A, Beyer P, Al-Babili S (2009) Characterization of the rice carotenoid cleavage dioxygenase 1 reveals a novel route for geranial biosynthesis. FEBS J 276:736–747
43. Bruno M, Beyer P, Al-Babili S (2015) The potato carotenoid cleavage dioxygenase 4 catalyzes a single cleavage of β-ionone ring-containing carotenes and non-epoxidated xanthophylls. Arch Biochem Biophys 572:126–133
44. Bruno M, Koschmieder J, Wuest F, Schaub P, Fehling-Kaschek M, Timmer J et al (2016) Enzymatic study on AtCCD4 and AtCCD7 and their potential to form acyclic regulatory metabolites. J Exp Bot 67:5993–6005
45. Giuliano G, Al-Babili S, Von Lintig J (2003) Carotenoid oxygenases: cleave it or leave it. Trends Plant Sci 8:145–149
46. Bouvier F, Isner JC, Dogbo O, Camara B (2005) Oxidative tailoring of carotenoids: a prospect towards novel functions in plants. Trends Plant Sci 10:187–194
47. Ilg A, Yu Q, Schaub P, Beyer P, Al-Babili S (2010) Overexpression of the rice carotenoid cleavage dioxygenase 1 gene in Golden Rice endosperm suggests apocarotenoids as substrates in planta. Planta 232:691–699
48. Alder A, Holdermann I, Beyer P, Al-Babili S (2008) Carotenoid oxygenases involved in plant branching catalyse a highly specific conserved apocarotenoid cleavage reaction. Biochem J 416:289–296
49. Scherzinger D, Ruch S, Kloer DP, Wilde A, Al-Babili S (2006) Retinal is formed from apo-carotenoids in Nostoc sp. PCC7120: in vitro characterization of an apo-carotenoid oxygenase. Biochem J 398:361–369

50. Ilg A, Bruno M, Beyer P, Al-Babili S (2014) Tomato carotenoid cleavage dioxygenases 1A and 1B: relaxed double bond specificity leads to a plenitude of dialdehydes, mono-apocarotenoids and isoprenoid volatiles. FEBS Open Bio 4:584–593
51. Collins P (1992) The role of annatto in food colouring. Foods Ingred Process Int 13:23–27
52. Timberlake CF, Henry BS (1986) Plant pigments as natural food colours. Endeavour 10:31–37
53. Palmer LS (1934) The biological and chemical nomenclature for the carotenoids. Science 79:488–490
54. Bouvier F, Odette D, Bilal C (2003) Biosynthesis of the food and cosmetic plant pigment bixin (annatto). Science 300:2089–2091
55. Mi J, Jia KP, Wang JY, Al-Babili S (2018) A rapid LC-MS method for qualitative and quantitative profiling of plant apocarotenoids. Anal Chim Acta 1035:87–95
56. Tie C, Hu T, Jia ZX, Zhang JL (2016) Derivatization strategy for the comprehensive characterization of endogenous fatty aldehydes using HPLC-multiple reaction monitoring. Anal Chem 88:7762–7768

Chapter 14

Fractionation of Tomato Fruit Chromoplasts

Karel De Pourcq and Albert Boronat

Abstract

Chromoplast differentiation involves an active synthesis of carotenoids associated with the remodeling of the preexisting plastid membrane systems to form specialized structures involved in the sequestration and storage of the synthesized carotenoids. These subplastidial structures show remarkable morphological differences and seem to be adapted to the accumulation of particular carotenoids in some plant species and organs. At present, very little is known about chromoplast biogenesis and the role of the different suborganellar structures in the synthesis and storage of carotenoids. The combination of classical fractionation methods with the use of biochemical and -omics techniques represents an attractive approach to unravel novel aspects related with the biochemical and cellular mechanisms underlying the biogenesis of the structures involved in the biosynthesis and storage of carotenoids during chromoplast differentiation. Here we describe a combined protocol for the isolation, lysis and fractionation of tomato fruit chromoplast. The fractions obtained are suitable for metabolomics and proteomics analysis.

Key words Chromoplast, Carotenoid, Proteomics, Fractionation, Tomato fruit

1 Introduction

Chromoplasts are plastids specialized in the synthesis and accumulation of carotenoids found in many fruits, flowers, seeds and roots. Chromoplasts derive from preexisting plastids (such as proplastids, chloroplasts, amyloplasts, or leucoplasts) in a process characterized by a massive biosynthesis of carotenoids along with a remodeling of the preexisting membrane systems and the formation of suborganellar structures involved in carotenoid sequestration and storage [1–3]. Because of their hydrophobic nature, carotenoids are usually stored in the form of lipoproteic complexes [4]. Based on the pigment storage substructures, chromoplasts are classified into five major types: globular, crystalline, membranous, fibrillar, and tubular [1, 2]. This variety of suborganellar structures has been associated with the large diversity of carotenoid profiles found in the chromoplasts present in different plant species, tissues, organs, or developmental stages [5]. Interestingly, different chromoplast

Manuel Rodríguez-Concepción and Ralf Welsch (eds.), *Plant and Food Carotenoids: Methods and Protocols*, Methods in Molecular Biology, vol. 2083, https://doi.org/10.1007/978-1-4939-9952-1_14,

types and substructures can coexist in a same species [1, 6]. Studies related with chromoplast differentiation have mainly focused on the synthesis of carotenoids using biochemical and molecular approaches [3, 5, 7]. In contrast, the biochemical, molecular and cellular mechanisms underlying carotenoid sequestration and storage during chromoplast biogenesis remain largely unknown [1, 3].

During tomato fruit ripening, photosynthetically active chloroplasts differentiate into photosynthetically inactive chromoplasts through a complex process distinguished by the breakdown of the photosynthetic apparatus and a massive synthesis and deposition of lycopene, the major carotenoid present in tomato fruit [8, 9]. The structures found in a typical tomato fruit chromoplast include large plastoglobules, an extensive laminar membranous system and long osmiophilic filaments reported to accumulate lycopene in a crystalline form [10]. Plastoglobules are lipoprotein particles present in plastids that have received an increasing interest during the last few years after the discovery that they contain enzymes involved in the synthesis of tocopherols and carotenoids [11]. Although chloroplast plastoglobules have been studied in detail, both at the structural and functional level, much less is known on chromoplast plastoglobules [11]. The proteomic analysis of plastoglobules isolated from red pepper chromoplasts has revealed the presence of carotenogenic enzymes [12]. These results suggest an active role of plastoglobules in carotenoid biosynthesis in addition to their function in storage. The role of the other subplastidial structures in carotenoid metabolism and storage during chromoplast differentiation in tomato fruit has not yet been characterized.

In recent years proteomics has become an efficient tool to study the protein composition of chromoplasts as well as the dynamic protein changes associated to their differentiation [13]. In tomato fruit, chromoplast differentiation has mainly been associated with a strong decrease in the abundance of proteins involved in photosynthesis and carbohydrate metabolism and an increase in the level of carotenoid biosynthetic enzymes and stress-related proteins [14]. In contrast, only a few proteins potentially involved in carotenoid sequestration and storage have been identified [15]. It is likely that proteomic and metabolomic analysis targeting particular chromoplast subfractions could provide new insights into biochemical and molecular processes related not only with carotenoid sequestration and storage but also with the organization and regulation of carotenoid biosynthesis during tomato fruit ripening.

In the present protocol two procedures used for the isolation of Arabidopsis plastoglobules [12, 16] have been adapted and merged to fractionate tomato fruit chromoplasts isolated as described previously [17, 18]. The fractionation method described here allows the isolation of fractions highly enriched in plastoglobules and crystalline structures as well as other fraction containing different membranous structures showing particular protein and metabolite profiles.

2 Materials

All solutions have to be prepared using ultrapure water and analytical grade reagents. Unless otherwise stated, all reagents are prepared at room temperature. All waste disposal regulations are to be carefully followed upon the discarding of used materials and reagents.

2.1 Buffers for Chromoplast Isolation

1. Buffer A: 100 mM Tris–HCl, pH 8.2, 330 mM sorbitol, 2 mM $MgCl_2$, 10 mM KCl, 8 mM EDTA, 10 mM L-ascorbic acid, 5 mM L-cysteine, 0.2% BSA. This buffer can be stored at 4 °C. Before use add DTT to 1 mM final concentration and PVPP to 1% (*see* **Note 1**).
2. Buffer B: Same as buffer A, but without PVPP.
3. Sucrose solutions: 15%, 30%, 40%, and 50% (w/v) solutions prepared in 50 mM Tris–HCl (pH 7.4). These solutions can be stored at −20 °C. Add DTT to 1 mM final concentration before use (*see* **Note 2**).
4. 1 M DTT stock solution in water. Aliquots can be stored at −20 ° C (stable for up to 1 year).

2.2 Buffers for Chromoplast Fractionation

1. Lysis buffer: 50 mM HEPES, pH 8.0, 5 mM $MgCl_2$, 17% (w/v) sucrose. The stock can be stored at −20 °C for up to 6 months. Add DTT to 1 mM final concentration immediately before use.
2. Sucrose solutions (to be prepared immediately before use): 0%, 7%, 20%, 30%, 40% and 50% (w/v) solutions in 50 mM HEPES pH 8.0, 5 mM $MgCl_2$. These solutions can be stored at −20 °C. Add DTT to 1 mM final concentration (*see* **Note 3**).
3. PMSF protease inhibitor: 1 M solution in DMSO. Aliquots can be stored for up to 6 months at −20 °C.

3 Methods

Keep all reagents and samples on ice and carry out all procedures at 4 °C. Short handling of centrifuge tubes and samples outside ice or a cold room is acceptable.

3.1 Chromoplast Isolation

1. Cut about 1 kg of tomato fruits into halves and remove seeds and the rest of the gelatinous material of the locular cavities (*see* **Note 4**).
2. Cut fruits into small pieces and place in a cold Waring blender jar. Add approx. 2 volumes of buffer A per weight of tomato tissue (*see* **Note 5**) and homogenize at low speed (three pulses lasting for 1 s each once full speed is attained).

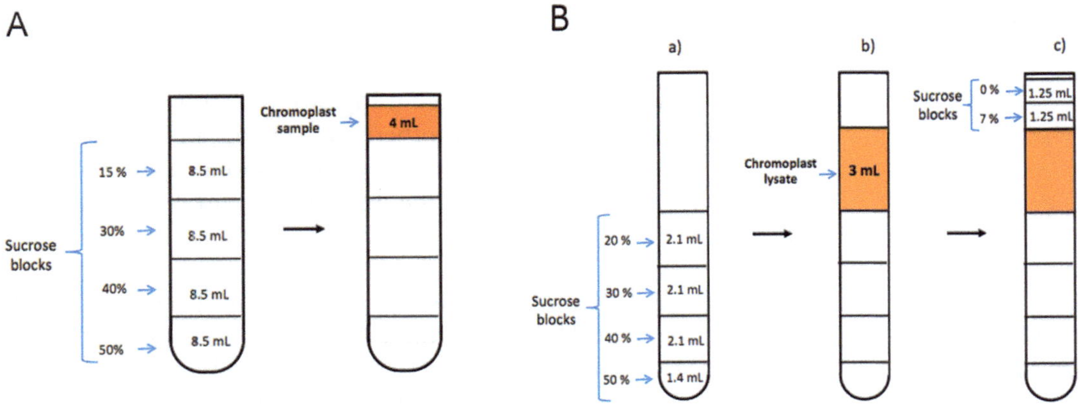

Fig. 1 Workflow illustrating the preparation of the sucrose density blocks used for chromoplasts isolation (A) and fractionation (B). Tubes sizes are not drawn to scale

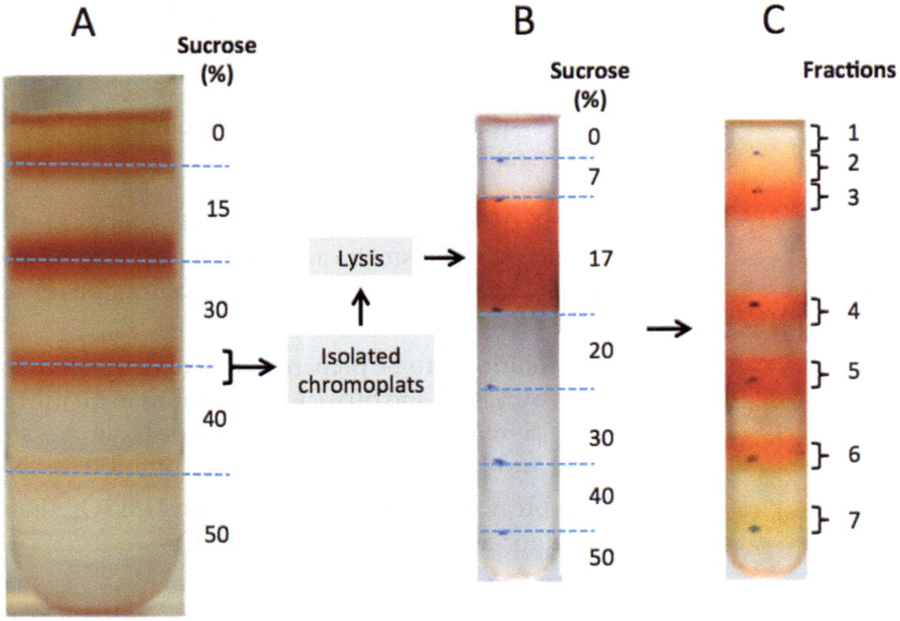

Fig. 2 Typical results obtained during the isolation of chromoplasts (A) and their fractionation (B and C) using ultracentrifugation in sucrose block gradients starting from 1 kg or ripe cherry tomato fruits. *See* **Note 14** for the main features of the isolated fractions. The relative intensity of the bands may differ when using fruits from other tomato varieties

3. Filter the homogenate through two layers of Miracloth (*see* **Note 6**).
4. Transfer the flow-through to centrifuge bottles (250 or 500 mL) and centrifuge at 5000 × *g* for 10 min, using rotors JA-10, JA-14, JLA 16.250 or equivalents.

5. Discard the supernatant and gently resuspend each pellet in 3 mL of buffer B using a 1 mL micropipette with the tip trimmed to approximately 2–3 mm of diameter to reduce chromoplast damage during handling (*see* **Note 7**).
6. Pool the resuspended pellets into a 30 mL polypropylene centrifuge tube using a 1 mL micropipette with the tip opening trimmed to approximately 2–3 mm of diameter. Complete to 25 mL with buffer B.
7. Centrifuge at 5000 × *g* for 10 min in rotors JA-17, JA-25.50, or similar.
8. Carefully discard the supernatant by decantation and gently resuspend the chromoplasts in 4 mL of buffer B using a 1 mL micropipette with the tip trimmed to approximately 2–3 mm of diameter.
9. Carefully load the 4 mL chromoplast sample on top of the sucrose density blocks (50%, 40%, 30% and 15%—bottom to top) prepared in advance in a 38.5 mL polyallomer ultracentrifuge tube as indicated in Fig. 1a (*see* **Note 8**). Add buffer B as required to equilibrate the tubes before ultracentrifugation.
10. Centrifuge at 100,000 × *g* for 60 min in a Beckman swinging bucket SW 28 Ti rotor (or equivalent) using mild brake settings.
11. Carefully withdraw the chromoplasts present in the 30–40% interface (Fig. 2a) using a wide-bore plastic Pasteur pipette and transfer into a polypropylene 30 mL centrifuge tube (*see* **Note 9**).
12. Add one volume of buffer B and centrifuge at 5000 × *g* for 10 min.
13. Carefully remove the supernatant by decantation and remove the remaining buffer droplets using a 200 μL micropipette (*see* **Note 10**).
14. Continue with the chromoplast fractionation steps as indicated in Subheading 3.2.

3.2 Chromoplast Fractionation

1. Resuspend the pellet in 3 mL of lysis buffer using a 1 mL micropipette and complete the final volume to 6 mL. Add PMSF to 1 mM final concentration and mix with vigorous agitation (*see* **Note 11**).
2. Lyse chromoplasts by sonication keeping the tube in an ice-filled beaker. Three 30 s pulses with pauses of at least 15–20 s are recommended (*see* **Note 12**).
3. Remove remaining unbroken chromoplasts and debris by centrifugation at 5000 × *g* for 10 min.

4. Load 3 mL of the chromoplast lysate on top of the sucrose density blocks (50%, 40%, 30% and 20% solutions, bottom to top) prepared in advance in two 13.2 mL polyallomer ultracentrifuge tubes as indicated in Fig. 1b (**steps a** and **b**). Afterward, load the 7% and 0% sucrose gradient blocks on top of the chromoplast lysate as indicated in Fig. 1b (**step c**) (*see* **Note 8**).
5. Centrifuge at 220,000 × *g* for 3 h using a Beckman swinging bucket rotor SW41Ti (or equivalent), using minimal but non-null brake settings (*see* **Note 13**). A pattern of bands similar to that shown in Fig. 2c should be obtained (*see* **Note 14**).
6. Harvest fractions using a 200 μL micropipette with the tip trimmed to approximately 1–1.5 mm of diameter doing gentle circular movements over the surface of the band. Undertake harvesting sequentially from the upper fraction. The interband sucrose solutions can be discarded carefully with a 1 mL micropipette.
7. The isolated fractions can be analyzed immediately or stored at −20 °C or −80 °C, depending on the downstream use. Samples for proteomic studies can be stored at −20 °C for short-term usage and at −80 °C for longer-term storage. Since lycopene is very prone to degradation by epoxidation [19] samples required for carotenoid profiling should be protected from light, frozen in liquid nitrogen and stored at −80 °C for the shortest time needed. When possible, provide an anoxic atmosphere inside the tube (nitrogen is suggested) to further prevent carotenoid breakdown.

4 Notes

1. It is recommended to keep this buffer in constant stirring before use to ensure that PVPP remains in suspension at the time of use. This will enhance the effectiveness of polyphenol removal from the homogenate.
2. Sucrose solutions can alternatively be prepared in the following way: For 100 mL mix the amount of sucrose needed in each case (15 g, 30 g 40 g or 50 g) with 10 mL of 0.5 M Tris–HCl (pH 7.4) buffer and 100 μL of 1 M DTT. Complete to 100 mL with water and use immediately.
3. Sucrose solutions can be prepared using the alternative method indicated in **Note 2**.
4. Extractions from 0.5 to 2.5 kg of starting material have been successfully performed by scaling up or down the volume of the isolation buffers and the number and volume of centrifuge tubes. Chromoplast yield may differ depending on the tomato variety used.

5. A typical extraction starting from 1 kg of fruits can yield around 600–700 g of tissue, which will entail the use of about 1.2–1.4 L of buffer A.
6. For preparations starting from large amounts of fruit a set of several filter funnels can be used in parallel to minimize the handling time.
7. Note that the upper layer of the pellets can be readily resuspended, while the lower part is more compact. Full resuspension can be facilitated by covering the remnant of the pellets with a small volume of buffer B (about 3 mL) and waiting for a few minutes before continuing. Carry on with this procedure until all pellets are resuspended.
8. Loading of the sucrose density blocks can be done in two ways: (a) using a fine-bore glass Pasteur pipette or (b) using a peristaltic pump coupled to a capillary tube. The exact volume of each sucrose block can be dispensed in advance into separate assay tubes. (a) If using a Pasteur pipette, very gently let flow a narrow stream of the sucrose solution on top of the previous (denser) block. Dispense the solutions against the tube wall from a close distance to the surface. Start with the 50% sucrose block and continue in an ascending way to the 40, 30, and 15% sucrose blocks. (b) If using a peristaltic pump, the gradient is prepared in an inverse way. The pump tubes are connected to a long capillary tube that has to be placed in contact with the bottom of the centrifuge tube. The pump will be fed sequentially with the right volumes of the 15%, 30%, 40%, and 50% sucrose solutions. The gradient blocks will be thus built from the bottom to the top. When removing the capillary tube, do it slowly and ensuring that no air bubbles leave from the capillary (which would disturb the gradient).
9. This fraction contains fully developed chromoplasts [17]. To minimize cross-contamination it is strongly recommended to withdraw the fractions present at the 0–15% and 15–30% interfaces in advance. Harvest the chromoplasts while gently moving the tip over the surface of the band. This makes the harvesting more effective (otherwise, inter-band material can easily be taken up).
10. The chromoplast pellet is very unstable at this particular stage. To minimize chromoplast losses perform decantation of the supernatant immediately after removing the tube from the centrifuge by carefully pouring the supernatant in the direction opposed to the pellet.
11. Chromoplast resuspension is more effective if initially done into a smaller volume before completing to the full volume. This forces any chromoplast clumps to pass more frequently through the pipette tip opening, favoring their disaggregation.

The PMSF stock needs vigorous agitation for proper dissolution in a cold aqueous solvent. However, the next sonication step helps to increase its solubilization.

12. The equipment used in this protocol was the Branson Ultrasonic Cell Disruptor SFX550 equipped with a 3 mm tip. Sonicators may vary notably in their performance and lysis efficiency. So optimal sonication conditions have to be determined for every particular equipment. Use as a reference the power settings routinely used for the sonication of E. coli cells. Monitor the lysis by measuring the size of the pellet after successive sonication steps (a microcentrifuge can be used). In our experience, the pellet size reaches a plateau (almost invariant pellet size) after a few sonication pulses.
13. Tools for the conversion of centrifugation settings among Beckman rotors are available at https://www.beckman.com/centrifuges/rotors/calculator. Note that increasing centrifugation times may reduce band resolution.
14. Electron microscopy observations and preliminary proteomics and metabolomics analysis have shown the following main features for each one of the isolated fractions. Fraction 1: contains colorless plastoglobules, its protein content is extremely low and is relatively rich in carotenoid precursors (phytoene and phytofluene). Fraction 2: is enriched in plastoglobules and crystals. Has a low protein content and relatively high levels of β-carotene. Fraction 3: contains plastoglobules and crystals and has a low protein content. Is relatively enriched in lycopene and β-carotene. Fraction 4: contains plastoglobule-like particles and other membranous components. Shows the presence of lycopene and β-carotene. Fraction 5: Contains membranous and vesicle-like structures. Shows the presence of lycopene and β-carotene and is enriched in MGDG and DGDG. Inner-membrane proteins have been identified. Fraction 6: it contains membranous structures and has a high protein content; is enriched in carotenoid precursors (phytoene and phytofluene) and tocopherols; and contains relatively high levels of MGDG, DGDG, and phospholipids. Outer-membrane proteins have been identified. Fraction 7: it contains membranous components and vesicle-like structures and has a relatively high protein content.

Acknowledgments

We acknowledge the financial support of AGAUR-Generalitat de Catalunya (Grant 2017 SGR 710), the CERCA Programme of the Generalitat de Catalunya and the Severo Ochoa Programme for Centres of Excellence in R&D 2016–2019 to CRAG (SEV-2015-

0533). AB is member of the Spanish Carotenoid Network (CaRed) funded by the Spanish Ministry of Economy and Competitiveness (Grants BIO2015-71703-REDT and BIO2017-90877-REDT).

References

1. Egea I, Barsan C, Bian W, Purgatto E, Latché A, Chervin C, Bouzayen M, Pech JC (2010) Chromoplast differentiation: current status and perspectives. Plant Cell Physiol 51:1601–1611
2. Li L, Yuan H (2013) Chromoplast biogenesis and carotenoid accumulation. Arch Biochem Biophys 539:102–109
3. Li L, Yuan H, Zeng Y, Xu Q (2016) Plastids and carotenoid accumulation. Subcell Biochem 79:273–293
4. Vishnevetsky M, Ovadis M, Vainstein A (1999) Carotenoid sequestration in plants: the role of carotenoid-associated proteins. Trends Plant Sci 4:232–235
5. Rodriguez-Concepcion M, Avalos J, Bonet ML, Boronat A, Gomez-Gomez L, Hornero-Mendez D, Limon MC, Meléndez-Martínez AJ, Olmedilla-Alonso B, Palou A, Ribot J, Rodrigo MJ, Zacarias L, Zhu C (2018) A global perspective on carotenoids: metabolism, biotechnology, and benefits for nutrition and health. Prog Lipid Res 70:62–93
6. Lado J, Zacarías L, Rodrigo MJ (2016) Regulation of carotenoid biosynthesis during fruit development. Subcell Biochem 79:161–198
7. Sun T, Yuan H, Cao H, Yazdani M, Tadmor Y, Li L (2018) Carotenoid metabolism in plants: the role of plastids. Mol Plant 11:58–74
8. Bathgate B, Purton ME, Grierson D, Goodenough PW (1985) Plastid changes during the conversion of chloroplasts to chromoplasts in ripening tomatoes. Planta 165:197–204
9. Cheung AY, McNellis T, Piekos B (1993) Maintenance of chloroplast components during chromoplast differentiation in the tomato mutant green flesh. Plant Physiol 101:1223–1229
10. Rosso SW (1968) The ultrastructure of chromoplast development in red tomatoes. J Ultrastruct Res 25:307–322
11. Van Wijk KJ, Kessler F (2017) Plastoglobuli: plastid microcompartments with integrated functions in metabolism, plastid developmental transitions, and environmental adaptation. Annu Rev Plant Biol 68:253–289
12. Ytterberg AJ, Peltier J-B, van Wijk KJ (2006) Protein profiling of plastoglobules in chloroplasts and chromoplasts. A surprising site for differential accumulation of metabolic enzymes. Plant Physiol 140:984–997
13. Zeng Y, Du J, Wang L, Pan Z, Xu Q, Xiao S, Deng X (2015) A comprehensive analysis of chromoplast differentiation reveals complex protein changes associated with plastoglobule biogenesis and remodeling of protein systems in sweet orange flesh. Plant Physiol 168:1648–1665
14. Barsan C, Sanchez-Bel P, Rombaldi C, Egea I, Rossignol M, Kuntz M, Zouine M, Latché A, Bouzayen M, Pech JC (2010) Characteristics of the tomato chromoplast revealed by proteomic analysis. J Exp Bot 61:2413–2431
15. Suzuki M, Takahashi S, Kondo T, Dohra H, Ito Y, Kiriiwa Y, Hayashi M, Kamiya S, Kato M, Fujiwara M, Fukao Y, Kobayashi M, Nagata N, Motohashi R (2015) Plastid proteomic analysis in tomato fruit development. PLoS One 10(9): e0137266
16. Vidi PA, Kanwischer M, Baginsky S, Austin JR, Csucs G, Dörmann P, Kessler F, Bréhélin C (2006) Tocopherol cyclase (VTE1) localization and vitamin E accumulation in chloroplast plastoglobule lipoprotein particles. J Biol Chem 281:11225–11234
17. Angaman DM, Petrizzo R, Hernández-Gras F, Romero-Segura C, Pateraki I, Busquets M, Boronat A (2012) Precursor uptake assays and metabolic analyses in isolated tomato fruit chromoplasts. Plant Methods 8:1
18. Hernández-Gras F, Petrizzo P, Pateraki I, Renato M, Angaman M, Azcón-Bieto J, Boronat A (2014) Isolation of tomato fruit chromoplasts and determination of ATP levels. Bio-Protocol 4(15):e1192. http://www.bio-protocol.org/e1192
19. Rodríguez-Amaya DB (2010) Quantitative analysis, in vitro assessment of bioavailability and antioxidant activity of food carotenoids—a review. J Food Compos Anal 23:726–740

Chapter 15

Quantification of Strigolactones

Carlos Rial, Rosa M. Varela, José M. G. Molinillo, Alexandra G. Durán, and Francisco A. Macías

Abstract

Strigolactones (SLs) are a family of natural products produced by the plants as shoot branching factors and responsible for the induction of hyphal branching in arbuscular mycorrhizal (AM) fungi. They have been also used by parasitic plant seeds as stimulators of their germination as a strategy to ensure the presence of a host in the environment. For all these bioactivities, SLs have kept the attention of the researchers in the last years, increasing the number of published papers, and have opened new areas of research in the multiple roles that they play in the rhizosphere and as plant hormones. However, the low amount of them produced by plants and their rapid degradability make it crucial to develop fast analytical methods with very low limits of quantification. Herein, it is described a protocol for the development of an LC-MS/MS method for the quantification of SLs, using GR24 as IS, in roots exudates and extracts.

Key words Strigolactones, Quantification method, LC-MS/MS, Carotenoids, Parasitic weeds, Arbuscular mycorrhizal fungi

1 Introduction

Strigolactones (SLs) are a family of compounds derived from carotenoids, [1, 2] which were firstly identified as stimulators of the germination of parasitic plants seeds [3, 4]. Parasitic plants are some of the most damaging agricultural pests, due to they reduce the yield of very important crops such as, maize, rice, legumes, tomato, sunflower, etc. [5, 6]. In 2005, they were also identified as branching factor for arbuscular mycorrhizal (AM) fungi [7, 8] and in 2008 two research groups described simultaneously that SLs have an endogenous role as plant hormone, regulating shoot branching [9, 10]. Due to all these results, SLs have kept the attention of the researchers in the last years, increasing the number of published papers, and opening new areas of research in the multiple roles that they play in the rhizosphere and as plant hormones [11].

Since the discovery of strigol as the first SLs isolated, 23 canonical and 6 noncanonical SLs have been isolated (Fig. 1). However,

Manuel Rodríguez-Concepción and Ralf Welsch (eds.), *Plant and Food Carotenoids: Methods and Protocols*, Methods in Molecular Biology, vol. 2083, https://doi.org/10.1007/978-1-4939-9952-1_15,

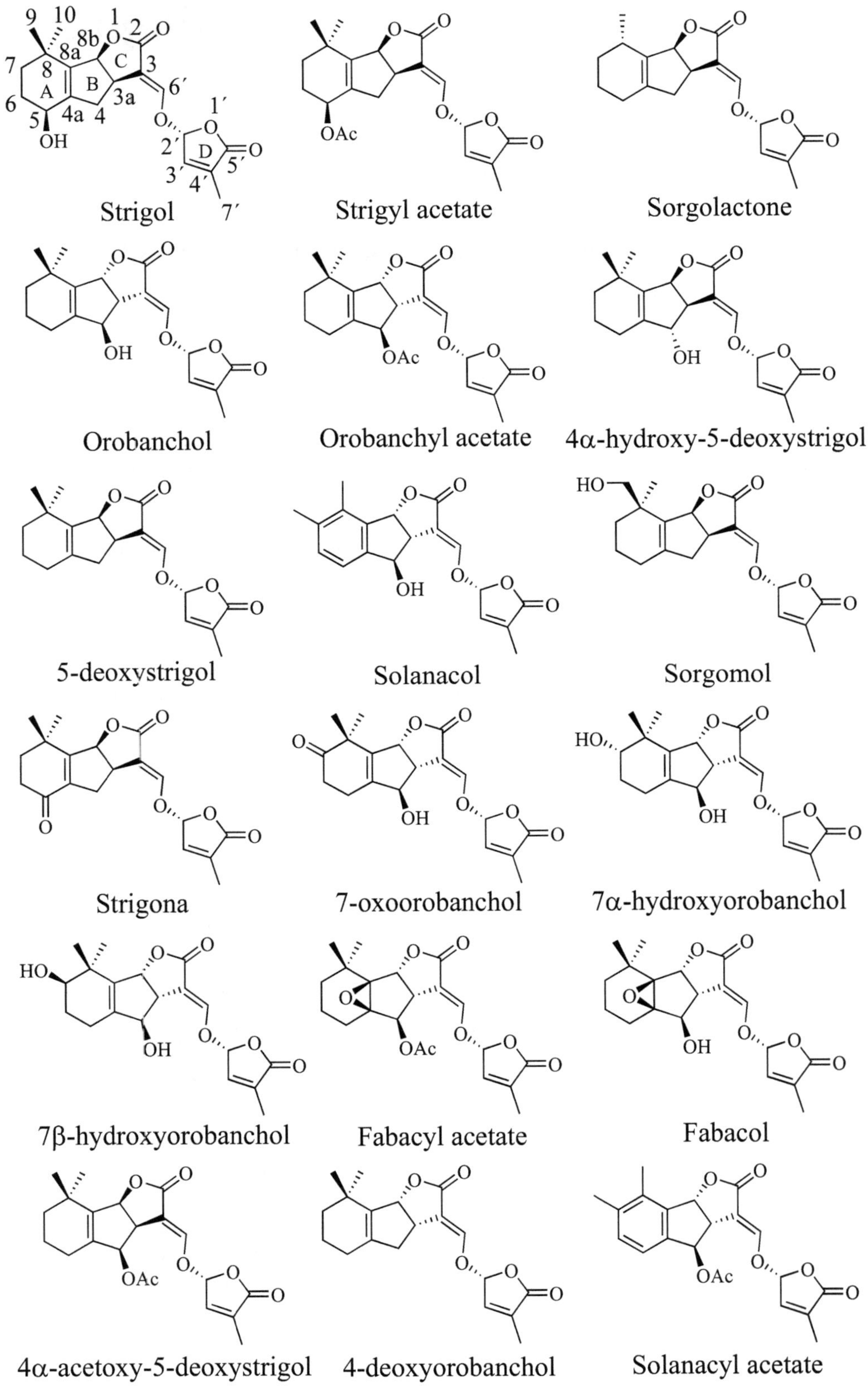

Fig. 1 Structure of the natural canonical and noncanonical SLs

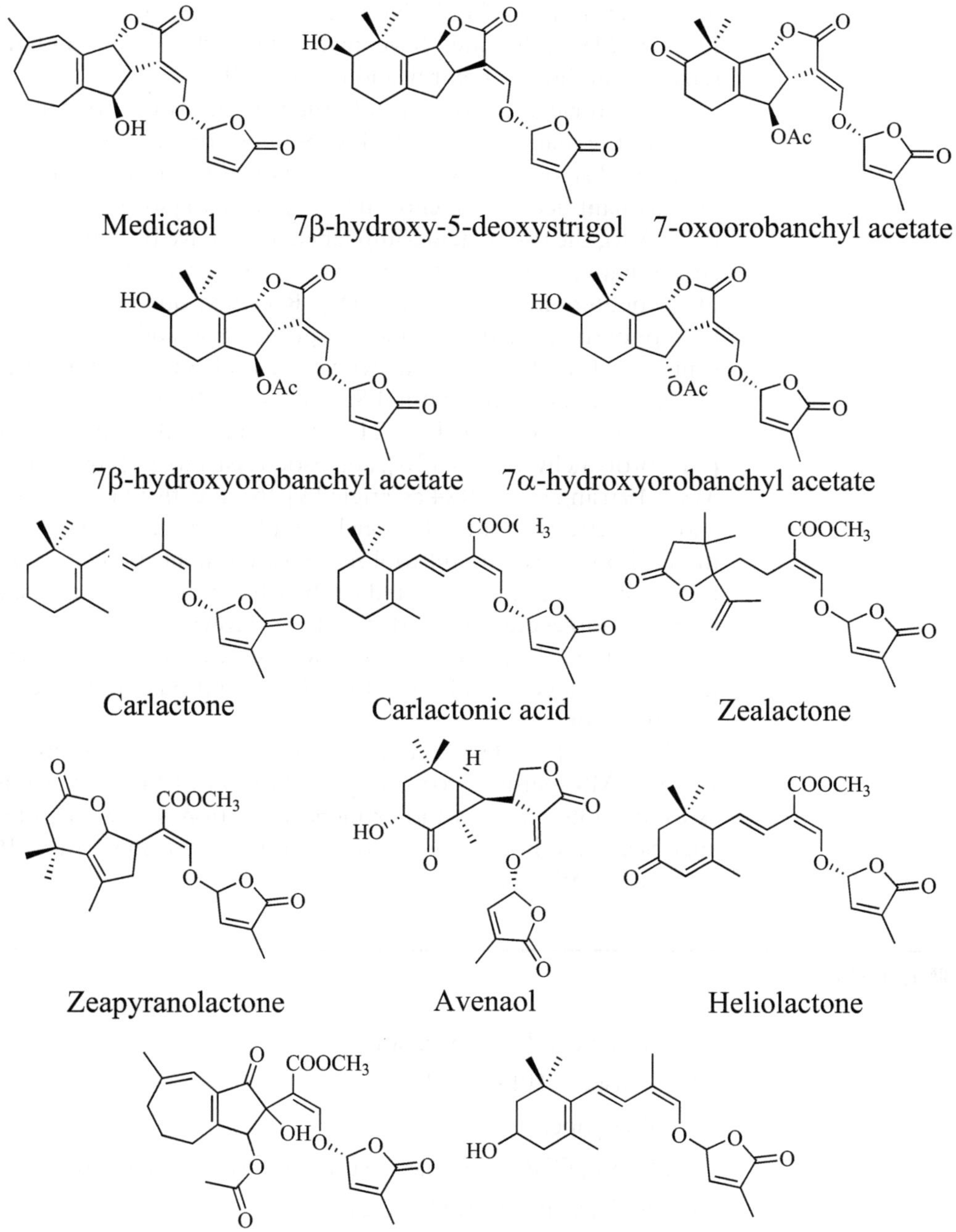

Fig. 1 (continued)

the low amount of stimulants produced by plants, together with their rapid degradability make it crucial to develop fast analytical methods with very low limits of quantification. In the last years several quantification methods have been used. First, the use of the parasitic seed germination bioassay was an indirect method for the

SLs quantification [12]. Nowadays, the improvement of the LC-MS/MS techniques have made them the fastest and most sensitive method to identify and quantify SLs.

Nevertheless, almost all published methods do not take into account the matrix effect, which should be corrected by the use of an internal standard (IS) [1, 13–16]. For instance, Jamil and coworkers quantified orobanchol and 2′-epi-5-deoxystrigol using an IS [17]. Also, the use of deuterium labeled SLs have been proposed as an alternative, however, it is limited because of the limited availability of deuterium labeled SLs [18]. Recently, our research group have published the development, optimization and validation of a simple, fast, and sensitive analytical method for the simultaneous quantification of seven natural SLs (7-oxoorobanchyl acetate [19], solanacol [20], orobanchol [21], strigol [22, 23], fabacyl acetate [24], orobanchyl acetate [25], and 5-deoxystrigol [7]) by UHPLC-MS/MS using (±)-GR24 as single IS [26] directly in root exudates and extracts, without additional sample preparation [27]. This method provides a useful tool for research in all the fields related to SLs, both for studies related to their function as hormones, and signaling molecules in the rhizosphere. However, due to the low availability of SLs standards just few research groups in the world have developed validated analytical methods for the SLs quantification.

Herein, it is described a protocol for the development of LC-MS/MS methods for the quantification of SLs, using GR24 as IS, in roots exudates and extracts. Validation of the quantitative analytical method was carried out following the ICH recommendations [28].

2 Materials

1. Distilled and ultrapure water.
2. Acetone (HPLC grade).
3. Filter paper.
4. 200 mg C18 solid phase extraction (SPE) cartridges.
5. Ethyl acetate (HPLC grade).
6. PTFE syringe filter (0.22 μm).
7. MeOH (LC-MS grade or HPLC grade filtered by a 0.22 μm membrane filter).
8. Solvent A: water, 0.1% formic acid (*see* **Note 1**).
9. Solvent B: MeOH, 0.1% formic acid.

3 Methods

3.1 Sample Collection

3.1.1 Exudate Collection and Purification

1. Once the plants are between 4 and 6 weeks old at least, rinse the pots with distilled water to collect between 100 and 200 mL of exudate from at least three different plants (Fig. 2) (*see* **Note 2**).
2. Filter the exudate by filter paper in a Buchner funnel with vacuum (Fig. 2).
3. Activate C18 SPE cartridges with 3 mL of MeOH.
4. Wash it with 6 mL of water.
5. Load the exudate into the cartridge.
6. Wash it with 3 mL of water and 3 mL of water:acetone 60:40 (v/v).
7. Elute SLs with 3 mL of acetone.
8. Evaporate the acetone in a rotary evaporator and store the sample at −80 °C until their analysis.

3.1.2 Root Extract

1. Once roots are collected, dry them in an oven at 40 °C for 48 h.
2. Powder the dry roots in an industrial mill.
3. Extract 100 mg of roots with 1 mL of ethyl acetate in an ultrasonic bath for 10 min.
4. Centrifuge the samples for 10 min at 5000 rpm.

Fig. 2 Exudate collection and filtration by rinsing the pots with distilled water

5. Transfer organic phase to a glass vial.
6. Repeat **steps 3–5** with the same roots three times and join the organic phases.
7. Evaporate the organic solvent in a rotary evaporator and store the samples at −80 °C until their analysis.

3.2 LC-MS/MS Analysis

This method was optimized using a Bruker EVOQ Triple Quadrupole Mass Spectrometer, an ACE Excel 1.7 column (Advanced Chromatography Technologies Ltd., Aberdeen, Scotland) and following the ICH recommendations [28]. The parameters described herein should be optimized for others mass spectrometers and columns.

3.2.1 LC-MS/MS Method Calibration and Optimization

1. Dissolve the SL with MeOH to achieve a concentration of 10 mg/L.
2. By direct infusion on the mass spectrometer in positive mode, determine the most intense SL adduct ($[M + H]^+$ or $[M + Na]^+$), their most stable fragments and the collision energy to achieve the maximum signal for them (*see* **Note 3**). When more than one fragments are obtained, the most intense is used as quantifier ion and the second as qualifier ion to improve the identification of the SLs (*see* **Notes 4** and **5**).
3. Prepare a diluted stock solution of 100 μg/L for each SL with 10 μg/L of GR24 in all of them (it is used as internal standard (IS)).
4. The calibration curve of each SL is prepared by serial dilution of the 10 mg/L stock solution prepared above with a solution of 10 μg/L of GR24 in MeOH.
5. Filter all the samples through a PTFE syringe filter (0.22 μm) (*see* **Note 6**).
6. Inject 5 μL of each level of the calibration curve onto the UHPLC-MS/MS three times with an electrospray ionization source (ESI) in positive mode and a C18 (100 × 2.1 mm, 1.7 μm particle size) UHPLC column maintained at 40 °C. Use as mobile phase solvent A and B with the flow rate set to 0.3 mL/min, with linear gradient as follows: 0–0.5 min, 50% B; 0.5–5 min, to 100% B; 5–7 min, 100% B; 7–7.5 min, to 50% B; 7.5–10.5 min, 50% B. Set the instrument parameters as follows: spray voltage +4500 V, cone temperature 300 °C, cone gas flow 15 psi, heated probe temperature 400 °C, heated probe gas flow 15 psi, nebulizer gas flow 55 psi, and collision pressure 2.0 mTorr (*see* **Note 7**).
7. Construct each calibration curve by plotting the peak area ratio (y = area of the standard/area of the IS) versus the ratio of their concentrations (x = concentration of the calibration curve level/10 μg/L) and fitting the curves to a linear function

with a weight of 1/nx ($R^2 > 0.99$), being "n" the calibration level.

8. Calculate the quantification (LOQ) and detection (LOD) limits for each SL, analyzing nine replicates and considering ten and three times the standard deviation of baseline noise respectively.
9. Study the precision of the method by analyze a sample with all the SLs in an intra- and interday assay ($n = 9$). Repeatability analyzing nine replicates of the same sample in the same day and the intermediate precisions injecting the same sample on three different days. They are expressed using the relative standard deviation percentage (RSD %) of the area and the retention times and they are calculated as (relative standard deviation/mean) × 100.

3.2.2 Analysis of Samples

1. Dissolve the exudates or the roots extracts obtained above with MeOH to achieve a ratio of 1/1 g/L.
2. Filter all the samples through a PTFE syringe filter (0.22 μm).
3. Transfer 1 ml to a 1.5 ml glass vial for LC-MS.
4. Add the IS GR24 to each sample to give a final concentration of 10 μg/L.
5. Inject 5 μL of each sample onto the UHPLC-MS/MS three times using the same parameters described above.
6. Calculate the concentration of the SLs present in each sample introducing the ratio of the area of the SL/area of IS in the calibration curves.
7. Calculate the RSD % of the concentration for each SL in each sample.

4 Notes

1. MeOH is used for the analysis because it helps to the proton adduct formation. However, SLs are more stable on acetonitrile (ACN). For this reason, it is recommended prepared a stock solution of SLs at 1000 or 100 mg/L in ACN.
2. Hydroponic or semihydroponic culture is recommended to reduce the interferences produced by the soil, for instance, use as support a mixture of vermiculite, sand, and clay (1:1:1) and half-strength Hoaglands as nutrient solution.
3. In the cases where SLs amounts are too small for the direct infusion in the mass spectrometer, the optimization of the mass transitions could be done by using the product scan mode in the mass spectrometer by injection of 5 μL aliquot in the LC-MS/MS. In this mode the mass spectrometer filters the

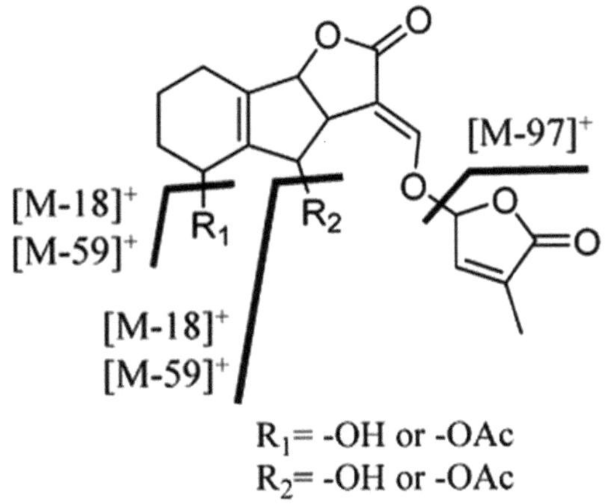

Fig. 3 Fragmentation pattern of SLs

precursor ion in the Q1 and study the fragments obtained in the collision cell by a full scan in the Q3. The collision energy for each transition must be optimized.

4. The use of the proton adduct as the precursor ion is always recommended (when it is produced), even if the sodium adduct is more intense, for two reasons. Firstly, the identification of the SL is improved and on the other hand the sodium adduct are normally more stable being the energy required for the fragmentation higher.
5. SLs fragments expected could be summarized as follow: Firstly, the loss of a water molecule (−18) or the loss of an acetate group (−59), when a hydroxyl or an acetate group is present in the structure, and then the loss of the D-ring moiety (−97) (Fig. 3).
6. Ambar vials are recommended to preserve SLs of the light degradation.
7. Set the autosampler with a temperature as lower than possible to preserve the samples.

Acknowledgments

This research was supported by the Ministerio de Economía, Industria y Competitividad (MINEICO) (Project AGL2017-88-083-R).

References

1. López-Ráez JA, Charnikhova T, Gómez-Roldán V et al (2008) Tomato strigolactones are derived from carotenoids and their biosynthesis is promoted by phosphate starvation. New Phytol 178:863–874
2. Delaux PM, Xie X, Timme RE et al (2012) Origin of strigolactones in the green lineage. New Phytol 195:857–871
3. Siame B, Weerasuriya Y, Wood K et al (1993) Isolation of strigol, a germination stimulant for *Striga asiatica*. J Agric Food Chem 41:1486–1491
4. Butler LG (1995) Chemical communication between the parasitic weed *Striga* and its crop host. A new dimension in allelochemistry. In: Inderjit KM, Dakshini M, Enhelling FA (eds) Allelopathy, organisms, processes and applications. American Chemical Society, Washington, DC, pp 158–166
5. Shen H, Ye W, Hong L et al (2006) Progress in parasitic plant biology: host selection and nutrient transfer. Plant Biol 8:175–185
6. Press MC, Scholes JD, Riches CR (2001) Current status and future prospects for management of parasitic weeds (*Striga* and *Orobanche*). World's Worst Weeds Proc:71–88
7. Akiyama K, Matsuzaki K, Hayashi H (2005) Plant sesquiterpenes induce hyphal branching in arbuscular mycorrhizal fungi. Nature 435:824–827
8. Hayashi H, Akiyama K (2008) Strigolactones: host recognition signals for arbuscular mycorrhizal fungi. New Phytol 2008:55–61
9. Gomez-Roldan V, Fermas S, Brewer PB et al (2008) Strigolactone inhibition of shoot branching. Nature 455:189–194
10. Umehara M, Hanada A, Yoshida S et al (2008) Inhibition of shoot branching by new terpenoid plant hormones. Nature 455:195–200
11. Screpanti C, Yoneyama K, Bouwmeester HJ (2016) Strigolactones and parasitic weed management 50 years after the discovery of the first natural strigolactone strigol: status and outlook. Pest Manag Sci 72:2013–2015
12. Yoneyama K, Arakawa R, Ishimoto K et al (2015) Difference in striga-susceptibility is reflected in strigolactone secretion profile, but not in compatibility and host preference in arbuscular mycorrhizal symbiosis in two maize cultivars. New Phytol 206:983–989
13. Sato D, Awad AA, Chae SH et al (2003) Analysis of strigolactones, germination stimulants for *Striga* and *Orobanche*, by high-performance liquid chromatography/tandem mass spectrometry. J Agric Food Chem 51:1162–1168
14. Sato D, A a A, Takeuchi Y et al (2005) Confirmation and quantification of strigolactones, germination stimulants for root parasitic plants *Striga* and *Orobanche*, produced by cotton. Biosci Biotechnol Biochem 69:98–102
15. Yoneyama K, Xie X, Sekimoto H et al (2008) Strigolactones, host recognition signals for root parasitic plants and arbuscular mycorrhizal fungi, from Fabaceae plants. New Phytol 179:484–494
16. Yoneyama K, Yoneyama K, Takeuchi Y et al (2007) Phosphorus deficiency in red clover promotes exudation of orobanchol, the signal for mycorrhizal symbionts and germination stimulant for root parasites. Planta 225:1031–1038
17. Jamil M, Charnikhova T, Cardoso C et al (2011) Quantification of the relationship between strigolactones and *Striga hermonthica* infection in rice under varying levels of nitrogen and phosphorus. Weed Res 51:373–385
18. Boutet-Mercey S, Perreau F, Roux A et al (2018) Validated method for Strigolactone quantification by ultra high-performance liquid chromatography – electrospray ionisation tandem mass spectrometry using novel deuterium Labelled standards. Phytochem Anal 29:59–68
19. Xie X, Yoneyama K, Kurita J et al (2009) 7-Oxoorobanchyl acetate and 7-Oxoorobanchol as germination stimulants for root parasitic plants from flax (*Linum usitatissimum*). Biosci Biotechnol Biochem 73:1367–1370
20. Xie X, Kusumoto D, Takeuchi Y et al (2007) 2′-Epi-orobanchol and Solanacol, two unique Strigolactones, germination stimulants for root parasitic weeds, produced by tobacco. J Agric Food Chem 55:8067–8072
21. Yokota T, Sakai H, Okuno K et al (1998) Alectrol and Orobanchol, germination stimulants for *Orobanche minor*, from its host red clover. Phytochemistry 49:1967–1973
22. Cook CE, Whichard LP, Turner B et al (1966) Germination of Witchweed (*Striga Lutea* Lour)—isolation and properties of a potent stimulant. Science 154:1189–1190
23. Cook C, Whichard L, Wall M et al (1972) Germination stimulants. II. Structure of strigol, a potent seed germination stimulant for witchweed (*Striga lutea*). J Am Chem Soc 94:6198–6199
24. Xie X, Yoneyama K, Harada Y et al (2009) Fabacyl acetate, a germination stimulant for root parasitic plants from *Pisum sativum*. Phytochemistry 70:211–215

25. Xie X, Yoneyama K, Kusumoto D et al (2008) Isolation and identification of alectrol as (+)-orobanchyl acetate, a germination stimulant for root parasitic plants. Phytochemistry 69:427–431
26. Zwanenburg B, Pospíšil T (2013) Structure and activity of strigolactones: new plant hormones with a rich future. Mol Plant 6:38–62
27. Rial C, Varela RM, Molinillo JMG et al (2019) A new UHPLC-MS/MS method for the direct determination of strigolactones in root exudates and extracts. Phytochem Anal 30:110–116
28. ICH (2005) ICH Topic Q2 (R1) Validation of analytical procedures: text and methodology, Geneva

Chapter 16

Carotenoid and Apocarotenoid Analysis by SFE-SFC-QqQ/MS

Daniele Giuffrida, Mariosimone Zoccali, and Luigi Mondello

Abstract

Carotenoids and carotenoids oxidative and enzymatic cleavage products called apocarotenoids are very important bioactive molecules in plants and humans, with different biological functions. Both central and noncentral carotenoid cleavage products have been reported to occur in food and in humans, where they may act as bioactive molecules with functions that were previously attributed to the parent carotenoid. However, relatively few studies are available in the literature on the apocarotenoid occurrence in food and biological fluids which were mainly based on liquid chromatographic separation approaches and even fewer reports are available on the carotenoid and apocarotenoid separation by a direct online supercritical fluid extraction-supercritical fluid chromatography with triple-quadrupole mass spectrometry detection (SFE-SFC-QqQ/MS) methodology. In comparison with offline approaches the online nature of the system drastically reduces the extraction time required in traditional solid/liquid extraction, which may require a few hours. Moreover, it reduces the analysis run time, as well as the risks of sample contamination and the possible occurrence of operator errors, improves run-to-run precision, and enables the setting of batch-type applications. The purpose of this contribution was to provide an updated description of the SFE-SFC-QqQ/MS methodology applied to carotenoid and apocarotenoid analysis in various matrices, although with a particular focus on the apocarotenoid applications.

Key words Carotenoids, Apocarotenoids, Supercritical fluid extraction, Supercritical fluid chromatography, Mass spectrometry, SFE-SFC-QqQ/MS

1 Introduction

Analytical methods for detecting carotenoids and their derivatives in food and biological samples are mainly based on liquid chromatography approaches with relatively long analytical times and considerable organic solvent consumption [1–4].

Supercritical carbon dioxide (CO_2) offers peculiar features, like low viscosity, high density, and high diffusion coefficient that make it suitable for both the carotenoid supercritical fluid extraction (SFE) and chromatography (SFC). Supercritical CO_2 is considered particularly suitable for carotenoid extraction and separation

Manuel Rodríguez-Concepción and Ralf Welsch (eds.), *Plant and Food Carotenoids: Methods and Protocols*, Methods in Molecular Biology, vol. 2083, https://doi.org/10.1007/978-1-4939-9952-1_16,

because of its low polarity. In both SFE and SFC quite often a proportion of an organic solvent is added to the mobile phase as modifier in order to widen the affinity of the mobile phase for the different compounds. Moreover little variations in the density of the fluid are achieved by small changes in pressure or temperature which can further ameliorate the extraction or separation. Moreover, additional benefits in SFC compared to LC, are the use of a much more ecological mobile phase with the reduction of organic solvent utilization and costs.

Although recently, supercritical fluid chromatography (SFC) coupled to mass spectrometry has gained attention as a green, fast and useful technology applied to the carotenoid analysis [5–11], and reports are available on the supercritical fluid extraction of carotenoids [12–16], only three reports are available in the literature on the direct online extraction and determination of carotenoids and apocarotenoids by a supercritical fluid extraction–supercritical fluid chromatography–mass spectrometry (SFE-SFC-MS) methodology in food samples and biological fluids [17–19]. The use of a triple-quadrupole mass spectrometer detector (QqQ/MS) online with the SFE-SFC system is mandatory especially for the detection and identification of apocarotenoids. This is caused by their usual very low abundance in food matrices (usually in the order of 1–2% of the parent carotenoid) and the possibility to perform multiple reaction monitoring (MRM) experiments which greatly help in compounds identifications. Moreover, usually, in the carotenoid field the mass spectrometer is operated using an atmospheric pressure chemical ionization (APCI) source with both the positive and negative ionization modes. The purpose of this contribution was to provide an updated description of the SFE-SFC-QqQ/MS methodology applied to the carotenoid and apocarotenoid analysis in various matrices, although with a particular focus on apocarotenoid applications.

2 Materials

1. Carbon dioxide (CO_2), purity 99.9% or more.
2. Methanol (MeOH) and chloroform ($CHCl_3$), both LC-MS grade.
3. 0.2 mL extraction vessel
4. Digital homogenizer.

2.1 Standards

1. Carotenoid commercial standards: β-carotene, zeaxanthin, lutein, β-cryptoxanthin, capsanthin, capsorubin, β-apo-8′-carotenal, and physalein. Standards are prepared at a concentration level of 10,000 mg L^{-1} in a solution of 90% MeOH and 10% $CHCl_3$ and stored in dark vials at −20 °C.

2. Intermediate stock internal standards: All intermediate internal standards are prepared by diluting the commercial stock in MeOH, and stored in dark vials at −20 °C.
3. Apocarotenoid standards were also generated by oxidative cleavage of the parent carotenoid by potassium permanganate following the methodology reported by Rodriguez and Rodriguez-Amaya [20]. A series of β-apo-carotenals were generated by oxidation of β-carotene, a series of apo-zeaxanthinals and apo-capsorubinals were generated by oxidation of zeaxanthin and capsorubin, respectively, and a series of ε-apo-luteinals were generated by oxidation of lutein. Briefly, in a round bottom amber vial, 1.0 M equivalent of the parent carotenoid was dissolved in 1 mL of $CHCl_3$ and treated with $KMnO_4$ (2.6 M equivalents) at room temperature for 12 h. When the reaction was complete, the formed apocarotenoid mixture was filtered through celite 535 coarse (Fluka) to remove MnO_2 and the filtrate was washed several times with water, dried over Na_2SO_4 and concentrated in controlled vacuum.
4. Adsorbent powder "Miyazaki Hydro-Protect," Patented in Japan no. 3645552 (Shimadzu Corporation, Kyoto, Japan).

2.2 Instrumentation for the SFE-SFC-MS Analysis

The SFE-SFC-MS analyses were performed on a Shimadzu Nexera UC system (Shimadzu, Japan), consisting of a CBM-20A controller, an SFE-30A module for supercritical fluid extraction, two LC-20ADXR dual-plunger parallel-flow pumps, an LC-30ADSF CO_2 pump, two SFC-30A back pressure regulator, a DGU degasser, a CTO-20 AC column oven, an SIL-30 AC autosampler, an LCMS-8050 mass spectrometer equipped with an APCI source. The entire system was controlled by the LabSolution ver. 5.8.

2.3 SFE-SFC-QqQ MS Solutions

1. Mobile Phase A: CO_2.
2. Mobile Phase B: MeOH.
3. Makeup Pump Mobile Phase: MeOH.
4. Wash 1: CO_2.
5. Wash 2: MeOH.

2.4 Supplies

Ascentis Express fused-core C30 column, (150 mm × 4.6 mm × 2.7 μm d.p., Merck Life Science (Merck KGaA, Darmstadt, Germany) or equivalent.

3 Methods

All procedures are carried out at room temperature; allow all standards and samples to equilibrate to room temperature before using.

1: Grind the sample — 2: Mix the sample with the adsorbent powder — 3: Put the mix into the extraction vessel

Fig. 1 The main three different steps involved in the solid sample preparation

3.1 Sample Preparation

1. 1 g of solid sample was ground by using the digital homogenizer (*see* **Note 1**).
2. The grinded sample (10 mg) was homogenized with an adsorbent powder (1 g) "Miyazaki Hydro-Protect". The sample homogenization using the digital homogenizer reduced the particle size to almost pseudo-liquid material, which was then mixed with the adsorbent material to remove the sample water content thus generating granules having dimension below 1 mm of diameter.
3. A 0.2 mL extraction vessel was used (the inner diameter of the extraction chamber was 6 mm and the length was 12 mm), loaded with 100 mg of sample/adsorbent. All the three steps described are shown in Fig. 1.
4. Liquid samples (10 μL) were directly loaded into the extraction vessel.

3.2 SFE-SFC-QqQ MS Analysis

1. The SFC-SFE-QqQ MS system scheme consists of different configurations in the three modes: (1) the static extraction mode, (2) the dynamic extraction mode or transfer mode, and (3) the analysis mode (Fig. 2). In order to better understand the operations of the instrument a brief description is reported here. During the static extraction mode (Fig. 2a), the CO_2/modifier flow is split by the use a T-Union through the analytical column and inside the extraction vessel in order to pressurize it (*see* **Note 2**). During the dynamic extraction mode or transfer mode (Fig. 2b) the CO_2/modifier flows continuously through the extraction vessel in opposite direction (from

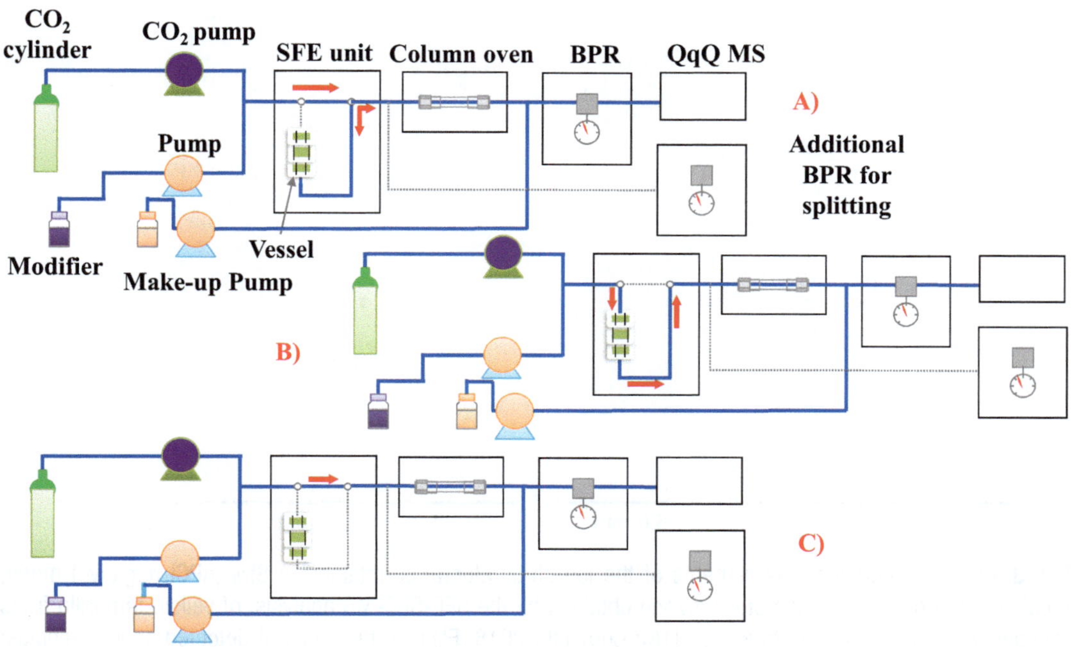

Fig. 2 Scheme of the system: (**a**) Static extraction mode, (**b**) Dynamic extraction mode, and (**c**) Analysis mode

the top to the bottom) and carries the extracted analytes into the analytical column. Of course, during this step also a dynamic extraction takes place (*see* **Note 3**). Finally, during the analysis mode (Fig. 3c), the mobile phase flows only through the analytical column.

2. SFE-SFC parameters are shown in Table 1; the total extraction time was 4 min (3 min static extraction and 1 min dynamic extraction) with a flow rate of 2 mL min^{-1} thus consuming only 0.5 mL of MeOH. Regarding the chromatography side, the same flow rate was used with a gradient increasing up to 40% of MeOH in 8 min, with an initial isocratic step of 2 min in order to focus the extracted analytes on the top of the analytical column. Another important parameter to take care of is the amount of makeup solvent, in this case also MeOH (0.5 mL min^{-1}) (*see* **Note 4**).

   The QqQ MS (triple-quadrupole mass detector) parameters are shown in Table 2, and apocarotenoid-specific parameters are shown in Table 3. The QqQ MS detector works in both Atmospheric Pressure Chemical Ionization (APCI) positive and negative ionization modes and SCAN, SIM (Selective Ion Monitoring), and MRM (Multiple Reaction Monitoring) modes in the same run. By using this approach is possible to perform both targeted and untargeted analysis, allowing also subsequent data treatment to look for more analytes. The transitions used in the MS/MS experiments were selected on

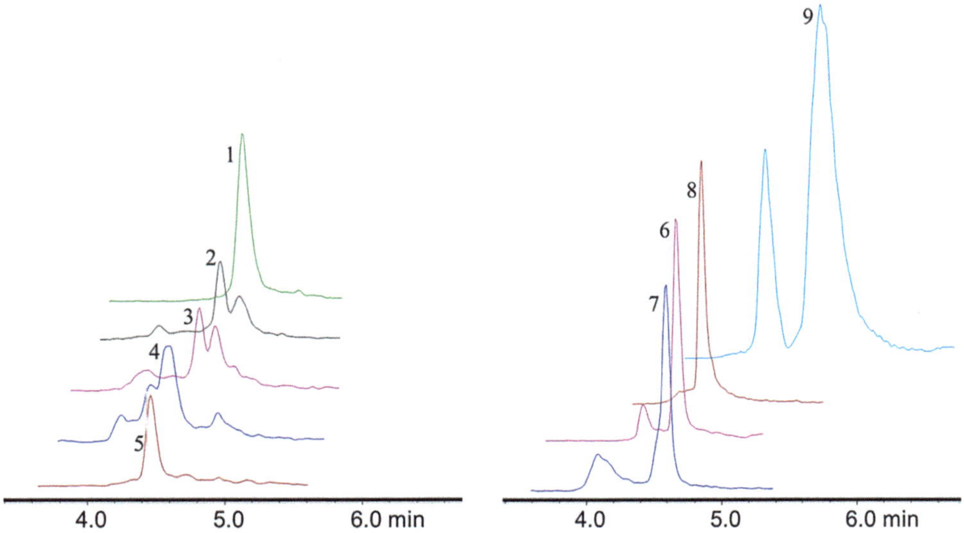

Fig. 3 The corresponding enlargements of the ion chromatograms obtained in SIM (APCI negative) mode, relative to the detected apocarotenoid zone obtained by the SFE-SFC-MS analysis, of yellow tamarillo fruits (reproduced with permission from Ref. [18]; Copyright 2018, Elsevier Inc.). List of detected apocarotenoids: (1) Apo-8′-Zeaxanthinal; (2) apo-10′-zeaxanthinal; (3) apo-12′-zeaxanthinal; (4) apo-14′-zeaxanthinal; (5) apo-15′-zeaxanthinal; (6) apo-12′-carotenal; (7) apo-14′-carotenal; (8) apo-10′-zeaxanthinal-c4:0; (9) apo-8′-zeaxanthinal-C12:0

Table 1
SFE-SFC conditions

SFE						
	Flow ($mL\ min^{-1}$)	**Time (min)**	**CO_2 (%)**	**MeOH (%)**	**Extraction vessel Temperature (°C)**	**Backpressure (bar)**
Static extraction	2	3	95	5	80	150
Dynamic extraction	2	1	90	10	80	150
SFC						
Time (min)	CO_2 (%)	MeOH (%)	Flow ($mL\ min^{-1}$)	Temperature (°C)	Backpressure (bar)	Makeup pump MeOH ($mL\ min^{-1}$)
0	100	0				
2	100	0	2	35	150	0.5
10	60	40				
12	60	40				

This methodology was applied to the carotenoid and apocarotenoid analysis of lyophilized yellow tamarillo samples [18]

Table 2
QqQ MS general conditions

Interface	APCI (+/− mode)
Interface temperature (°C)	350
DL temperature (°C)	200
Block heater temperature (°C)	200
Nebulizing gas flow (N_2) L min^{-1}	3
Drying gas flow (N_2) L min^{-1}	5
Acquisition mode	Scan (200–1200 amu) SIM MRM
Event time (s)	0.05

the basis of the Product Ion Scan (PIS) experiments carried out on the various available standards using various collision energies both in positive and negative modes before the MRM experiments in order to further confirm the reported compounds identifications.

3. After a run is completed, the results are analyzed and quantified. Each peak is automated integrated by the software and eventually further reviewed by a technologist. Quantification is done using LabSolution *ver.* 5.8. software (Shimadzu Japan).

An example with samples of yellow tamarillo fruits analyzed by the SFE-SFC-MS methodology (Reproduced with permission from Ref. [18]; Copyright 2018, Elsevier Inc.) is shown in Fig. 3. It shows the corresponding enlargements of the ion chromatograms obtained in SIM (APCI negative) mode relative to the detected apocarotenoid zone. Interestingly, in this study nine apo-carotenoids including also two apocarotenoid fatty acids esters were detected in less than 6 min including extraction time. Another example is shown in Fig. 4 with different positions of eccentric zeaxanthin oxidative cleavages sites leading to different apo-zeaxanthinals (Reproduced with permission from Ref.[4]; Copyright 2018, American Chemical Society Publications).

3.3 Quantitative Analysis

Quantitative analysis is performed by the preparation of appropriate calibration curves when the necessary standards are available; however, the complete carotenoid and apocarotenoid quantifications are very often challenging due to the lack of reference material and the high cost of the standards. A possible alternative to perform quantitative analysis is the calculation of the extraction yield performing multiple extractions until depletion on the same

Table 3
Optimized multiple reaction monitoring (MRM) with molecular ion (*m/z*), quantifier (Q) and qualifier (q) transitions, and Collision Energy (CE) for apocarotenoid analysis

Compound	(+/−)[a]	*m/z*	Quantifier	CE (V)	Qualifier	CE (V)
Apo-8′-Zeaxanthinal	(+)	433	433 > 119	−30	433 > 105	−35
Apo-10′-Zeaxanthinal	(+)	393	393 > 119	−25	393 > 105	−35
Apo-12′-Zeaxanthinal	(+)	367	367 > 105	−35	367 > 119	−30
Apo-14′-Zeaxanthinal	(+)	327	327 > 105	−35	327 > 119	−30
Apo-15′-Zeaxanthinal	(+)	301	301 > 173	-15	301 > 105	−30
β-Apo-8′-Carotenal	(+)	417	417 > 119	−25	417 > 105	−35
β-Apo-10′-Carotenal	(+)	377	377 > 105	−35	377 > 119	−30
β-Apo-12′-Carotenal	(+)	351	351 > 119	−25	351 > 105	−35
β-Apo-14′-Carotenal	(+)	311	311 > 105	−30	311 > 119	−25
ε-Apo-8′-Luteinal	(−)	415	415 > 119	40	415 > 91	50
ε-Apo-10′-Luteinal	(−)	375	375 > 105	40	375 > 91	50
ε-Apo-12′-Luteinal	(−)	349	349 > 105	40	349 > 91	50
ε-Apo-14′-Luteinal	(−)	309	309 > 91	50	309 > 105	40
ε-Apo-15′-Luteinal	(−)	283	283 > 105	40	283 > 91	50
Apo-8′-Capsorubinal	(+)	449	449 > 109	−25	449 > 127	−15
Apo-10′-Capsorubinal	(+)	409	409 > 109	−20	409 > 127	−15
Apo-12′-Capsorubinal	(+)	383	383 > 109	−20	383 > 127	−15
Apo-14′-Capsorubinal	(+)	343	343 > 109	−20	343 > 127	−15
Apo-15′-Capsorubinal	(+)	317	317 > 109	−20	317 > 127	−15
Apo-8′-Capsorubinal	(−)	448	448 > 239	25	448 > 342	20
Apo-10′-Capsorubinal	(−)	408	408 > 199	25	408 > 302	20
Apo-12′-Capsorubinal	(−)	382	382 > 276	20	382 > 173	25
Apo-14′-Capsorubinal	(−)	342	342 > 173	25	342 > 153	25
Apo-15′-Capsorubinal	(−)	316	316 > 169	20	316 > 147	20

[a](+) APCI positive; (−) APCI negative

sample. For all the compounds considered in the fourth extraction only some traces were eventually present [17]. The multiple extraction process could be compared to a traditional solid–liquid extraction method. Here, the recording of reliable quantitative data requires the extraction to be repeated several times until sample decoloration, thus consuming a very high amount of organic solvent and sample in a time consuming process.

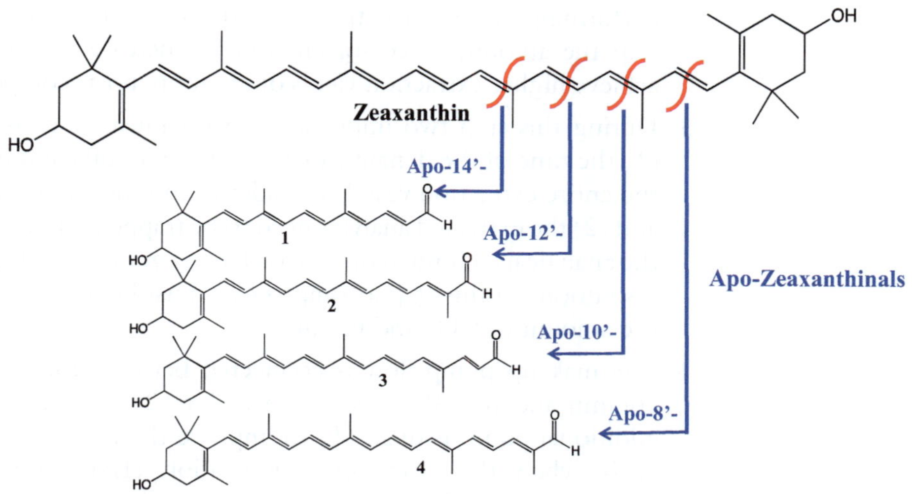

Fig. 4 Different positions of eccentric zeaxanthin oxidative cleavages sites leading to different apo-zeaxanthinals. (1) apo-14′-zeaxanthinal; (2) apo-12′-zeaxanthinal; (3) apo-10′-zeaxanthinal; (4) apo-8′-zeaxanthinal. Reproduced with permission from Ref. [4]; Copyright 2018, American Chemical Society Publications

3.4 Conclusions

The occurrence of apocarotenoids in food and in biological matrices should be much further investigated and the here reported methodology provides a tool for a fast and reliable determination of these bioactive metabolites in various matrices. Detection of oxidative or cleavage products of carotenoids requires highly sensitive techniques without laborious preparation steps, therefore the hyphenated here described automated online SFE-SFC-QqQ/MS technique will provide a useful tool for their identification. The online nature of the system, compared to offline approaches, drastically reduces the extraction time (compared to the traditional solid/liquid extraction, which may require a few hours), reduces the analysis run time, reduces the risks of sample contamination and the possible occurrence of operator errors, improves run-to-run precision, and enables the setting of batch-type applications.

4 Notes

1. By using the SFE-SFC instrument the main goal of the "manual" sample preparation is to achieve an accurate homogenization of the sample and sampling in order to achieve accurate and reproducible results.
2. During this step it is very important to completely fill the extraction vessel in order to reach the same pressure of the back pressure regulator (BPR). The split ratio of the mobile phase between the analytical column and the extraction vessel depends on the analytical conditions, but can be calculated

performing only static extraction with the empty vessel. In this step the amount of co-solvent can be increased in order to achieve higher extraction yields of the more polar compounds.

3. During this step two important aspects must be considered: (1) the time of the dynamic extraction mode sufficient to flush the entire extraction vessel (consider both time and flow rate), and (2) the extracted analytes need to be trapped on the head of the analytical column in order to achieve a good peak shape and resolution. In this step a compromise must be found between the amount of CO_2 and modifier.
4. The makeup pump flow is connected between the analytical column and the BPR (as shown in Fig. 2). The aim of this additional pump is dual: (1) to improve the ionization, especially when the electrospray ionization (ESI) interface is employed, and (2) to improve the peak shape especially if a low amount of organic modifier is employed. In fact, when a low amount of organic solvent is employed the expansion of the CO_2 after the BPR can cause the clogging of the transfer tube to the MS. An additional trick in order to improve the peak shape is to limit both the internal diameter of the transferring tube between the BPR and the MS system and the use of unions than can cause an expansion of the CO_2.

References

1. Rivera SM, Canela-Garayoa R (2012) Analytical tools for the analysis of carotenoids in diverse materials. J Chromatogr A 1224:1–10
2. Amorim-Carrolho KT, Cepeda A, Fente C, Regal P (2014) Review of methods for analysis of carotenoids. Trends Anal Chem 56:49–73
3. Mercadante AZ, Rodrigues DB, Petri FC, Mariutti LRB (2017) Carotenoid esters in foods—a review and practical directions on analysis and occurrence. Food Res Int 99:830–850
4. Giuffrida D, Donato P, Dugo P, Mondello L (2018) Recent analytical techniques advances in the carotenoids and their derivative determination in various matrices. J Agric Food Chem 66:3302–3307
5. Jumaah F, Plaza M, Abrahamsson V, Turner C (2016) A fast and sensitive method for the separation of carotenoids using ultra-high performance supercritical fluid chromatography-mass spectrometry. Anal Bioanal Chem 408:5883–5894
6. Li B, Zhao H, Liu J, Liu W, Fan S, Wu G, Zhao GR (2015) Application of ultra-high performance supercritical fluid chromatography for the determination of carotenoids in dietary supplements. J Chromatogr A 1425:287–292
7. Berger TA, Berger BK (2013) Separation of natural food pigments in saponified and un-saponified paprika oleoresin by ultra-high-performance supercritical fluid chromatography (UHPSFC). Chromatographia 76:591–601
8. Matsubara A, Uchikata T, Shinohara M, Nishiumi S, Yoshida M, Fukusaki E, Bamba T (2012) Highly sensitive and rapid profiling method for carotenoids and their epoxidized products using supercritical fluid chromatography coupled with electrospray ionization-triple quadrupole mass spectrometry. J Biosci Bioeng 113:782–787
9. Wada Y, Matsubara A, Uchikata T, Iwasaki Y, Morimoto S, Kan K, Okura T, Fukusaki E, Bamba T (2011) Metabolic profiling of β-cryptoxanthin and its fatty acid esters by supercritical fluid chromatography coupled with triple quadrupole mass spectrometry. J Sep Sci 34:3546–3552
10. Matsubara A, Bamba T, Ishida H, Fukusaki E, Hirata K (2009) Highly sensitive and accurate profiling of carotenoids by supercritical fluid chromatography coupled with mass spectrometry. J Sep Sci 32:1459–1464

11. Giuffrida D, Zoccali M, SV G, Dugo P, Mondello L (2017) Apocarotenoids determination in *Capsicum chinense* Jacq. cv habanero, by supercritical fluid chromatography-mass spectrometry. Food Chem 231:316–323
12. Durante M, Lenucci MS, Mita G (2014) Supercritical carbon dioxide extraction of carotenoids from pumpkin (*Cucurbita* spp.): a review. Int J Mol Sci 15:6725–6740
13. Herrero M, Mendolia JA, Cifuentes A, Ibanez E (2010) Supercritical fluid extraction: recent advances and applications. J Chromatogr A 1217:2495–2511
14. Filho GL, De Rosso VV, Meireles MAA, Rosa PTV, Oliveira AL, Mercadante AZ, Cabral FA (2008) Supercritical CO_2 extraction of carotenoids from Pitanga fruits (*Eugenia uniflora* L.). J Supercrit Fluids 46:33–39
15. Machmudah S, Kawahito Y, Sasaki M, Goto M (2008) Process optimization and extraction rate analysis of carotenoids extraction from rosehip fruit using supercritical CO_2. J Supercrit Fluids 44:308–314
16. Daood HG, Illes V, Gnayfeed MH, Meszaros B, Horvath G, Biacs PA (2002) Extraction of pungent spice paprika by supercritical carbon dioxide and subcritical propane. J Supercrit Fluids 23:143–152
17. Zoccali M, Giuffrida D, Dugo P, Mondello L (2017) Direct online extraction and determination by supercritical fluid extraction with chromatography and mass spectrometry of targeted carotenoids from habanero peppers (*Capsicum chinense* Jacq.). J Sep Sci 40:3749–3922
18. Giuffrida D, Zoccali M, Arigò A, Cacciola F, Osorio-Roa C, Dugo P, Mondello L (2018) Comparison of different analytical techniques for the analysis of carotenoids in tamarillo (*Solanum betaceum* Cav.). Arch Biochem Biophys 646:161–167
19. Zoccali M, Giuffrida D, Salafia F, Giofrè SV, Mondello L (2018) Carotenoids and apocarotenoids determination in intact human blood samples by online supercritical fluid extraction-supercritical fluid chromatography-tandem mass spectrometry. Anal Chim Acta 1032:40–47
20. Rodriguez EB, Rodriguez-Amaya DB (2007) Formation of apocarotenals and epoxycarotenoids from β-carotene by chemical reactions and by autoxidation in model system and processed foods. Food Chem 101:563–572

Part IV

Imaging

Chapter 17

Elucidating Carotenoid Biosynthetic Enzyme Localization and Interactions Using Fluorescent Microscopy

Maria Shumskaya, Rena F. Quinlan, and Eleanore T. Wurtzel

Abstract

Carotenoids are essential for survival of all plants, where these colorful pigments and derivatives are biosynthesized, as well as for humans and other species that obtain plant-derived carotenoids in their diets and rely upon them for vitamin biosynthesis or antioxidant actions. The plant carotenoid biosynthetic pathway consists of nuclear encoded enzymes that are imported into chloroplasts and other plastids. The pathway structural genes are known and have been targeted for metabolic engineering to improve carotenoid profiles or content. However, results are not always as expected because there remain fundamental gaps in understanding how the pathway is physically organized. Many of the enzymes have been found in high molecular weight complexes which are poorly described. Elucidation of enzyme localization as well as enzyme interactions in vivo are needed for advancing the carotenoid field and facilitating our understanding of the three-dimensional organization of this important pathway. Fluorescent protein fusions with carotenoid enzymes can provide in vivo information when these fusions are introduced and transiently expressed in plant cells. Current advances in fluorescent microscopy, especially confocal microscopy, provide the resolution needed to localize fluorescently tagged carotenoid enzymes within suborganellar locations of plastids. Interactions between carotenoid biosynthetic enzymes can be determined using bimolecular fluorescence complementation (BiFC), a method whereby genes of interest are fused with sequences encoding nonfluorescent N- and C-terminal halves of YFP (yellow fluorescent protein), and then introduced into plant protoplasts to allow expression and visualization by fluorescence microscopy. The YFP fluorescence is restored only if the N and C-terminal regions are brought together by interacting fusion partners. Here we describe the methodology, with extensive tips and notes, for determining in vivo carotenoid enzyme localization and enzyme interactions by transient expression of enzyme–fluorescent protein fusions.

Key words Carotenoid enzymes, Biosynthetic pathway, Plants, Maize, Bimolecular fluorescence, Plastid localization, Protein interaction, Microscopy

1 Introduction

Carotenoids constitute a large class of lipophilic pigments that are biosynthesized by plants and other organisms [1]. In plants, carotenoids and their bioactive derivatives have manifold roles, ranging from photosynthetic light capture and photoprotection to signaling needed to promote proper growth and development.

Manuel Rodríguez-Concepción and Ralf Welsch (eds.), *Plant and Food Carotenoids: Methods and Protocols*, Methods in Molecular Biology, vol. 2083, https://doi.org/10.1007/978-1-4939-9952-1_17,

The importance of carotenoids, both in plants and organisms that consume plants in their diets, has fueled efforts in metabolic engineering of the biosynthetic pathway [2] and instilled future interest in use of synthetic biology to capitalize on these multifunctional molecules [3]. The long-standing efforts to manipulate the pathway have not always led to expected results [4] because fundamental understanding of pathway regulation, in particular the organization of the multienzyme complex, is poorly understood [5]. The carotenoid biosynthetic pathway enzymes are nuclear-encoded and localize to chloroplasts and other plastids via targeting mediated by N-terminal transit peptides that are cleaved upon plastid import. Research to date indicates that the carotenoid biosynthetic enzymes can be found in various locations within plastids [5]. The carotenoid biosynthetic pathway enzymes are part of poorly described, high molecular weight complexes, and the pathway shows evidence of substrate channeling, as expected for a multienzyme pathway structure. Enzymes such as phytoene synthase (PSY) require membrane association for activity and PSY isozymes show varying locations within the plastid, while other downstream pathway enzymes are widely distributed within plastids [6]. However, there is a dearth of information on how the complete biosynthetic machinery is assembled to form a fully functional complex. This gap in knowledge severely limits opportunities for predictable engineering of the pathway, especially in the variable genotypes of plants.

Early studies to explore carotenoid enzyme localization entailed polyclonal antibodies raised against the biosynthetic enzymes (for examples, *see* Refs. [7, 8]). Use of antibodies for localizing the carotenoid enzymes requires high-level protein expression in bacteria, followed by purification of large quantities of the carotenoid enzymes needed to raise antibodies in animals. This immunological approach is time consuming and fraught with problems, including variable antigenic specificities. Furthermore, antibodies may reveal location, but not enzyme–enzyme interactions in vivo. Recently, we utilized fluorescence microscopy and transient expression in protoplasts for in vivo localization of carotenoid enzymes in maize and other plants [6, 8, 9] and tested potential interactions between carotenoid enzymes [10]. This approach skips the problematic step of high level expression of carotenoid enzymes needed for immunolocalization experiments, and also allows for the rapid (as compared to immunolocalization or stable transformation) in vivo examination of carotenoid enzyme interactions. In this method, plasmids encoding a fluorescent protein, most commonly Green Fluorescent Protein (GFP), that is C-terminally fused to the carotenoid enzymes, are delivered to isolated plant protoplasts. The sequence encoding the carotenoid enzyme includes the native transit peptide needed for plastid targeting and import. After a period of incubation to allow protein expression and plastid import, the fluorescence is visualized by confocal microscopy. For testing interactions between two carotenoid enzymes, each construct encodes an enzyme, whose C-terminus is fused in frame to the N-terminus of one of two halves of a fluorescent protein, such as Yellow Fluorescent Protein (YFP). If the enzymes interact in vivo, the full structure of the fluorescent protein is restored and can be visualized by the known technique of bimolecular fluorescence [11]. The use of protoplasts for transient expression of fluorescently tagged carotenoid enzymes reveals enzyme location and potential interactions in vivo. Detailed below is the methodology, with careful explanations and cautionary notes, used to examine carotenoid enzyme localization and enzyme–enzyme interactions by creation of test plasmids that are introduced into freshly isolated protoplasts that are then visualized by fluorescence microscopy.

2 Materials

Distilled water is recommended. All manipulations for the isolation and transformation of protoplasts should be carried out at room temperature. The procedure is not sterile, no flow hood is required and all manipulations with protoplasts can be performed on a lab bench. Transformed protoplasts survive no longer than 24–48 h, thus fluorescence should be examined the day after transformation. Here, a C4 plant, *Zea mays* (maize), is the choice for protoplast preparation, since most carotenoid biosynthetic enzymes studied in our laboratory were of maize origin. In maize, etiolated leaves give protoplasts with the best transformation efficiency of 80–90%. However, if the carotenoid enzymes to be tested are thought to interact with a developed thylakoid membrane, green leaves should be used for protoplasts preparation. If a C3 plant is preferred as a source of protoplasts, then *Vigna unguiculata* subsp. *unguiculata* (black-eyed peas) is recommended, however, transformation efficiency in protoplasts isolated from these leaves is lower and peas do not grow well as etiolated. The methodology detailed here, with extensive tips and notes, is based on the protocols used to examine localization and interaction of carotenoid enzymes [6, 8–10].

1. Cellulase from *Trichoderma viridae* or *Trichoderma* sp. (preferably plant cell culture-tested).
2. Pectinase from *Rhizopus* sp.
3. Reagents #1 and #2 can be substituted with Onozuka R10 (cellulase+pectinase mix).
4. BSA (bovine serum albumin).
5. Mannitol.
6. MES (2-(N-morpholino)ethanesulfonic acid) $CaCl_2$.

7. $MgCl_2$.
8. $Ca(NO_3)_2$.
9. β-mercaptoethanol.
10. PEG6000.
11. Nylon filter mesh, 60 micron (Fisher Scientific).
12. 5 mL pipettes.
13. Round bottom polystyrene tubes, 14 mL.
14. 500-mL sterile side-arm flask.
15. Double-sided thin razor blades.
16. 24-well flat bottomed microplate for fluorescent microscopy.
17. 500 μL tips with a cut end to widen the opening (cut with a razor blade or scissors and autoclave).
18. Vermiculite.
19. Perlite.
20. Fluorescent microscope (e.g., Leica Microsystems) equipped with lasers or lamp filters for detection of the fluorescent signal of choice.
21. Buffer A. Basic protoplast buffer: 10 mM $CaCl_2$, 0.6 M mannitol, 20 mM MES-KOH (pH 5.7), autoclaved and stored at room temperature.
22. Buffer B. Isolation buffer: to 50 mL of Buffer A, add 1% cellulase (500 mg), 0.3% pectinase (160 mg), 5 mM (18 μL) β-mercaptoethanol, and 0.1% BSA (50 mg). Alternatively, use 300–500 mg of Onozuka R10 mix, containing cellulase and pectinase). This buffer must be made immediately before use. Keep at room temperature. Do not store.
23. Buffer C. PEG solution: in a 50 mL plastic tube mix 6 mL of a 0.1 M $Ca(NO_3)_2$ solution with 0.91 g mannitol (0.5 M final concentration). Dissolve the mannitol by swirling, then add 4 g of PEG6000 (40% final concentration). Tightly close the tube and microwave on high (~800 watts) for exactly 6 s so that the solution does not boil and change concentration. Gently mix, microwave again for 6 s, and gently mix again. Let the solution cool down. The solution must appear clear with no visible particles. This solution must be prepared fresh before use and cooled down to room temperature. Do not store.
24. Buffer D. Incubation buffer: 15 mM $MgCl_2$, 0.5 M mannitol, 5 mM MES-KOH (pH 5.5), autoclaved and stored at room temperature.

3 Methods

3.1 Guidelines for Creating Test Plasmids

For carotenoid enzyme localization experiments, plasmid constructs should be assembled in such a way that a gene for a fluorescent protein, such as GFP, Red Fluorescent Protein (RFP) or YFP is fused in frame and downstream of the open reading frame or cDNA sequence encoding the carotenoid enzyme; the carotenoid enzyme transit sequence should not be altered since it is essential for plastid import. A plant promoter for constitutive expression, such as the 35S Cauliflower Mosaic virus promoter is recommended (see examples of constructs used in Fig. 1a). It is preferable to insert a Kozak sequence [12] to support translation, if the vector used does not possess this translation start sequence before the insert site. For testing interactions between two carotenoid enzymes, each construct encodes an enzyme (with its native transit sequence) fused in-frame at and upstream of one of two halves of a fluorescent protein. For examples, *see* [6, 10] and Fig. 1b.

3.2 Protoplast Isolation and DNA Transformation

1. Place maize kernels in a 50:50 mix of vermiculate and perlite and keep in the dark at 25 °C so the leaves that emerge are completely etiolated (*see* **Note 1**). Water the seeds and then water as necessary using tap water (*see* **Note 2**).

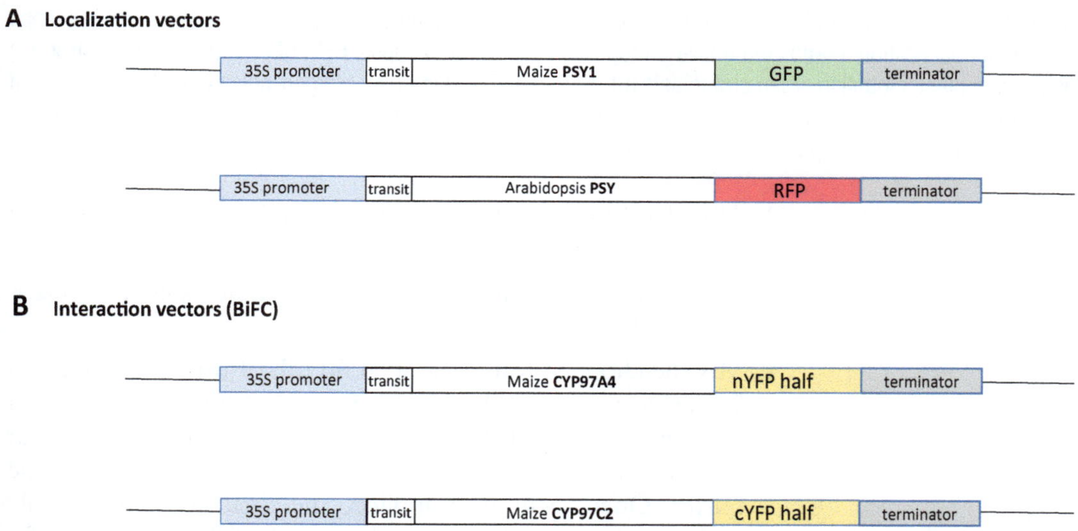

Fig. 1 Examples of plasmid constructs used for localization or interaction experiments. (**a**) Elucidation of carotenoid enzyme localization by fusing sequences of carotenoid enzymes with GFP (pUC35S-Zm-PSY1-sGFP-Nos for maize PSY1) [6] or RFP (pSAT-At-PSY-RFP for Arabidopsis PSY) [6]. (**b**) Constructs required for carotenoid enzyme interaction studies using the BiFC method. Vectors shown are for CYP97A4 and CYP97C2 hydroxylases, which were fused in-frame and upstream of either the N-terminal or C-terminal halves of YFP [10], by use of vectors pSAT4(A)-nEYFP-N1 (encoding the N-terminal half of YFP) and pSAT6-cEYFP-N1 (encoding the C-terminal half of YFP) [11]. Transit, refers to the native sequences encoding the cleavable transit peptides of the corresponding carotenoid enzymes; terminator refers to transcriptional terminators contained within the backbone vectors

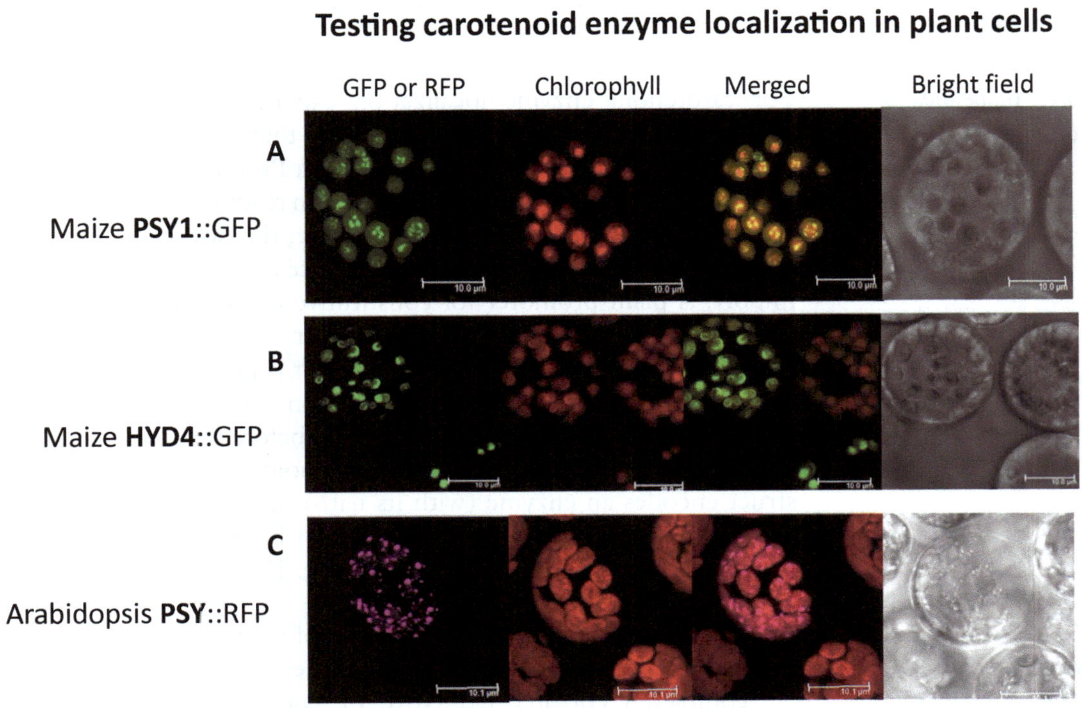

Fig. 2 Localization of carotenoid biosynthetic proteins in different compartments of chloroplasts. (**a**) Maize PSY1::GFP (*see* Fig. 1a construct) is localized to stroma and associated with prolamellar bodies in etiolated maize protoplasts [6]. (**b**) Maize HYD4::GFP localizes as a half-moon or circular pattern typical for envelope membrane localization in etiolated maize protoplasts [10]. (**c**) Arabidopsis PSY::RFP (*see* Fig. 1a construct) is localized to plastoglobuli in green protoplasts [6]. *GFP or RFP, Chlorophyll, Merged* indicate the fluorescence of the marker protein, autofluorescence of chlorophyll, or superposition of both signals, respectively. Images here are similar to others published [6, 10]. Bar: 10 μm

2. After 10–14 days, collect the middle section of the second true leaves from ~20 plants.
3. Chop leaves perpendicularly with a double-sided thin razor blade in 1 mm pieces.
4. Mix the chopped leaf tissue with 50 mL of Buffer B (Isolation buffer) in a sterile 500 mL side-arm flask. Close the flask with a stopper and connect the arm to a vacuum pump. Apply vacuum in 2–3 short, 1 s bursts to substitute air in leaf tissues with the buffer. Once vacuum is applied, the tissue becomes visibly darker and chopped pieces sink instead of float in the buffer (*see* **Note 3**).
5. Incubate the flask on a rotating platform, 60 rpm, at 25 °C for 2–3 h in the dark.
6. Immediately after initiating this incubation, prepare **Buffer C** (PEG solution) so that there is sufficient time for the buffer to cool down to room temperature before it is needed for the later steps. The plasmid DNA to be used for transformation should be defrosted on ice (*see* **Note 4**).

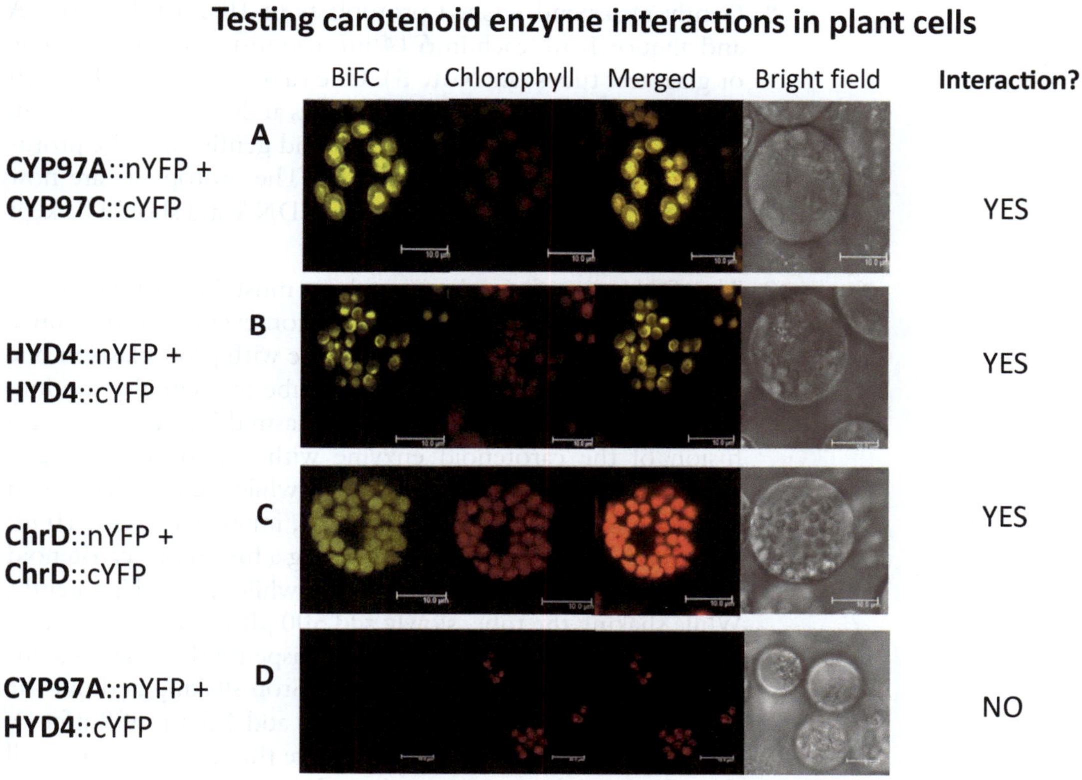

Fig. 3 Protein interaction shown with BiFC in etiolated maize protoplasts. BiFC constructs encoding carotenoid enzymes or controls (*see* Fig. 1b) are introduced into protoplasts and expressed proteins are visualized by fluorescence microscopy. Fluorescence only occurs if the interacting proteins bring the YFP N and C terminal halves in proximity to restore a complete protein structure as needed to produce fluorescence. (**a**) BiFC fusions show that maize carotenoid enzymes CYP97A and CYP97C interact in vivo as seen by the restoration of YFP fluorescence. The protein complex is localized to stroma. (**b**). Maize HYD4 forms a homodimer in vivo as evidenced by restored fluorescence; the homodimer is localized in the plastid envelope (consistent with the GFP fusion localized in Fig. 2b). (**c**) BiFC positive control, ChrD plastid protein from *Cucumis sativus*, forms a homodimer as evidenced by restored fluorescence. (**d**) CYP97A and HYD4 do not interact in vivo, as evidence by the absence of YFP fluorescence. *BiFC, Chlorophyll, Merged* correspond to restored fluorescence of YFP, autofluorescence of chlorophyll which is a marker for the chloroplast, and a superposition of both signals, respectively. Images here are similar to others published [6, 10]. Bar: 10 μm

7. After the 2–3 h incubation period, remove the flask from the rotator platform to a lab bench. Shake vigorously by hand for 3–5 min in a circular motion to completely release protoplasts from their cell walls and plant tissues. Collect the released protoplasts into a 50 mL plastic tube by pouring through a 60 μm nylon filter folded twice, then centrifuge at 80 × *g* for 10 min at 20 °C, discard the supernatant (be careful not to disturb the protoplast pellet, it is very fragile), gently resuspend the protoplasts in 50 mL of Buffer A (Basic protoplast buffer), centrifuge at 80 × *g*, discard the supernatant and repeat the washing procedure again using 50 mL of Buffer A. Never vortex protoplasts, only gently mix!

8. Gently resuspend washed protoplasts in 10–20 mL Buffer A and aliquot 1 mL each into 14 mL round-bottom polystyrene or glass test tubes (*see* **Note 5**). One tube will be used for each transformation. Centrifuge protoplasts at 80 × *g*, 20 °C, 5 min. Remove 850 μL of the supernatant and gently swirl the protoplasts in the remaining supernatant. The protoplasts are now ready to be transformed with plasmid DNA and should be kept at RT and used immediately.
9. The DNA transformation reactions must be carried out in about a minute for each transformation event, without interruptions (*see* **Note 6**). Take a test tube with protoplasts in one hand and start to gently shake the tube in a circular motion. With the other hand, add 10 μL of plasmid DNA (encoding a fusion of the carotenoid enzyme with a fluorescent-tagged protein, or with one half of YFP) while gently shaking in circular motions. In the case of BiFC, immediately add 10 μL of the second plasmid DNA (encoding a fusion of a carotenoid enzyme with the second half of YFP), while still shaking gently. While shaking the tube, slowly add 500 μL Buffer C, measured using a cut pipette tip to accurately dispense this viscous solution, and continue shaking for 10 s. Stop shaking and put the tube in a rack. Immediately and gently add 4.5 mL of Buffer D, cap the tube and do not shake or move the tube. Repeat for all the other transformation reactions. Leave the tubes in a rack for 20–30 min undisturbed.
10. After the incubation is over, add another 5 mL of Buffer A and gently invert the tubes upside down several times. Centrifuge at 80 × *g*, 20 °C, 5 min. Discard the supernatant, and wash with 5 mL of Buffer A. Discard the supernatant and gently resuspend each protoplast sample in 1 mL of the Buffer A using cut tips to gently resuspend the protoplasts. Transfer the transformed protoplasts into the wells of a 24-well plate if an inverted microscope is used (*see* **Note 7**), and incubate overnight at 21–25 °C in the dark or dim light (*see* **Note 8**).
11. Samples are then visualized by fluorescent microscopy and compared with controls to determine where the proteins are localized or if they interact. For these experiments, a confocal fluorescent microscope, such as the inverted DM16000B with TCS SP8 system and LAS AF software (Leica Microsystems CMS, Germany) is recommended. Confocal laser microscopes provide visualization at Z-planes (e.g., several "layers" of the image that can be fused together by the software into a 3D image for better visualization), which is especially important for analysis and detailed localization of the enzymes in plastids. An oil or water immersion objective 63–100× is required for optimal resolution. In confocal microscopy, a 488 nm argon laser is used to excite fluorescence of GFP and chlorophyll; the

514 nm line of an argon laser is used to excite the fluorescence of YFP, and the 543 nm laser is used to excite RFP. Chlorophyll autofluorescence is detected between 664 and 696 nm, GFP fluorescence is detected between 500 and 539 nm, YFP between 524 and 575 nm, and RFP between 600 and 650 nm. One of the advantages of the Leica system is the ability to perform wavelength scanning (lambda scan) to obtain the emission spectrum of the excited molecules. The analyzed emission spectra should be identical to the spectra of fluorescent molecules as described in the reference database (included in the Leica LAS AF software) to exclude potential false signals caused by multiple autofluorescent components of plant cells. Typical localization results can be seen for test carotenoid enzymes such as PSY from maize and Arabidopsis, and β-carotene hydroxylase 4 (HYD4) from maize (*see* Fig. 2). The red chlorophyll autofluorescence provides a good internal marker for plastid identification and therefore it is useful to merge the red fluorescence signal with the fluorescent tag (GFP, YFP, or RFP). Note that the PSY enzymes from maize and Arabidopsis do not localize in the same place, as previously reported [6]. Figure 3 shows typical results for protein-protein interactions, as evidenced by the restored fluorescence seen when coexpressing BiFC fusions of the P450 carotene hydroxylases from maize, CYP97C and CYP97A (*see* Fig. 3a) [10]. BiFC fusion pairs using the nonheme diiron enzyme HYD4 also restores YFP fluorescence, evidence for HYD4 forming a dimeric complex (Fig. 3b); this complex localization is consistent with the GFP-tagged localization in Fig. 2b. Figure 3c shows restored fluorescence indicating interaction for a known homodimeric plastid protein, ChrD (carotenoid-associated protein D), which is commonly used as a positive control for BiFC experiments. Figure 3d shows no restoration of fluorescence, indicating that while CYP97C can interact with the P450 carotene β-ring hydroxylase CYP97A, it cannot interact with the nonheme diiron enzyme β-ring hydroxylase HYD4.

4 Notes

1. Any maize variety can be used, preferably with a known genotype and/or one for which there is known sequence information such as the B73 cultivar. Maize mutants defective in carotenoid biosynthesis may be useful as long as the phenotype does not affect leaf photosynthesis [13]. Many maize genetic stocks can be obtained from the Maize Genetic Cooperation Stock Center (University of Illinois; http://maizecoop.cropsci.uiuc.edu) which is part of the National Plant Germplasm

System in the USA. Maize germplasm is also available from the North Central Regional Plant Introduction Station (NCRPI; https://www.ars.usda.gov/midwest-area/ames/plant-introduction-research/).

2. Do not use fertilizer, as the seed endosperm contains sufficient nutrients to promote growth.
3. Buffer B is used to facilitate degradation of the cell wall and release of the fragile protoplasts. Therefore, from this point onward, care should be taken in using the gentlest of manipulations and avoiding hard shaking of the sample.
4. Purified plasmids for transformation should be of a concentration >1000 ng/μL in water or 10 mM Tris/HCl buffer (pH 8.5). A preparation kit that allows large-scale purification of plasmid DNA with the final concentration of at least 1 μg/μ L should be used instead of manual preparation since impurities might affect efficiency of transformation.
5. Polystyrene or glass material is important, because during the PEG-initiated transformation the protoplasts might stick to polyethylene or other plastics. Use 1 mL tips for which the ends are cut off for a wider opening to make sure the protoplasts are not destroyed by pipetting.
6. All of the materials needed for DNA transformations of the isolated protoplasts should be preassembled so that the transformations can be done quickly. Place in front of you the following items: Buffer C (PEG solution), Buffer D (Incubation buffer), purified plasmids on ice, pipette for 500 μL with cut tips, pipette for 20 μL with tips, and a pipet for 5 mL with the dispenser attached. If introducing two plasmids, prepare either 10 μL of 1 mg/μL of each plasmid, or premix the two plasmids and dispense a total of 20 μL to the protoplasts. To work quickly, shake the sample with one hand, and add everything with the other hand, one by one. After Buffer D is added, shaking stops and the PEG in Buffer C becomes diluted. The protoplasts will die if you stop shaking after adding Buffer C (PEG solution) and before adding Buffer D, or shaking after adding Buffer C (PEG solution) takes too long before addition of Buffer D. Therefore, the entire process for one protoplast sample, including addition of plasmids, Buffer C and Buffer D, should be completed within a couple of minutes. If you cannot work with two hands, then a partner can help with gentle shaking of the protoplast-containing tube while you add everything step by step. When testing multiple enzymes and controls for localization or interactions, each protoplast sample must be completely processed before going on to the next DNA transformation.

7. An inverted fluorescent microscope is recommended, since after transformation the cells can be incubated overnight in glass-bottom cell culture plates while expressing the fusion genes, and then observed directly. Transferring protoplasts to microscopic slides after incubation is complete should be avoided because protoplasts can be destroyed during pipetting. If an inverted microscope is not available, leave the samples in their tubes for overnight incubation. For upright microscopy, on the following day use cut tips to transfer the samples to concave glass slides, instead of standard slides, to avoid breakage of the large and fragile protoplasts.
8. Light does not affect transformation efficiency, but bright light is not recommended since etiolated protoplasts do not have developed photosystems. A controlled temperature chamber is recommended in order to avoid any unexpected temperature changes that might occur if the samples are left on a lab bench overnight.

Acknowledgments

Research in the Wurtzel laboratory has been funded by the National Institutes of Health (grant GM081160), National Science Foundation, American Cancer Society, Rockefeller Foundation International Rice Biotechnology Program, McKnight Foundation, USDA, DOD, New York State, The City University of New York, and Lehman College. MS acknowledges support from Kean University, NJ.

References

1. Moise AR, Al-Babili S, Wurtzel ET (2014) Mechanistic aspects of carotenoid biosynthesis. Chem Rev 114(1):164–193
2. Giuliano G (2017) Provitamin A biofortification of crop plants: a gold rush with many miners. Curr Opin Biotechnol 44:169–180. https://doi.org/10.1016/j.copbio.2017.02.001
3. Wurtzel ET (2018) Changing form and function through carotenoids and synthetic biology. Plant Physiol 179(3):830–843. https://doi.org/10.1104/pp.18.01122
4. Schaub P, Al-Babili S, Drake R, Beyer P (2005) Why is golden rice golden (yellow) instead of red? Plant Physiol 138(1):441–450. https://doi.org/10.1104/pp.104.057927
5. Shumskaya M, Wurtzel ET (2013) The carotenoid biosynthetic pathway: thinking in all dimensions. Plant Sci 208(1):58–63
6. Shumskaya M, Bradbury LMT, Monaco RR, Wurtzel ET (2012) Plastid localization of the key carotenoid enzyme phytoene synthase is altered by isozyme, allelic variation, and activity. Plant Cell 24:3725–3741. https://doi.org/10.1105/tpc.112.104174
7. Li F, Vallabhaneni R, Yu J, Rocheford T, Wurtzel ET (2008a) The maize phytoene synthase gene family: overlapping roles for carotenogenesis in endosperm, photomorphogenesis, and thermal stress-tolerance. Plant Physiol 147:1334–1346
8. Beltrán J, Kloss B, Hosler JP, Geng J, Liu A, Modi A, Dawson JH, Sono M, Shumskaya M, Ampomah-Dwamena C, Love JD, Wurtzel ET (2015) Control of carotenoid biosynthesis through a heme-based cis-trans isomerase. Nat Chem Biol 11(8):598–605
9. Ampomah-Dwamena C, Driedonks N, Lewis D, Shumskaya M, Chen X, Wurtzel ET,

Espley RV, Allan AC (2015) The phytoene synthase gene family of apple (Malus x domestica) and its role in controlling fruit carotenoid content. BMC Plant Biol 15(1):185

10. Quinlan RF, Shumskaya M, Bradbury LMT, Beltrán J, Ma C, Kennelly EJ, Wurtzel ET (2012) Synergistic interactions between carotene ring hydroxylases drive lutein formation in plant carotenoid biosynthesis. Plant Physiol 160:204–214. https://doi.org/10.1104/pp.112.198556
11. Citovsky V, Lee L-Y, Vyas S, Glick E, Chen M-H, Vainstein A, Gafni Y, Gelvin SB, Tzfira T (2006) Subcellular localization of interacting proteins by bimolecular fluorescence complementation *in planta*. J Mol Biol 362(5):1120–1131
12. Joshi CP, Zhou H, Huang X, Chiang VL (1997) Context sequences of translation initiation codon in plants. Plant Mol Biol 35(6):993–1001
13. Wurtzel ET, Cuttriss A, Vallabhaneni R (2012) Maize provitamin A carotenoids, current resources and future metabolic engineering challenges. Front Plant Sci 3:29. https://doi.org/10.3389/fpls.2012.00029

Chapter 18

Visualization of Carotenoid-Storage Structures in Fruits by Transmission Electron Microscopy

Joanna Lado, Jaime Zacarias, María Jesús Rodrigo, and Lorenzo Zacarías

Abstract

Plastids are cell organelles that, beside other functions, have the capability to store carotenoids in specialized structures, which may vary among the different plant species, tissues or according to the carotenoid complement. Fruits are an important source of carotenoids, and during ripening, chloroplasts differentiate into chromoplasts that are able to accumulate large amounts of carotenoids, rendering then the characteristic fruit coloration. Whereas lycopene or β-carotene may accumulate as crystal in the chromoplasts of some fruit, other xanthophyll-accumulating fruits differentiate plastoglobuli as a preferred system to enhance carotenoids stability and storage. Visualization of plastid ultrastructure and their transformation during ripening or in fruit of contrasting coloration are fundamental objectives within carotenoids research in fruits. Therefore, in this chapter, we describe a protocol for the visualization and analysis of plastid ultrastructure by transmission electron microscopy (TEM), specially designed and adapted to fruit tissues.

Key words Carotenoids, Citrus fruits, Chloroplast, Chromoplast, Plastids, Plastoglobuli, Transmission electron microscopy

1 Introduction

Carotenoids are a group of tetraterpenoid pigments synthetized in the plastids of plants and in some algae, fungi, and bacteria [1]. They are responsible for the attractive coloration of flowers, fruits, and other tissues in many plant species and also play critical roles in human nutrition and health [2–4].

Accumulation of carotenoids in plants is the result of a fine balance between biosynthesis, degradation, and their accumulation in specialized structures, the plastid [5]. During plant development and tissue differentiation, there are substantial changes in the number of plastids and in their structure and functionality that may determine the capacity for carotenoid storage. Various types of plastids can be identified in plant cells, such as proplastids, amyloplasts, etioplasts, chloroplasts, and chromoplasts [6, 7]. All these types of plastids, except proplastids, possess the capacity for

Manuel Rodríguez-Concepción and Ralf Welsch (eds.), *Plant and Food Carotenoids: Methods and Protocols*, Methods in Molecular Biology, vol. 2083, https://doi.org/10.1007/978-1-4939-9952-1_18,

carotenoid production [8, 9]. Amyloplasts are starch-storing plastids found in seed, roots and tubers (wheat, maize, potato, cassava) and in general contain relatively low levels of carotenoids, mainly xanthophylls (i.e., lutein, zeaxanthin, and violaxanthin) [8, 10]. Etioplasts are developed in dark-grown plants and usually contain small amounts of photosynthetic carotenoids, primarily lutein and violaxanthin, as well as the chlorophyll precursor protochlorophyllide [11]. Chloroplasts are the typical plastids found in green photosynthetic tissues (leaves and immature fruits) while the synthesis of carotenoids is believed to occur in their envelop and in the thylakoid membrane [12–14]. The formation of a lipoprotein complex between chlorophyll and carotenoids allows the accumulation of large amounts of carotenoids in the chloroplasts. Lutein, β-carotene, violaxanthin and neoxanthin are the most abundant carotenoids in chloroplast [14]. Finally, chromoplast are specialized plastid with large capacity to synthetize and accumulate massive amounts of carotenoids, and then providing particular coloration to specialized tissue, such as flowers or fruits [9, 15, 16].

Chromoplasts in most tissues contain plastoglobuli, which are lipid body structures accumulating carotenoids. Moreover, other types of carotenoid-accumulating structures may also be found inside the chromoplast, which determine their classification as globular, tubular, fibrillary, crystalline, and membranous [9, 15]. Globular chromoplasts are characterized by abundant plastoglobule, and are characteristics of many fruits such as mango (*Mangifera indica*), yellow papaya (*Carica papaya*), tomato (*Solanum lycopersicum*) [17], saffron stigma [18] and citrus [19, 20]. Globular chromoplasts usually contain lutein, β-cryptoxanthin, and β-carotene as major carotenoids [21]. Crystalline chromoplasts typically accumulate large amount of lycopene and β-carotene, as red/orange crystals and are abundant in tomato [17] and carrot (*Daucus carota*) [22]. Interestingly, while *all-trans*-β-carotene is deposited as crystalline, its *cis* isomers are in globular chromoplasts [23]. Membranous chromoplasts are typified by the large amount of condensed multi-layer membrane structures, as shown in orange cauliflower [24]. Fibrillar chromoplasts contain spindle-shaped structures with numerous tubules arranged in bundles, as in yellow papaya [21]. In general, fruits with low carotenoid content do not differentiate plastids of define any structure, whereas fruit accumulating uncommon carotenoids develop uncommon structures. The high heterogeneous nature or plasticity of the carotenoid-sequestering substructures in the chromoplasts is determinant of the diversity and the levels of carotenoids accumulating in different plant tissues.

The diversity in coloration and in the carotenoid content and composition in the peel and pulp of citrus fruits, provide an excellent system for the morphological and structural analysis of plastid diversity into the chromoplast and the relation with the carotenoid

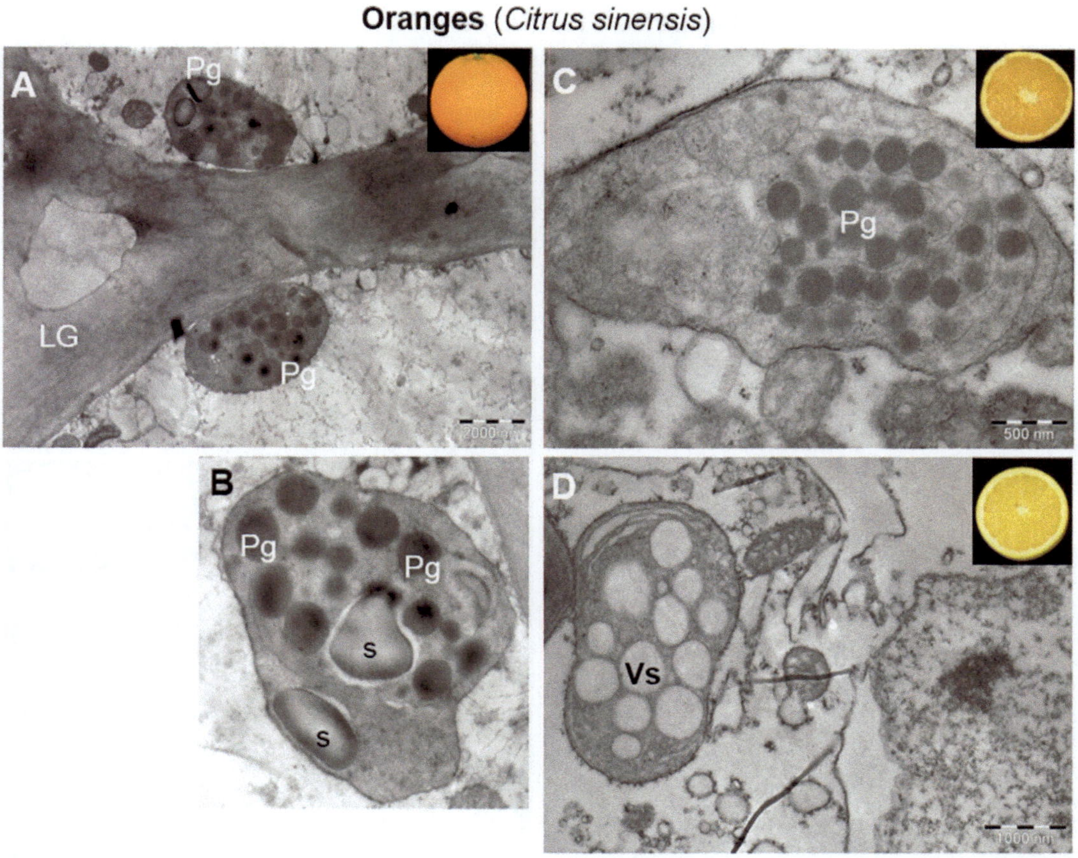

Fig. 1 Image of the ultrastructure of chromoplast from orange (*Citrus sinensis*) fruit with distinctive coloration. Peel (**a**) and pulp (**c**) of the ordinary mature Navel orange fruit. Magnification of a chromoplast from the peel (**b**). Pulp of the yellow-mutant Pinalate (**d**). Structures for carotenoid accumulation are indicated with abbreviations: Pg, plastoglobuli; LG, lipid grains; s, starch grains; Vs, round vesicles typical in the mutant Pinalate

complement [25–27]. Citrus plastids have an extraordinary plasticity to rearrange carotenoids inside different compartments to allow massive pigment accumulation, probably as a mechanism to avoid potential harmful effects [28]. Transmission electron microscopy (TEM) observations of peel and pulp sections of oranges (Fig. 1) and grapefruits (Fig. 2) of contrasting coloration, reveal striking differences in chromoplast structure, potentially related to their particular colors and to their unique carotenoid complement. The peel and pulp of ordinary orange fruit that accumulate high amounts of xanthophylls contained even and typical round plastoglobuli, which are easily visualized as electron-dense bodies of variable shape and size, coexisting in certain cases with starch grains (Fig. 1 a–c). The pulp of the yellow orange mutant Pinalate accumulates large amounts of unusual carotenoids, such as phytoene, phytofluene and ζ-carotene, and interestingly, their chromoplasts

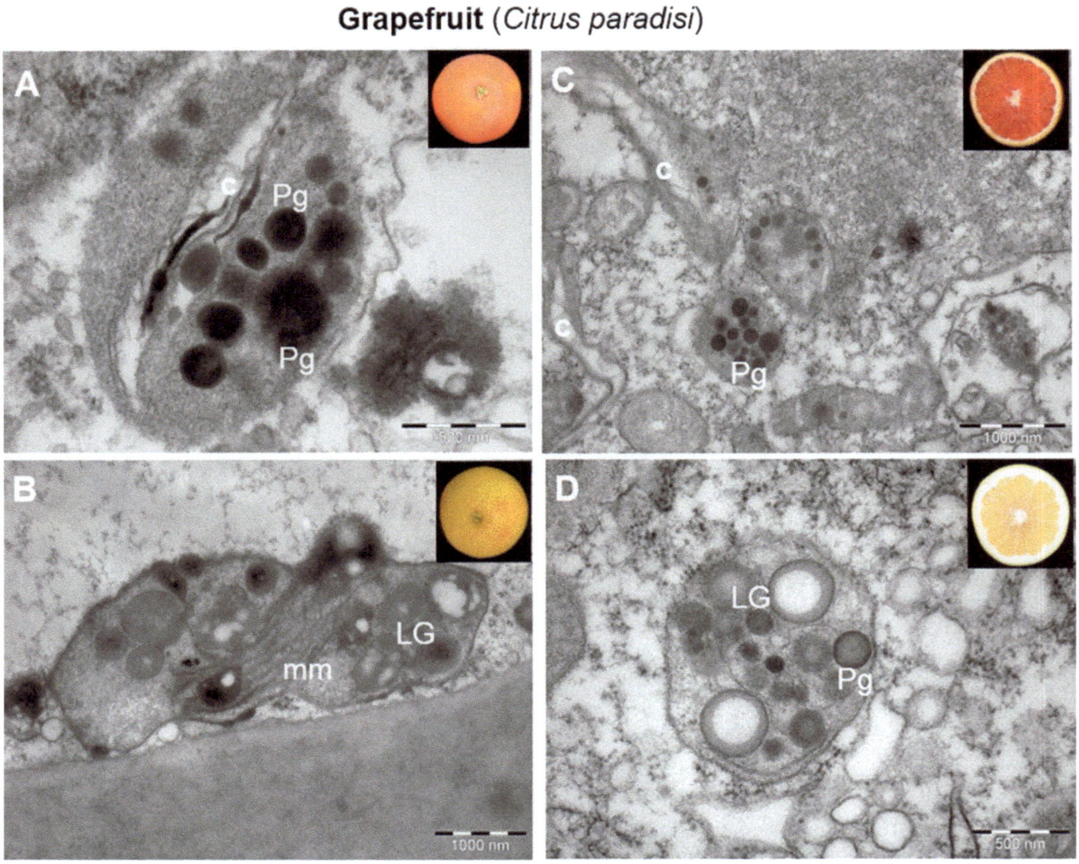

Fig. 2 Image of the ultrastructure of chromoplast from peel and pulp of red and yellow mature grapefruits (*Citrus paradisi*). Red peel (**a**) and pulp (**c**), and yellow peel (**b**) of Star Ruby grapefruit, and pulp of the yellow Marsh grapefruit (**d**). The intense red coloration of the peel in Star Ruby grapefruit was achieved by avoiding light exposure of fruit on the field, whereas the yellow fruit was directly exposed to sunlight on the field [19]. Structures for carotenoid accumulation are indicated with abbreviations: Pg, plastoglobuli; LG, lipid grains; s, starch grains; mm, achlorophyllous membranes; c, lycopene crystals with membrane remnants inside (after osmium fixation the crystalloids are largely dissolved during the dehydration procedure and their expanded envelopes are shrunken into an undulating shape)

contain novel even round vesicles, instead of the characteristics plastoglobuli of the ordinary oranges (Fig. 1d). Mature fruit of red grapefruit, such as Star Ruby, accumulate lycopene in the pulp and also in the peel (Fig. 2) [28]. Chromoplasts of these tissues contained characteristic plastoglobuli with electron-dense bodies of variable size and number. In these cells, the presence of lycopene is evidenced by the observation of long crystals (Fig. 2a, c). In yellow sectors of the peel and in the pulp of the white Marsh grapefruit, that contain very low carotenoid concentration, typical plastoglobuli are merely seen, showing instead lipid grains of variable size and achlorophyllous membranes in the peel (Fig. 2b, d). These observations support the motion that a differential carotenoid complement among plant cells determinates the type of plastid substructures for pigments storage [28].

Then, in order to understand the biosynthesis and accumulation of carotenoids in plant cell, and in fruits in particular, it is essential a reliable visualization of the carotenoids-containing structures in cells and plastids ultrastructure. Moreover, carotenoids storage has been also demonstrated to be determinant for their bioaccessibility. In this chapter, we provide a detailed protocol for transmission electron microscopy (TEM) analysis of plastids accumulating carotenoids, specially focused on fruit tissue, since these cells have high potential for carotenoid storage, and exhibit large carotenoid and plastid diversity.

2 Materials

2.1 Reagents and Solutions

1. PIPES buffer: Mix piperazine-NN′-bis-2-ethanesulfonic acid (PIPES)- 60.48 g with 10 M sodium hydroxide (NaOH) to bring the pH up to 7.2. Make up to 1 L with distilled water to produce a 0.2 M solution of PIPES buffer. The buffer can be divided up into suitable volumes and frozen until required.
2. Fixative solution (4% formaldehyde): Mix 4 g of paraformaldehyde with 20 mL distilled water in a small beaker. Heat to 85 °C on a hotplate in the fume cupboard. Add three drops of 0.1 M NaOH and stir until the liquid clears. Make up to 50 mL with distilled water. This will produce 8% formaldehyde and it should be then diluted 50/50 in PIPES buffer with glutaraldehyde (*see* **Note 1**).
3. Reynolds lead stain: Weight 1.33 g lead nitrate ($Pb(NO_3)_2$) and 1.76 g Tri-sodium citrate and place both into 30 mL distilled water in a very clean glass container. Shake forl min and then intermittently for 30 min to produce a milky suspension. Add 8 mL of freshly prepared 1 M sodium hydroxide and shake until clear. Make up to 50 mL with fresh distilled water. The stain can be divided up into suitable volumes and stored in sealed containers (*see* **Note 2**).
4. Uranyl acetate stain: Weight 1–2 g of uranyl acetate (*see* **Note 3**) and place it in a centrifuge tube capable of withstanding repeated centrifugation (i.e. polypropylene). Fill the tube with 50 mL of ethanol 50% and leave to stand in the fridge overnight. Solution reacts with light forming a precipitate, therefore should be stored in the dark. The tube must be centrifuged before use (*see* **Note 4**).

2.2 Equipment

1. Scalpels.
2. Petri dishes.

3. Tubes and racks for sample preparation.
4. Microcentrifuge.
5. Nickel grids for TEM samples.
6. Microtome.
7. Glass or diamond knife.
8. Transmission electron microscope (TEM) Hitachi H7000, equipped with a SIS Megaview III digital camera.

3 Methods

3.1 Sample Fixation, Preparation and Inclusion in Resin

1. The fruit tissue sample should be cut into small squares sections 3–5 mm long and 1 mm wide and fixed in the main fixative solution (3% glutaraldehyde +4% formaldehyde in 0.1 M PIPES buffer at pH 7.2). The tissue sample should be fixed for at least 3 h. Be aware of a fast inclusion in the fixative in order to avoid long contact with oxygen.
2. Rinse the sample with buffer (PIPES 0.1 M) for 10 min.
3. Postfix in buffered 1% osmium tetroxide prepared in 0.1 M PIPES buffer. This step should not exceed more than 1 hour. The sample could not become black in color. Move/rotate the sample during the treatment.
4. Rinse the section with buffer (PIPES 0.1 M) 10 min.
5. Rinse the section briefly in water.
6. Place the fruit section in 2% uranyl acetate for 20 min (not longer than 30 min) and move/rotate the sections during the treatment. Uranyl acetate staining could be done at this step or[f] afterward (*see* Subheading 3.3).
7. Place the tissue for dehydration sequentially in the following solutions: ethanol 30% for 10 min, ethanol 50% for 10 min, ethanol 70% for 10 min, ethanol 95% for 10 min, absolute ethanol for 20 min.
8. Place the section on acetonitrile for 10 min.
9. Place the section in 1 mL acetonitrile–Spurr resin (50:50) overnight.
10. Place the fruit section in the spur resin solution in continuous rotation during 6 h.
11. Embedding the fruit section in fresh epoxy resin (Spurr) allowing to polymerize in oven at 60 °C for a period between 20 and 24 h.
12. Collect the rising blocks and store them property labelled in dry environment.

3.2 Cutting Procedure

A microtome should be used; a diamond knife is highly recommended, but a glass knife could be also useful, which may be replaced after 2 or 3 cuttings.

1. Place the sample in the correct orientation.
2. Prepare and cut the resin end with a knife or scalpel.
3. Put water in the knife collector/recipient in order to allow the thin sections to launch on the water surface and to be easily recovered.
4. The thin sections could be first observed by optic microscopy in order to guarantee preservation of the tissue structure and quality.
5. A chloroform drop should be included in the water for section stretching before picking.
6. Fish sections with an instrument containing the selected grid. Tissue to be observed should be localized in the center of the grid. Sections should be put in nickel grids (and not copper).
7. Leave the sections to dry for 24 h before staining.

3.3 Grid Staining

1. Place a piece of wax-paper into a petri dish.
2. Invert grids section side down onto drops of uranyl acetate solution for 15 min in the dark (*see* **Notes 5** and **6**).
3. Wash gently in a stream of distilled water and blot dry.
4. Invert grids section side down onto drops of Reynolds lead stain (5 min) in the presence of NaOH pellets (in order to absorb CO_2). A petri dish with cover is a suitable place to avoid gas exchange. Lead stain produces a precipitate on exposure to carbon dioxide. Be careful and avoid breathing into samples during staining (*see* **Note 7**).
5. Wash gently in a stream of distilled water and blot dry (*see* **Notes 8** and **9**).
6. Store the staining grids in appropriate racks, properly identified, until microscopic observation (*see* **Note 9**).

3.4 TEM Observations

1. To initiate TEM observation of the different grids, the microscope should be property adjusted following the Microscopy Unit instructions and working protocols that may vary for each particular microscope.
2. As a general procedure, the grid should be mounted over the sample holder, properly adjusted to the microscope allowing then perform the vacuum.
3. Selection of appropriate voltage and brightness, and searching for observation cell sections of interest usually at low magnifications to get a wider image of the sections. Increase

magnification for an in deep observation discrete of cell sections and organelles ultrastructure.

4. Take pictures of the sections of interest using a SIS Megaview III digital camera.

4 Notes

1. Paraformaldehyde and formaldehyde are toxic by inhalation, ingestion and skin absorption and corrosive on skin contact. Wear gloves and use a cupboard.
2. It should be protected from CO_2 since produces a precipitate. It should not be used if it becomes cloudy.
3. Lead nitrate is toxic by inhalation and harmful by ingestion and skin absorption. Avoid breathing and use gloves.
4. Uranyl acetate is very toxic by ingestion and inhalation. Wear gloves and prepare the solution in a fume cup-board.
5. Centrifuge both stains before using (put an aliquot of 0.5 mL in an Eppendorf tube) and centrifuge at maximum speed.
6. Uranyl acetate should be made always before lead staining.
7. Uranyl acetate produce an undesired precipitate on exposure to light, therefore grids should be stained and stored in darkness afterward.
8. Both stains are harmful by inhalation and ingestion. Avoid breathing dust and wear gloves. In case of spillage wash with abundant water.
9. Leave samples to dry 24 h before TEM visualization. Store in the dark.

Acknowledgments

Financial support by research grants AGL-2015-70218 (Ministerio Economía y Competitividad (Spain), and FEDEGENT/2018/007 (Generalitat Valenciana, Spain) are gratefully acknowledged. MJR and LZ are members of Eurocaroten (COST_Action CA15136) and CaRed (Spanish Carotenoid Network, BIO2015-71703-REDT and BIO2017-90877-REDT).

References

1. Rodriguez-Concepcion M, Avalos J, Bonet ML, Boronat A, Gomez-Gomez L, Hornero-Mendez D, Limon MC, Meléndez-Martínez AJ, Olmedilla-Alonso B, Palou A, Ribot J, Rodrigo MJ, Zacarias L, Changfu Z (2018) A global perspective on carotenoids: metabolism, biotechnology, and benefits for nutrition and health. Prog Lipid Res 70:62–93

2. Fasset RG, Coombes JS (2012) Astaxanthin in cardiovascular health and disease. Molecules 17:2030–2048
3. Mordente A, Guantario B, Meucci E, Silvestrini A, Lombardi E, Martorana GE (2011) Lycopene and cardiovascular diseases: an update. Curr Med Chem 18:1146–1163
4. Rao AV, Rao LG (2007) Carotenoids and human health. Pharmacol Res 55:207–216
5. Li L, Yuan H (2013) Chromoplast biogenesis and carotenoid accumulation. Arch Biochem Biophys 539:102–109
6. Lopez-Juez E, Pyke KA (2004) Plastids unleashed: their development and their integration in plant development. Int J Dev Biol 49:557–577
7. Jarvis P, Lopez-Juez E (2013) Biogenesis and homeostasis of chloroplasts and other plastids. Nat Rev Mol Cell Biol 14:787–802
8. Howitt CA, Pogson BJ (2006) Carotenoid accumulation and function in seeds and non-green tissues. Plant Cell Environ 29:435–445
9. Li L, Yuan H, Zeng Y, Qiang X (2016) Carotenoids in nature biosynthesis, regulation and function. In: Stange C (ed) Plastids and carotenoids accumulation, 1st edn, Switzerland, pp 273–293
10. Wurtzel E, Cuttriss A, Vallabhaneni R (2012) Maize provitamin a carotenoids, current resources, and future metabolic engineering challenges. Front Plant Sci 3:29
11. Rodrıguez-Villalon A, Gas E, Rodrıguez-Concepcion M (2009) Phytoene synthase activity controls the biosynthesis of carotenoids and the supply of their metabolic precursors in dark-grown Arabidopsis seedlings. Plant J 60:424–435
12. Niyogi KK, Truong TB (2013) Evolution of flexible nonphotochemical quenching mechanisms that regulate light harvesting in oxygenic photosynthesis. Curr Opin Plant Biol 16:307–314
13. Ruban AV (2016) Nonphotochemical chlorophyll fluorescence quenching: mechanism and effectiveness in protecting plants from photodamage. Plant Physiol 170:1903–1916
14. Rodriguez-Concepcion M, Ruiz-Sola MA (2012) Carotenoid biosynthesis in Arabidopsis: a colorful pathway. Arabidopsis Book 10: e0158
15. Egea I, Barsan C, Bian W, Purgatto E, Latche A, Chervin C, Bouzayen M, Pech JC (2010) Chromoplast differentiation: current status and perspectives. Plant Cell Physiol 51:1601–1611
16. Lado J, Zacarias L, Rodrigo MJ (2016) Carotenoids in nature biosynthesis, regulation and function. In: Stange C (ed) Regulation of carotenoid biosynthesis during fruit development, 1st edn, Switzerland, pp 161–198
17. Jeffery J, Holzenburg A, King S (2012) Physical barriers to carotenoid bioaccessibility. Ultrastructure survey of chromoplast and cell wall morphology in nine carotenoid-containing fruits and vegetables. J Sci Food Agric 92:2594–2602
18. Gómez-Gómez L, Parra-Vega A, Rivas-Sendra JM, Seguí-Simarro R, Molina V, Pallotti C, Rubio-Moraga A, Diretto G, Prieto A, Ahrazem O (2017) Unraveling massive crocins transport and accumulation through proteome and microscopy tools during the development of saffron stigma. Int J Mol Sci 18:76
19. Lado J, Cronje P, Alquezar B, Page A, Manzi M, Gomez-Cadenas A, Stead AD, Zacarias L, Rodrigo MJ (2015) Fruit shading enhances peel color, carotenes accumulation and chromoplast differentiation in red grapefruit. Physiol Plant 154:469–484
20. Lu P, Wang C, Yin T, Zhong S, Grierson D, Chen K, Xu C (2017) Cytological and molecular characterization of carotenoid accumulation in normal and high-lycopene mutant oranges. Sci Rep 7:761
21. Schweiggert RM, Steingass CB, Heller A, Esquivel P, Carle R (2011) Characterization of chromoplasts and carotenoids of redand yellow-fleshed papaya (*Carica papaya* L.). Planta 234:1031–1044
22. Kim JE, Rensing KH, Douglas CJ, Cheng KM (2010) Chromoplasts ultrastructure and estimated carotene content in root secondary phloem of different carrot varieties. Planta 231:549–558
23. Vasquez-Caicedo AL, Heller A, Neidhart S, Carle R (2006) Chromoplast morphology and beta-carotene accumulation during postharvest ripening of mango cv. 'Tommy Atkins'. J Agric Food Chem 54:5769–5776
24. Paolillo DJ, Garvin DF, Parthasarathy MV (2004) The chromoplasts of or mutants of cauliflower (Brassica oleracea L. var. botrytis). Protoplasma 224:245–253
25. Gross J (1987) Pigments in fruits. In: Schweigert BS (ed) Food science and technology: a series of monographs. Academic Press. London Sci Technol
26. Alquézar B, Rodrigo MJ, Zacarías L (2008) Regulation of carotenoid biosynthesis during fruit maturation in the red-fleshed orange mutant Cara Cara. Phytochemistry 69:1997–2007

27. Rodrigo MJ, Alquézar B, Alos E, Lado J, Zacarias L (2013) Biochemical bases and molecular regulation of pigmentation in the peel of Citrus fruit. Sci Hortic 163:46–62

28. Lado J, Zacarías L, Gurrea L, Page A, Stead A, Rodrigo MJ (2015) Exploring the diversity in Citrus fruit colouration to decipher the relationship between plastid ultrastructure and carotenoid composition. Planta 242:645–661

Chapter 19

Light Microscopy and Raman Imaging of Carotenoids in Plant Cells In Situ and in Released Carotene Crystals

Tomasz Oleszkiewicz, Marta Z. Pacia, Ewa Grzebelus, and Rafal Baranski

Abstract

Light microscopy with a bright field mode offers an easy and fast examination of plant specimen for carotenoid presence in its cells. Using basic techniques such as hand sectioned or squashed preparations, carotenoid-rich chromoplasts can be identified without applying any staining procedure and their localization within the cell, their shape and number can be assessed. More detailed information can be obtained by using Raman spectroscopy which is suitable for the analysis of carotenoids due to their unique Raman spectra and allows semiquantification of their contents. Raman imaging (mapping) can be additionally used to show the distribution of carotenoids within the sample. Raman spectra can be taken from extracted carotenoids but can be also obtained directly from plant tissues or cells as Raman measurements are nondestructive for the sample. Here we describe preparations of intact tissue samples, monolayer cell samples, isolated protoplasts as well as carotene crystals released from chromoplasts that are suitable for subsequent observations using light microscopy and for analysis using Raman spectroscopy.

Key words Carrot, Chromoplasts, Bright-field microscopy, Nondestructive measurement, Protoplast, Raman mapping, Spectroscopy

1 Introduction

Carotenoid biosynthesis, accumulation and metabolism vary among species, plant organs and tissues. The content and composition of carotenoids is additionally modulated during ontogenesis and affected by environmental factors [1–3]. Carotenoids are components of chloroplasts and are also sequestered in nongreen plant tissues in chromoplasts. Four main types of chromoplasts are identified in plant tissues, that is, crystalloid, globular, tubular, and membranous [4]. One or more types of chromoplasts may occur in the same cell or tissue which is related to plastid differentiation and the transition between different plastid types [5]. Crystalline sub-structures may have different shapes, that is, needle-like, broad ribbon-like, and tube-like. They can be identified by their orange-red color and regular shape. Globular chromoplasts contain lipid-

Manuel Rodríguez-Concepción and Ralf Welsch (eds.), *Plant and Food Carotenoids: Methods and Protocols*, Methods in Molecular Biology, vol. 2083, https://doi.org/10.1007/978-1-4939-9952-1_19,

dissolved carotenoids in plastoglobuli of various sizes and are more often found in tissues exposed to light [4, 6, 7]. Both types can be identified using a light microscopy. Due to their absorbance in the visible light spectrum, carotenoids can be recognized in bright field mode without any staining. To avoid destruction of these pigments (dissolving in washing buffers), microscopic preparations should be prepared from a nonfixed specimen. A low resolution of light microscopy allows only for indicating the localization of crystalloid and globular chromoplasts in the cells, but their high number or close localization may restrict unambiguous determination of their shape, size, and number. Thus, chromoplast isolation from cells may be required for detailed characterizations. The other two types of chromoplasts—tubular and membranous—are less frequently observed; they have a fine ultrastructure and can be identified mainly by using transmission electron microscopy (TEM). However, this technique is destructive to carotenoids which are dissolved during sample fixation and usually only traces of preserved carotenoid remnants are finally visualized [6].

Carotenoids can be detected and identified nondestructively using Raman spectroscopy. This technique allows for measurements of carotenoids present in either intact living tissues or single cells; thus, neither chemical nor structural changes to carotenoids are expected, as neither exogenous chemicals nor physical forces are applied during sample preparation. A high resolution Raman spectroscopy enables also the measurements of individual carotene crystals both when present in the cell or after their extraction. The Raman spectra of the samples can be collected from a single point or by measuring several points of the defined area. The latter approach, Raman mapping or Raman imaging, can be used to assess the distribution of carotenoids and the homogeneity of the sample. Such information is retrieved by calculating the integral intensity of the carotenoid marker band at every point of the mapped sample and visualized using a color 2D surface plot.

Carotenoid molecules contain a chromophore, which is a polyene chain with alternating double and single bonds. When the wavenumber of the laser excitation used for Raman measurements coincides with an electronic transition of the individual carotenoid, that is, it is in the visible range (or near-infrared range, NIR), the resonance (or preresonance) Raman effect is observed [8]. Due to this effect, Raman spectra of carotenoids can be recorded even when these compounds are present at very low concentrations (down to 10^{-8} M) and when they are embedded in a complex biological matrix. The most intense and characteristic bands in the Raman spectra of carotenoids are located at 1500–1550 cm^{-1} (ν_1) and 1150–1170 cm^{-1} (ν_2), and are assigned to the C=C and C–C stretching modes, respectively. Additionally, in-plane rocking modes of CH_3 groups attached to the polyene chain can be observed as a medium intense band at 1000–1020 cm^{-1} [9]. The band location strongly depends on the number of the conjugated double bonds, length of the polyene chain, terminal moieties and their interaction with the environment. Thus, the band position in the Raman spectrum and the band shape can be used for identification of different carotenoids present in the sample, including biological material.

Carotenoids in plant tissues were thoroughly investigated using Raman spectroscopy. For fresh samples containing mainly β-carotene the Raman spectra exhibited the symmetric ν_1 band located at ca. 1520 cm^{-1}, while for samples containing a higher amount of lutein and α-carotene that marker band was asymmetric and exhibited a shoulder at 1527 cm^{-1}. Lycopene rich samples provided Raman spectra with a band at 1510 cm^{-1} [10]. Raman imaging applied to leaf, fruit and root sections revealed different content and distribution of carotenoids, depending on the tissue. Point measurements and high resolution imaging of carotenoid crystals performed directly in the plant cell as well as after crystal isolation indicated that the crystals may have different compositions depending on their morphology, that is, rhomboidal, needle-like, or ribbon-like shape. Moreover, the distribution of various carotenoids within an individual crystal was visualized [11, 12].

Here we present protocols that enable sample preparation for both light microscopy and Raman spectroscopy of carotenoids. We describe the preparation of intact macroscopic biological samples as well as microscopic slides with a monolayer of cells [11] and isolated protoplasts [13]. The protocol extends to the procedure of carotenoid crystals isolation from protoplasts and is a modification of the method reported by Wang et al. [14]. The protocols are optimized for carrot samples but can be applied to other species as well, in particular when intact tissue samples or monolayer cell specimens are desired.

2 Materials

2.1 Sample Preparation

2.1.1 Plant Material

1. Any of the following carrot material:
   (a) Carrot storage root.
   (b) Callus tissue maintained in vitro.

2.1.2 Protoplast Isolation

1. Centrifuge capable to achieve at least 150 × *g* and equipped with an angle-tilt rotor for 15 ml tubes.
2. 26 °C rotary incubator with a flat table to hold petri dishes.
3. Laminar flow cabinet.
4. Autoclave.
5. pH meter.

6. Laboratory scale.
7. Automatic pipettes for 1–5 ml and 100–1000 µl.
8. Glass or plastic funnels (Ø 6 cm).
9. Nylon sieve fitting the funnel (80 µm pore, Ø 8 cm).
10. Tweezers.
11. Scalpel.
12. Sterile 15 ml centrifuge tubes.
13. Sterile 3 ml Pasteur pipettes.
14. Sterile petri dishes Ø 90 mm or 2 ml Eppendorf tubes (*see* **Note 1**).
15. 100 ml and 250 ml glass or plastic beakers.
16. 100–1000 µl and 1–5 ml sterile pipette tips.
17. Preplasmolysis solution (PS) (0.5 M mannitol): dissolve 22.77 g mannitol in dH_2O, adjust volume to 250 ml with distilled water (dH_2O) and pH to 5.6–5.8. Sterilize by autoclaving at 0.1 MPa and 121 °C for 15 min. Store at room temperature (RT).
18. Enzyme mixture solution (ES): dissolve 1 g cellulase Onozuka R10, 0.1 g pectolyase Y-23, 10.93 g mannitol (0.6 M final conc.), 0.39 g 2-(N-morpholino)ethanesulfonic acid (MES) (20 mM final conc.), 0.1 g $MgCl_2 \times 6H_2O$ (5 mM final conc.) in dH_2O (*see* **Note 2**). Adjust volume to 100 ml with dH_2O and pH to 5.6. Sterilize by filtering using 0.22 µm syringe filter. Store at 4 °C. For long-term storage, place at −20 °C.
19. Sucrose (0.5 M)/MES (1 mM) solution: dissolve 42.79 g sucrose and 0.05 g MES in dH_2O. Adjust volume to 250 ml with dH_2O and pH to 5.8. Sterilize by autoclaving at 0.1 MPa and 121 °C for 15 min. Store at RT.
20. W5 solution: dissolve 2.25 g NaCl (154 mM final conc.), 4.59 g $CaCl_2 \times 2H_2O$ (125 mM final conc.), 0.09 g KCl (5 mM final conc.) and 0.23 g glucose (5 mM final conc.) in dH_2O. Adjust volume to 250 ml with dH_2O and pH to 5.8. Sterilize by autoclaving at 0.1 MPa and 121 °C for 15 min. Store at RT.

2.1.3 Extraction of Carotenoid Crystals

1. Centrifuge with cooling to 4 °C capable to achieve 26,000 × *g* and equipped with a rotor for 2 ml tubes.
2. Automatic 1000 µl pipette.
3. 1000 µl pipette tips.
4. 1.5 ml Eppendorf tubes.
5. 2 ml Eppendorf tubes with round shaped bottom.

6. 25 and 50 ml glass or plastic beakers
7. Crushed ice.
8. 2× Extraction Buffer (2× EB): dissolve 1.19 g 4-(2-hydroxyethyl)-1-piperazineethanesulfonic acid (HEPES) (100 mM final conc.), 0.06 g ethylenediaminetetraacetic acid disodium salt (EDTA) (4 mM final conc.), 6.01 g sorbitol (660 mM final conc.) in dH_2O. Adjust volume to 50 ml with dH_2O and pH to 7.5 using NaOH (*see* **Note 3**). Store at 4 °C.
9. Sucrose gradient solutions: dissolve 5.0 g, 3.0 g, and 1.7 g sucrose in 5 ml of 2× EB each, and adjust volume with dH_2O to 10 ml each to acquire 50%, 30%, and 17% sucrose in 1× EB, respectively. Store at 4 °C.
10. 2-Mercaptoethanol.

2.1.4 Microscopic Slide Preparation

1. Dry block thermostat up to 60 °C.
2. 2 ml Eppendorf tubes.
3. Glass petri dish or microscopic slide.
4. Automatic pipette for 100–1000 μl.
5. 100–1000 μl pipette tips.
6. SuperFrost glass microscope slides (75 × 25 × 1 mm) or Raman grade calcium fluoride (CaF_2) slides (*see* **Note 4**).
7. Coverslips (22 × 22 mm or 22 × 32 mm).
8. Small pieces of filter paper (3 × 2 cm).
9. Wooden pencil with round-shaped end.
10. Adhesive tape.
11. Preparation needle.
12. Tweezers.
13. Knife.
14. Deep freezer for −80 °C (Optional).
15. 1 N HCl: take 0.1 mol of HCl and add to dH_2O, make up to 100 ml.

2.2 Light Microscopy and Raman Spectroscopy

1. Bright-field microscope for routine tasks (10×–60× objective magnification).
2. Raman spectrometer equipped with a mapping stage and suitable software (*see* **Note 5**).

3 Methods

3.1 Sample Preparation

3.1.1 Intact Organ Specimen

1. Cut a thin (0.1–2.0 mm) slice of carrot root or other tissue and place it on a microscope slide. Cover with another microscopic slide and press gently to flatten the specimen. Protect the movement of slides by joining their both ends using an adhesive tape that will protect also any specimen movement.
2. Use directly for Raman measurements.

3.1.2 Cell Monolayer Specimen

1. Using a knife make manual cross-sections of the storage root of ca. 1 mm thickness or less. Then cut into 2 × 5 mm fragments containing the secondary phloem. Place into an Eppendorf tube and add 1 N HCl to completely cover the tissue. Incubate on dry block thermostat at 60 °C for 40 min (*see* **Note 6**).
2. Using forceps gently transfer small pieces of softened tissue to the microscope slide. Absorb the excess of HCl with filter paper and add 1–2 drops of dH_2O.
3. Attach the edge of the coverslip to the side of the water drop and slowly lower the coverslip over the specimen to prevent forming air bubbles.
4. Gently apply pressure to the coverslip using the wooden pencil with round-shaped end to flatten the specimen and to finally spread individual cells (*see* **Note 7**).
5. Check the quality of the preparation under the light microscope. If a multilayer specimen is visible apply additional pressure to the coverslip with the pencil until cells are dispersed in monolayers.
6. Start observation under the bright-field microscope immediately as water used to close the preparations evaporates quickly und samples might shrink due to drought (*see* **Note 8**).
7. For Raman spectroscopy remove the coverslip. For this purpose put the water-closed slides to the deep freezer (−80 °C) for 48 h. Then lever the coverslip with the scalpel blade and tear it out vigorously (*see* **Note 9**) and let the slides air dry for 3–5 min.
8. Use the open slide directly for Raman measurements.

3.1.3 Protoplast Specimen

1. Place 1–3 g of callus or root pieces in a petri dish (*see* **Note 10**) and add 8 ml of PS solution to cover completely the plant tissue.
2. Incubate plant material at 26 °C for 1 h in the dark on a rotary shaker with gentle rotation (20–50 rpm).
3. Carefully remove as much as possible PS solution by pipetting while avoiding moving the plant tissue.

4. Add 8 ml ES solution and incubate the plant material at 26 °C for 3–5 h in the dark on a rotary shaker with gentle rotation (20–50 rpm) to release protoplasts (*see* **Note 11**).
5. Mount a 80 μm nylon sieve in the funnel placed in the 15 ml centrifuge tube.
6. Separate undigested tissue by transferring the whole suspension from the petri dish to the sieve using a Pasteur pipette and collect the released protoplasts in the 15 ml centrifuge tube (*see* **Note 12**). Pipet slowly to avoid damaging (cell wall-free) protoplasts.
7. Centrifuge the filtered suspension of protoplasts and cell debris at 100 × *g* for 5 min.
8. Remove supernatant and resuspend the pellet in 8 ml of sucrose/MES solution.
9. Carefully pipette 2 ml of W5 solution onto the top of the sucrose/MES solution with resuspended protoplasts. Two distinct phases should be visible (*see* **Note 13**). Centrifuge at 145 × *g* for 10 min.
10. After centrifugation a thin layer of protoplasts between two phases is visible. Using a Pasteur pipette gently collect the protoplasts and transfer them to a 2 ml tube with a round-shaped bottom (*see* Fig. 1).
11. Centrifuge at 100 × *g* for 5 min.
12. Remove supernatant and use the pellet for carotenoid crystal extraction (*see* Subheading 3.1.4).
13. Alternatively add a few drops of W5 and transfer 1–2 drops of protoplast suspension to a microscopic slide using a Pasteur pipette and cover with the coverslip.
14. Prepare slides for light microscopy or Raman spectroscopy as described in the **steps 6** or **7** in the Subheading 3.1.2, respectively.

3.1.4 Carotenoid Crystals Specimen

The centrifuge and solutions should be cooled down to 4 °C before crystal isolation. During the whole procedure all solutions and samples should be kept on ice.

1. Prepare 1× EB by mixing 2× EB and dH_2O in 1:1 ratio. Add 0.35 μl 2-mercaptoethanol for each 1 ml of 1× EB and for each 1 ml of sucrose gradient solutions (50%, 30%, 17%).
2. Add 1 ml of dH_2O to the protoplast pellet in 2 ml tube (**step 12** in the Subheading 3.1.3).
3. Mix tubes with protoplasts for at least 1 min by inverting (*see* **Note 14**).
4. Centrifuge at 350 × *g* for 5 min.

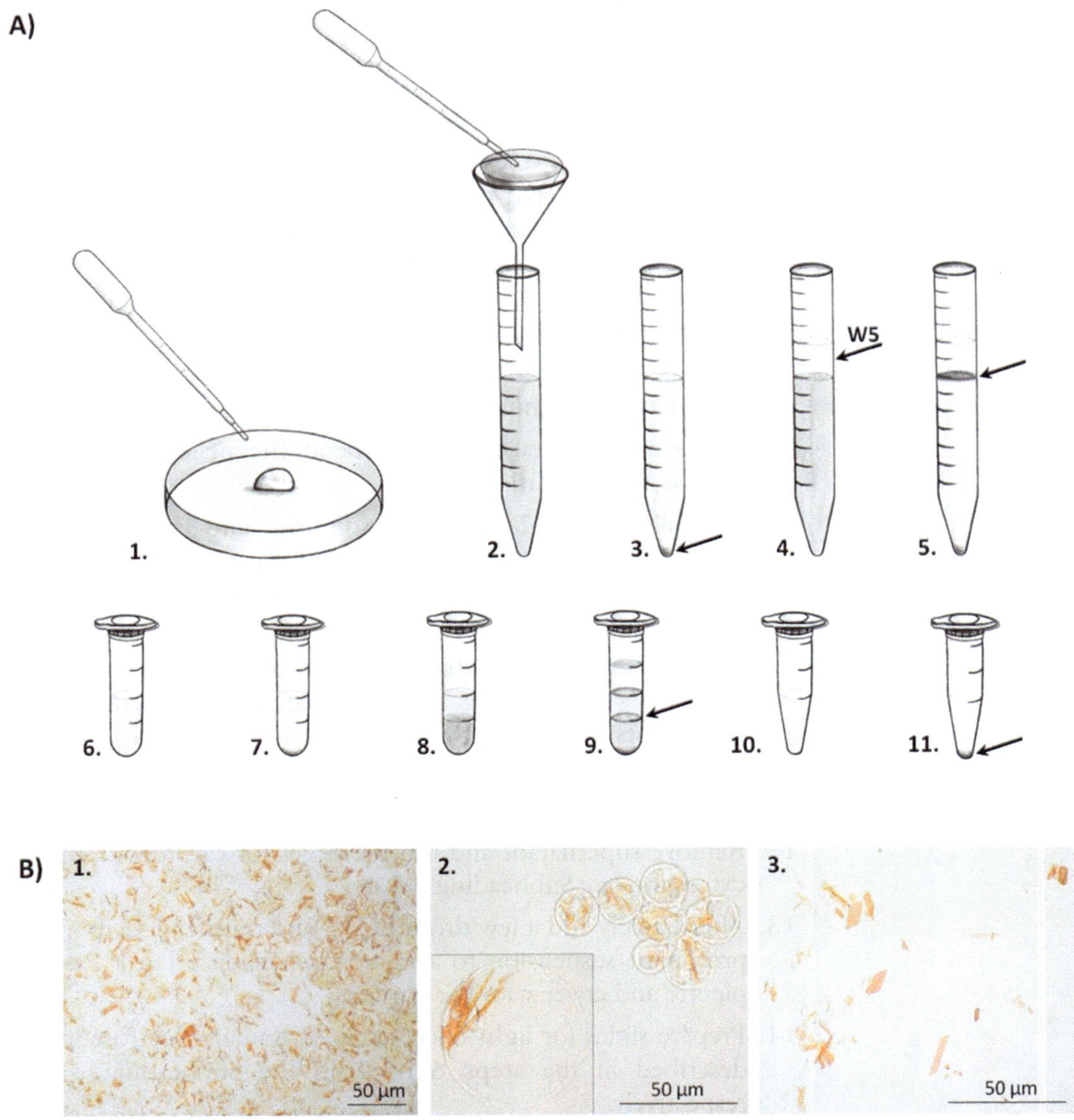

Fig. 1 Isolation of protoplasts and carotenoid crystals (**a**). (1) Callus tissue incubation in enzymatic solution. (2) Separation of digested and undigested tissue. (3) Centrifuged protoplasts at the bottom of tube (*arrow*). (4) Resuspended protoplasts overlaid with W5 solution. (5) Layer of protoplasts between two phases after centrifugation (*arrow*). (6, 7) Collected protoplast suspension before and after centrifugation. (8) Sucrose gradient with protoplasts resuspended in the bottom 50% sucrose layer. (9) Chromoplasts arrested in the ring between 50% and 30% sucrose layers (*arrow*). (10, 11) Purification of carotenoid crystals and the final pellet of crystals (*arrow*). Microscopic images (**b**). (1) Monolayer specimen of callus cells after treatment with 1 N HCl at 60 °C for 5 min. (2) Isolated protoplasts from callus rich in carotenoids. (3) Carotenoid crystals released from protoplasts (Photos: Dariusz Kadłuczka)

5. Remove supernatant and carefully resuspend the pellet in 50% sucrose in 1× EB.
6. Carefully pipette 30% sucrose in 1× EB on the inner side of the tube to create a second layer of the sucrose gradient (*see* **Note 15**).
7. Repeat the previous step by pipetting 17% sucrose in 1× EB to create the third layer of the sucrose gradient.
8. Centrifuge at 26,000 × *g* for 45 min in the centrifuge cooled down to 4 °C.
9. Collect intact chromoplasts localized between the 50% and 30% sucrose layers using an automatic or Pasteur pipette (*see* Fig. 1).
10. Resuspend the pellet in 1× EB and centrifuge at 5000 × *g* for 5 min.
11. Remove supernatant and resuspend the pellet in 1 ml dH_2O.
12. Centrifuge at 20,000 × *g* for 3 min.
13. Repeat **steps 11** and **12** two times more to wash crystals from the remaining EB solution.
14. Resuspend the pellet in 0.2–0.5 ml dH_2O (*see* **Note 16**).
15. Transfer one drop of carotenoid crystals onto a microscope slide and smear very gently (to avoid breaking the crystals) using a coverslip (*see* **Note 17**). Let the slides air dry for 3–5 min in the darkness.
16. Use directly for light microscopy or Raman measurements.

3.2 Light Microscopy

1. Put the slide onto microscopic stage and observe using 10× objective at first. 40×–60× magnifications are recommended to observe the details of carotenoid crystals both located in the cell and extracted from the cell (*see* Fig. 1b).
2. Adjust light intensity and aperture to visualize crystals, whereas orange colour of carotenoid crystals should contrast with non-colored cell organelles.

3.3 Raman Spectroscopy

1. Place the slide with the sample onto the instrument table under the microscope and focus on the sample.
2. For single point Raman measurements select the following measurement conditions:
   (a) Laser wavelength (*see* **Notes 18** and **19**).
   (b) Laser power (*see* **Note 20**).
   (c) Integration time (*see* **Note 21**).
   (d) Number of accumulations (*see* **Note 22**).
   (e) Spectral resolution (*see* **Note 23**).

   The scheme of single point Raman measurement is presented in Fig. 2a.

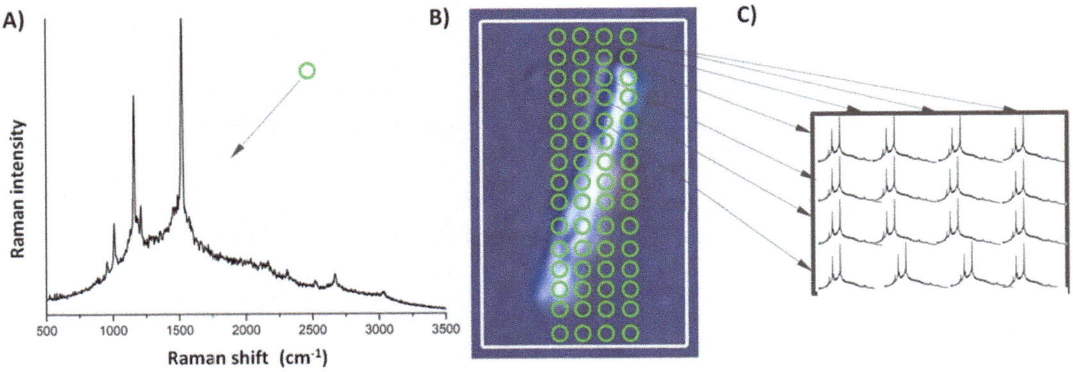

Fig. 2 Raman spectrum obtained as a result of a single-point measurement of the carotenoid crystal (**a**). Positioning of the measurement points during the Raman imaging. From each point a single Raman spectrum is acquired. The number and location of measurement points is determined by the sampling density (**b**). The array of Raman spectra in the whole wavenumber range recorded from selected points in the defined sample area (**c**)

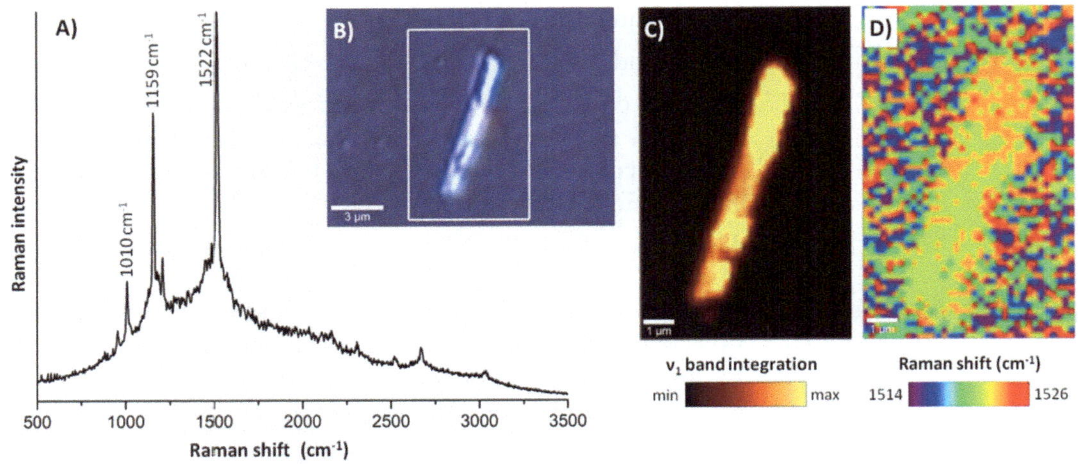

Fig. 3 An exemplary Raman spectrum of the isolated carotenoid crystal (**a**) and a microscopic image (100× magnification) of this crystal (**b**). Distribution of carotenoids visualized based on the intensity of the marker band integrated in the 1495–1540 cm^{-1} range (**c**). Distribution of carotenoids visualized based on the positions of the marker band in the 1514–1526 cm^{-1} range (**d**)

3. For Raman imaging select additionally the size of samples to be imaged and sampling density (*see* **Note 24**). The scheme of Raman imaging measurement is presented in Fig. 2b, c.
4. Optimize all the parameters (*see* **Note 25**) and perform measurement. If necessary, repeat the measurements by choosing the same area within a sample and optimize the measurement condition.
5. Read the band positions and assign the bands in single Raman spectra of carotenoids (*see* **Note 26**) (Fig. 3a).

6. For Raman imaging carry out the analyses of the selected marker bands:
   (a) Distribution of carotenoids (*see* **Note 27**) (Fig. 3c).
   (b) Homogeneity of carotenoid composition (*see* **Note 28**) (Fig. 3d).

4 Notes

1. It is recommended to use Ø 90 mm petri dish, however a smaller size can be used, but then the volume of solutions should be downscaled proportionally. Protoplast isolation can also be done in 2 ml Eppendorf tubes using one-eighth volume of recommended solutions, however in this case protoplast yield may be too low for carotenoid crystals isolations. Use glass petri dishes if it is necessary to cut the tissue.
2. Start preparing the ES solution with dissolving the salts, then pectolyase, and then cellulase by continuous low-speed stirring as enzymes require long time to dissolve.
3. The pH adjusts very slowly. To avoid changing the volume of extraction buffer use high concentrated solution of NaOH or solid NaOH in powder or tablets.
4. Raman measurements of carotenoids can be done using glass microscopic slides or CaF_2 slides of Raman grade. The Raman signal coming from microscopic glass is significantly lower than the signal coming from carotenoids. Glass microscope slides can be safely used when measurements are focused only on carotenoids (without measuring biological matrix, e.g., in protoplasts). Nevertheless, it is recommended to perform Raman measurements with the use of CaF_2 slides. Raman signal coming from CaF_2 substrate (one single Raman band at ca. 320 cm^{-1}) do not interfere with Raman signal originating from samples.
5. To localize small objects, a microscope (or at least a camera) is required. To optimize measurement parameters (e.g., spatial or spectral resolution vs. signal enhancement) the possibility of using various lasers is convenient.
6. When using a soft tissue like callus, place small pieces (0.3 cm × 0.3 cm) into an Eppendorf tube with ca. 1 ml 1 N HCl (to cover the tissue) and incubate on a dry block thermostat at 60 °C for 5–7 min to soften the specimen.
7. Apply slight and uniform pressure to avoid breaking the coverslip.
8. If microscopic observations are to be done later, seal the coverslip with colorless nail polish for longer storage of the slide.

9. The critical point of this step is to prevent thawing the coverslip, otherwise the specimen will be removed along with the coverslip.
10. For more efficient enzyme maceration it is recommended to cut the tissue in PS solution into fine pieces using, for example, a scalpel. For carotenoid crystals isolation more tissue (ca. 3 g) is needed.
11. The duration of the enzyme treatment depends on tissue type and on the desired protoplast yield. For carrot callus tissue small number of protoplasts can be produced already after 1 h incubation time. For root tissues overnight digestion is recommended.
12. To isolate protoplasts from undigested tissue, prepare a prefilter column by placing nylon sieve in the plastic funnel inserted into a centrifuge tube. The nylon sieve will retain undigested tissue.
13. To achieve distinct phases between two solutions, add the W5 solution drop by drop along the tube wall over sucrose/MES solution to avoid mixing of the solutions.
14. After water addition the cell membranes break due to osmotic pressure and carotenoid crystals are released from the chromoplasts into the solution.
15. It is very important to avoid mixing of solutions. This step should be done carefully, with smoothly working automatic pipette.
16. Pellet of crystals can be stored at 4 °C in the dark until slide preparation.
17. Insert the coverslip edge into the drop with crystals and drag it along the slide. Alternatively, use an inoculating needle instead of the coverslip.
18. The intensity of Raman signal is proportional to the fourth power of the frequency of the incident light. This means that the strongest Raman spectra are obtained when a measurement is performed with a laser emitting in the UV region while the weakest spectra are obtained in the infrared region. However, the risk of side effect (e.g., fluorescence, photodamage of a sample) is higher when using the UV-Vis lasers than the NIR lasers. Spectrometers may be equipped with several lasers of different wavelengths. Dispersive Raman spectrometers are usually equipped with lasers emitting UV, visible and infrared (IR) light while Fourier transformed (FT, also called interferometric) spectrometers are coupled with a Nd:YAG laser (yttrium aluminum garnet doped with neodymium Nd^{3+} ions) operating in NIR region. So, the shorter the laser wavelength, the stronger Raman signal can be obtained, and

detection of the analyte is possible at lower concentrations, but the risk of side effects is higher. Raman measurements using a FT-spectrometer with the NIR laser offer spectra of lower intensity and resolution, but much lower sample fluorescence which can be the only option for the analysis of many plant sample.

19. For samples containing a chromophore it is possible to perform resonance Raman measurements. When the wavenumber of the laser excitation used for Raman measurements coincides with an electronic transition of the individual molecule, the resonance (or preresonance) Raman effect is observed. Due to this effect, Raman spectra of the analytes can be recorded even when these compounds are present in the sample at very low concentrations (down to 10^{-8} M).
20. The laser power varies depending on the laser type and needs to be adjusted for each sample. The higher the laser power, the stronger the Raman signal observed. Too high power can cause sample heating and water evaporation thus resulting in sample damage (Table 1).
21. The higher the integration time, the stronger the Raman signal. Integration time is a parameter set for dispersive Raman spectrometers. It is the exposure time of the sample to laser light as well as the time of recording signals during a single-point Raman measurement. Too long integration time can cause damage to the sample.
22. Number of accumulations is the number of individual spectra registered at the same point (number of scans per point), averaged over scanned points. The higher the number of accumulations, the better the Raman spectrum quality (higher signal to noise ratio).
23. Spectral resolution set for interferometric spectrometers is usually 2 or 4 cm^{-1}. In dispersive Raman spectrometers, spectral resolution depends on the chosen type of grating (e.g., 600 and 1800 grooves per mm providing spectral resolution of 3 and 1 cm^{-1}, respectively).
24. The sampling density (an increment, a step size) defines how many points are measured in the selected sample area. The lower the increment, the more points per line and lines per image are measured, and the better is the spatial resolution of the Raman image obtained. A measurement of more points requires longer acquisition time, so the compromise between these two criteria, the spatial resolution and the acquisition time, is necessary. It may be useful to measure the whole area with a lower spatial resolution (small number of points) first and then to repeat the measurement of a selected smaller area with a higher spatial resolution.

Table 1
Exemplary Raman imaging parameters used for measurements of intact carrot root slices, carotenoid crystals present in carrot cells and isolated carotenoid crystals

Sample	Type of spectrometer[a]	Laser wavelength (nm)	Laser power (mW)	Integration time (s) per spectrum	No. of accumulations (scans)	Spectral resolution (cm^{-1})	Size of the image (μm)	Sampling density (μm)	Reference
Root	I	1064	150	–	8	4	10,500 × 12,000	250	[15]
Cells	D	532/488	40/30	0.3	1	3/1	10 × 10	0.2	[11]
Isolated crystal	D	533/488	20/0.01	0.1/0.3	1	3/1	20 × 20	0.2/0.5	[12]

[a] *I* interferometric (FT), *D* dispersive spectrometer

25. The Raman imaging parameters including laser wavelength, laser power, integration time (for dispersive Raman spectrometers), size of the image and sampling density, are interdependent. For example, by increasing the laser power the integration time can be reduced. Lower laser power or integration time can be applied when the measurement is performed using a laser of a shorter wavelength (i.e., from UV-Vis region). Optimization of spectrometer parameters is crucial for successful Raman measurements. Measurement parameters are different for various samples and always must be determined individually taking into account which spectrometer type is used. Sample parameters for measuring specimens containing carotenoids are presented in Table 1.
26. Generally, Raman spectrum of carotenoids contain three intense bands located at 1500–1550 cm^{-1}, 1150–1170 cm^{-1}, and 1000–1020 cm^{-1} assigned to the C=C stretching modes, C–C stretching modes, and in-plane rocking modes of CH_3 groups attached to the polyene chain, respectively. The exact positions of these bands highly depend on the number of the conjugated double bonds in the structure of carotenoid molecule (i.e., length of the polyene chain), terminal moieties as well as on the interaction between carotenoids and other compounds in the sample. The band located in the 1500–1550 cm^{-1} wavenumber range is the most discriminating one.
27. Distribution of carotenoids can be visualized by plotting intensities of the selected marker band. Generally, the intensity can be calculated as the band height or band integral intensity (area under the band). As the result a 3-dimensional matrix is obtained (Fig. 2c), where the x and y dimensions indicate the location of sample point within the measured area where Raman signal was acquired, and the z dimension provides information about the intensity of the marker band in the Raman spectrum at individual points. The intensity is proportional to the carotenoid amount occurring in the sample at that point.
28. Homogeneity of carotenoid compositions can be visualized by plotting the intensities of the marker bands characteristic for various carotenoid compounds. Positions of the marker bands in Raman spectra are shown in colors corresponding to the chosen range of band wavelengths. Carotenoids occurring in specific places of the sample can be identified by referring to the color scale indicating the position of the marker Raman band (Fig. 2d).

Acknowledgments

The authors thank Prof. M. Baranska for valuable comments on the manuscript and Mr. W. Wesolowski for consultations on chromoplast preparation. The protocol is partially the result of the research project supported by the National Science Centre in Poland (UMO-2013/09/B/NZ9/02379).

References

1. Egea I, Barsan C, Bian W, Purgatto E, Latche A, Chervin C, Bouzayen M, Pech J (2010) Chromoplast differentiation: current status and perspectives. Plant Cell Physiol 51 (10):1601–1611
2. Schaub P, Rodriguez-Franco M, Cazzonelli CI, Alvarez D, Wust F, Welsch R (2018) Establishment of an *Arabidopsis* callus system to study the interrelations of biosynthesis, degradation and accumulation of carotenoids. PLoS One 13(2):e0192158
3. Sun T, Yuan H, Cao H, Yazdani M, Tadmor Y, Li L (2018) Carotenoid metabolism in plants: the role of plastids. Mol Plant 11(1):58–74
4. Schweiggert R, Carle R (2017) Carotenoid deposition in plant and animal foods and its impact on bioavailability. Crit Rev Food Sci Nutr 57(9):1807–1830
5. Pyke K (2007) Plastid biogenesis and differentiation. In: Bock R (ed) Cell and molecular biology of plastids, vol 19. Springer-Verlag, Berlin, pp 1–28
6. Kim JE, Rensing KH, Douglas CJ, Cheng KM (2010) Chromoplasts ultrastructure and estimated carotene content in root secondary phloem of different carrot varieties. Planta 231:549–558
7. van Wijk K, Kessler F (2017) Plastoglobuli: plastid microcompartments with integrated functions in metabolism, plastid developmental transitions, and environmental adaptation. Annu Rev Plant Biol 68:253–289
8. Ozaki Y, Cho R, Ikegaya K, Muraishi S, Kawauchi K (1992) Potential of near-infrared Fourier transform Raman spectroscopy in food analysis. Appl Spectrosc 46(10):1503–1507
9. Schulz H, Baranska M, Baranski R (2005) Potential of NIR-FT-Raman spectroscopy in natural carotenoid analysis. Biopolymers 77:212–221
10. Baranska M, Baranski R, Schulz H, Nothnagel T (2006) Tissue-specific accumulation of carotenoids in carrot roots. Planta 224 (5):1028–1037
11. Roman M, Marzec K, Grzebelus E, Simon P, Baranska M, Baranski R (2015) Composition and (in)homogeneity of carotenoid crystals in carrot cells revealed by high resolution Raman imaging. Spectrochim Acta A Mol Biomol Spectrosc 136:1395–1400
12. Rygula A, Oleszkiewicz T, Grzebelus E, Pacia MZ, Baranska M, Baranski R (2018) Raman, AFM and SNOM high resolution imaging of carotene crystals in a model carrot cell system. Spectrochim Acta A Mol Biomol Spectrosc 197:47–55
13. Grzebelus E, Szklarczyk M, Baranski R (2012) An improved protocol for plant regeneration from carrot leaf- and hypocotyl-derived protoplasts. Plant Cell Tissue Organ Cult 109:101–109
14. Wang Y-Q, Yang Y, Fei Z, Yuan H, Fish T, Thannhauser TW, Mazourek M, Kochian LV, Wang X, Li L (2013) Proteomic analysis of chromoplasts from six crop species reveals insights into chromoplast function and development. J Exp Bot 64:949–961
15. Baranski R, Baranska M, Schulz H (2005) Changes in carotenoid content and distribution in living plant tissue can be observed and mapped *in situ* using NIR-FT-Raman spectroscopy. Planta 222:448–457

Part V

Dynamics

Chapter 20

Carotenoid Isotopolog Profiling in ^{13}C-Labeled Leaf Extracts by LC-MS and LC-FTICR-MS

Björn Thiele and Shizue Matsubara

Abstract

Mass spectrometry (MS)-based metabolite analysis combined with stable isotope labeling offers a powerful tool to study dynamic regulation of metabolic pathways and metabolite fluxes in biological systems. Here we describe a method to analyze the composition of carotenoid isotopologs in ^{13}C-labeled leaf extracts by using liquid chromatography (LC)-MS and LC-Fourier transform ion cyclotron resonance (FTICR)-MS. High mass resolution of the latter enables unambiguous assignment of observed mass to a unique chemical formula. Based on peak intensity the relative abundance and the degree of ^{13}C labeling are calculated for individual carotenoid isotopologs.

Key words ^{13}C labeling, Carotenoid metabolism, Degree of labeling, FTICR-MS, Isotopolog profiling, LC-MS

1 Introduction

Plants are a rich source of naturally occurring carotenoids. They accumulate the yellow to orange-reddish pigments in fruits and flowers to attract pollinators and seed dispersers. In leaves carotenoids play a dual role, collecting sunlight in the blue wavelength region to drive photosynthesis, while at the same time protecting the photosynthetic membrane against photooxidation. Furthermore, some carotenoids can serve as precursors to produce plant hormones, signaling agents, and fragrances. Owing to these important functions in plants, biosynthesis and degradation of carotenoids have been studied intensively by researchers [1, 2], aided by analytical tools to identify and quantify pathway products and intermediates.

Stable isotope labeling was a key technique in the early days of biochemical pathway analysis, including the biosynthesis of isoprenoids and carotenoids [3, 4]. Besides using specifically labeled organic tracers, carbon fluxes in plants can be studied in vivo by

Manuel Rodríguez-Concepción and Ralf Welsch (eds.), *Plant and Food Carotenoids: Methods and Protocols*, Methods in Molecular Biology, vol. 2083, https://doi.org/10.1007/978-1-4939-9952-1_20,

$^{13}CO_2$ labeling. The heavy carbon of $^{13}CO_2$ is quickly incorporated in various primary metabolites in photosynthesizing leaves [5]. In species, which emit isoprene from leaves, rapid ^{13}C labeling of isoprene via the plastidic methylerythritol phosphate (MEP) pathway is closely linked to photosynthetic CO_2 fixation in the Calvin-Benson cycle [6, 7]. Part of carbon fluxes through these pathways also maintain the levels of photosynthetic pigments (carotenes and chlorophyll *a*) which undergo continuous turnover in illuminated leaves [8]. Insights into dynamic pathway regulation and interaction in different genotypes and environments can be gained by quantitative analysis of isotope incorporation in metabolites of the corresponding pathways.

Mass spectrometry (MS) can reliably differentiate isotopically labeled metabolites based on their mass. Accurate mass determination by high resolution MS, such as Fourier transform ion cyclotron resonance (FTICR)-MS and Orbitrap-MS, enables unambiguous annotation of metabolites by empirical formula calculation [9]. Today MS-based technologies coupled with separation techniques, such as liquid chromatography (LC) and gas chromatography, have become an essential platform to study metabolic pathways. Especially, they provide powerful tools for nontargeted or multitargeted metabolite analysis (metabolite profiling and fingerprinting) in complex biological samples, thus complementing the fast advancement in genomic, transcriptomic and proteomic approaches [9–11]. When combined with stable isotope labeling, metabolite snapshots obtained from a time course experiment can provide dynamic information about pathway fluxes in a metabolic network [12, 13].

Here we describe a method for carotenoid profiling in ^{13}C-labeled leaf extracts using LC- MS and LC-FTICR-MS. Pigments are extracted from leaves of plants that are cultivated in a $^{13}CO_2$ environment [14, 15]. After separation by LC and detection by (photo)diode array detector (PDA or DAD), ionization is performed by electrospray ionization (ESI) or atmospheric pressure chemical ionization (APCI) in the positive mode [16–18]. Mass assignment of product ions to specific carotenoid isotopologs is achieved by FTICR-MS. Based on peak intensity, the relative abundance and the degree of ^{13}C labeling are calculated for individual carotenoid isotopologs.

2 Materials

Prepare all aqueous solutions with ultrapure deionized water and analytical grade (LC-MS grade) reagents. Discard remaining extracts and solvents following waste disposal regulations.

2.1 Pigment Extraction

1. Optional for grinding of frozen leaf material: Ball mill with stainless steel balls (3 mm diameter) and adaptors to fit microcentrifuge tubes (Retsch MM 300 or equivalent).
2. Liquid nitrogen.
3. Acetone: Precool a necessary amount in a glass bottle on ice.
4. 2 mL microcentrifuge tubes.
5. Refrigerated microcentrifuge (Eppendorf centrifuge 5415R or equivalent).
6. 3 mL or 5 mL disposable syringes.
7. Nylon syringe filters (0.45 μm pore size, 3 mm diameter).
8. Amber glass vials (MS certified).

2.2 LC-PDA-MS

1. 500 mM ammonium acetate (NH_4Ac) stock solution: Weigh 3.85 g NH_4Ac in a 100 mL volumetric flask and make up to 100 mL with water (*see* **Note 1**).
2. Waters ACQUITY UPLC system (comprising a binary pump and autosampler) or equivalent.
3. Waters PDA detector or equivalent.
4. Waters Xevo TQ-S triple quadrupole mass spectrometer or equivalent.
5. Bischoff ProntoSil 200–3-C30 column (250 × 4.6 mm, 3 μm particle size) and a corresponding guard column (10 × 4.0 mm, 3 μm) or equivalent.
6. Elution solvent A: Mix 499 mL methanol (MeOH) and 1 mL 500 mM NH_4Ac stock solution in a glass bottle to make 1 mM NH_4Ac in MeOH/water (99.8:0.2, v/v).
7. Elution solvent B: Methyl tert-butyl ether (MTBE).
8. Standards of plant photosynthetic carotenoids (Table 1) purchased from DHI (Hørsholm, Denmark) (*see* **Note 2**).

2.3 LC-DAD-FTICR-MS

1. Agilent 1200 series HPLC system consisting of a binary pump, autosampler, column oven and DAD or equivalent.
2. Hybrid linear ion trap FTICR mass spectrometer LTQ FT Ultra (Thermo Fisher Scientific) equipped with a 7 T supraconducting magnet or equivalent.
3. For HPLC column, elution solvents and pigment standards, *see* Subheading 2.2, **items 5–8**.

3 Methods

Carry out pigment extraction under a fume hood and wearing safety glasses. Dim the light and keep samples on ice while extracting pigments to avoid degradation. Amber glass vials should be used for LC-MS for protection against light.

Table 1
Chemical formulae and monoisotopic mass of photosynthetic carotenoids in plants

Carotenoid	Abbreviation	Chemical formula	Monoisotopic mass[a] [u]
Violaxanthin	Vio	$C_{40}H_{56}O_4$	600.417869
Neoxanthin	Neo	$C_{40}H_{56}O_4$	600.417869
Antheraxanthin	Anth	$C_{40}H_{56}O_3$	584.42295
Lutein	Lut	$C_{40}H_{56}O_2$	568.42803
Zeaxanthin	Zea	$C_{40}H_{56}O_2$	568.42803
α-Carotene	α-Car	$C_{40}H_{56}$	536.43820
β-Carotene	β-Car	$C_{40}H_{56}$	536.43820

[a]Values are for the lightest (most abundant) isotopologs

3.1 Pigment Extraction

1. Grinding of leaf samples: Precool a mortar and a pestle. Pour a small amount of liquid N_2 into the mortar and grind ^{13}C-labeled (or nonlabeled) leaf material (*see* **Note 3**) carefully to fine powder. If you use a ball mill, precool stainless steel balls and mill adaptors before use. The vibrational frequency and milling duration should be optimized for different plant leaves. You can continue with pigment extraction directly, or collect the frozen leaf powder in a precooled centrifuge tube, freeze it in liquid N_2 and store at −80 °C until extraction.
2. Weigh 15–20 mg of frozen leaf powder quickly; leaf powder should not thaw while weighing. Write down the exact weight of each sample (*see* **Note 4**).
3. Transfer the weighed frozen leaf powder into a mortar precooled with a small amount of liquid N_2. Add 1 mL of chilled acetone and homogenize the sample quickly but thoroughly until no leaf particles are recognized. Collect the homogenate into a 2-mL microcentrifuge tube using a pipette (*see* **Note 5**) and put it on ice.
4. Rinse the mortar and pestle with 0.5 mL of chilled acetone. Do not discard this acetone; it should be collected and added to the sample. Adjust the total volume of the sample to 2 mL. The optimal volume for extraction is different for different plant leaves.
5. Repeat **steps 2–4** for all samples.
6. Centrifuge the samples at 15,700 × *g*, 4 °C for 5 min.
7. Collect the green supernatant into a fresh microcentrifuge tube and put it on ice. Check the color of the pellets remaining at the bottom of the first microcentrifuge tubes. If the pellets are still greenish or yellowish, extraction is not complete. Repeat the **steps 3**, **4**, and **6** until the pellets become white (*see* **Note 6**).

8. Filter the extract into an amber glass vial through a nylon syringe filter by gently pressing down on the plunger of a disposable syringe (*see* **Note 7**).
9. Place the open vial containing the extract briefly under a gentle stream of N_2 gas (*see* **Note 8**) to blow away the air (O_2) above the extract. Close the vial with a cap under the N_2 stream.

3.2 Obtaining Mass Spectra with LC-PDA-MS

1. Optimize the ionization parameters of MS by infusing 0.3–0.4 μg/mL solutions of β-carotene and lutein (representatively for carotenes and xanthophylls) dissolved in 1 mM NH_4Ac in MeOH/water (99.8, 0.2). The optimal parameters will be different for different instruments. In the case of Xevo TQ-S (Waters) with ESI in the positive mode, the capillary voltage is set to 2.5 kV, cone voltage 30 V, desolvation temperature and source temperature to 500 °C and 150 °C, respectively. The desolvation gas flow is 1000 L/h and the cone gas flow 150 L/h, both using N_2.
2. Chromatographic separation is achieved with a C30 column using the following LC parameters: column temperature 25 °C, injection volume 50 μL and flow rate 0.5 mL/min. Start a gradient program with a linear gradient from 85 to 70% solvent A over 12 min, isocratic separation at 70% A for 6 min, then a linear gradient to 15% A over 5 min, isocratic at 15% A for 7 min, return to the initial condition of 85% A over 5 min followed by equilibration at 85% A for 5 min (total run time: 40 min).
3. Measure absorption spectra with PDA detector to identify carotenoid species. To estimate the retention times and determine the concentrations of different carotenoids, set the PDA detector to 450 nm and calibrate the system using the pigment standards (*see* **Note 9**).
4. Subsequent to the PDA detection, mass spectra are obtained for individual carotenoid species in the full scan mode in the range of mass-to-charge ratio (m/z) between 350 and 1000 (*see* **Note 10**). Adjust the dwell time and gain of the MS detection to minimize background signal (*see* **Note 11**).
5. Process data using MassLynx software version 4.1 (Waters) or equivalent.

3.3 Confirmation of ^{13}C Isotopologs by LC-DAD-FTICR-MS

1. Optimize the ionization parameters of the MS in the same way as described for LC-MS in Subheading 3.2, **step 1**. The optimal parameters will be different for different instruments. We use LTQ FT Ultra (Thermo Fisher Scientific) with APCI operated in the positive mode at 4.0 μA ion spray current. Nitrogen serves as both the sheath gas (40 arb) and the auxiliary gas (10 arb). The transfer capillary temperature is set at 350 °C and

the vaporizer temperature at 400 °C. 40 V and 130 V are applied to capillary and tube lens, respectively.

2. Set up the LC and DAD as described for LC-MS in Subheading 3.2, **steps 2** and **3**.
3. Measure mass spectra in the full scan mode from *m/z* 350 to 1000 (*see* **Note 10**) with a mass resolution of 100,000 (full width at half maximum) at *m/z* 400. Set the automatic gain control to provide a constant ion population in the ICR cell. For LTQ FT Ultra in the full scan mode, the automatic gain control is set at $5e^5$ with the maximum ion trap fill time of 10 ms and the maximum ICR cell fill time of 500 ms.
4. Process data using Xcalibur software version 2.0.7 (Thermo Fisher Scientific) or equivalent.

3.4 Processing of MS Data and Calculation of the Degree of ^{13}C Labeling

1. Chromatograms obtained by LC-MS: Place the 450 nm and total ion current (TIC) chromatograms of the same sample above each other (Fig. 1) to identify the peak of interest in the TIC chromatogram. Figure 1b shows lutein and zeaxanthin peaks in the extracted ion chromatogram at *m/z* 568 ($^{12}C_{40}H_{56}O_2$).
2. Extract mass spectrum of the selected pigment and run background subtraction (*see* **Note 11**). Set the mass range to cover all possible isotopologs and their quasi-molecular ions (*see* **Note 12**). Four types of quasi-molecular ions are found in mass spectra of leaf carotenoids by using our method: $[M]^+$ and $[M + H]^+$ for carotenes and additionally two dehydration products $[M + H - H_2O]^+$ and $[M + H - 2\ H_2O]^+$ for xanthophylls (Table 2). The mass spectrum in Fig. 2 belongs to lutein extracted from partially ^{13}C-labeled leaves.
3. Proceed with the chromatograms of FTICR-MS in the same way as described for LC-MS in Subheading 3.4, **steps 1** and **2**.
4. In the XCalibur software, set the constraints of isotope numbers (Table 3) to calculate elemental compositions of carotenoid isotopologs. Select the option to show deviations (delta; in ppm) from the expected mass of isotopologs (Fig. 3). The high mass resolution of FTICR-MS allows mass assignment to a unique chemical formula by distinguishing ^{12}C-^{13}C substitution (change in mass = 1.003355 u) from ^1H addition/subtraction (change in mass = 1.007825 u).
5. Export the high accuracy spectrum list of FTICR-MS to Excel. Inspect the list carefully and delete peaks if their *m/z* values do not match the expected mass of any isotopologs of the carotenoid species analyzed. Once all carotenoid isotopolog peaks are identified by FTICR-MS in representative samples of an experiment, the same annotation can be applied to other samples of the same experiment.

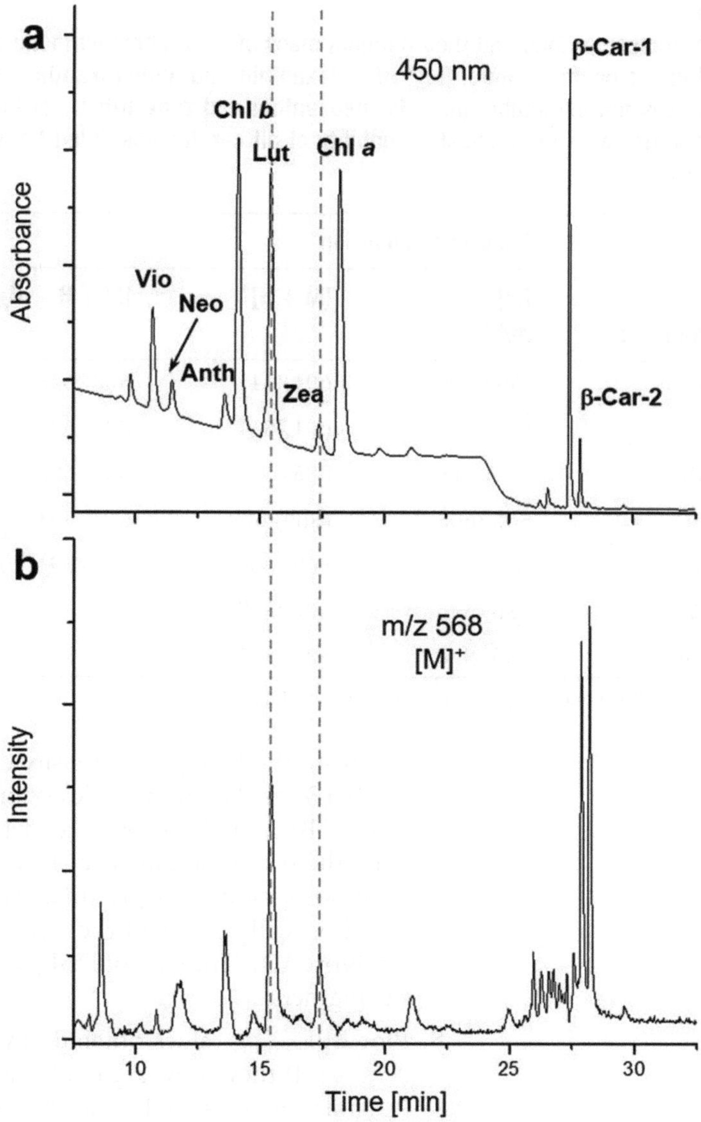

Fig. 1 Chromatograms of a leaf pigment extract obtained by LC-PDA-MS. (**a**) Pigment separation monitored at 450 nm. Vio, violaxanthin; Neo, neoxanthin; Anth, antheraxanthin; Chl *b*, chlorophyll *b*; Lut, lutein; Zea, zeaxanthin; Chl *a*, chlorophyll *a*; β-Car, β-carotene. (**b**) Total ion current chromatogram of the same sample extracted at *m/z* 568 in tandem with PDA. The positions of Lut and Zea peaks ($^{12}C_{40}H_{56}O_2$) are indicated by *dotted lines*

6. Export the spectrum list of LC-MS to Excel. Compare the *m/z* values of the two lists and delete no-match peaks also from the LC-MS list. Apply the mass assignment of FTICR-MS to all spectrum lists of LC-MS.

Table 2
Quasi-molecular ions and their nominal mass of plant photosynthetic carotenoids observed by ESI or APCI in the positive mode. $[M]^+$ of violaxanthin and antheraxanthin is detected at very low levels while it is not detectable (n.d.) for neoxanthin and zeaxanthin. For lutein, which, upon protonation, loses water with the highest probability of all carotenoids listed here, only a trace of $[M + H]^+$ is found

	Quasi-molecular ion			
Carotenoid	$[M]^+$ m/z^a	$[M + H]^+$	$[M + H - H_2O]^+$	$[M + H - 2\ H_2O]^+$
Vio	(600/640)	601/641	583/623	565/605
Neo	(n.d.)	601/641	583/623	565/605
Anth	(584/624)	585/625	567/607	549/589
Lut	568/608	(569/609)	551/591	533/573
Zea	(n.d.)	569/609	551/591	533/573
α-Car	536/576	537/577		
β-Car	536/576	537/577		

[a]Values are for monoisotopic isotopologs $^{12}C_{40}/^{13}C_{40}$

7. Calculate the base peak intensity of each isotopolog (BPI_i; *see* **Note 13**) as shown in Tables 4 and 5. Normalize BPI_i to the sum of BPI_i of all isotopologs detected in the sample to estimate the relative abundance ($BPI_{i(\mathrm{norm})}$). The ^{13}C enrichment in an isotopolog (or degree of labeling, DoL_i) can be calculated based on $BPI_{i(\mathrm{norm})}$ and the number of ^{13}C in the isotopolog (*see* **Note 13**). The sum of DoL_i of all isotopologs gives DoL of the pigment.
8. Plot $BPI_{i(\mathrm{norm})}$ values obtained by LC-MS versus FTICR-MS (Fig. 4). If they show a good linear correlation with a slope of 1, similar samples can be analyzed by LC-MS alone. The correlation must be checked for each pigment.

4 Notes

1. A common problem with NH_4Ac solutions is rapid growth of microalgae especially at room temperature. To avoid this problem, we recommend amber glass bottles or bottles wrapped with aluminum foil to store NH_4Ac solutions at 4 °C.
2. Open the ampules containing the standards directly before use. Once opened, the standards cannot be stored as they are prone to degradation by oxygen and light.

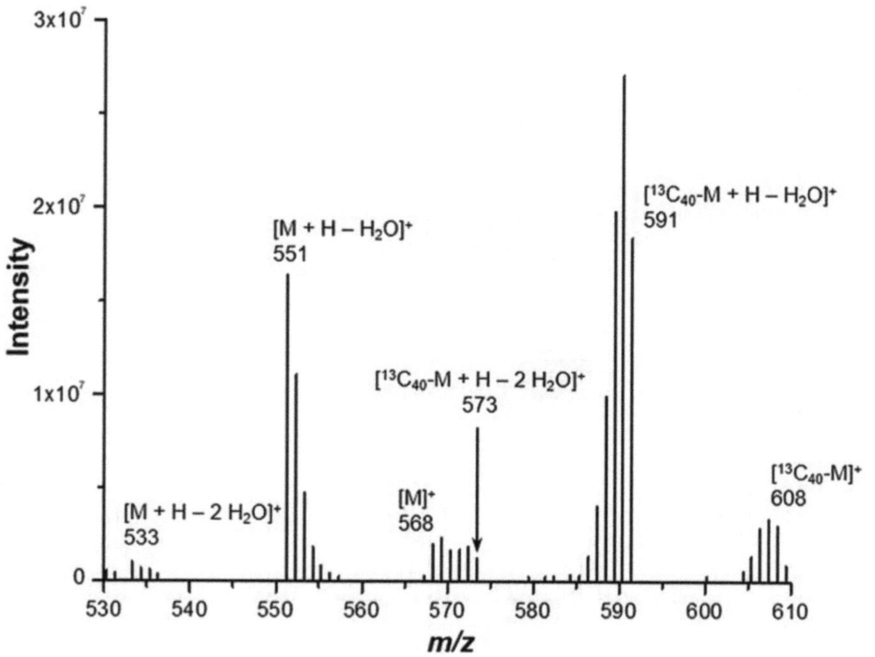

Fig. 2 Mass spectrum of lutein from a partially ^{13}C-labeled leaf sample. Shown are data in the mass range between *m/z* 530 and *m/z* 610. Three types of quasi-molecular ions ($[M]^+$, $[M + H - H_2O]^+$ and $[M + H - 2 H_2O]^+$) are observed for lutein, with a minor peak of $[M + H]^+$ at *m/z* 609. Note that the peaks of nonlabeled $[M]^+$ and ^{13}C-labeled $[M + H - 2 H_2O]^+$ are partly overlapping at *m/z* ~570. These peaks cannot be separated by triple quadrupole MS

Table 3
The setting of isotope numbers used for calculation of elemental compositions of carotenoid isotopologs

Isotope	Natural abundance [%]	Mass [u]	Min. number	Max. number
^{1}H	99.9885	1.007825	1	57
^{12}C	98.93	12.000000	0	40
^{13}C	1.07	13.003355	0	40
^{16}O	99.757	15.994914	0	4

3. You can use fresh leaf material which is taken from plants directly before extraction. In many cases, however, leaves are collected beforehand in centrifuge tubes or aluminum bags, shock-frozen in liquid N_2 and stored at −80 °C until extraction.
4. You need to measure sample weight to calculate concentrations of different carotenoids per unit leaf mass. For concentrations per unit leaf area, there is no need to determine the weight but the area of the sample must be known.

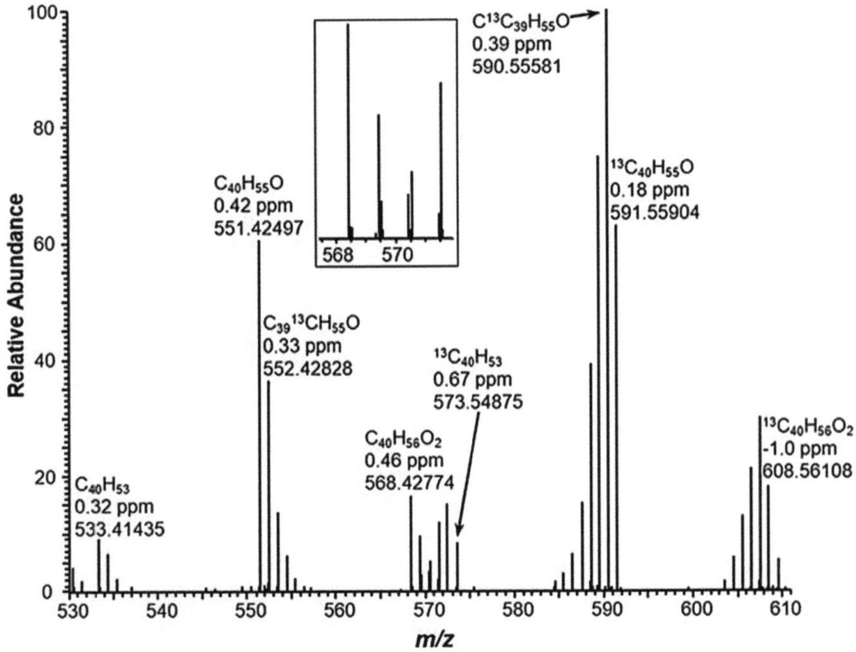

Fig. 3 High-resolution mass spectrum of lutein. Data are from the same sample as shown in Fig. 2. High mass accuracy of FTICR-MS allows assignment of each mass to a unique empirical formula. Deviations from the expected mass of isotopologs are given in parts per million (ppm). The mass differences between nonlabeled $[M]^+$ and ^{13}C-labeled $[M + H - 2\ H_2O]^+$ at *m/z* ~570 are resolved by FTICR-MS (inset)

5. If acetone has evaporated during homogenization in the mortar, add some (e.g., 0.5 mL) chilled acetone to the green slurry so that you can collect it with a pipette.
6. You can check the effectiveness of extraction by eyes. After complete extraction of pigments (carotenoids and chlorophylls) pellets have a white color (or brownish in the case of anthocyanin accumulation). Use less volume of acetone when you do reextraction from pellets. All extracts from the same sample must be collected together in a larger centrifuge tube. Write down the total volume of the extract for calculation of pigment concentration.
7. Very rarely, leaves of some plant species contain polymers which block nylon membrane. If there is a strong back pressure, do not apply excessive pressure to the syringe as this could cause the filter housing to burst.
8. Brief application of N_2 gas stream is sufficient to blow away the air out of the sample vial. Longer exposure to dry N_2 gas will evaporate acetone and alter the pigment concentration.
9. The concentrations of carotenoids in your samples must be within the range of calibration.

Table 4
Calculation of base peak intensity (BPI_i), normalized BPI ($BPI_{i(norm)}$) and degree of ^{13}C labeling (DoL_i) for individual isotopologs. Data are for lutein obtained by FTICR-MS

m/z	Intensity	Delta (ppm)	Empirical formula	Quasi-molecular ion	BPI_i	$BPI_{i(norm)}$ (%)	*i*	DoL_i (%)
533.41435	34052.9	0.32	$C_{40}H_{53}$	$[M + H - 2 H_2O]^+$	7.09	1.14	0	0.00
534.41755	27257.4	0.03	$C_{39}{}^{13}CH_{53}$		5.68	0.92	1	0.02
535.42062	7322.5	−0.49	$C_{38}{}^{13}C_2H_{53}$		1.53	0.25	2	0.01
551.42497	299994.0	0.42	$C_{40}H_{55}O$	$[M + H - H_2O]^+$	62.48	10.07	0	0.00
552.42828	176169.1	0.33	$C_{39}{}^{13}CH_{55}O$		36.69	5.92	1	0.15
553.43165	70194.0	0.35	$C_{38}{}^{13}C_2H_{55}O$		14.62	2.36	2	0.12
554.43528	31908.2	0.86	$C_{37}{}^{13}C_3H_{55}O$		6.65	1.07	3	0.08
555.43866	14361.3	0.89	$C_{36}{}^{13}C_4H_{55}O$		2.99	0.48	4	0.05
556.44183	5491.6	0.57	$C_{35}{}^{13}C_5H_{55}O$		1.14	0.18	5	0.02
568.42774	84386.7	0.46	$C_{40}H_{56}O_2$	$[M]^+$	17.57	2.83	0	0.00
569.43038[a]	49129.1	−0.81	$C_{39}{}^{13}CH_{56}O_2$		10.23	1.65	1	0.04
570.53868[a]	35126.0	0.66	$C_3{}^{13}C_{37}H_{53}$	$[M + H - 2 H_2O]^+$	7.32	1.18	37	1.09
571.54186[a]	50284.3	0.36	$C_2{}^{13}C_{38}H_{53}$		10.47	1.69	38	1.60
572.54516	65877.7	0.25	$C^{13}C_{39}H_{53}$		13.72	2.21	39	2.16
573.54875	57541.9	0.67	$^{13}C_{40}H_{53}$		11.98	1.93	40	1.93
584.53577	4676.7	0.55	$C_7{}^{13}C_{33}H_{55}O$	$[M + H - H_2O]^+$	0.97	0.16	33	0.13
585.53728	10401.5	−2.6	$C_6{}^{13}C_{34}H_{55}O$		2.17	0.35	34	0.30
586.54286	34141.2	1.18	$C_5{}^{13}C_{35}H_{55}O$		7.11	1.15	35	1.00
587.54588	88589.6	0.62	$C_4{}^{13}C_{36}H_{55}O$		18.45	2.97	36	2.68
588.54931	193427.3	0.75	$C_3{}^{13}C_{37}H_{55}O$		40.28	6.49	37	6.01
589.55250	377630.9	0.46	$C_2{}^{13}C_{38}H_{55}O$		78.65	12.68	38	12.05
590.55581	480154.9	0.39	$C^{13}C_{39}H_{55}O$		100.00	16.12	39	15.72
591.55904	328182	0.18	$^{13}C_{40}H_{55}O$		68.35	11.02	40	11.02
603.54477	4244.8	−0.22	$C_5{}^{13}C_{35}H_{56}O_2$	$[M]^+$	0.88	0.14	35	0.12
604.54849	25521.7	0.39	$C_4{}^{13}C_{36}H_{56}O_2$		5.32	0.86	36	0.77
605.55171	52490.3	0.16	$C_3{}^{13}C_{37}H_{56}O_2$		10.93	1.76	37	1.63
606.55496	108882.4	−0.01	$C_2{}^{13}C_{38}H_{56}O_2$		22.68	3.66	38	3.47
607.55812	150582.8	−0.33	$C^{13}C_{39}H_{56}O_2$		31.36	5.06	39	4.93
608.56108	86775.4	−0.98	$^{13}C_{40}H_{56}O_2$		18.07	2.91	40	2.91
609.57005	23360.5	0.9	$^{13}C_{40}H_{57}O_2$	$[M + H]^+$	4.87	0.78	40	0.78
					Σ	100.00		70.80

[a]As shown in the inset of Fig. 3, there are at least six peaks detected between *m/z* 569 and *m/z* 572. The above calculation takes into account the larger ones of the overlapping peaks. Recalculation including the three smaller peaks results in a decrease in the total DoL by 0.3

10. The scanning mass range can be narrowed to *m/z* 530–610 for analysis of carotenoids. By extending the scan to *m/z* 350–1000 you have the possibility to analyze other compounds in the same extracts, such as chlorophylls and tocopherols. Triple quadrupole mass spectrometer can then be used to obtain fragment spectra of those compounds. Note that tocopherols must be analyzed by PDA/DAD at 295 nm.

Table 5
Calculation of base peak intensity (BPI_i), normalized BPI ($BPI_{i(norm)}$) and degree of ^{13}C labeling (DoL_i) for individual isotopologs. Data are for lutein from the same sample as in Table 4 but obtained by LC-MS

m/z	Intensity	Quasi-molecular ion	BPI_i	$BPI_{i(norm)}$ (%)	*i*	DoL_i (%)
533.3	1.04E+06	$[M + H - 2\ H_2O]^+$	3.20	0.63	0	0.00
534.3	8.19E+05		2.53	0.50	1	0.01
535.3	3.72E+05		1.15	0.22	2	0.01
551.3	1.67E+07	$[M + H - H_2O]^+$	51.47	10.08	0	0.00
552.3	1.19E+07		36.71	7.19	1	0.18
553.3	5.30E+06		16.37	3.21	2	0.16
554.3	2.44E+06		7.52	1.47	3	0.11
555.3	1.24E+06		3.83	0.75	4	0.07
556.3	5.56E+05		1.72	0.34	5	0.04
568.3	2.19E+06	$[M]^+$	6.77	1.33	0	0.00
569.3[a]	2.55E+06		7.87	1.54	1	0.04
570.3[a]	1.90E+06	$[M + H - 2\ H_2O]^+$	5.86	1.15	37	1.06
571.4[a]	2.22E+06		6.85	1.34	38	1.27
572.4	2.33E+06		7.19	1.41	39	1.37
573.4	1.47E+06		4.53	0.89	40	0.89
584.4	5.78E+05	$[M + H - H_2O]^+$	1.78	0.35	33	0.29
585.4	3.89E+05		1.20	0.24	34	0.20
586.4	1.55E+06		4.80	0.94	35	0.82
587.4	4.98E+06		15.36	3.01	36	2.71
588.4	1.21E+07		37.33	7.31	37	6.76
589.4	2.38E+07		73.33	14.36	38	13.64
590.4	3.24E+07		100.00	19.58	39	19.09
591.4	2.10E+07		64.87	12.70	40	12.70
604.4	7.68E+05	$[M]^+$	2.37	0.46	36	0.42
605.4	1.82E+06		5.61	1.10	37	1.02
606.4	3.78E+06		11.67	2.28	38	2.17
607.4	4.67E+06		14.42	2.82	39	2.75
608.4	3.81E+06		11.77	2.31	40	2.31
609.4	8.50E+05	$[M + H]^+$	2.62	0.51	40	0.51
			Σ	100.00		70.61

[a]These peaks appear as three single peaks in the mass spectra of LC-MS instead of six as detected by FTICR-MS (*see* Figs. 2 and 3 inset)

11. Mass spectra of plant extracts are often spoiled by unwanted matrix peaks even after selective extraction and chromatographic separation. Try to reduce background signal by adjusting detection parameters of MS and performing background subtraction with data processing software. For Waters Xevo TQ-S, for example, background signal can be suppressed by lowering the gain to 0.1 albeit at the expense of sensitivity. Matrices may also interfere with detection of your analytes.

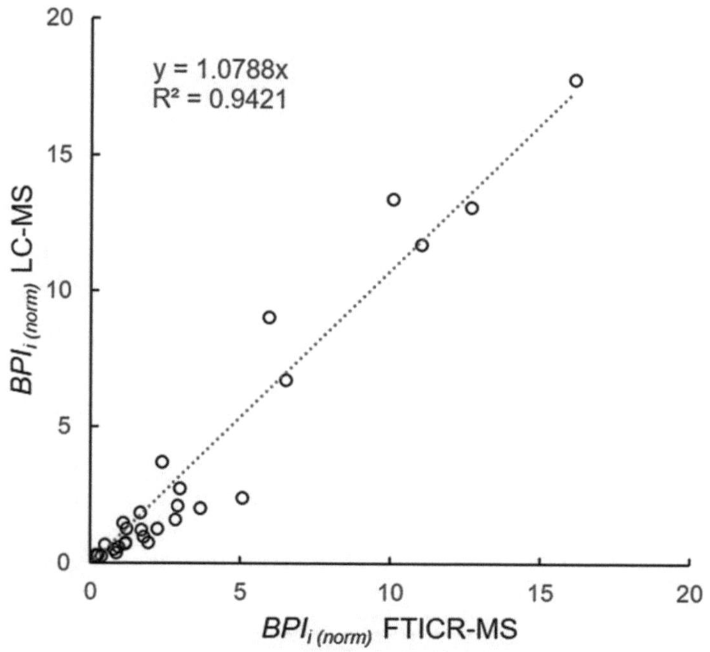

Fig. 4 Correlation between $BPI_{i(norm)}$ of lutein isotopologs obtained by FTICR-MS and LC-MS. The values were taken from Table 4 to Table 5 for FTICR-MS and LC-MS, respectively

Check matrix effects by obtaining mass spectra of known amounts of ^{13}C-labeled and nonlabeled carotenoids added to a blank (blank spike) and to a plant sample during extraction (matrix spike). ^{13}C-labeled standards of lutein and β-carotene are commercially available.

12. To find out the mass range of isotopologs, you can analyze two contrasting samples with the lowest and the highest degree of ^{13}C labeling when these are known. In the case of ^{13}C pulse-chase labeling experiment, for instance, you can use the samples taken at the beginning and at the end of ^{13}C labeling or subsequent chase.
13. Equations for calculation of BPI and DoL:

$$BPI_i = \frac{I_i}{I_{i(\max)}} \cdot 100$$

BPI_i: Base peak intensity of an isotopolog with i ^{13}C atoms.

i: Number of ^{13}C atoms.

I_i: Intensity of an isotopolog with i ^{13}C atoms.

$I_{i(\max)}$: Highest intensity of all isotopologs

$$\mathrm{BPI}_{i(norm)} = \frac{\mathrm{BPI}_i}{\sum_{i=0}^{n} \mathrm{BPI}_i} \cdot 100$$

$\mathrm{BPI}_{i(\mathrm{norm})}$: Base peak intensity of an isotopolog normalized to the sum of BPI of all isotopologs

$$\mathrm{DoL}_i = \frac{\mathrm{BPI}_{i(norm)} \cdot i}{n}$$

DoL_i: Degree of ^{13}C labeling of an isotopolog with i ^{13}C atoms.

n: Number of C atoms in the molecule (40 for carotenoids)

$$\mathrm{DoL} = \sum_{i=0}^{n} \mathrm{DoL}_i$$

DoL: Total degree of ^{13}C labeling of a metabolite.

Acknowledgments

We thank Ingar Janzik and Andrea Neuwohner for ^{13}C-labeled leaf material of peppermint.

References

1. Cazzonelli CI, Pogson BJ (2010) Source to sink: regulation of carotenoid biosynthesis in plants. Trends Plant Sci 15:266–274
2. Ruiz-Sola MÁ, Rodríguez-Conceptión M (2012) Carotenoid biosynthesis in Arabidopsis: A colorful pathway. In: The Arabidopsis Book, vol. 10. American Society of Plant Biologists, p e0158
3. Britton G (1985) Stable isotopes in carotenoid biochemistry. Pure Appl Chem 57:701–708
4. Lichtenthaler HK (1999) The 1-deoxy-D-xylulose-5-phosphate pathway of isoprenoid biosynthesis in plants. Annu Rev Plant Physiol Plant Mol Biol 50:47–65
5. Szecowka M, Heise R, Tohge T, Nunes-Nesi A, Vosloh D, Huege J, Feil R, Lunn J, Nikoloski Z, Stitt M, Fernie AR, Arrivault S (2013) Metabolic fluxes in an illuminated *Arabidopsis* rosette. Plant Cell 25:694–714
6. Delwiche CF, Sharkey TD (1993) Rapid appearance of ^{13}C in biogenic isoprene when $^{13}CO_2$ is fed to intact leaves. Plant Cell Environ 16:587–591
7. Ghirardo A, Wright LP, Bi Z, Rosenkranz M, Pulido P, Rodríguez-Concepción M, Niinemetz Ü, Brüggeman N, Gershenzon J, Schnitzler J-P (2014) Metabolic flux analysis of plastidic isoprenoid biosynthesis in poplar leaves emitting and nonemitting isoprene. Plant Physiol 165:37–51
8. Beisel KG, Jahnke S, Hofmann D, Köppchen S, Schurr U, Matsubara S (2010) Continuous turnover of carotenes and chlorophyll *a* in mature leaves of Arabidopsis revealed by $^{14}CO_2$ pulse-chase labeling. Plant Physiol 152:2188–2199
9. Aharoni A, de Vos CHR, Verhoeven HA, Maliepaard CA, Kruppa G, Bino R, Goodenowe DB (2002) Nontargeted metabolome analysis by use of Fourier transform ion cyclotron mass spectrometry. OMICS 6:217–234
10. Fiehn O (2002) Metabolomics—the link between genotypes and phenotypes. Plant Mol Biol 48:151–171
11. Lei Z, Huhman DV, Sumner LW (2011) Mass spectrometry strategies in metabolomics. J Biol Chem 286:25435–25442
12. Heise R, Arrivault S, Szecowka M, Tohge T, Nunes-Nesi A, Stitt M, Nikoloski Z, Fernie AR (2014) Flux profiling of photosynthetic carbon

metabolism in intact plants. Nat Protoc 9:1803–1824

13. Allen DG (2016) Quantifying plant phenotypes with isotopic labeling & metabolic flux analysis. Curr Opin Biotechnol 37:45–52
14. Gleichenhagen M, Zimmermann BF, Herzig B, Janzik I, Jahnke S, Boner M, Stehle P, Galensa R (2013) Intrinsic isotopic ^{13}C labeling of polyphenols. Food Chem 141:2582–2590
15. Dersch LM, Beckers V, Rasch D, Melzer G, Bolten C, Kiep K, Becker H, Bläsing OE, Fuchs R, Ehrhardt T, Wittmann C (2016) Novel approach for high-throughput metabolic screening of whole plants by stable isotopes. Plant Physiol 171:25–41
16. van Breemen RB, Huang C-R, Tan Y, Sander LC, Schilling AB (1996) Liquid chromatography/mass spectrometry of carotenoids using atmospheric pressure chemical ionization. J Mass Spectrom 31:975–981
17. Kurilich AC, Britz SJ, Clevidence BA, Novotny JA (2003) Isotopic labeling and LC-APCI-MS quantification for investigating absorption of carotenoids and phylloquinone from kale (*Brassica oleracea*). J Agric Food Chem 51:4877–4888
18. Stöggl W, Huck C, Wongyai S, Scherz H, Bonn G (2005) Simultaneous determination of carotenoids, tocopherols, and gamma-oryzanol in crude rice bran oil by liquid chromatography coupled to diode array and mass spectrometric detection employing silica C30 stationary phases. J Sep Sci 28:1712–1718

Chapter 21

Quantification of Carotenoid Pathway Flux in Green and Nongreen Systems

Julian Koschmieder and Ralf Welsch

Abstract

Metabolite accumulation in plant tissues represents the transient net result of their constant biosynthesis and degradation. For carotenoids, degradation might occur enzymatically by carotenoid cleavage producing plant hormones and volatiles or by nonenzymatic oxidation, both depending on environmental and developmental conditions. Carotenoid biosynthesis is therefore constantly regulated at various levels to attain sufficient carotenoid accumulation, mainly for photosynthesis and photoprotection. Due to the plenitude of carotenoids and their degradation products, it is not feasible to investigate overall carotenoid biosynthetic activity and its regulation by the quantification of all carotenoids including their derivatives. This is an issue encountered in investigations on many other highly branched pathways. We therefore present protocols to determine carotenoid biosynthesis flux in a given plant tissue by HPLC quantification of phytoene, the first pathway-specific intermediate and precursor of all carotenoids synthesized by phytoene synthase (PSY). For this purpose, enzymatic metabolization of phytoene in the tissue under investigation is prevented by treatment with the bleaching herbicide norflurazon, blocking the carotenogenic pathway downstream of PSY. As phytoene is more resistant to oxidation than desaturated carotenoids, the rate of phytoene biosynthesis serves as a good measure for total carotenogenic flux in a given tissue. The method is described for Arabidopsis for two photosynthetically active sample types, namely, seedlings and leaves, as well as for seed-derived callus as nongreen tissue. It should be realizable using only a relatively simple experimental setup and is applicable to other plant tissues as well as to different plant species. Additionally, similar experimental setups could be a useful tool to investigate total flux and turnover rates in other biosynthetic pathways.

Key words Phytoene desaturase, Phytoene, Quantitative HPLC, Norflurazon, Seedlings, Callus, Arabidopsis

1 Introduction

Carotenoids accumulate in all tissues and throughout all developmental stages of a plant's life. In chloroplasts of photosynthetically active tissues, carotenoids are constituents of light-harvesting proteins and photosynthetic reaction center while in chromoplasts, carotenoids accumulate at high levels and are protein-bound, integrated into proliferated membranes, accumulate as crystals or

Manuel Rodríguez-Concepción and Ralf Welsch (eds.), *Plant and Food Carotenoids: Methods and Protocols*, Methods in Molecular Biology, vol. 2083, https://doi.org/10.1007/978-1-4939-9952-1_21,

are dissolved in lipid droplets [1–3]. However, once synthesized, carotenoid amounts are not static. In contrast, they are subject to constant degradation and their continuous biosynthesis is required to maintain carotenoid levels present in various tissues, as is characteristic of homeostatic systems [4, 5].

Degradation occurs to a minor proportion through carotenoid-cleaving enzymes (carotenoid cleavage dioxygenases, CCDs and nine-*cis* epoxycarotenoid dioxygenases, NCEDs) generating plant hormones like strigolactones and abscisic acid, volatiles like β-cyclocitral and β-ionone, and apocarotenoids like in saffron or apocarotenoids which are glycosidically bound [5–8]. However, the majority of carotenoids are degraded nonenzymatically through oxidation, producing apocarotenoids of various chain lengths as primary oxidation products which are truncated further through continued oxidation [9–12]. The terminal fate of the products of continued apocarotenoid degradation is not known but subject of current investigations of numerous research groups in the field of carotenoid research.

Continuous carotenoid biosynthesis is required to compensate for carotenoid degradation which varies between developmental stages, tissue types and environmental conditions, each characterized by different rates of phytohormone biosynthesis and carotenoid oxidation. Therefore, carotenoid pathway flux takes place permanently and is dynamically regulated. However, determination of pathway flux through integrative quantification of all carotenoid-derived products is impossible due to their sheer diversity. One possible approach for the quantification of carotenoid biosynthetic flux is to determine the turnover of carotenoids by ^{14}C pulse chase assay [13, 14]. The high turnover rate observed for, for example, β-carotene in these experiments strongly supports a very high carotenogenic pathway flux in leaves.

Alternatively, inhibition of early pathway enzymes can be used to determine the amounts of carotenoid biosynthesis intermediates, these accumulating upon pathway inhibition and mirroring the overall carotenoid biosynthesis rate through the rate-limiting enzyme phytoene synthase (PSY). A common approach to achieve this is to inhibit phytoene desaturase (PDS) positioned downstream of PSY by the application of bleaching herbicides, mainly fluridone and norflurazon [15–17]. This results in the accumulation of phytoene, an intermediate which is usually instantly further metabolized in the pathway and thus absent in most tissues. As shown previously, phytoene is practically not metabolized by carotenoid cleavage enzymes and is also resistant toward oxidation as oxidative cleavage targets higher desaturated carotenoids such as xanthophylls and β-carotene [10]. Accordingly, the phytoene amount under such inhibitory conditions is a good measure for the total carotenoid amount that would have been synthesized during the time the sample is exposed to the bleaching herbicides [4, 18–20].

In this article, we summarize current protocols for the application of this technique to different plant tissues and include a detailed protocol on the identification and quantification of phytoene by HPLC analysis. Detectable amounts of phytoene accumulate within a few hours in herbicide-treated photosynthetically active leaves, equivalent to the level of carotenoid biosynthesis. Moreover, the pathway activity is dependent on light intensity and increases significantly after exposure to high light conditions [5]. We also include a protocol on the treatment of Arabidopsis seedlings with norflurazon which allows determination of pathway flux alterations in photomorphogenic mutants [20, 21]. Finally, we include a method for determining the carotenogenic pathway flux in callus generated from germinating Arabidopsis seedlings [10, 22]. Remarkably, pathway flux determined matched with expected pathway alterations, for example, in mutants or upon overexpression of enzymes positioned upstream to phytoene formation. This supports the notion that this system as a useful tool for basic pathway research. We recently identified a number of apocarotenoid-degradation products in callus, corroborating the strong impact of oxidative destruction on carotenoid stability.

The carotenoid extraction protocol includes α-tocopheryl acetate as internal standard to correct for extraction errors. Quantification of phytoene is performed without requiring purified phytoene as authentic standard compound. The amount of phytoene in each sample is calculated using a response factor of the diode array detector for β-carotene after applying a conversion factor obtained from molecular extinction coefficients of β-carotene and phytoene.

2 Materials

2.1 Carotenoid Pathway Flux in Leaves

1. 4-week-old Arabidopsis plants grown in soil under long-day conditions.
2. Small petri dishes (100 mm diameter).
3. Two solutions with 10 μM and 70 μM norflurazon in water. For this, prepare a 100 mM norflurazon stock solution in isopropanol and further dilute in water. Norflurazon stocks can be stored at 4 °C for several years.
4. Forceps.
5. Phytochamber or light source with 100 μmol photons m^{-2} s^{-2}.

2.2 Norflurazon Treatment Arabidopsis Seedlings

1. Arabidopsis seeds.
2. Sterilization solution (6% (v/v) sodium hypochlorite, 0.02% (v/v) Triton X-100), sterile water.
3. Tabletop centrifuge.

4. Sterile bench.
5. 0.4% (w/v) agar-agar.
6. Arabidopsis growth agar plates. Add per liter: 4.33 g Murashige and Skoog medium, 0.5 g 2-(N-morpholino)ethanesulfonic acid/KOH, pH 5.7, 4 g Phytagel. After autoclaving, poor into large petri dishes (145 mm diameter). Prepare as many plates without and with 5 μM norflurazon. Norflurazon is added from stock solution (see above) to the autoclaved medium before pouring the plates.
7. Rectangular stainless steel nets (e.g., V2A-315/200) of 3 × 3 cm in size with a mesh size of 0.315 mm and 0.2 mm wire thickness. Wrap in small batches of, for example, six pieces in aluminum foil and sterilize by autoclaving. Place in drying oven to evaporate water before use.
8. Cold chamber or fridge at 4 °C.
9. Phytochamber or light source with 100 μmol photons m^{-2} s^{-2}.

2.3 Carotenoid Pathway Flux in Seed-Derived Callus

1. Arabidopsis seeds.
2. Sterilization solution (see above), sterile water.
3. Table top centrifuge.
4. Seed-derived callus medium (SDC medium). For 1 L, dissolve the following components: 4.33 g Murashige and Skoog medium, 30 g sucrose, 0.5 mg 2,4-dichlorophenoxyacetic acid (50 μl of 10 mg/mL stock in 1 M KOH), 2 mg indole-3-acetic acid (20 μl of 100 mg/ml stock in 1 M KOH), 0.5 mg 2-isopentenyladenine (50 μl of 10 mg/ml stock in 1 M KOH). Adjust pH to 5.8 with KOH. Add 4 g Phytagel per L medium and autoclave, let cool to about 40 °C, add 1000 μl 1000× Gamborg B5 vitamins per liter medium and poor into big petri dishes (145 mm diameter). Prepare as many plates without and with 1 μM norflurazon. Norflurazon is added from stock solution (see above) to the autoclaved medium before pouring the plates.
5. 0.4% (w/v) agar-agar.
6. Sterile filter paper, forceps.

2.4 Carotenoid Extraction and Analysis

1. Liquid nitrogen, mortar, pistil (alternatively: microdismembrator), aluminum foil.
2. Freeze dryer, speed vacuum evaporator, sonicator, table top centrifuge, centrifuge for 15 ml plastic tubes, vortex evaporator or nitrogen steam.
3. 15 ml plastic reaction tubes.
4. 100 mM Tris (pH not adjusted).

5. Acetone, petroleum ether–diethyl ether (2:1, v/v), chloroform–methanol (2:1, v/v).
6. α-Tocopheryl acetate (2 mg/ml in acetone).

2.5 Quantitative HPLC Carotenoid Analysis

1. β-carotene.
2. Chloroform–methanol (2:1, v/v), petroleum ether.
3. Photometer, 15 ml plastic reaction tubes (*see* **Note 1**).
4. C18 column (e.g., Hypersil Gold, 150 × 2.1 mm i.d., 1.9 μm, Thermo Scientific, Erlangen, Germany) or equivalent.

3 Methods

3.1 Carotenoid Pathway Flux in Leaves

This method is adopted from Ref. [14].

1. Collect 20 rosette leaves from 4-week-old Arabidopsis plants and place them with the bottom side facing downward in a small petri dish containing 25 ml of 70 μM norflurazon. Control leaves are placed on water (*see* **Note 2**).
2. Incubate for 2 h in the dark at room temperature, for example, in a cardboard box.
3. Transfer leaves to 25 ml 10 μM norflurazon in small petri dishes using a forceps, again with the abaxial side in contact with the liquid.
4. Illuminate for 4 h with 100 μmol photons m^{-2} s^{-2} (*see* **Note 3**).
5. Collect leaves with forceps on paper wipe to remove adhering liquid. Put harvested leaves in small aluminum foil envelopes, freeze in liquid nitrogen immediately after harvest and store for carotenoid extraction (*see* Subheading 3.4) and HPLC analysis (Subheading 3.5).

3.2 Norflurazon Treatment Arabidopsis Seedlings

This method is adopted from Ref. [20].

1. Weigh 10 mg Arabidopsis seeds per large petri dish (145 mm diameter) into reaction tube, add 1 ml sterilization solution and incubate for 10 min with gentle shaking at room temperature (*see* **Note 4**).
2. Centrifuge seeds for 2 min at 3000 × *g* in a table top centrifuge, remove supernatant under the sterile hood, wash three times by adding 1 ml of sterile water followed by centrifugation. After the last washing, resuspend seeds in 1 ml 0.4% (w/v) agar-agar (*see* **Note 5**).
3. Distribute seeds onto three metal grids per big petri dish containing Arabidopsis growth agar. Let the plates with lids closed

for about 1 h in the sterile bench with the ventilation switched on to allow water drops from the seeds to evaporate.

4. Seal the plates with tape and place in closed cardboard boxes.
5. Stratify over night at 4 °C.
6. Light-germinate for 10 days at long-day conditions.
7. Transfer nets to Arabidopsis growth agar supplemented with 5 μM norflurazon. Close lids and seal with tape.
8. Illuminate for 3 h with 100 μmol photons m^{-2} s^{-2} (growth chamber conditions).
9. Collect seedlings from the metal grids with forceps. Put harvested seedlings in small aluminum foil squares folded like envelopes, freeze in liquid nitrogen immediately after harvest and store for carotenoid extraction (*see* Subheading 3.4) and HPLC analysis (Subheading 3.5).

3.3 Carotenoid Pathway Flux in Seed-Derived Callus

This method is adopted from Ref. [10].

1. Weight 7.5 mg Arabidopsis seeds per big petri dish into a reaction tube, add 1 ml of sterilization solution and incubate 10 min with gentle shaking at room temperature (*see* **Note 6**).
2. Centrifuge seeds for 2 min at 3000 × *g* in a table top centrifuge, remove supernatant under the sterile hood, wash three times by adding 1 ml of sterile water followed by centrifugation. After the last washing, resuspend seeds in 2 ml 0.4% (w/v) agar-agar (*see* **Note** 7).
3. Place one sterile filter paper on a big petri dish containing SDC medium and incubate briefly to soak some liquid. Place sterilized seeds on the filter paper by dribbling agar-agar solution containing the seeds. Seal the plates with tape and place in closed cardboard boxes.
4. Stratify over night at 4 °C.
5. Place in growth chamber on white paper to reduce appearance of condense water drops on lids. Germinate under long day conditions at 100 μmol photons m^{-2} s^{-2} for 5 days (16 h light/8 h dark, 26 °C).
6. For norflurazon treatment transfer the filter paper including germinated seeds on a fresh petri dish containing SDC medium supplemented with 1 μM norflurazon. For the control, transfer filter paper including germinated seeds on a fresh petri dish containing fresh SDC medium without norflurazon (*see* **Note 8**).
7. Reseal the plates with tape and place them into dark, for example, in closed cardboard boxes. Continue incubation for 14 days at 26 °C.

8. Collect seed-developed calli from filter paper either with forceps or by pushing the plates upside down on a table for several times to collect seed-derived calli on the lid. Put harvested seed-derived calli in small aluminum foil squares folded like envelopes, freeze in liquid nitrogen immediately after harvest and store for carotenoid extraction (*see* Subheading 3.4) and HPLC analysis (Subheading 3.5).

3.4 Carotenoid Extraction

1. Put harvested leaves, seedlings or seed-derived calli in small aluminum foil envelopes. Freeze envelopes with samples in liquid nitrogen immediately after harvest. When all samples are collected, grind plant material with a mortar and pistil in liquid nitrogen, return powder into aluminum envelope and lyophilize overnight (*see* **Note 9**).
2. Weigh lyophilized powder in a 15 ml plastic tube: 5 mg for leaves, 20 mg for seed-derived calli and 10 mg for seedlings (dry weight). Note the exact weight.
3. Add 100 μl of 10 mM Tris (pH not adjusted), vortex (*see* **Note 10**).
4. Add 2 ml of acetone and exactly 100 μl of internal standard (α-tocopheryl acetate, 2 mg/ml in acetone) to each sample. Prepare three aliquots with 100 μl each from same standard dilution in reaction tubes, dry in a SpeedVac and dissolve in 100 μl chloroform/methanol (2:1, v/v), like samples (see below). Store at −20 °C until start of HPLC analysis. These samples serve as external controls to determine standard recovery during extraction and HPLC analysis (*see* **Note 11**).
5. Apply short pulses of sonication to mix and extract sample.
6. Centrifuge for 5 min at 3000 × *g*.
7. Transfer supernatant to a new 15 ml tube.
8. Add another 2 ml of acetone to the pellet and reextract sample by sonication.
9. Centrifuge for 5 min at 3000 × *g*.
10. Combine supernatant with supernatant from **step** 7.
11. Repeat **steps 8** to **10**.
12. Add 2 ml PE:DE (2:1; v:v) to combined supernatants (~ 6 ml) and mix by vortexing.
13. Add water up to 14 ml.
14. Close tube tightly and mix samples by inverting tube several times.
15. Centrifuge for 5 minutes at 3000 × *g*.
16. Transfer upper organic phase to a new 15 ml tube.

17. Add another 2 ml PE:DE (2:1; v:v) to sample and repeat **steps 13–15** to repartition.
18. Transfer upper organic phase to 15 ml tube and dry combined organic phases in a vacuum evaporator or under a nitrogen stream.
19. Transfer extract into 2 ml tube using 2 × 100 μl chloroform/methanol (2:1, v/v).
20. Dry organic phase using a SpeedVac.
21. Dissolve dried extract in exactly 100 μl of chloroform/methanol (2:1, v/v), vortex.
22. Transfer complete extract into a HPLC vial and inject 2 μl for HPLC analysis (*see* Subheading 3.5).

3.5 Phytoene Quantification by HPLC Analysis

The quantification procedure described here for phytoene can in principle be applied for the quantification of all carotenoids if chromatographic conditions allow for sufficient resolution of analytes and if molar extinction coefficients are known.

1. Separation of samples is performed on a C18 column (e.g., Hypersil Gold C18 UPLC-column; 150 × 2.1 mm i.d., 1.9 μm, Thermo Scientific), using the solvent system A, 0.1% formic acid in water and B, 0.1% formic acid in acetonitrile. The mobile gradient at a constant flow rate of 0.5 mL min^{-1} and at 20 °C is as follows: 70% B for 1 min, linear increase to 100% B within 4 min and separation at 100% B for 20 min. The column is then reequilibrated to 70% B (*see* **Note 12**).
2. Prepare β-carotene solution of about 5 pmol/μl in petroleum ether (*see* **Note 13**).
3. Analyze 1, 2, 5, 10, 25, and 50 pmol β-carotene on column by HPLC. Use the same injection volume in each run (e.g., 2 μl from 50 μl β-carotene solutions of appropriate concentration). Perform all concentrations in independent triplicates.
4. Integrate the β-carotene peak area in a chromatogram at 450 nm and plot the peak areas for all measurements against the amount of β-carotene on column. Generate a calibration curve to infer the ratio β-carotene area per pmol β-carotene, that is, the response factor for β-carotene (RF_{car}).
5. Analyze 2 μl of plant extracts (total volume 100 μl, *see* Subheading 3.4) by HPLC.
6. Identify phytoene and the internal standard α-tocopheryl acetate using a chromatogram at 285 nm which is λ_{max} for both analytes (*see* Fig. 1 for norflurazon-treated leaves and Fig. 2 for norflurazon-treated calli). Determine the phytoene peak area A_{Phyt} and the α-tocopheryl acetate peak area A_{IST}.

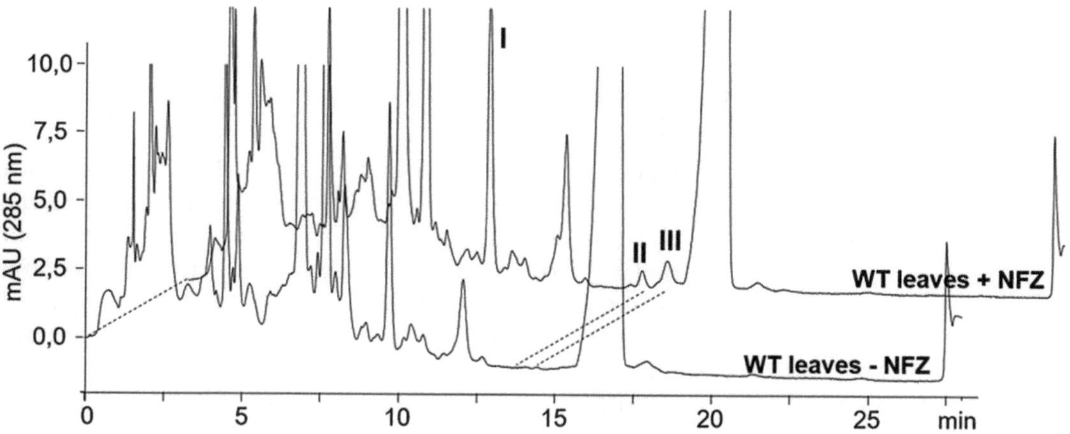

Fig. 1 Identification of phytoene in norflurazon-treated wild-type leaves. Chromatograms at 285 nm from a C18 HPLC analysis of carotenoids in Arabidopsis wild-type leaves, generated with or without norflurazon supplementation (+/− NFZ). UV/VIS spectra of α-tocopheryl acetate as internal standard (I), 15-*cis*-phytoene (II) and all-*trans*-phytoene (III) are given as insets in Fig. 2. Only in the presence of norflurazon, phytoene is detected in wild-type Arabidopsis leaves and its amount upon norflurazon treatment represents carotenogenic pathway flux

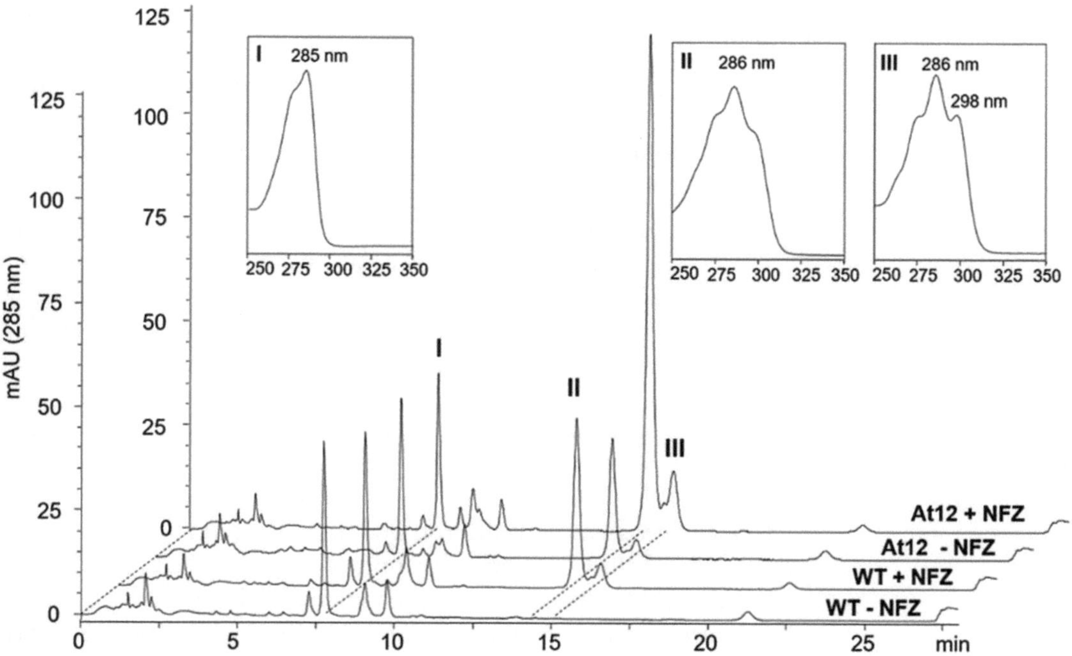

Fig. 2 Identification of phytoene in norflurazon-treated seed-derived calli. Chromatograms at 285 nm from a C18 HPLC analysis of carotenoids in seed-derived calli of Arabidopsis wild-type and one line constitutively overexpressing *AtPSY* (At12), generated with or without norflurazon supplementation (+/− NFZ). UV/VIS spectra of α-tocopheryl acetat as internal standard (I), 15-*cis*-phytoene (II) and all-*trans*-phytoene (III) are given as insets. *AtPSY* overexpression in At12 as well as norflurazon treatment leads to accumulation of phytoene. Phytoene amounts upon norflurazon treatment represent carotenogenic pathway flux

7. Analyze 2 μl of each external standard sample by HPLC and determine the α-tocopheryl acetate peak area A_{EST}.

8. In order to calculate phytoene amounts in the samples, use the following conversions:

   (a) First, convert peak area of phytoene A_{Phyt} virtually into β-carotene, which the HPLC was calibrated for, by multiplication with factor 1.97. This factor originates from molar extinction coefficients ε for β-carotene (134,000 L × mol^{-1} × cm^{-1}) and phytoene (68,125 L × mol^{-1} × cm^{-1}) [23]. Second, compensate for sample losses during extraction as well as fluctuations in HPLC injection volumes as follows. Calculate the mean of all three A_{EST} (A_{EST_MEAN}) and divide A_{EST_MEAN} by A_{IST} for every individual sample to determine a sample-specific tocopheryl acetate (TocAc) correction factor. Third, divide the converted phytoene peak area A_{Phyt} by the TocAc correction factor. In summary, this provides the corrected phytoene peak area $A_{PhytCORR}$ (= A_{Phyt} × 1.97/ TocAc correction factor) (*see* **Note 14**).

   (b) Convert corrected phytoene peak area $A_{PhytCORR}$ into pmol of phytoene ($amount_{Phyt}$) by division by the β-carotene response factor RF_{car}. This yields the $amount_{Phyt}$ [pmol] in the 2 μl injected from your sample. Multiply by factor 50 (= 100 μl extract sample/2 μl injection volume) to determine the total phytoene amount in your 100 μl extract sample [pmol], that is, $amount_{Phyt\_tot}$.

   (c) Convert $amount_{Phyt\_tot}$ into total mass of phytoene [ng] using the molecular weight of phytoene of 545 g mol^{-1}. Further divide this by the sample's dry weight to determine total mass of phytoene per dry weight [ng phytoene mg^{-1} DW) (*see* **Note 15**).

4 Notes

1. HPLC system with photodiode array detector (e.g., our Shimadzu HPLC system consisting of degasser DGU-20A3, two pumps LC-20AD XR, autosampler SIL-20 AC XR, thermostatted column compartment CTO-20AC, photodiode array detector SPD-M20A, and system controller unit CBM-20A (Shimadzu, Duisburg, Germany)).

2. Plant health has great impact on carotenoid flux. Use nonsenescent, healthy leaves for experiments. If several different lines are compared, grow them simultaneously under same conditions and make sure plant health and leaf quality is comparable. We usually use leaves from Arabidopsis grown at long day conditions.

3. We found that light intensity directly affects carotenoid pathway flux in detached leaves. For standard experiments, use 100 μmol photons m^{-2} s^{-2} which corresponds to standard growth chamber conditions.
4. If higher number of seeds needs to be sterilized, up to 55 mg seeds can be treated at once with the same amount of sterilization solution and resuspended in accordingly increased final medium.
5. This low concentrated agar-agar solution is viscous and can easily be pipetted. Arabidopsis seeds resuspended in this solution do not settle and distribute equally after mixing which allows to distribute them evenly on several metal grids.
6. If higher number of seeds need to be sterilized, up to 55 mg seeds can be treated at once with the same procedure and resuspended in equivalently increased final medium.
7. Seeds should not settle, *see* **Note 5**.
8. A convenient way to do this is to place both plates in a row (fresh plate in front), lift the filter paper with sterile forceps and drag it over to the front plate touching the fringe of the rear petri dish.
9. Grinding to a fine powder is essential as carotenoids will not be quantitatively extracted from a sample if grinding is omitted or not sufficient. Grinding lyophilized plant material is less effective than grinding fresh frozen material. Aluminum foil envelopes should not be closed too tight to allow water to sublimate during lyophilization. Avoid thawing of powder throughout the whole process.
10. Addition of basic Tris neutralizes acids in samples to prevent acid-catalyzed furanoid rearrangement of epoxy groups [24].
11. While slight volume errors during extraction process are tolerable, addition of exact and identical volumes of internal standard is crucial. In dispenser-type pipettes (e.g., Eppendorf multipette) piston is in direct contact with solvent, in contrast to standard pipettes in which piston is filled with air. Absence of vapor pressure in dispenser-type pipettes avoids dripping of solvents and results in exact and precise pipetting volumes. Use these pipettes for addition of internal and external standards.
12. These chromatographic conditions are only sufficient for separation and quantification of α-tocopheryl acetate and phytoene in the presence of other carotenoids and chlorophylls. If quantification of other carotenoids in the presence of chlorophylls is required, chromatography can be performed on a 250 mm C30 reverse phase column, for instance as described in Hoa et al. [25]. If quantification of other carotenoids in the absence

of chlorophylls is required, separation on a 150 mm C30 reverse phase column is suitable (e.g., YMC carotenoid (150 mm × 3 mm, 5 μm) with solvent system A: methanol–*tert*-Butyl methyl ether (1:1, v/v) and B: methanol–*tert*-Butyl methyl ether–water (5:1:1, v/v/v), 0.75 ml min^{-1}; 40 °C; increase linearly from 0% A to 100% A within 20 min, maintain final conditions for 4 min).

13. Dissolve some β-carotene crystals in 20 μl chloroform in a 15 ml tube and add PE up to 10 ml. Determine concentration photometrically at λ_{max} (β-carotene) = 450 nm by Beer's law using the molar extinction coefficient ε 134,000 L × mol^{-1} × cm^{-1} [23] and dilute accordingly to obtain a 5 pmol/μl solution in PE.
14. If enough care was taken during extraction and sample preparation, the TocAc correction factor ranges between 0.75 and 1.0.
15. Phytoene and α-tocopheryl acetate are quantified at 285 nm using the deuterium lamp (D2) signals, whereas all other carotenoids, such as β-carotene for calibration, are quantified using the Tungsten lamp (W) signal. For reliable phytoene quantification with the method described, it is necessary to monitor sensitivity of both lamps over time. That is, β-carotene calibration to evaluate W lamp sensitivity should be repeated in regular intervals. Similarly, deuterium lamp (D2) sensitivity should be evaluated by monitoring the α-tocopheryl acetate peak area at 285 nm in external standard samples in repeated experiments, that is, signals should remain fairly constant over a longer period of time. In this context, lamp lifetime as recommended by the manufacturer should not be exceeded.

Acknowledgments

This work was supported in part by the HarvestPlus research consortium (grant 2014H6320.FRE) to R.W. J.K. was funded by Grant WE 4731/3-1 from the Deutsche Forschungsgemeinschaft. The authors thank the present and former members of the laboratory of Prof. Peter Beyer for their contributions to establishing this method.

References

1. Schweiggert RM, Mezger D, Schimpf F et al (2012) Influence of chromoplast morphology on carotenoid bioaccessibility of carrot, mango, papaya, and tomato. Food Chem 135:2736–2742
2. Li L, Yuan H (2013) Chromoplast biogenesis and carotenoid accumulation. Arch Biochem Biophys 539:102–109
3. Sun T, Yuan H, Cao H et al (2017) Carotenoid metabolism in plants: the role of plastids. Mol Plant 11:58–74

4. Simkin AJ, Zhu C, Kuntz M, Sandmann G (2003) Light-dark regulation of carotenoid biosynthesis in pepper (Capsicum annuum) leaves. J Plant Physiol 160:439–443
5. Lätari K, Wüst F, Hübner M et al (2015) Tissue-specific Apocarotenoid glycosylation contributes to carotenoid homeostasis in Arabidopsis leaves. Plant Physiol 168:1550–1562
6. Havaux M (2013) Carotenoid oxidation products as stress signals in plants. Plant J 79:597–606
7. Al-Babili S, Bouwmeester HJ (2015) Strigolactones, a novel carotenoid-derived plant hormone. Annu Rev Plant Biol 66:161–186
8. Ahrazem O, Rubio-Moraga A, Nebauer SG et al (2015) Saffron: its Phytochemistry, developmental processes, and biotechnological prospects. J Agric Food Chem 63:8751–8764
9. Schaub P, Wuest F, Koschmieder J et al (2017) Non-enzymatic β-carotene degradation in (Provitamin A-biofortified) crop plants. J Agric Food Chem. https://doi.org/10.1021/acs.jafc.7b01693
10. Schaub P, Rodriguez-Franco M, Cazzonelli CI et al (2018) Establishment of an Arabidopsis callus system to study the interrelations of biosynthesis, degradation and accumulation of carotenoids. PLoS One 13:e0192158
11. Ramel F, Mialoundama AS, Havaux M (2013) Nonenzymic carotenoid oxidation and photooxidative stress signalling in plants. J Exp Bot 64:799–805
12. Walter MH, Strack D (2011) Carotenoids and their cleavage products: biosynthesis and functions. Nat Prod Rep 28:663–692
13. Beisel KG, Jahnke S, Hofmann D et al (2010) Continuous turnover of carotenes and chlorophyll a in mature leaves of Arabidopsis thaliana revealed by 14CO2 pulse-chase labeling. Plant Physiol 152:2188–2199
14. Beisel KG, Schurr U, Matsubara S (2011) Altered turnover of β-carotene and Chl a in Arabidopsis leaves treated with lincomycin or norflurazon. Plant Cell Physiol 52:1193–1203
15. Breitenbach J, Zhu C, Sandmann G (2001) Bleaching herbicide norflurazon inhibits phytoene desaturase by competition with the cofactors. J Agric Food Chem 49:5270–5272
16. Koschmieder J, Fehling-Kaschek M, Schaub P et al (2017) Plant-type phytoene desaturase: functional evaluation of structural implications. PLoS One 12:e0187628
17. Brausemann A, Gemmecker S, Koschmieder J et al (2017) Structure of Phytoene Desaturase provides insights into herbicide binding and reaction mechanisms involved in carotene desaturation. Structure 25(8):1222–1232.e3
18. Welsch R, Zhou X, Yuan H et al (2017) Clp protease and OR directly control the proteostasis of phytoene synthase, the crucial enzyme for carotenoid biosynthesis in Arabidopsis. Mol Plant 11:149–162
19. Zhou X, Welsch R, Yang Y et al (2015) Arabidopsis OR proteins are the major posttranscriptional regulators of phytoene synthase in controlling carotenoid biosynthesis. Proc Natl Acad Sci 112:3558–3563
20. Rodríguez-Villalón A, Gas E, Rodríguez-Concepción M (2009) Phytoene synthase activity controls the biosynthesis of carotenoids and the supply of their metabolic precursors in dark-grown Arabidopsis seedlings. Plant J 60:424–435
21. Rodríguez-Villalón A, Gas E, Rodríguez-Concepción M, Rodriguez-Villalon A (2009) Colors in the dark: a model for the regulation of carotenoid biosynthesis in etioplasts. Plant Signal Behav 4:965–967
22. Álvarez D, Voß B, Maass D et al (2016) Carotenogenesis is regulated by 5'UTR-mediated translation of Phytoene synthase splice variants. Plant Physiol 172:2314–2326
23. Britton G, Liaaen-Jensen S, Pfander H (1994) Carotenoids—Volume 1 B: Spectroscopy, 1st edn. Birkhäuser, Basel
24. Britton G, Liaaen-Jensen S, Pfander H (1994) Carotenoids—volume 1 a: isolation and analysis, 1st edn. Birkhäuser, Basel
25. Hoa TTC, Al-Babili S, Schaub P et al (2003) Golden Indica and japonica rice lines amenable to deregulation. Plant Physiol 133:161–169

Chapter 22

Histochemical and Molecular Quantification of Arbuscular Mycorrhiza Symbiosis

Juan M. García, María J. Pozo, and Juan A. López-Ráez

Abstract

Arbuscular mycorrhizae (AM) are one of the most widespread and studied plant associations with beneficial microorganisms. Indeed, more than 80% of land plants, including most agricultural and horticultural crop species, are able to establish this mutualistic symbiosis with AM fungi. Through this association the fungus helps the plant in the acquisition of water and mineral nutrients, especially under stress conditions. AM symbiosis affects other ecologically and economically important traits such as plant architecture, flowering, and fruit quality but also tolerance against biotic and abiotic stresses. As a consequence, AM fungi have a great potential as biofertilizers and bioprotection agents in sustainable agriculture. However, in order to take advantage of all these benefits, a good and functional symbiosis is required. Here we present methods for reliable quantification of colonization levels which should be useful not only for research but also from the agronomic point of view.

Key words Arbuscular mycorrhiza symbiosis, Mycorrhizal quantification, Histochemical staining, Apocarotenoids, Liquid chromatography, Roots

1 Introduction

Among the beneficial microorganisms present in the rhizosphere, AM fungi deserve special interest as they establish mutualistic associations with the roots of more than 80% of land plants, including most agricultural and horticultural crop species [1]. This association dates back more than 400 million years ago, and it is considered a key component of the plant microbiome and crucial in plant evolution [2, 3]. AM fungi are obligate biotrophs that depend on plant carbon to complete their life cycle. In "compensation," the fungus helps the plant in the acquisition of water and mineral nutrients, mainly phosphorus (Pi) and to a lesser extent nitrogen, and other inorganic nutrients. In addition to a better mineral nutrition and the corresponding increased plant growth, AM symbiosis affects other ecologically and economically important traits such as plant architecture, flowering, fruit quality, tolerance against

Manuel Rodríguez-Concepción and Ralf Welsch (eds.), *Plant and Food Carotenoids: Methods and Protocols*, Methods in Molecular Biology, vol. 2083, https://doi.org/10.1007/978-1-4939-9952-1_22,

biotic and abiotic stresses, and ecosystem biodiversity [2, 4]. Indeed, inoculation with AM fungi can reduce the need for fertilizers, while increasing crop resilience against diverse stresses under lab and field conditions [5]. Because its benefits for the host plant, AM symbiosis has an immediate potential for application in modern and sustainable agriculture as biofertilizers and bioprotection agents. However, to take advantage of all these benefits, a good and functional symbiosis is required. Therefore, reliable quantification of colonization levels is key, not only for research but also from the agronomic point of view.

AM symbiosis is a very dynamic system, which establishment and functioning depends on a high degree of coordination between the two partners. The success of the association relies on a continuous molecular dialogue that orchestrates the complex symbiotic programs that are regulated by certain apocarotenoids, such as strigolactones, ABA, and C_{13} α-ionol glycosides (blumenols) and C_{14} mycorradicin derivatives [6, 7]. Apocarotenoids are a class of compounds that play important roles in nature, derived from carotenoids by oxidative cleavage. In AM symbiosis, strigolactones are important during the presymbiotic stage by enhancing spore germination and hyphal development [8, 9]. Blumenols and mycorradicin (known as the yellow pigments) seems to be more important at later stages of the symbiosis. Indeed, their accumulation correlates with arbuscule occurrence within the roots, being associated with a well-established and functional symbiosis [7, 10]. Therefore, these compounds constitute a reliable and quantitative metabolic marker for a functional AM symbiosis. In the present work, we present two different methods for the quantification of mycorrhization levels. The first one is a traditional method based on the histochemical analysis of the fungal structures within the roots, by means of ink staining. The second one consists in the analytical quantification by liquid chromatography associated to mass spectrometry of the apocarotenoids C_{13} α-ionol glycosides (blumenols) and C_{14} mycorradicin derivatives (Fig. 1).

2 Materials

2.1 Histochemical Quantification

This method, known as the gridline intersect method [11], is the most frequently used to quickly evaluate the presence of mycorrhiza and quantify the percentage of colonized roots within a microscope field (Fig. 2). It is a reliable, simple, and nonexpensive method. Here, we show a slightly modified protocol described by Vierheilig et al. [12] , which avoids the use of the toxic compound trypan blue [12]. All solutions should be prepared using deionized water and analytical grade reagents., and they should be stored at room temperature.

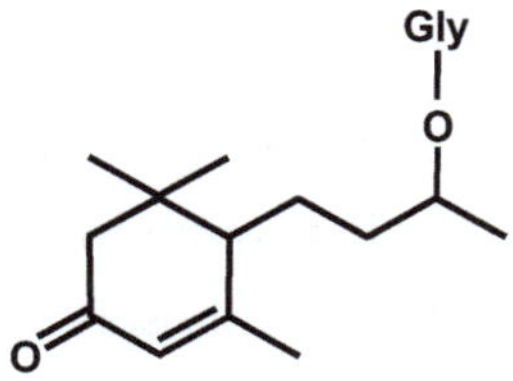

Fig. 1 General structure of the apocarotenoids C_{13} α-ionol glycosides (blumenols) and C_{14} mycorradicin derivatives, specifically induced in mycorrhizal plants

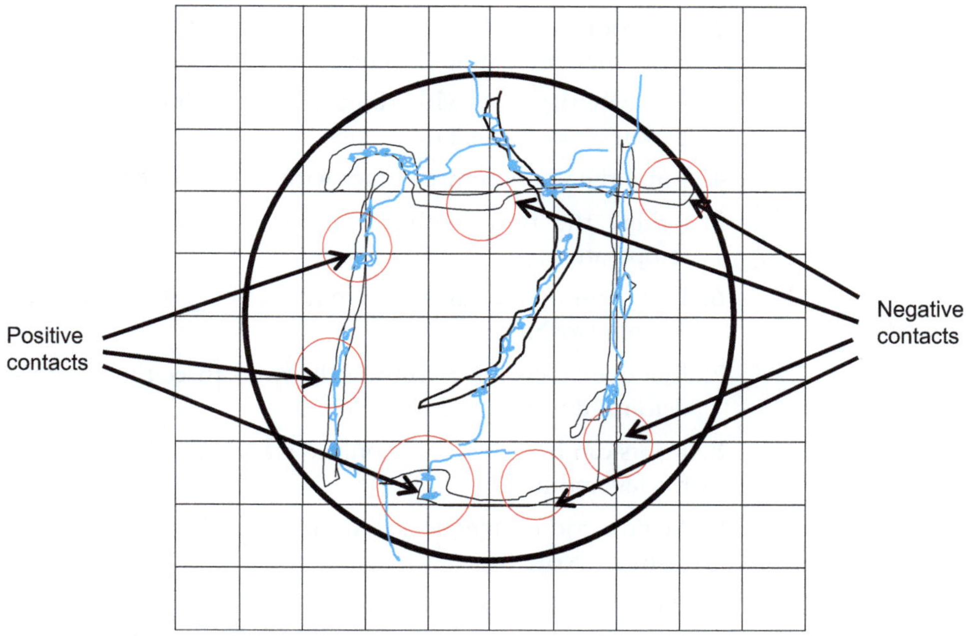

Fig. 2 Scheme showing the gridline intersect method for quantification of AM fungal colonization in mycorrhizal roots. Based on the method described by Giovannetti and Mosse in 1980 [11]

1. Clearing solution 1: 10% (w:v) potassium hydroxide (KOH).
2. Clearing solution 2: 2% (v:v) acetic acid in water.
3. Staining solution: 5% Lamy black ink (*see* **Note 1**) and 2% acetic acid.

2.2 HPLC Quantification of Molecular Markers (Apocarotenoids)

This method is more accurate than the previous one, but it is also more expensive and requires more specific analytical equipment. The methodology is based on the protocol described by Fester et al. [13]. All solutions should be prepared using deionized water and analytical grade reagents., and they should be stored at room temperature.

1. 80% (v:v) methanol (MeOH).
2. 2 mg/ml ribitol.
3. Potassium hydroxide (KOH), pure (pellets).
4. 1.5% phosphoric acid (H_3PO_4)
5. Acetonitrile (MeCN), pure.

3 Methods

3.1 Histochemical Quantification

1. Rinse harvest roots with tap water to remove solid particles from the substrate (*see* **Notes 2** and **3**).
2. Dry roots with filter paper, cut them into 1 cm long pieces and place them in a 50 ml Falcon tube.
3. Incubate the roots with 40 ml 10% KOH for 2 days at 4 °C in darkness (*see* **Note 4**).
4. Rinse the samples three times with 40 ml deionized water.
5. Acidify the roots with 2% acetic acid for 5 min at room temperature.
6. Dye the roots by immersion in the staining solution for at least 30 min (*see* **Note 5**).
7. Remove excess ink by rinsing the roots three times with 40 ml deionized water (*see* **Note 6**).
8. Roots can be stored in deionized water or 70% ethanol at 4 °C for few months.
9. Mark a grid of lines of 1 cm on the bottom of a 90 mm Petri dish (*see* **Note 7**).
10. Randomly disperse the stained roots (about 200) in the Petri dish with gridlines using forceps and a dissecting needle.
11. Assess mycorrhizal colonization under a dissecting microscope by following all horizontal and vertical lines.
12. Count intersects with roots and mycorrhiza (positive) separately. Do, at least, two sets of quantification, recording 100–200 root/gridline intersections.
13. Calculate the colonization percentage by counting all the positives contacts. Divide this number by the total number of contacts (positives and negatives), and multiply by 100.

3.2 HPLC Quantification of Apocarotenoids

1. Weight 50 mg of lyophilized root material and put it in a 5 ml glass vial (*see* **Notes 8** and **9**).
2. Add 50 μl of ribitol, used as internal standard to further yield the polar fraction (*see* **Note 10**).
3. Extract with 400 μl of 80% MeOH.
4. Sonicate the mixture for 2 min at room temperature.
5. Centrifuge at 13,000 × *g* for 5 min at room temperature and collect the supernatant.
6. Repeat **steps 3–5** two times more and put together the three supernatants, obtaining a fraction of about 1.2 ml (*see* **Note 11**).
7. Take and aliquot of 300 μl of the supernatant and adjust it to a final concentration of 0.5 M with KOH and incubate for 1 h at room temperature (*see* **Note 12**).
8. Centrifuge at 13,000 × *g* for 10 min and collect the supernatant. Samples are ready for their analysis by liquid chromatography (HPLC).
9. Perform HPLC with a system including a separation module. Normally a system from Waters is used. The liquid chromatograph must be equipped with a 5 μm Nucleosil C_{18} column (250 × 4 mm) or similar.
10. Use as solvent A: 1.5% H_3PO_4; and solvent B: MeCN.
11. The sample injection volume is 20 μl.
12. For an optimal separation of the compounds use the following gradient: a first step from 5% B to 25% B within 40 min, and then from 25% B to 80% within 20 min.
13. Use a flow rate of ml min^{-1}.
14. Detect the apocarotenoids photometrically at 250 and 380 nm, using a photodiode array detector (*see* **Note 13**).
15. Calculate quantitative values after standardization with ribitol (used as internal standard), and using an appropriate software (*see* **Note 14**).

4 Notes

1. We observed that with this ink, a better contrast was obtained. However, other brands, such as Sheaffer or Pelikan can also be used.
2. Washing tap water gives better results than with deionized water. Moreover, the later can disrupt root membranes with the corresponding loss of metabolites.

3. Roots for at least five independent mycorrhizal plants should be harvest and analyzed. It is important to include also negative controls without AM fungal inoculation to discard the presence of contamination.
4. For a faster processing, roots can be also cleared by boiling in 10% KOH for 10 min.
5. Longer incubation times (2–4 h) results in better staining. Therefore mycorrhizal colonization can be assessed easier under de microscope.
6. Elimination of excess ink can be done also by rinsing the roots with 2% acetic acid. This treatment will keep samples in better condition for longer when stored at 4 °C.
7. The dimension of the grid squares is important for the measurement of total root length (colonized or noncolonized), but not for measuring the percentage of colonized roots, since the gridlines only act as a device for the systematic selection of observation points. Grids of 1 cm^2 are normally used.
8. Fresh material can be also used. If this is the case, at least 200 mg of material is required. However, larger quantities give better signals in the chromatography, and therefore better results.
9. Roots for at least five independent mycorrhizal plants should be extracted and analyzed. Although the apocarotenoids C_{13} α-ionol glycosides (blumenols) and C_{14} mycorradicin derivatives are specific of mycorrhizal plants, it is important to include also negative controls without AM fungal inoculation to compare the different profiles.
10. Ribitol is used as internal standard to normalize extraction efficiency of apocarotenoids. Ribitol is a crystalline pentose alcohol ($C_5H_{12}O_5$) formed by the reduction of ribose.
11. A total of three extractions are required to ensure the obtaining 99% of the polar compounds.
12. Pure KOH (pellets) is used to adjust (increase) the pH of the samples previous their analysis by HPLC. It allows an easier detection of polar compounds.
13. The software for the analysis of the HPLC data is normally supplied by the company of the equipment.
14. The integrated values are normalized to the mean value of the internal standard ribitol calculated for the whole set of experiments per extract.

Acknowledgments

This work was supported by grants AGL2015-64990-C2-1-R and BIO2017-90877-REDT from the Ministry of Economy and Competitiveness (MINECO) and 201640I040 from CSIC.

References

1. Barea JM, Pozo MJ, Azcon R, Azcon-Aguilar C (2005) Microbial co-operation in the rhizosphere. J Exp Bot 56:1761–1778
2. Gianinazzi S et al (2010) Agroecology: the key role of arbuscular mycorrhizas in ecosystem services. Mycorrhiza 20:519–530
3. Tkacz A, Poole P (2015) Role of root microbiota in plant productivity. J Exp Bot 66:2167–2175
4. Pozo MJ, López-Ráez JA, Azcón C, García-Garrido JM (2015) Phytohormones as integrators of environmental signals in the regulation of mycorrhizal symbioses. New Phytol 205:1431–1436
5. Duhamel M, Vandenkoornhuyse P (2013) Sustainable agriculture: possible trajectories from mutualistic symbiosis and plant neodomestication. Trends Plant Sci 18:597–600
6. López-Ráez JA et al (2015) Differential spatiotemporal expression of carotenoid cleavage dioxygenases regulates apocarotenoid fluxes during AM symbiosis. Plant Sci 230:59–69
7. Walter MH, Stauder R, Tissier A (2015) Evolution of root-specific carotenoid precursor pathways for apocarotenoid signal biogenesis. Plant Sci 233:1–10
8. López-Ráez JA, Shirasu K, Foo E (2017) Strigolactones in plant interactions with beneficial and detrimental organisms: the yin and yang. Trends Plant Sci 22:527–537
9. Waters MT, Gutjahr C, Bennett T, Nelson DC (2017) Strigolactone signaling and evolution. Annu Rev Plant Biol 68:291–322
10. Wang M et al (2018) Blumenols as shoot markers of root symbiosis with arbuscular mycorrhizal fungi. Elife 7:pii: e37093
11. Giovannetti M, Mosse B (1980) Evaluation of techniques for measuring vesicular arbuscular mycorrhizal infection in roots. New Phytol 84 (3):489–500
12. Vierheilig H, Coughlan AP, Wyss U, Piché Y (1998) Ink and vinegar, a simple staining technique for arbuscular-mycorrhizal fungi. Appl Environ Microbiol 64:5004–5007
13. Fester T et al (2002) Stimulation of carotenoid metabolism in arbuscular mycorrhizal roots. Planta 216:148–154

Part VI

Biotechnology

Chapter 23

Use of Potyvirus Vectors to Produce Carotenoids in Plants

José-Antonio Daròs

Abstract

Potyviruses are plus-strand RNA viruses that can be easily transformed into expression vectors to quickly express one carotenogenic enzyme or transcription factor, or more, in plant tissues. Unlike the technically challenging and time-consuming process of plant transformation, manipulation of a roughly 10,000 nt-long viral genome is rather straightforward via common molecular biology techniques. Here I describe how to insert the cDNAs of the proteins of interest into two particular positions of the cDNA of a *Tobacco etch virus* (TEV) mutant that lacks the viral NIb cistron and only infects the plants in which this protein is expressed. This deletion increases the space to harbor foreign sequences. The selection of the expression site must be made according to subcellular localization requirements. The recombinant virus is then inoculated into *Nicotiana benthamiana* plants by means of *Agrobacterium tumefaciens*. The expression of the viral genome entails the production of carotenogenic proteins in the plant tissues with a consequent effect on the plant carotenoid pathway.

Key words Viral vector, Potyvirus, Transient expression, Protein expression, Polyprotein processing, Carotenoid biosynthesis

1 Introduction

Plant viruses, if properly manipulated, can be converted into vectors to produce compounds of interest in plant tissues [1, 2]. In contrast to the complex genomes of host plants, which are technically difficult and time-consuming to manipulate, the small genomes of most plant viruses can be cloned in a single plasmid, or a few, and can be easily manipulated by common molecular biology techniques [3]. Once engineered and converted into expression vectors, most recombinant viruses can be quickly and efficiently delivered to plant tissues by means of *Agrobacterium tumefaciens*. In essence, this is no different from the most general method of the *A. tumefaciens*-mediated transient expression of RNAs and proteins in plant tissues [4], a technique that also frequently uses elements of a viral origin, such as promoters, suppressors of RNA silencing pathways, and other regulatory elements. However, the use of

Manuel Rodríguez-Concepción and Ralf Welsch (eds.), *Plant and Food Carotenoids: Methods and Protocols*, Methods in Molecular Biology, vol. 2083, https://doi.org/10.1007/978-1-4939-9952-1_23,

recombinant viruses may add some advantages to a particular expression project as all the regulatory elements that the virus provides have been shaped by evolution to work coordinately to optimize the expression of the viral proteins and, expectantly, foreign proteins. In our recent research, we found that potyviruses are particularly helpful for expressing carotenogenic enzymes in plant tissues, since a substantial accumulation of products of this metabolic pathway can take place in infected tissues a few days after inoculation [5, 6]. Potyviruses (genus *Potyvirus*) compose the largest group of RNA viruses that infect plants. They consist of an approximately 10,000 nt-long RNA genome of plus polarity that is attached to a viral protein genome-linked (VPg) at the 5′ end, containing a poly(A) tail at the 3′ end, and encapsidated into an elongated and flexuous virion [7]. Potyviruses have a unique strategy to express their genomes. Basically, genomic RNA encodes a large polyprotein that is processed after translation into about 10 mature proteins by three viral-encoded proteases [8]. Conversion of a potyvirus into an expression vector is quite a straight forward process that only requires the insertion of the cDNAs corresponding to one protein of interest, or more, in some particular intercistronic positions of the viral genome, provided that these cDNAs are flanked by the right sequences to mediate the release of the heterologous proteins from the viral polyprotein [9, 10]. In practice, nuclear inclusion *a* (NIa) protease (NIaPro), which recognizes a −6/+1 seven amino-acid motif with high specificity, is used for this purpose [11]. Attention to signal peptides and the right position of the heterologous proteins in the viral polyprotein must be also paid if these proteins need to be targeted to any particular subcellular compartments [12].

Here I describe a method to express one or more carotenogenic enzymes or transcription factors in *Nicotiana benthamiana* plants using *Tobacco etch virus* (TEV; genus *Potyvirus*) (Fig. 1). More specifically, in this method we use a TEV version that lacks the NIb cistron, which encodes for the viral RNA-dependent RNA polymerase and is only able to infect the *N. benthamiana* plants that stably express NIb [9, 13, 14]. Deletion of viral NIb cistron increases the space to harbor foreign RNA, the same strategy herein described can also be used with a full-length TEV or any other potyvirus vector. After construction of the recombinant virus (Figs. 2 and 3), plants are agroinoculated. To replicate its genome in infected tissues, the recombinant virus expresses a polyprotein from which foreign proteins are released (Fig. 2). This method can be useful for producing a particular carotenoid in *N. benthamiana* tissues (Fig. 4), but to also quickly analyze the effect of the expressions of native, heterologous or engineered proteins in the carotenoid biosynthesis pathway.

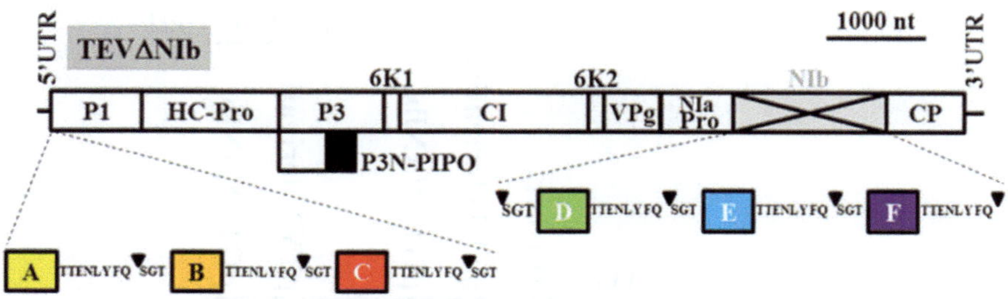

Fig. 1 Schematic representation of recombinant viruses derived from TEVΔNIb in which proteins of interest are expressed from the amino terminal end (proteins A, B, and C) and from the inside (replacing NIb, proteins D, E, and F) of viral polyprotein. Boxes represent the TEV cistrons P1, HC-Pro, P3, P3N-PIPO, 6 K1, CI, 6 K2, VPg, NIaPro, NIb, and CP, and lines the viral 5′ and 3′ untranslated regions (UTR). Note that the release of the foreign proteins from the viral polyprotein is mediated by the NIaPro cleavage site (TTENLYFQ/SGT). The arrowheads point to the exact cleavage sites

```
  Proteolytic cleavage site:  T   T   E   N   L   Y   F   Q / S   G   T
   Wild type NIaPro NIb/CP:  ACG ACT GAG AAT CTT TAT TTT CAG/AGT GGC ACT
      Artificial NIaPro a1:  ACG ACT GAA AAC TTG TAC TTC CAA/AGC GGC ACT
      Artificial NIaPro a2:  ACT ACA GAG AAC CTC TAC TTT CAA/TCA GGT ACA
      Artificial NIaPro a3:  ACA ACC GAA AAT TTA TAT TTC CAG/TCT GGA ACA
      Artificial NIaPro a4:  ACC ACA GAA AAC TTA TAC TTT CAA/AGT GGG ACC
      Artificial NIaPro a5:  ACG ACC GAA AAC CTG TAT TTC CAA/TCT GGC ACC
```

Fig. 2 Alternative artificial TEV NIaPro cleavage sites (−8/+3) with different nucleotide sequences that have been experimentally tested. Silent mutations with respect to the sequence that splits NIb from CP in wild-type TEV are indicated in red

2 Materials

2.1 Plasmids

1. Plasmids pTV1 (Addgene plasmid # 122528) and pTV3 (Addgene plasmid # 122529) contain fragments of TEV (GenBank accession number DQ986288, including silent mutations G273A and A1119G) cDNA from position 1 to 3198, and from position 6517 to 9539, respectively.
2. Plasmid pGTEVΔNIb (Addgene plasmid # 122530) contains the cDNA of TEVΔNIb (DQ986288, G273A, A1119G, NIb deletion 6982-8517) under the control of the *Cauliflower mosaic virus* (CaMV) 35S promoter and terminator in a binary plasmid derived from pCLEAN-G181 [15]. pTV1 and pTV3 also contain the CaMV 35S promoter and terminator, respectively.

2.2 Culture Media

1. Liquid Luria–Bertani (LB) medium: weigh 10 g of tryptone, 5 g of yeast extract, and 10 g of NaCl, dissolve in 1 L of water and sterilize by autoclaving. Add the appropriate antibiotics.
2. Solid LB agar medium: add 15 g of agar to 1 L of liquid LB medium before sterilizing and autoclave. Adjust the solution to approximately 55 °C, add the appropriate antibiotics and plate.

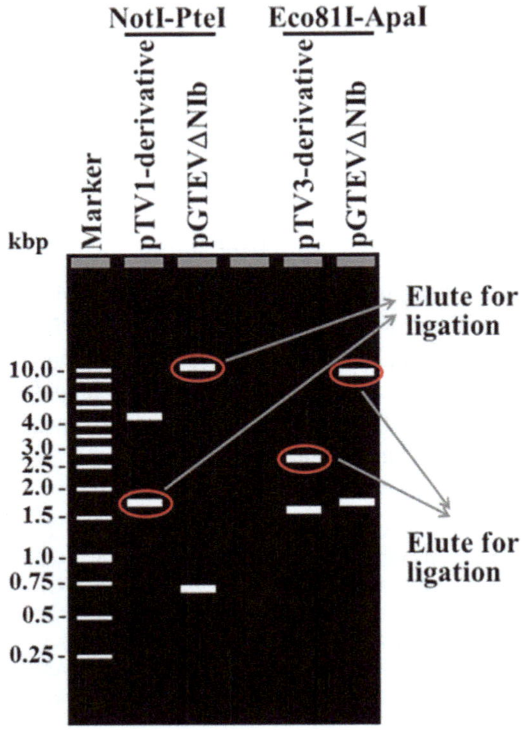

Fig. 3 Schematic representation of an electrophoresis gel run to separate the DNA fragments obtained in the NotI–PteI digestion of the pTV1-derivative and pGTEVΔNIb, and the Eco81I–ApaI digestion of the pTV3-derivative and pGTEVΔNIb. The pTV1 derivative and the NotI–PteI digestion, and the pTV3 derivative and the Eco81I–ApaI digestion are used to build recombinant viruses that express the foreign proteins from the amino terminal end and from the internal part of the viral polyprotein, respectively. Migration of the schematic DNA fragments have been arbitrarily adjusted for a foreign cDNA of 1000 bp. DNA fragments marked with a red ellipse must be cut from the gel and the DNA eluted for ligation

3. Chloramphenicol stock solution: weigh 340 mg of chloramphenicol and dissolve in 10 mL ethanol. Store at −20 °C. The final concentration is 34 mg/mL chloramphenicol.
4. Rifampicin stock solution: weigh 0.5 g of rifampicin and dissolve in 10 mL of dimethyl sulfoxide. Store frozen aliquots at −20 °C. The final concentration is 50 mg/mL rifampicin.
5. Tetracycline stock solution: weigh 75 mg of tetracycline and dissolve in 10 mL of ethanol. Store at −20 °C. The final tetracycline concentration is 7.5 mg/mL.
6. Kanamycin stock solution: weigh 0.5 g of kanamycin, dissolve in 10 mL of water and sterilize by filtration. Store frozen aliquots at −20 °C. The final kanamycin concentration is 50 mg/mL.

Fig. 4 Pictures of *Nicotiana benthamiana* leaves infected with TEV recombinant clones that express carotenogenic proteins. (A) Expression of *Pantoea ananatis* phytoene synthase (crtB). Upper noninoculated leaf photographed at 7 dpi. (B) Expression of *P. ananatis* crtE: geranylgeranyl pyrophosphate synthase (crtE), crtB and phytoene desaturase (crtI). Agroinoculated leaf photographed at 6 dpi

7. Growth media with antibiotics: prepare the final media by adding 1 μL of the corresponding antibiotic stock per 1 mL of medium under sterile conditions. The final concentrations are 34 μg/mL chloramphenicol, 50 μg/mL rifampicin, 7.5 μg/mL tetracycline (*see* **Note 1**), and 50 μg/mL kanamycin.

2.3 Electrophoresis

1. TAE electrophoresis buffer: 40 mM Tris, 20 mM sodium acetate, 1 mM sodium ethylenediaminetetraacetate (EDTA), pH 7.2. Prepare (10×) concentrated by weighing 97 g of Tris, 54 g of sodium acetate ($\cdot 3H_2O$) and 7 g of $Na_2EDTA \cdot 2H_2O$, and dissolve in water. Adjust pH to 7.2 with acetic acid and bring to a final volume of 1 L with water. There is no need to use this solution under sterile conditions, although autoclaving is a common practice to improve conservation at room temperature.
2. 1% agarose electrophoresis gel: weigh 1 g of agarose and dissolve in 100 mL (×1) of TAE buffer by heating. Adjust the solution to approximately 65 °C and cast the agarose gel. Melted agarose can be stored for a few days at 65 °C.

2.4 **Agrobacterium tumefaciens** *Induction Media*

1. Acetosyringone stock solution (100 mM): weigh 0.59 g of acetosyringone and dissolve in 30 mL of dimethyl sulfoxide. Prepare aliquots and store them frozen at −20 °C.
2. *A. tumefaciens* induction solution: 10 mM 2-(N-morpholino) ethanesulfonic acid (MES)-NaOH, pH 5.6, 10 mM $MgCl_2$, 150 μM acetosyringone. Weigh 1.07 g of $MES \cdot H_2O$ and 1.02 g of $MgCl_2 \cdot 6H_2O$, and dissolve with water. Adjust pH to 5.6 with 0.1 M NaOH and bring to 500 mL. Sterilize the

solution by autoclaving. Prepare the working solution by adding 1.5 μL of the 100 mM acetosyringone stock per 1 mL of the final solution to reach an acetosyringone final concentration of 150 μM.

3 Methods

3.1 Construction of the Recombinant Virus

To build a recombinant TEV that expresses one carotenogenic protein or more, the cDNAs corresponding to these proteins are first assembled in intermediate plasmids pTV1 or pTV3, depending on these heterologous proteins are intended to be expressed from the amino terminal end or internally in the viral polyprotein, respectively (Fig. 1) (*see* also **Note 2**). The combination of both expression positions is possible.

1. Obtain the cDNA or cDNAs corresponding to the recombinant proteins with the appropriate NIaPro cleavage sites (Fig. 2) by PCR (*see* **Note 3**) or gene synthesis. Flank these cDNAs by appropriate homologous sequences to assemble them between positions 454 and 455 of pTV1, or between 474 and 1978 of pTV3 (*see* **Note 4**).
2. Amplify by PCR the DNAs corresponding to pTV1 or/and pTV3 opened between positions 454 and 455, and between 474 and 1978, respectively. Use primers 5′-GGCTATCGTTCGTAAATGG-3′ and 5′-ATGGCACTCATCTTTGGC-3′ to open pTV1, and primers 5′-CTTCTCCCCTTGCGAGTAC-3′ and 5′-ACGACTGAGAATCTTTATTTTCAG-3′ to open pTV3.
3. Mix all the DNA fragments at a vector–insert 1:3 molar ratio and perform the Gibson assembly reaction for 1 h at 50 °C [16] (*see* also **Note 5**).
4. Purify the DNA from the assembly reaction using silica gel columns and use it to electroporate *Escherichia coli* DH5α.
5. Select the transformed *E. coli* clones in the LB agar plates containing chloramphenicol.
6. Miniprep the plasmids from some of the colonies and analyze them by electrophoresis in 1% agarose gel to select one with the expected migration according to the inserted cDNAs (*see* **Note 6**).
7. For the expression from the amino terminal end of the polyprotein, digest both the pTV1 derivative and pGTEVΔNIb with *Not*I and *Pte*I, separate the DNA fragments by electrophoresis in 1% agarose gel and elute the appropriate fragments from the gel using silica gel columns. For the expression from inside the viral polyprotein, digest the pTV3 derivative and

pGTEVΔNIb with *Eco*81I and *Apa*I, and follow the same procedure (Fig. 3).

8. Mix the appropriate amounts of the DNA fragments to set up a DNA ligation reaction with T4 DNA ligase for 1 h at 22 °C.
9. Purify the products of ligation reaction using a silica gel column and employ them to electroporate *E. coli* DH5α.
10. Select the transformed bacteria in the LB agar plates containing kanamycin.
11. Miniprep the plasmids from some colonies and analyze them by electrophoresis in a 1% agarose gel.
12. Choose a plasmid with the expected migration (*see* **Note 7**) and confirm the incorporation of the desired DNA fragment first by a restriction analysis and then, if necessary, by sequencing.

3.2 Transformation of *A. tumefaciens*

1. Using the pGTEVΔNIb derivative that contains the TEV recombinant clone, electroporate an *A. tumefaciens* C58C1 clone that harbors helper plasmid pCLEAN-S48 [15] (*see* **Note 8**).
2. Select the transformed bacteria in the LB agar plates containing rifampicin, tetracycline and kanamycin. Incubate plates for 5–7 days at 28 °C (*see* **Note 9**).

3.3 Plant Agroinoculation

1. Pick an *A. tumefaciens* colony from the plate with three antibiotics and inoculate 10 mL of the liquid LB media containing kanamycin (*see* **Note 10**).
2. Grow the liquid culture for approximately 24 h at 28 °C with shaking (200 revolutions per minute, rpm).
3. Measure the optic density of the culture at 600 nm and harvest bacteria at the exponential grow rate (*see* **Note 11**).
4. Based on the actual optic density at which *A. tumefaciens* are harvested, spin down an aliquot of the bacterial culture to prepare 4.5 mL of cells in agroinoculation media (*see* **Note 12**).
5. Induce the *A. tumefaciens* culture for 3 h at 28 °C with shaking (200 rpm).
6. Use a 1-mL needleless syringe to infiltrate the induced *A. tumefaciens* culture into the abaxial side of an *N. benthamiana* plant leaf that expresses TEV NIb. The middle leaves of the plant (youngest among those fully expanded) usually give the best results (*see* **Note 13**).

3.4 Carotenoid Production

1. Cultivate plants in a growth chamber at 25 °C with a common day-night cycle.
2. Symptoms of infection should be visible in young tissue around 1 week after inoculation (*see* **Note 14**).

3. Harvest the symptomatic tissue 2 days later for the carotenoid analysis (Fig. 4).

4 Notes

1. The tetracycline concentration to select pCLEAN-S48 (see below) is a slightly lower than those normally used with the plasmids containing tetracycline resistance selection markers.
2. From the amino terminal end of the polyprotein, heterologous proteins can be targeted to any subcellular location, provided that they harbor an appropriate signal peptide. From inside the polyprotein, heterologous proteins can be efficiently targeted only to the nucleus and the cytosol [12].
3. For the PCR amplifications, the use of a high-fidelity DNA polymerase, such as Phusion DNA polymerase (Thermo Scientific), is recommended.
4. The use of a program or web tool, such as the NEBuilder Assembly Tool from New England BioLabs (https://nebuilder.neb.com/#!/), facilitates the design of the flanking sequences that allow DNA assembly during the Gibson reaction. Separate the cDNAs corresponding to the different proteins by the −8/+3 cleavage peptides recognized by TEV NIaPro. To avoid undesired recombinations during the replication of the recombinant virus, while maintaining the same amino acid sequence (TTENLYFQ/SGT), use different nucleotide sequences. Some examples are found in Fig. 2. Remember that the recombinant protein or block of recombinant proteins expressed from the amino terminal end of the polyprotein must end with a NIaPro proteolytic site to separate the last protein from the viral P1 (Fig. 1). If the recombinant protein or proteins are expressed from inside the polyprotein, take advantage of the native NIaPro/NIb and NIb/CP cleavage sites that remain in pTV3 for the first and last cleavages.
5. For the Gibson reaction, we successfully used the NEBuilder HiFi DNA Assembly Master Mix from New England Biolabs.
6. Plasmid sequencing around the region where the insertion takes place is recommended.
7. Due to the size of the plasmids at this point, do not expect to observe a clear difference in migration between those containing the recombinant virus and those with the unmodified TEVΔNIb clone. A restriction analysis and sequencing are recommended for confirmation purposes.
8. pGTEVΔNIb is a derivative of pCLEAN-G181 (GenBank accession number EU186083.1) [15] and requires the presence of a helper plasmid of the pSoup series to replicate in

A. tumefaciens. We obtained good results with pCLEAN-S48, but any other helper plasmid that carries the pSA replicase gene intended to act *in trans* upon the pSA replication origin should work just as well.

9. Discard any colonies that appear in plates too early because they will most probably contain plasmids with deleted versions of the virus. Remember that the pGTEVΔNIb derivative is a plasmid with a large insert (more than 10 kbp in most clones), and good colonies will necessarily grow slowly in the plates with three antibiotics. The grow rate depends on the length of the recombinant clone, which is reason why 5–7-day window is given.
10. Kanamycin is the critical antibiotic to select the recombinant *A. tumefaciens*. There is no need to add rifampicin and tetracycline to the liquid culture.
11. Optic density is not critical, although an interval between 0.2 and 1 is recommended to harvest bacteria under the best physiological conditions.
12. We found that 5 mL Eppendorf tubes were most convenient in this step. Centrifuge for 5 min at 10,000 rpm to sediment *A. tumefaciens*.
13. If the recombinant TEV clone is infectious, a single infiltration in one leaf is enough. However, we recommend infiltrating two or three leaves per plant until efficient infectivity is confirmed.
14. Once again, the exact day in which symptoms appear will depend on the complexity of the recombinant virus.

Acknowledgments

This work was supported by the Ministerio de Ciencia, Innovación y Universidades (Spain) grant BIO2017-83184-R (AEI/FEDER UE).

References

1. Pogue GP, Lindbo JA, Garger SJ, Fitzmaurice WP (2002) Making an ally from an enemy: plant virology and the new agriculture. Annu Rev Phytopathol 40:45–74. https://doi.org/10.1146/annurev.phyto.40.021102.150133
2. Hefferon K (2017) Plant virus expression vectors: a powerhouse for Global Health. Biomedicine 5(3). https://doi.org/10.3390/biomedicines5030044
3. Pasin F, Menzel W, Daròs JA (2019) Harnessed viruses in the age of metagenomics and synthetic biology: an update on infectious clone assembly and biotechnologies of plant viruses. Plant Biotechnol J. https://doi.org/10.1111/pbi.13084
4. Krenek P, Samajova O, Luptovciak I, Doskocilova A, Komis G, Samaj J (2015) Transient plant transformation mediated by *Agrobacterium tumefaciens*: principles, methods and applications. Biotechnol Adv 33(6 Pt 2):1024–1042. https://doi.org/10.1016/j.biotechadv.2015.03.012
5. Majer E, Llorente B, Rodríguez-Concepción M, Daròs JA (2017) Rewiring

carotenoid biosynthesis in plants using a viral vector. Sci Rep 7:41645. https://doi.org/10.1038/srep41645

6. Aragonés V, Pérez-de-Castro A, Cordero T, Cebolla-Cornejo J, López C, Picó B, Daròs JA (2019) A *Watermelon mosaic virus* clone tagged with the yellow visual maker phytoene synthase facilitates scoring infectivity in melon breeding programs. Eur J Plant Pathol 153:317. https://doi.org/10.1007/s10658-018-01621-x
7. Wylie SJ, Adams M, Chalam C, Kreuze J, López-Moya JJ, Ohshima K, Praveen S, Rabenstein F, Stenger D, Wang A, Zerbini FM, Ictv Report C (2017) ICTV Virus Taxonomy Profile: *Potyviridae*. J Gen Virol 98 (3):352–354. https://doi.org/10.1099/jgv.0.000740
8. Revers F, García JA (2015) Molecular biology of potyviruses. Adv Virus Res 92:101–199. https://doi.org/10.1016/bs.aivir.2014.11.006
9. Bedoya L, Martínez F, Rubio L, Daròs JA (2010) Simultaneous equimolar expression of multiple proteins in plants from a disarmed potyvirus vector. J Biotechnol 150 (2):268–275. https://doi.org/10.1016/j.jbiotec.2010.08.006
10. Majer E, Salvador Z, Zwart MP, Willemsen A, Elena SF, Daròs JA (2014) Relocation of the NIb gene in the tobacco etch potyvirus genome. J Virol 88(8):4586–4590. https://doi.org/10.1128/JVI.03336-13
11. Carrington JC, Haldeman R, Dolja VV, Restrepo-Hartwig MA (1993) Internal cleavage and trans-proteolytic activities of the VPg-proteinase (NIa) of tobacco etch potyvirus in vivo. J Virol 67(12):6995–7000
12. Majer E, Navarro JA, Daròs JA (2015) A potyvirus vector efficiently targets recombinant proteins to chloroplasts, mitochondria and nuclei in plant cells when expressed at the amino terminus of the polyprotein. Biotechnol J. https://doi.org/10.1002/biot.201500042
13. Li XH, Carrington JC (1995) Complementation of tobacco etch potyvirus mutants by active RNA polymerase expressed in transgenic cells. Proc Natl Acad Sci U S A 92(2):457–461
14. Martí M, Diretto G, Aragonés V, Ahrazem O, Gómez-Gómez L, Daròs JA (2019) Efficient production of saffron crocins and picrocrocin in *Nicotiana benthamiana* using a virus-driven system. Submitted
15. Thole V, Worland B, Snape JW, Vain P (2007) The pCLEAN dual binary vector system for *Agrobacterium*-mediated plant transformation. Plant Physiol 145(4):1211–1219
16. Gibson DG, Young L, Chuang RY, Venter JC, Hutchison CA 3rd, Smith HO (2009) Enzymatic assembly of DNA molecules up to several hundred kilobases. Nat Methods 6 (5):343–345. https://doi.org/10.1038/nmeth.1318

Chapter 24

Agrobacterium tumefaciens-Mediated Stable Transformation of *Daucus carota*

Christian Gonzalez-Calquin and Claudia Stange

Abstract

Daucus carota L. (carrot) is one of the ten most important vegetables cultivated and consumed worldwide and is a main source of provitamin A. Carrot storage root is rich in dietary fiber, antioxidants, and other nutrients but especially in carotenoids. It has been also used as plant model for studding embryogenesis, as well as the genetic and genomic evolution of carrots and for carotenoid synthesis regulation, among others. Research in carrot often needs genetic transformation. Here we describe a step-by-step protocol on the nuclear and stable transformation of carrot through *Agrobacterium tumefaciens* and somatic embryogenesis in vitro culture. Somatic embryos, induced by supplementation of Murashige–Skoog medium with the 2,4D hormone, develop into seedlings after 6 months approximately when plants are ready to be transferred to a greenhouse. The protocol has over 85% of transformation efficiency.

Key words Carrot, Transformation protocol, 2,4-dichlorophenoxiacetic acid, Somatic embryogenesis, *Agrobacterium tumefaciens*, In vitro culture

1 Introduction

Carrot (*Daucus carota* L., $2n = 18$) belongs to the Umbelliferae group together with parsley *(Petroselinum hortense)*, anise (*Pimpinella anisum*), celery (*Apium graveolens*), caraway (*Carum carvi*), dill (*Anethum graveolens)*, and other thousands of species. Carrot is a biennial plant, consumed worldwide, whose storage root accumulates high level of carotenoids. The orange carrot was stated recently until the sixteenth century, being the first emerged domesticated carrot root yellow and purple [1, 2]. The production of carrots has quadrupled in the last four decades, principally due to the consumption as fresh vegetable and other high-value products [3]. Carrot storage root is rich in dietary fiber, antioxidants, and other nutrients [2], but the most important nutritional value of carrots lies in the high quantities of α- and β-carotene, making carrots the richest source of provitamin A in the diet [4].

Manuel Rodríguez-Concepción and Ralf Welsch (eds.), *Plant and Food Carotenoids: Methods and Protocols*,
Methods in Molecular Biology, vol. 2083, https://doi.org/10.1007/978-1-4939-9952-1_24,

It is also used as a plant model for tissue culture and embryogenesis studies [5–8], as well as genetic and genomic studies looking for understanding the evolution of domesticated carrots [9–11]. Other researches have been centered on studying gene function and regulation of carotenoid synthesis in the storage root through over expressing or down regulating carotenogenic genes [12–16]. Genetic engineering of carrots has been also successful for producing astaxanthin [17] and human interferon alpha-2b protein as medical treatment for several virus diseases [18]. On the other hand, carrots expressing disease resistance genes have provided defense against abiotic and biotic stresses [19, 20]. Recently, the CRISPR/Cas technology was established in carrot targeting the flavanone-3-hydroxylase (F3H) [21]. For most of these researches, carrot genetic transformation is required. Among the diverse techniques, those based in *Agrobacterium tumefaciens* are the most common methods used elsewhere since the first transformation protocol was reported [22]. Thereafter, many optimized transformation procedures have been established using roots, cotyledons, hypocotyls, and petioles as explants as well as different cocultivation time and culture medium [19, 23–25]. In this work we present a detailed method for carrot nuclear transformation through somatic embryogenesis for basic and applied research.

2 Materials

All culture media should be prepared with deionized water and analytical grade reagents, then, sterilized (150 °C for 15 min in autoclave) before use. Glass petri dishes, tweezers, and scalpels should be sterilized in autoclave as well.

2.1 Carrot Transformation

1. Liquid MS 1× media: 1× Murashige and Skoog medium with vitamins (MS), 2% Sucrose, 0.05% myoinositol, pH 5.8. Weigh 2.215 g of MS, 10 g of sucrose, 50 mg of myoinositol and add 400 mL of deionized water. Stir the solution at room temperature until all components are dissolved, then adjust pH with KOH (*see* **Note 1**). Make up to 500 mL with deionized water and sterilized.
2. Solid MS 1× media: 1× MS, 2% sucrose, 0.05% myoinositol, 0.7% agar, pH 5.8. The procedure is the same as the previous culture media, but after adjusting pH and making up the volume, 7 g/L agar is added and then sterilized.
3. Coculture plates: Solid MS 1× media supplemented with 1 mg/L 2.4-dichlorophenoxyacetic acid (2.4-D). Melt the solid MS 1× media and let it cool (*see* **Note 2**), then measure 30 mL of media inside a 50 mL, sterile plastic tube, and add 2.4-D to a final concentration of 1 mg/L (use a stock solution

of 1 mg/mL 2.4-D). Mix the media very gently to avoid generating bubbles, and deposit 25 mL of media in each plastic, sterile petri dish and let solidify. 90 mm Petri dishes are recommended.

4. Sterile absorbent paper: In an autoclavable flask, deposit squares of absorbent paper of approximately 4 cm^2. Sterilize by autoclave and then dry in an oven at 60 °C by 2 days minimum.
5. Sterile quirurgic and glass equipment: use, preferably, a 773A tweezers, as it is better for handling the small and thin carrot explants. A small-blade scalpel is recommended as well. The equipment should be sterilized at 150 °C for 15 min and dried in an oven for 2 days minimum before use.

2.2 Somatic Embryogenesis Induction

1. Media I: Solid MS 1× media supplemented with 0.5 mg/L 2.4-D.
2. Media II: Solid MS 1× media supplemented with 0.25 mg/L 2.4-D.
3. Media III plates: Solid MS 1× media without hormones.

Melt the solid MS 1× media and let it cool, then measure 30 mL of media inside a 50 mL, sterile plastic tube. When correspond, add 2.4-D to a final concentration of 0.5 mg/L or 0.25 mg/L. Mix the media very gently to avoid generating bubbles, and deposit the media in a plastic, sterile petri dish and let solidify.

3 Methods

3.1 Carrot Explant Preparation

1. Cut 2 cm-long carrot epicotyls from 3 weeks old in vitro-germinated plants and deposit in the coculture plates.
2. Incubate the explants for 3 days in dark at 24 °C.

3.2 Agrobacterium tumefaciens Suspension Preparation

1. Inoculate 30 mL of liquid Luria–Bertani broth supplemented with the required selection agents of the *A. tumefaciens* strain being used (*see* **Note 3**) and let the culture grown in agitation at 28 °C overnight.
2. Centrifuge the cells at 4000 × *g* for 15 min at 4 °C. Discard the media and add Liquid MS 1× until an Optical Density at 600 nm of 0.3 is reached.
3. Resuspend the pellet very gently by pipetting (*see* **Note 4**).
4. Add acetosyringone to a final concentration of 1 μM and incubate the suspension at room temperature for 20 min (*see* **Note 5**).

3.3 Explant Transformation

1. Deposit the *Agrobacterium tumefaciens* suspension media in a sterilized glass Petri dish.
2. Deposit the carrot explants inside the cell suspension and proceed to cut the explants into 1 cm pieces (*see* **Note 6**) with the help of a sterile scalpel. Procure to generate little cuts in the tissue to facilitate the Agrobacterium cells to penetrate the tissue.
3. Incubate the explants inside the cell suspension for 15 min (*see* **Note 7**).
4. Deposit the sterile absorbent paper squares in a sterilized glass petri dish, with the help of tweezers.
5. When the incubation time of the explants is over, deposit them on the paper for removing the excess cell suspension (*see* **Note 8**).
6. After removing the excess cell suspension, deposit the explants in a coculture plate.
7. Incubate the explants for 3 days at 24 °C in darkness.

3.4 Somatic Embryogenesis Induction

1. After the third day, change the explants from the coculture plates to the MI plates. Incubate the explants for 14 days in darkness. During this process, the tissue will swallow and little cell cumuli (preembryogenic cells) will appear (Fig. 1).
2. After the 14th day, change the explants to MII plates. The explants should be incubated at 24 °C in a 16/8 h photoperiod for at least 30 days, refreshing the media every 14 days (*see* **Note 9**). During this period, the tissue will continue to swallow, and little protuberances will form, which correspond to globular-stage somatic embryos (Fig. 2). When these cells start to fall away from the mother tissue, proceed to the next step.

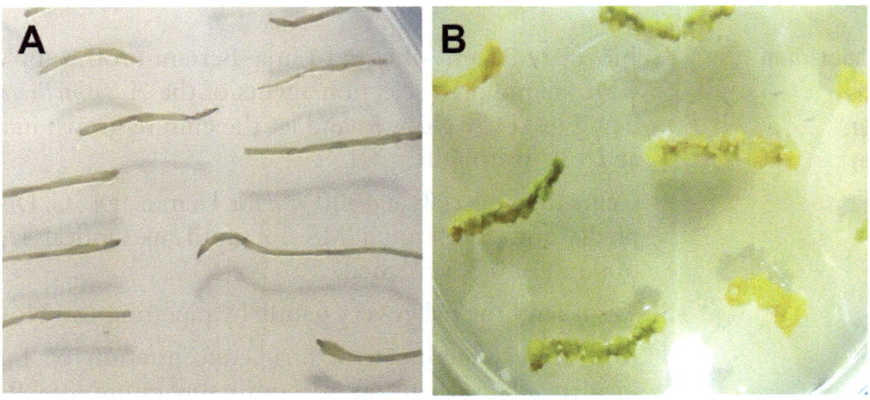

Fig. 1 Initial stages of carrot explants transformation. (**a**) Initial explants in coculture media plates. (**b**) Explants after 14 days in MI

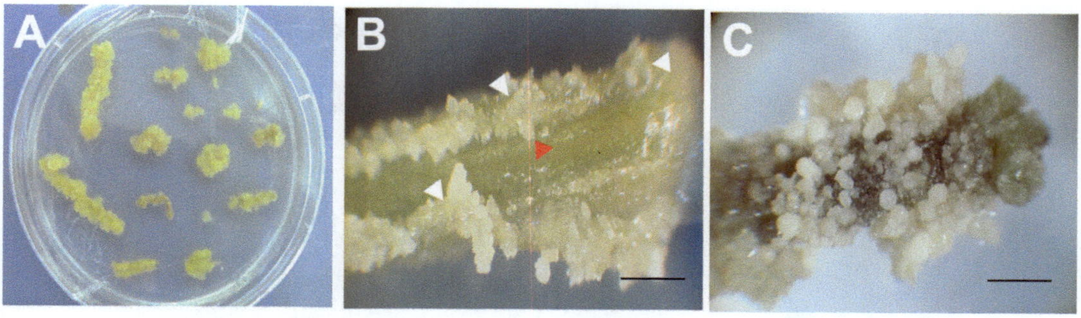

Fig. 2 Somatic embryo development from preembryonic cells. (**a**) Explants after 30 days in MII plates. (**b**) Detail of immature embryo development that emerges from the mother explant. (**c**) Detail of globular embryos. *White arrowheads* indicate somatic embryos; *red arrowhead* indicates the mother explant

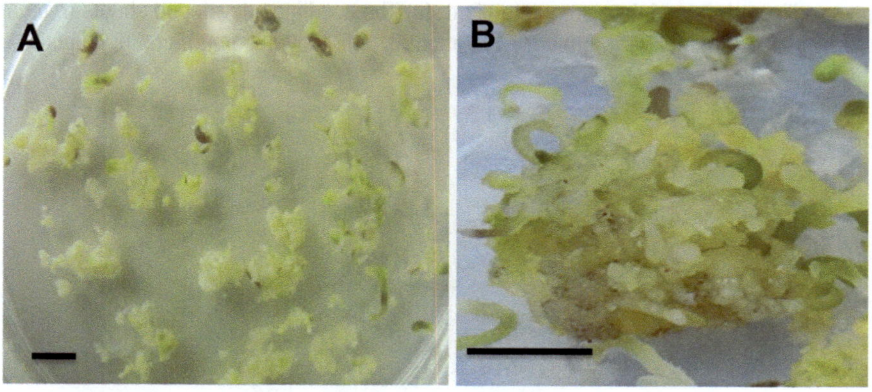

Fig. 3 Somatic embryos develop into seedlings. (**a**) In MIII, the somatic embryos will develop until they produce a new seedling. (**b**) Amplification of an explant that produces several embryos that are developing into a complete seedling. Line: 1 cm

3. When these cells start to fall apart from the mother explant, change to MIII plates and incubate them at 24 °C with a photoperiod of 16/8 h. In this media, the somatic embryos will start to differentiate into a plantlet, process that takes around 3 months (Fig. 3). Procure to transfer the globular somatic embryos to new MIII each month, handling them with care as this type of cells are very fragile and can be damaged very easily through manipulation with the tweezers.
4. When leaves and roots are developed (Fig. 4a), transfer the seedlings to Magenta GA-7 Vessels with 30 mL of MIII to allow shoot extension and plant development (Fig. 4b).
5. When plants are big enough to manipulate (3 cm long), change them to long pots with sterilized soil. Very carefully, remove the plant from the culture media and remove any traces of the solid media that could be on the root.

Fig. 4 Carrot seedlings develop into a complete plant. (**a**) The seedlings developed shoots (*blue arrows*) and roots (*red arrows*) and (**b**) continue to grow for 1–2 months until they become a complete plant

Fig. 5 Transgenic carrot plant of 6 months old. Picture of a transgenic line growing in the greenhouse for 6 months after transfer from in vitro culture

6. Place the plant on the soil and cover it with plastic foil to protect seedlings from the low relative humidity. Generate two or three little holes in the plastic to let the plant adapt to the water potential of the soil and the low relative humidity. The plants should be incubated in a greenhouse with long photoperiod (16 hours light and 8 hours dark at 24 °C).
7. Remove the plastic after 2 or 3 weeks and let the plant grow until they reach the size and age for molecular and functional analysis. Plants of 3–6 months (Fig. 5) are harvested when storage roots will be studied. Usually 3–5 transformed plants can be obtained from one induced epicotyl and over 85% of them present the transgene after genotyping.

4 Notes

1. The normal pH of the solution should oscillate between 3.8 and 4.2.
2. You can accelerate the cooling of the media by introducing the bottle inside an ice bucket with a little bit of ice and tap water. By doing this, you should be very careful of not overcool the media, or else the media at the bottom of the bottle will start to solidify.
3. Different strains of Agrobacterium need different selection agents: GV3101 strain needs gentamicin and rifampicin, alongside the selection agent that the vector confers; meanwhile, EHA101 only needs gentamicin and the selection agent that the vector confers.
4. Do not vortex the cells or the transformation efficiency will drop dramatically.
5. Longer incubation times do not affect transformation efficiency.
6. Avoid generating smaller explants. At early stages, the tissue is difficult to manipulate as is very fragile and desiccates very easily.
7. If the exposure time is shorter, it could affect the number of transformant explants.
8. Avoid exposing the explant too long on the paper to prevent desiccation of the tissue.
9. Depending on the incubator chamber, the media from the plates could evaporate more rapidly in some cases.

References

1. Vaughan J, Geissler C (2009) The new Oxford book of food plants. OUP, Oxford
2. Simpson K, Cerda A, Stange C (2016a) Carotenoid biosynthesis in *Daucus carota*. Subcell Biochem 79:199–217
3. FAOSTAT Database on Agriculture (2014). In: food and agriculture Organization of the United Nations, Rome, Italy
4. Simon P, Pollak L, Clevidence B, Holden J, Haytowitz D (2009) Plant breeding for human nutritional quality. Plant Breeding Rev 31:325–392
5. Kreuger M, Van Holst G (1993) Arabinogalactan proteins are essential in somatic embryogenesis of *Daucus carota L.* Planta 189:243–248
6. Quiroz-Figueroa F, Rojas-Herrera R, Galaz-Avalos R, Loyola-Vargas V (2006) Embryo production through somatic embryogenesis can be used to study cell differentiation in plants. Plant Cell Tissue Organ Cult 86:285–301
7. Shiota H, Ko S, Wada S, Otsu C, Tanaka I, Kamada H (2008) A carrot G-box binding factor-type basic region/leucine zipper factor DcBZ1 is involved in abscisic acid signal transduction in somatic embryogenesis. Plant Physiol Biochem 46:550–558
8. Guan Y, Ren H, Xie H, Ma Z, Chen F (2009) Identification and characterization of bZIP-type transcription factors involved in carrot (*Daucus carota L.*) somatic embryogenesis. Plant J 60:207–217

9. Xu Z, Tan H, Wang F, Hou X, Xiong A (2014) CarrotDB: a genomic and transcriptomic database for carrot. Database 2014
10. Iorizzo M, Ellison S, Senalik D, Zeng P, Satapoomin P, Huang J, Bowman M, Iovene M, Sanseverino W, Cavagnaro P, Yildiz M, Macko-Podgnórni A, Moranska E, Grzebelus E, Grzebelus D, Ashrafi H, Zheng Z, Cheng S, Spoonder D, Van Deynze A, Simon P (2016) A high-quality carrot genome assembly provides new insights into carotenoid accumulation and asterid genome evolution. Nat Genet 48:657–666
11. Ellison S, Luby C, Corak K, Coe K, Senalik D, Iorizzo M, Goldman I, Simon P, Dawson J (2018) Carotenoid presence is associated with the or gene in domesticated carrot. Genetics 210:1497–1508
12. Stange C, Fuentes P, Handford M, Pizarro L (2008) *Daucus carota* as a novel model to evaluate the effect of light on carotenogenic gene expression. Biol Res 41:289–301
13. Fuentes P, Pizarro L, Moreno J, Handford M, Rodriguez-Concepcion M, Stange C (2012) Light-dependent changes in plastid differentiation influence carotenoid gene expression and accumulation in carrot roots. Pant Mol Biol 79:47–59
14. Moreno J, Pizarro L, Fuentes P, Handford M, Cifuentes V, Stange C (2013) Levels of lycopene β-cyclase 1 modulate carotenoid gene expression and accumulation in *Daucus carota*. PLoS One 8:e58144
15. Simpson K, Quiroz L, Rodriguez-Concepción M, Stange C (2016b) Differential contribution of the first two enzymes of the MEP pathway to the supply of metabolic precursors for carotenoid and chlorophyll biosynthesis in carrot (*Daucus carota*). Front Plant Sci 7:1344
16. Simpson K, Fuentes P, Quiroz-Iturra L, Flores-Ortiz C, Contreras R, Handford M, Stange C (2018) Unraveling the induction of phytoene synthase 2 expression by salt stress and abscisic acid in *Daucus carota*. J Exp Bot 69:4113–4126
17. Jayaraj J, Devlin R, Punja Z (2008) Metabolic engineering of novel ketocarotenoid production in carrot plants. Transgenic Res 17:489–501
18. Luchakivskaya Y, Kishchenko O, Gerasymenko I, Olevinskaya Z, Simonenko Y, Spivak M, Kuchuk M (2011) High-level expression of human interferon alpha-2b in transgenic carrot (*Daucus carota L.*) plants. Plant Cell Rep 30:407–415
19. Chen W, Punja Z (2002) Transgenic herbicide- and disease-tolerant carrot (*Daucus carota L.*) plants obtained through agrobacterium-mediated transformation. Plant Cell Rep 20:929–935
20. Wally O, Punja Z (2010) Enhanced disease resistance in transgenic carrot (*Daucus carota L.*) plants over-expressing a rice cationic peroxidase. Planta 232:1229–1239
21. Klimek-Chodacka M, Oleszkiewicz T, Lowder L, Qi Y, Baranski R (2018) Efficient CRISPR/Cas9-based genome editing in carrot cells. Plant Cell Rep 37(4):575–586
22. Scott R, Draper J (1987) Transformation of carrot tissues derived from proembryogenic suspension cells: a useful model system for gene expression studies in plants. Plant Mol Biol 8:265–274
23. Pawlicki N, Sangwan R, Sangwan-Norreel B (1992) Factors influencing the agrobacterium tumefaciens-mediated transformation of carrot (*Daucus carota L.*). Plant Cell Tissue Organ Cult 31:129–139
24. Hardegger M, Sturm A (1998) Transformation and regeneration of carrot (*Daucus carota L.*). Mol Breeding 4:119–127
25. Baranski R (2008) Genetic transformation of carrot (*Daucus carota*) and other Apiaceae species. Trans Plant J 2:18–38

Chapter 25

CRISPR-Cas-Mediated Gene Knockout in Tomato

Gwen Swinnen, Thomas Jacobs, Laurens Pauwels, and Alain Goossens

Abstract

Loss-of-function mutants are crucial for plant functional genomics studies. With the advent of CRISPR-Cas genome editing, generating null alleles for one or multiple specific gene(s) has become feasible for many plant species including tomato (*Solanum lycopersicum*). An easily programmable RNA-guided Cas endonuclease efficiently creates DNA double-strand breaks (DSBs) at targeted genomic sites that can be repaired by nonhomologous end joining (NHEJ) typically leading to small insertions or deletions that can produce null mutations. Here, we describe how to utilize CRISPR-Cas genome editing to obtain stable tomato gene knockout lines.

Key words Genome editing, CRISRP-Cas, Gene knockout, Loss-of-function mutation, Null mutation, Site-directed mutagenesis, Targeted mutagenesis, Tomato, *Solanum lycopersicum*, *Solanaceae*

1 Introduction

In recent decades, studying genetic tomato (*Solanum lycopersicum*) mutants has allowed functional gene characterization through reverse genetics. This involved the generation and screening of mutations randomly introduced in the genome by physical (e.g., gamma radiation), chemical (e.g., ethyl methanesulfonate), or insertional (e.g., T-DNA or transposon insertion) mutagenesis [1]. The availability of the tomato genome sequence and the advent of genome editing using sequence-specific nucleases revealed the possibility of site-directed mutagenesis [2]. With the emergence of clustered regularly interspaced short palindromic repeats-CRISPR associated protein (CRISPR-Cas) systems, targeted introduction of a DNA double-strand break (DSB) to create a gene knockout has become feasible for *Solanaceae* species like tomato.

The ease and efficiency by which CRISPR-Cas can be applied rely on its simple composition and programmable versatility. An artificial guide RNA (gRNA) directs a generic Cas endonuclease to the genomic target site through base pairing of the gRNA with the genomic protospacer sequence, which neighbors a protospacer

Manuel Rodríguez-Concepción and Ralf Welsch (eds.), *Plant and Food Carotenoids: Methods and Protocols*, Methods in Molecular Biology, vol. 2083, https://doi.org/10.1007/978-1-4939-9952-1_25,

adjacent motif (PAM). Repair of the resulting DSB by error-prone nonhomologous end joining (NHEJ) typically produces small insertions or deletions (indels) at the cleavage site. Gene function can be disrupted by the introduction of an indel that generates a frameshift mutation (*see* Fig. 1).

In this chapter, we outline a comprehensive workflow for CRISPR-Cas genome editing in tomato that goes from gRNA target selection to the identification of stable gene knockout lines (*see* Fig. 2). We also provide variations in the experimental setup regarding the use of different *Agrobacterium*-mediated transformation systems and CRISPR-Cas vectors.

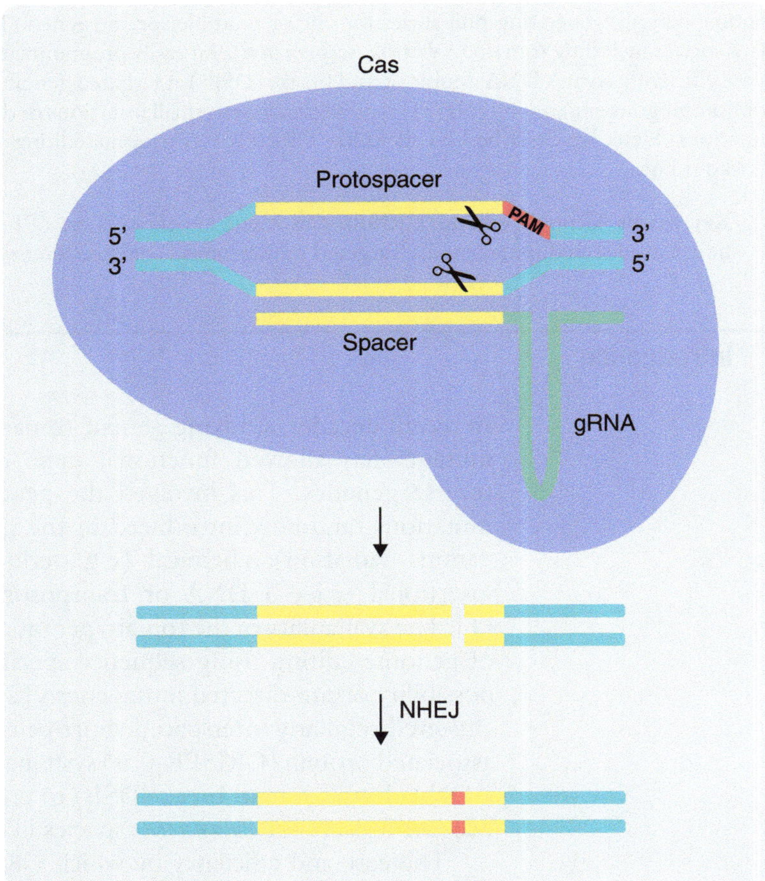

Fig. 1 Molecular basis of the CRISPR-Cas system. A generic Cas endonuclease is directed to the genomic target site through base pairing of an artificial guide RNA (gRNA) spacer sequence with the genomic protospacer sequence next to a protospacer adjacent motif (PAM). Repair of a DNA double-strand break at the target site by error-prone nonhomologous end joining (NHEJ) typically leads to small insertions or deletions at the cleavage site

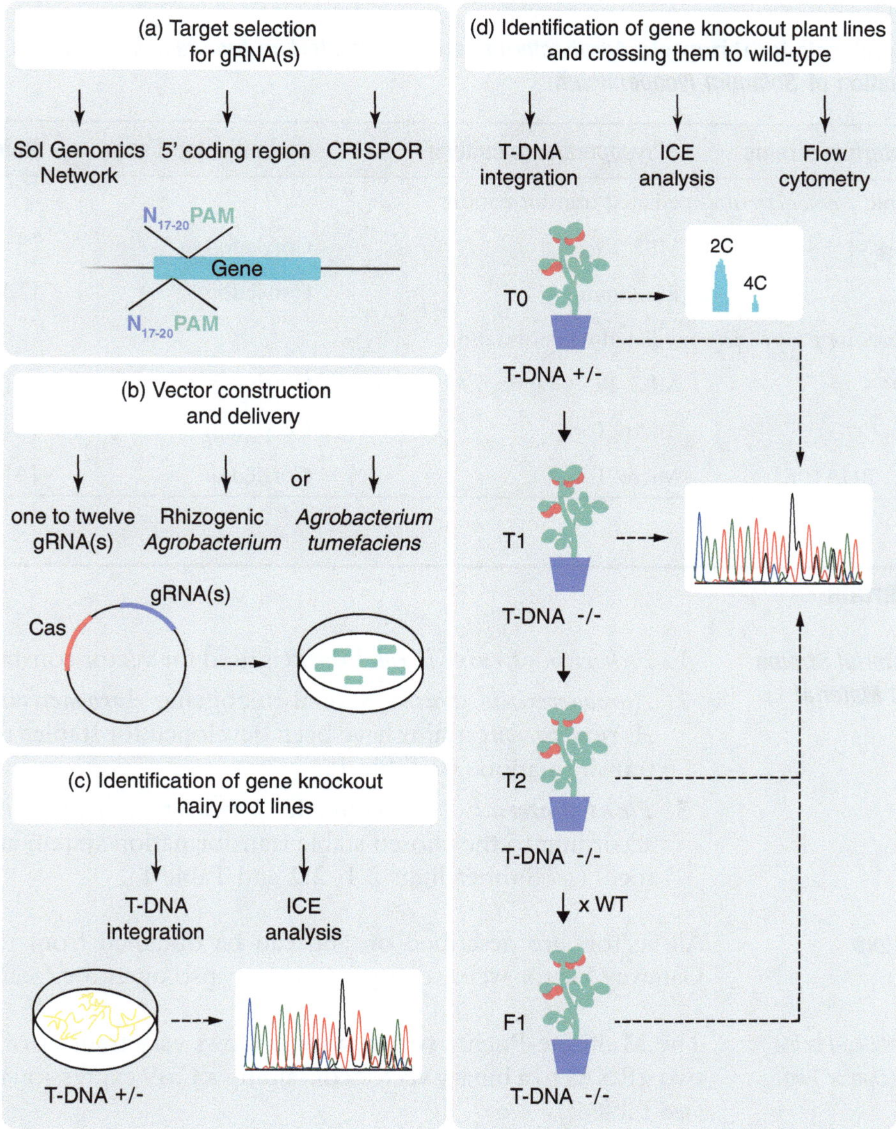

Fig. 2 A workflow for CRISPR-Cas genome editing to establish stable tomato gene knockout lines. (**a**) Target selection for guide RNAs (gRNAs). Select a protospacer of 17–20 nucleotides next to a protospacer adjacent motif (PAM) in the 5′ coding region of the genomic sequence of your gene of interest, retrieved from Sol Genomics Network, utilizing CRISPOR. (**b**) Vector construction and delivery. Construct a CRISPR-Cas binary vector containing one to 12 gRNAs and transfer it to cotyledon tissue by either rhizogenic *Agrobacterium*- or *Agrobacterium tumefaciens*-mediated transformation. (**c**) Identification of gene knockout tomato hairy root lines. Perform gene mutagenesis analysis on hairy root lines, containing the CRISPR-Cas T-DNA, using Sanger sequencing of the gRNA target region followed by Inference of CRISPR Editing (ICE) analysis. (**d**) Identification of gene knockout tomato plant lines and crossing them to wild-type (WT). Subject primary plant transformant (T0) lines, which contain the CRISPR-Cas T-DNA, to Sanger sequencing of the gRNA target region, followed by ICE analysis for gene mutagenesis analysis, and to ploidy determination by flow cytometry. Perform gene mutagenesis analysis, as described for plant T0 lines, on plant T1 and T2 lines not containing the CRISPR-Cas T-DNA. Cross obtained gene knockout plant lines to the WT to exclude somaclonal variation and/or to generate single-gene knockout plant lines from multiple-gene knockout plant lines. Gene mutagenesis analysis of F1 lines and their progeny can be performed as described for plant T0 lines

Table 1
Selected protocols for rhizogenic *Agrobacterium* and *Agrobacterium tumefaciens*-mediated transformation of *Solanum lycopersicum*

Agrobacterium strains	*S. lycopersicum* cultivar	Co-cultivated tissue	Reference
Rhizogenic *Agrobacterium*-mediated transformation:			
ATCC15834	M82	Cotyledon	[3]
ARqua1	Rio Grande	Cotyledon	[13]
Agrobacterium tumefaciens-mediated transformation:			
LBA4404	M82, Rio Grande, Castlemart	Cotyledon	[4]
GV2260	Micro-Tom	Cotyledon	[5]
GV3101, EHA105	Micro-Tom	Cotyledon	[6]

2 Materials

2.1 Bacterial Strains and Plant Material

1. *Escherichia strain*: *E. coli* DH5α is used for vector construction.
2. *Agrobacterium strains:* Several rhizogenic *Agrobacterium* and *A. tumefaciens* strains have been developed for stable (tomato) transformation (*see* Table 1).
3. *Plant material:* Transformable *S. lycopersicum* plant material according to the chosen stable transformation system and protocol (*see* Subheadings 2.1, 2.2 and Table 1).

2.2 Vectors

All vectors are described on and can be obtained from the PSB Gateway Vector website https://gateway.psb.ugent.be/search.

2.2.1 MultiSite-Puchta Vectors for One or Two gRNAs

The MultiSite-Puchta vector system allows easy transfer of one or two gRNAs to a binary vector containing a Cas9 expression cassette (*see* Table 2).

2.2.2 Modular Vectors for Up to 12 gRNAs

The modular vector system uses GreenGate technology to enable the construction of binary vectors, which contain a Cas9 expression cassette, with up to 12 gRNAs (*see* **Note 1**) [7]. Six Golden Gate entry modules, with each of them containing either one gRNA, two gRNAs or a linker, are assembled into one binary vector (*see* Table 3 and **Note 2**).

2.3 Web Tools and Software

1. Sol Genomics Network (https://solgenomics.net/).
2. Tomato 360 variants SL.50 dataset in JBrowse (https://solgenomics.net/jbrowse_solgenomics/?data=data/json/tomato_variants_2.5).

Table 2
Characteristics of MultiSite-Puchta vectors for one or two gRNAs

Vector	Gateway *att* sites	gRNA promoter	Cas promoter	Cas	PAM	*E. coli* Marker	Plant Marker	References
Entry vectors:								
pENC1.1	L1-L2	AtU6–26	—	—		Amp	—	[14]
pMR217	L1-R5	AtU6–26	—	—		Km	—	[15, 16]
pMR218	L5-L2	AtU6–26	—	—		Km	—	[15, 16]
Binary vectors:								
pDe-Cas9-PPT	R1-R2	—	PcUbi4–2	SpCas9[a]	NGG	Sp	PPT	[14]
pDe-Cas9-km	R1-R2	—	PcUbi4–2	SpCas9[a]	NGG	Sp	Km	[15, 16]
pDe-Cas9-km	R1-R2	—	PcUbi4–2	SpCas9 (VQR)[a]	NGA	Sp	Km	[17]

Abbreviations: gRNA, guide RNA; PAM, protospacer adjacent motif; Amp, ampicillin; Km, kanamycin; Sp, spectinomycin; PPT, phosphinothricin; Ref., reference
[a] *Arabidopsis thaliana* codon-optimized *Streptococcus pyogenes* Cas9

3. Tomato Genetics Resource Center Accessions (https://tgrc.ucdavis.edu/Data/Acc/dataframe.aspx?start=AccSearch.aspx&navstart=nav.html).
4. Sol Genomics Network BLAST (https://solgenomics.net/tools/blast/).
5. TomExpress (http://tomexpress.toulouse.inra.fr/).
6. InterPro (https://www.ebi.ac.uk/interpro/).
7. CRISPOR (http://crispor.tefor.net/).
8. Primer-BLAST (https://www.ncbi.nlm.nih.gov/tools/primer-blast/).
9. Inference of CRISPR Editing (ICE) CRISPR Analysis Tool (https://ice.synthego.com/#/).
10. Software for ploidy level analysis (e.g., FloMax software from Sysmex).

2.4 Equipment

1. Microtiter tubes (eight-strip format).
2. Stainless steel beads (diameter 3.0 mm).
3. Tissue lyser.

Table 3
Characteristics of modular vectors for up to 12 gRNAs

Vector	Golden Gate overlaps	gRNA promoter	Cas promoter	Cas	PAM	*E. coli* Marker	Plant Marker	Reference
PCR template vectors:								
pEN-2xAtU6	—	AtU6–26	—	—	—	Km	—	[10]
Entry vectors:								
pGG-A-AtU6ccdB-B	AB	AtU6–26	—	—		Amp	—	[10]
pGG-B-AtU6ccdB-C	BC	AtU6–26	—	—		Amp	—	[10]
pGG-C-AtU6ccdB-D	CD	AtU6–26	—	—		Amp	—	[10]
pGG-D-AtU6ccdB-E	DE	AtU6–26	—	—		Amp	—	[10]
pGG-E-AtU6ccdB-F	EF	AtU6–26	—	—		Amp	—	[10]
pGG-F-AtU6ccdB-G	FG	AtU6–26	—	—		Amp	—	[10]
pGG-B-linkerII-G	BG	—	—	—		Gm	—	[10]
pGG-C-linkerII-G	CG	—	—	—		Gm	—	[18]
pGG-D-linkerII-G	DG	—	—	—		Gm	—	[10]
pGG-E-linkerII-G	EG	—	—	—		Gm	—	[18]
pGG-F-linkerII-G	FG	—	—	—		Gm	—	[10]
Binary vectors:								
pFASTGK AtCas9A-ccdB-G	AG	—	PcUbi4–2	SpCas9[a]	NGG	Sp	Km, green seed	—
pFASTRK AtCas9A-ccdB-G	AG	—	PcUbi4–2	SpCas9[a]	NGG	Sp	Km, red seed	—

Abbreviations: *gRNA*, guide RNA; *PAM*, protospacer adjacent motif; *Km*, kanamycin; *Amp*, ampicillin; *Gm*, gentamycin; *PPT*, phosphinothricin; *Sp*, spectinomycin; *Ref.*, reference

[a]*Arabidopsis thaliana* codon-optimized *Streptococcus pyogenes* Cas9

4. Vortex.
5. PCR tubes (eight-strip format).
6. Multichannel pipette (eight channels).
7. Thermal cycler.
8. Tabletop centrifuge for PCR tubes (eight-strip format or 96-well plate).
9. Magnetic separator for PCR tubes (e.g., MagWell™ Magnetic Separator 96 from EdgeBio, *see* **Note 3**).
10. Razor blade.
11. Nylon filter of 50 μm.
12. Flow cytometer (e.g., CyFlow Space from Sysmex).
13. Ultrafine point forceps (Dumont, *see* **Note 4**).

2.5 Reagents

All solutions are prepared with Milli-Q water.

2.5.1 MultiSite-Puchta Vector Construction for One or Two gRNAs

1. BpiI (10 U/μL; e.g., from Thermo Scientific, *see* **Note 5**).
2. T4 DNA ligase (5 U/μL).
3. 10× T4 DNA ligase buffer (containing 10 mM ATP, *see* **Note 6**).
4. 1 mg/mL bovine serum albumin (BSA).
5. LR Clonase II.
6. LR Clonase II Plus.
7. Proteinase K (2 μg/μL).
8. 100 mg/mL carbenicillin, sterile-filtered.
9. 50 mg/mL kanamycin, sterile-filtered.
10. 50 mg/mL spectinomycin, sterile-filtered.
11. LB medium.
12. Plasmid miniprep kit.
13. Oligonucleotides for colony-PCR and Sanger sequencing can be found in Table 4.

2.5.2 Modular Vector Construction for Up to 12 gRNAs

1. BbsI-HF (20,000 U/mL (e.g., from New England Biolabs, *see* **Note 5**).
2. BsaI (20,000 U/mL).
3. NheI (20,000 U/mL).
4. T4 DNA ligase (400,000 U/mL, *see* **Note 6**).
5. 10× Recommended restriction enzyme buffer (containing 1 mg/mL BSA).
6. 10 mM ATP.
7. 100 mg/mL carbenicillin, sterile-filtered.
8. 50 mg/mL spectinomycin, sterile-filtered.

Table 4
Primers for MultiSite-Puchta and modular vector construction

Primer name	Sequence (5′-3′)	Orientation
MultiSite-Puchta vector construction for one or two gRNAs:		
pEn-F	TCCCAGGATTAGAATGATTAGG	Forward
pDe-F	TGCCCACACACCAGAATCCTAC	Forward
pDe-R	GACCTTAGGCGACTTTTGAACG	Reverse
Modular vector construction for up to 12 gRNAs:		
pGG-F	GTGAGCGGATAACAATTTCACA	Forward
pGG-R	CGACGGCCAGGTAATACGACT	Reverse

9. LB medium.
10. Gel extraction kit.
11. Plasmid miniprep kit.
12. Oligonucleotides for colony-PCR and Sanger sequencing can be found in Table 4.

2.5.3 Rapid Genomic DNA (gDNA) Extraction

1. Extraction buffer: 0.1 M tris(hydroxymethyl)aminomethane (Tris)–HCl, 0.25 M KCl, 0.01 M ethylenediaminetetraacetic acid (EDTA), pH 9.5.
2. 3% (w/v) BSA.

2.5.4 Gene Mutagenesis Analysis

1. Magnetic beads for PCR amplicon purification (e.g., High-Prep™ PCR reagent from MAGBIO, *see* **Note 7**).
2. 80% ethanol.
3. 10 mM Tris–HCl, pH 8.0.

2.5.5 Ploidy Level Analysis

1. Nuclei extraction buffer (e.g., CyStain UV Precise P Nuclei Extraction Buffer from Sysmex).
2. Staining buffer containing 1 mg/L 4,6-diamino-2-phenyl-indole (DAPI) (e.g., CyStain UV Precise P Staining Buffer from Sysmex).

3 Methods

Here, we describe the protocols to establish stable tomato loss-of-function mutant lines utilizing CRISPR-Cas genome editing (*see* Fig. 2).

3.1 Target Selection for gRNA(s)

Sequence, expression and phenotypic information on your gene of interest should be obtained before initiating gRNA(s) target selection.

3.1.1 Getting to Know Your Gene(s) of Interest

1. Search for your gene of interest on the Sol Genomics Network to retrieve the genomic sequence with annotated exon-intron structure, any alternative splicing variants (*see* **Note 8**) and the amino acid sequence.
2. Search for your gene of interest in the Tomato 360 variants SL.50 data set in JBrowse and select the track of your *S. lycopersicum* cultivar of interest (*see* **Note 9**) to check if there are any single nucleotide polymorphisms (SNPs) present in the protein coding sequence compared to the reference sequence.
3. Search the literature for your gene of interest. In particular, look for functional characterization, functionally important protein domains and redundant genes.
4. Identify paralogous genes using Sol Genomics Network BLAST and compare their gene expression pattern with that of your gene of interest on TomExpress to identify potentially redundant genes (*see* **Note 10**).
5. Predict protein domains important for your gene of interest's function using InterPro.

3.1.2 Target Region Selection for gRNA(s)

Select a target region (*see* **Note 11**) in the genomic sequence of your gene of interest using the information gathered in Subheading 3.1.1 to take the following considerations into account.

1. The target region should consist of an early (*see* **Note 12**) and constitutive (*see* **Note 13**) protein-coding exon.
2. The target region should be located downstream of (potential) alternative first exons and in-frame start codons (*see* **Note 14**).
3. The target region should be located upstream of regions coding for (predicted) functionally important protein domains (*see* **Note 15**).

3.1.3 Target Site Selection for gRNA(s)

Select a target site, called a protospacer, immediately preceding a PAM (*see* **Note 16**), within the target region selected in Subheading 3.1.2 and using the information gathered in Subheading 3.1.1.

1. Enter the sequence of your selected target region (*see* **Note 17**) as input for CRISPOR and select the latest *S. lycopersicum* genome version and PAM requirement for your chosen Cas nuclease from the drop-down menus.
2. Select a protospacer of 17–20 nucleotides for one or two gRNAs (*see* **Notes 18** and **19**) that (a) has good on- and

off-target scores (*see* **Notes 20** and **21**) and (b) does not contain a BpiI or BsaI site (*see* **Note 22**).

3. Design and synthesize a pair of PCR primers to amplify a genomic region of approximately 700 bp that includes the gRNA target site (*see* **Notes 23–25**). Extract genomic DNA from your wild-type cultivar following the steps in Subheading 3.5.1 and Sanger sequence an amplified PCR product as described in Subheading 3.5.3 to ensure the reference sequence is correct and the designed primers are accurate (*see* **Note 26**).

3.2 MultiSite-Puchta Vector Construction for One or Two gRNA(s)

This protocol has been adapted from Schiml et al. [8].

3.2.1 Annealing and Transfer of Oligonucleotides into the Entry Vector

1. For one gRNA target site, order two complementary oligonucleotides with 4-bp overhangs (*see* **Notes 27–29**):

   Forward: 5′-ATTG-protospacer-3′,

   Reverse: 5′-AAAC-reverse complement of protospacer-3′.
2. Mix 1 μL of each 100 μM complementary oligonucleotide with 48 μL of Milli-Q water in a PCR tube.
3. Place in a thermal cycler and use the following parameters: 95 °C for 5 min and 25 °C for 20 min.
4. Mix 1 μL of the annealed oligonucleotides with 1 μL of 100 ng/μL entry vector (*see* **Note 30**), 3 μL of 10× T4 DNA ligase buffer, 3 μL of 1 mg/mL BSA, 5 units of BpiI, 5 units of T4 DNA ligase and 20.5 μL of Milli-Q water in a PCR tube.
5. Place in a thermal cycler and run the following program to perform a Golden Gate reaction: 37 °C for 5 min, (20 °C for 5 min, 37 °C for 5 min) × 50 cycles, 50 °C for 10 min, 80 °C for 10 min, and hold at 12 °C.
6. Transform 5 μL of the Golden Gate reaction into competent *E. coli* DH5α and plate on LB medium with the appropriate antibiotic (*see* **Note 31**).
7. Set up a colony-PCR for four colonies using pEn-F and your protospacer reverse oligonucleotide as primers and anneal at 56 °C. The expected band size is 280 bp.
8. Isolate pEn-Chimera DNA from two colonies using a plasmid miniprep kit.
9. Conduct Sanger sequencing with primer pEn-F to verify the presence of your spacer.

3.2.2 Transfer of gRNA(s) into the Binary Vector

1. To transfer one gRNA, set up a Gateway reaction with 1 μL of 100 ng/μL pEn-Chimera, 1 μL of 100 ng/μL pDe-Cas9 (*see* **Note 32**) and 0.5 μL of LR Clonase II. Proceed to **step 3**.
2. To transfer two gRNAs, set up a MultiSite Gateway reaction with 1 μL of each pEn-Chimera (50 ng/μL), 4 μL of 50 ng/μL pDe-Cas9 (*see* **Note 32**), 2 μL of Milli-Q water, and 2 μL of LR Clonase II Plus.
3. Incubate at 25 °C overnight.
4. Add 1 μL of 2 μg/μL Proteinase K.
5. Incubate at 37 °C for 10 min.
6. Transform 2.5–5 μL of the Gateway reaction into competent *E. coli* DH5α and plate on LB medium with 75 mg/L spectinomycin.
7. Set up a colony-PCR for three colonies using pDe-F and pDe-R as primers and anneal at 55 °C. The expected band size is 1180 bp and 1660 bp if one or two gRNAs are transferred, respectively.
8. Isolate binary vector DNA from two colonies using a plasmid miniprep kit.
9. Conduct Sanger sequencing with primers pDe-F and pDe-R to verify the presence of your gRNA(s).

3.3 Modular Vector Construction for Up to 12 gRNAs

3.3.1 Golden Gate Entry Module for One gRNA

1. For one gRNA target site, order two complementary oligonucleotides with 4-bp overhangs (*see* **Notes 27**, **28** and **33**):

   Forward: 5′-ATTG-protospacer-3′,

   Reverse: 5′-AAAC-reverse complement of protospacer-3′.
2. Mix 1 μL of each 100 μM complementary oligonucleotide with 48 μL of Milli-Q water in a PCR tube.
3. Place in a thermal cycler and use the following parameters: 95 °C for 5 min and 25 °C for 20 min.
4. Mix 1 μL of the annealed oligonucleotides with 1 μL of 50 ng/μL entry vector (*see* **Note 34**), 5 units of BbsI-HF, 200 units of T4 DNA ligase, 1.5 μL of 10 mM ATP, 1.5 μL of 10× recommended restriction enzyme buffer and 9 μL of Milli-Q water in a PCR tube.
5. Place in a thermal cycler and run the following program to perform a Golden Gate reaction: (37 °C for 3 min, 16 °C for 3 min) × 30 cycles, 50 °C for 5 min, 80 °C for 5 min, hold at 12 °C.
6. Transform 5 μL of the Golden Gate reaction into competent *E. coli* DH5α and plate on LB medium with 100 μg/mL carbenicillin.

7. Set up a colony-PCR for four colonies using pGG-F and pGG-R as primers and anneal at 55 °C. The expected band size is 610 bp.
8. Isolate binary vector DNA from two colonies using a plasmid miniprep kit.
9. Conduct Sanger sequencing with pGG-R to verify the presence of your spacer.

3.3.2 Golden Gate Entry Module for Two gRNAs

1. For two gRNA target sites, order two oligonucleotides with 16-bp overhangs (*see* **Notes 27**, **28** and **33**):

   Forward: 5′-TTTTGAAGACAAATTG-protospacer (gRNA target site 1)- GTTTTAGAGCTAGAAATAGC-3′,

   Reverse: 5′-TTTTGAAGACAAAAAC-reverse complement of protospacer (gRNA target site 2)- CAATCACTACTTC-GACTC-3′.

2. Set up a PCR using pEN-2xAtU6 as template and the oligonucleotides for the two gRNA target sites as primers and anneal at 55 °C. The expected band size is 580 bp.
3. Isolate the PCR amplicon using gel extraction (*see* **Note 35**).
4. Mix 1 μL of purified PCR product (~5 ng) with 1 μL of 50 ng/μL entry vector (*see* **Note 34**), 5 units of BbsI-HF, 200 units of T4 DNA ligase, 1.5 μL of 10 mM ATP, 1.5 μL of 10× recommended restriction enzyme buffer and 9 μL of Milli-Q water in a PCR tube.
5. Place in a thermal cycler and run the following program to perform a Golden Gate reaction: (37 °C for 3 min, 16 °C for 3 min) × 30 cycles, 50 °C for 5 min, 80 °C for 5 min, hold at 12 °C.
6. Transform 5 μL of the Golden Gate reaction into competent *E.* coli DH5α and plate on LB medium with 100 μg/mL carbenicillin.
7. Set up a colony-PCR for four colonies using pGG-F and pGG-R as primers and anneal at 55 °C. The expected band size is 1140 bp.
8. Isolate binary vector DNA from two colonies using a plasmid miniprep kit.
9. Conduct Sanger sequencing with pGG-F and pGG-R to verify the presence of your spacers.

3.3.3 Transfer of gRNA(s) into the Binary Vector

Here, six Golden Gate entry modules, with each of them containing either one gRNA, two gRNAs or a linker, are assembled into a binary vector (*see* **Note 2**).

1. Mix 100 ng of each of the six Golden Gate entry modules with 100 ng of binary vector (*see* **Note 36**), 5 units of BsaI-HFv2, 200 units of T4 DNA ligase, 1.5 μL of 10 mM ATP, 1.5 μL of 10× recommended restriction enzyme buffer, and 9 μL of Milli-Q water in a PCR tube.
2. Place in a thermal cycler and run the following program to perform a Golden Gate reaction: (37 °C for 3 min, 16 °C for 3 min) × 30 cycles, 50 °C for 5 min, 80 °C for 5 min, and hold at 12 °C.
3. Transform 5 μL of the Golden Gate reaction into competent *E. coli* DH5α and plate on LB medium with 75 μg/mL spectinomycin.
4. Isolate binary vector DNA from two colonies using a plasmid miniprep kit.
5. Perform a diagnostic digest with NheI on ~1 μg of binary vector to verify the presence of your gRNA(s) and linker if applicable (*see* **Notes 37** and **38**).

3.4 Vector Delivery

The two predominant stable tomato transformation systems used to obtain gene knockouts using CRISPR-Cas are rhizogenic *Agrobacterium*- and *Agrobacterium tumefaciens*-mediated transformation (*see* Table 1). Rhizogenic *Agrobacterium*-induced hairy root cultures edited by CRISPR-Cas can be obtained within 2 months for rapid functional characterization of genes expressed in the root of *S. lycopersicum* [3]. Although generating CRISPR-Cas-mediated gene knockout plant lines using *A. tumefaciens* is more labor-intensive and time-consuming because of lower transformation efficiencies and long generation times [4–6], it allows for the investigation of gene function in any tissue and developmental stage.

1. Transform your binary vector into rhizogenic *Agrobacterium* or *A. tumefaciens.*
2. Stably transform *S. lycopersicum* via rhizogenic *Agrobacterium*- or *A. tumefaciens*-mediated transformation (*see* Table 1).

3.5 Identification of Gene Knockout Lines

In this section, we describe how to identify stable gene knockout lines obtained through rhizogenic *Agrobacterium*- or *A. tumefaciens*-mediated transformation of CRISPR-Cas vectors.

3.5.1 Rapid gDNA Extraction

1. Collect approximately 1.0 cm of hairy root material (*see* **Note 39**) or 1.0 cm^2 of leaf material (*see* **Note 40**) in microtiter tubes (eight-strip format) filled with two stainless steel beads (diameter 3.0 mm).
2. Dip in liquid nitrogen for flash freezing or store at −80 °C for at least 1 h.

3. Homogenize using a tissue lyser for 60 s at 30 oscillations s^{-1}.
4. Repeat **step 3**.
5. Add 50 μL of extraction buffer.
6. Mix by vortexing.
7. Transfer to PCR tubes (eight-strip format or 96-well plate).
8. Place in a thermal cycler and use the following parameters: 95 °C for 10 min and 4 °C for 5 min.
9. Add 50 μL of 3% (w/v) BSA.
10. Mix by vortexing.
11. Centrifuge at 1000 × *g* for 15 min.
12. Collect the supernatant, which contains the gDNA.

3.5.2 T-DNA Integration Analysis

1. Set up a PCR using Cas-specific primers and the extracted gDNA in order to select (a) hairy root and primary plant transformant (T0) lines in which the T-DNA is present or (b) plant T1 and T2 lines in which the T-DNA is absent (*see* **Note 41**).

3.5.3 Gene Mutagenesis Analysis

1. Set up a PCR using the primers designed in **step 3** of Subheading 3.1.3 to amplify the gRNA(s) target region(s) in the extracted gDNA (a) from hairy root and primary plant transformant (T0) lines containing the T-DNA or (b) from plant T1 and T2 lines not containing the T-DNA (*see* **Note 42**).
2. Mix the magnetic beads for PCR amplicon purification by vortexing to resuspend the magnetic beads.
3. Add 9 μL of magnetic beads for PCR amplicon purification to 5 μL of PCR reaction (*see* **Note 43**).
4. Mix by pipetting.
5. Incubate at room temperature for 5 min.
6. Place on the magnetic separator (*see* **Notes 44** and **45**).
7. Remove the cleared supernatant (*see* **Note 46**).
8. Add 200 μL of 80% ethanol.
9. Incubate at room temperature for 30 s.
10. Remove the cleared supernatant (*see* **Note 46**).
11. Repeat **steps 8–10**.
12. Incubate at room temperature for 10–15 min to dry the beads (*see* **Note 47**).
13. Remove from the magnetic separator.
14. Add 50 μL of 10 mM Tris–HCl.
15. Mix by pipetting.
16. Incubate at room temperature for 2 min.

17. Place on the magnetic separator.
18. Collect the cleared supernatant, which contains the purified PCR amplicon (*see* **Note 46**).
19. Conduct Sanger sequencing (*see* **Note 48**).
20. Identify hairy root or plant lines containing gene knockout mutation(s) using ICE to analyze obtained Sanger sequencing data (*see* **Notes 49** and **50**).

3.5.4 Ploidy Level Analysis

Regeneration of polyploid primary transformant (T0) plants is not unusual in plant tissue-culture dependent *A. tumefaciens*-mediated transformation. As an increase in tomato ploidy level can alter the development of vegetative and/or reproductive organs, it is necessary to select diploid primary transformant (T0) individuals [9].

1. Chop 1.0 cm^2 of leaf material from a primary transformant (T0) plant in 200 μL of chilled nuclei extraction buffer for 2 min using a razor blade.
2. Filter the suspension through a 50-μm nylon filter to remove cell fragments and large debris.
3. Add 800 μL of chilled staining buffer to the isolated nuclei.
4. Mix by pipetting.
5. Measure the DNA content of about 5000 to 10,000 nuclei per leaf sample using a flow cytometer (*see* **Note 51**).
6. Analyze with FloMax software to identify diploid primary transformant (T0) plant lines.

3.6 Crossing Gene Knockout Plant Lines to Wild-Type

Here, we describe a method to cross an obtained gene knockout plant line to the wild-type. This allows the exclusion of somaclonal variation and the generation of single-gene knockout plant lines from multiple-gene knockout plant lines. Alternatively, it can be used to develop heterozygous gene knockout plant lines from homozygous gene knockout plant lines that are low in seed set or sterile (*see* **Note 52**).

1. Select an immature bud (*see* **Note 53**) as female parent, which shows a little yellowing of the petals (*see* **Note 54**), from a T-DNA free gene knockout plant line (*see* **Note 55**).
2. Remove sepals from the female parent using ultrafine point forceps.
3. Emasculate the female parent (*see* **Note 56**) using ultrafine point forceps to carefully pinch the side of the petals and anther cone and pull them off thereafter (*see* **Note 57**).
4. Remove all other flowers present on the inflorescence to which the female parent belongs (*see* **Note 58**).
5. Select an open flower as male parent from a wild-type plant.

6. Make a slit in the anther cone of the male parent using ultrafine point forceps.
7. Drag the ultrafine point forceps upward through the inside of the anther cone to collect pollen (*see* **Note 59**).
8. Apply collected pollen from the male parent using the ultrafine point forceps to the stigma of the female parent (*see* **Note 60**).
9. Tag the crossed flower (*see* **Note 61**).
10. Repollinate the female parent one or 2 days later by repeating **steps 5–8** (*see* **Note 62**).
11. Ovary enlargement of the cross-pollinated flower can be observed within 4 to 5 days if fertilization was successful. Otherwise, the flower will abort (*see* **Note 63**).
12. F1 lines can be analyzed as described in Subheading 3.5 (*see* **Note 64**).

4 Notes

1. Because the modular system is based on GreenGate technology, it can easily be adapted to any new development (e.g., for alternative Cas endonucleases) [10].
2. For instance, to transfer three gRNAs to a binary vector of choice, you can combine three Golden Gate entry modules for one gRNA (with Golden Gate overlaps AB, BC and CD) together with the Golden Gate entry module pGG-D-linkerII-G. If a binary vector containing six gRNAs is desired, you can assemble six Golden Gate entry modules for one gRNA (with Golden Gate overlaps AB, BC, CD, DE, EF and FG). To obtain a binary vector with seven gRNAs, you can combine five Golden Gate entry modules for one gRNA (with Golden Gate overlaps AB, BC, CD, DE, and EF) together with one Golden Gate entry module for two gRNAs (with Golden Gate overlap FG) (*see* Table 3).
3. With the MagWell™ Magnetic Separator 96 (EdgeBio), the magnetic beads will be attracted to one side and up from the bottom of PCR tubes, allowing for easy recovery of your samples.
4. Ultrafine point forceps can be purchased from any supplier.
5. BpiI from Thermo Scientific and BbsI-HF from New England Biolabs are stable at −20 °C.
6. Store the T4 ligase buffer in aliquots of 10–20 μL to minimize degradation of DTT (and of ATP in case it is included in the buffer).

7. Although magnetic beads for PCR amplicon purification may be purchased from any supplier, **steps 2–19** in Subheading 3.5.3 apply specifically to the use of HighPrep™ PCR reagent from MAGBIO.
8. In the imaginary Solyc identifier Solyc01g001010.1.1, the second ".1" denotes the splice variant.
9. Look up the accession number of your cultivar of interest in the Tomato Genetics Resource Center Accessions database.
10. Paralogous genes with overlapping expression patterns might code for proteins with redundant functions.
11. If (potentially) redundant genes were identified in **steps 3** and/or **4** of Subheading 3.1.1, targeting a conserved region might be considered.
12. Mutagenesis in the 3′ end coding region or introns may not result in null mutations.
13. Do not target exons that are excluded in alternative splice variants to ensure complete gene function disruption, unless you want to impair the function of a protein encoded by a particular splice variant only.
14. N-terminal protein truncations, resulting from mutagenesis in a region upstream of an alternative first exon or in-frame start codon, may have little or no effect on gene function.
15. Mutagenesis downstream of functionally important protein domains may not lead to completely impaired protein function.
16. If (potentially) redundant genes were identified in **step 3** and/or **4** of Subheading 3.1.1, targeting a conserved site might be considered.
17. Adapt the sequence of your selected target region according to any SNPs that may be present in your *S. lycopersicum* cultivar of choice (*see* **step 2** of Subheading 3.1.1).
18. Using two gRNAs to target your gene of interest can cause a bigger genomic deletion, ensuring the interruption of gene function.
19. When selecting two target sites (for two gRNAs), the Cas-cleavage sites should be out of frame in the coding sequence. The cleavage site is specific for each Cas nuclease and is located 3–4 bp upstream of the PAM sequence for Cas9.
20. Mismatches of the protospacer within the target region distal to the PAM are generally tolerated, whereas complete complementarity with the 8–12 bp immediately preceding the PAM is crucial for efficient cleavage by the Cas nuclease.
21. In case there are no protospacers available with both good on- and off-target scores, consider what is most important for your project.

22. If the protospacer were to contain a BpiI or BsaI site, this would interfere with the MultiSite-Puchta vector construction in Subheading 3.2 and the modular vector construction in Subheading 3.3.
23. The Cas-cleavage site should be located 150–250 bp from the forward or reverse primer to enable downstream analysis using ICE in Subheading 3.5.3.
24. One PCR primer pair might be used to amplify two gRNA target sites if they are located near each other, with either Cas-cleavage site located 150–250 bp from each primer.
25. Use Primer-BLAST to ensure specific amplification of your target region.
26. This step should be performed prior to any vector construction. A reliable PCR amplicon is required to simplify downstream analysis. Possible problems with amplification and sequencing can be (a) that the material in your hands does not match the reference sequence, (b) the presence of highly homologous sequences, resulting in a mixture of PCR products and (c) the presence of homopolymer region(s) within the amplicon, which will result in overlapping chromatograms.
27. The PAM sequence is not included in the oligonucleotides.
28. The 5′ overhang of the forward oligonucleotide contains the G initiation nucleotide for the AtU6–26 polymerase III promoter.
29. The overhangs are complementary with the sticky ends generated by BpiI in the chosen entry vector in **steps 4–5** resulting in seamless cloning.
30. Use pENC1.1 to transfer one gRNA to a binary vector in Subheading 3.2.2. To transfer two gRNAs to a binary vector in Subheading 3.2.2, use pMR217 for your first gRNA and pMR218 for your second (*see* Table 2).
31. Use LB with 100 μg/mL carbenicillin for a cut-ligation reaction with pENC1.1 and LB with 75 μg/mL kanamycin for a cut-ligation reaction with pMR217 or pMR218.
32. Use a binary vector according to your preference (*see* Table 2).
33. The overhangs are complementary with the sticky ends generated by BbsI-HF in the chosen entry vector in **steps 4–5** resulting in seamless cloning.
34. Use an entry vector according to your preference (*see* Table 3).
35. In our experience, gel extraction of the PCR amplicon yields better results than PCR purification.
36. Use a binary vector according to your preference (*see* Table 3).

37. Colony-PCR and Sanger sequencing are not encouraged here because of the tandem array of gRNAs produced.

38. Fragment size after digestion will vary depending on the binary vector used in **step 1**.

39. As hairy root cultures obtained by rhizogenic *Agrobacterium*-mediated transformation of CRISPR-Cas are often chimeric, make sure to cut off approximately 1.0 cm from the distal end of a single, lateral root that was recently subcultured. This will ensure that the Sanger sequencing data obtained in Subheading 3.5.3 will be representative for that recent hairy root subculture.

40. In case leaf material needs to be collected from primary transformant (T0) plant lines, only collect it from well-growing regenerated explants.

41. Use gDNA from a wild-type plant as a negative control template and the used binary vector as a positive control template.

42. Use gDNA from a wild-type plant as a control template and Milli-Q water as a no-template control.

43. Use PCR-tubes (eight-strip format).

44. The beads will be pulled to the magnet.

45. Keep samples on the magnetic separator until **step 12**.

46. Do this without disturbing the beads.

47. All traces of ethanol need to be removed before proceeding to the next step, but beware not to over dry the beads to avoid cracking and subsequent yield loss.

48. Sequence with the PCR primer located 150–250 bp from the Cas-cleavage site used in **step 1** of Subheading 3.5.3.

49. Indels that are not a multiple of three bases will shift the open reading frame, which can result in the translation of a truncated nonfunctional protein and/or nonsense-mediated decay triggered by a premature stop codon [11].

50. Alternatively, CRISPR/Cas-induced mutations can be analyzed using cleaved amplified polymorphic sequence (CAPS), T7 endonuclease, next-generation sequencing and high-resolution melting curve analysis [12]. However, Sanger sequencing followed by ICE analysis is the most cost-effective option.

51. Use the 2C-peak of nuclei isolated from a wild-type leaf (diploid) for calibration.

52. Likewise, a primary transformant (T0) plant line that has difficulty setting seed can be rescued by crossing it to the wild-type.

53. A flower that is still closed is chosen, because it is unlikely to have self-pollinated.

54. The stigma of the selected female parent should be receptive to pollen.
55. Alternatively, select an immature bud as female parent from a primary transformant (T0) plant line that has difficulty setting seed.
56. Removal of the anthers is needed because tomato is a self-pollinating species.
57. Pull the petals and anther cone straight off and away from the flower axis to avoid (a) damaging the stigma and style and (b) pollen transfer.
58. Prevent pollen transfer from these flowers to the female parent.
59. Collected pollen should be visible to the naked eye on the tip of the ultrafine point forceps.
60. Provide the stigma with a visible amount of pollen to help assure fertilization.
61. Indicate both parents and the date of crossing.
62. This will increase the likelihood that fertilization of the female parent will occur.
63. It is recommended to perform multiple crosses, since not all of them might succeed.
64. To obtain single-gene knockout lines, identified heterozygous F1 lines can be selfed and their progeny can be analyzed as described in Subheading 3.5.

References

1. Rothan C, Bres C, Garcia V, Just D (2016) Tomato resources for functional genomics. In: Causse M, Giovannoni J, Bouzayen M, Zouine M (eds) The tomato genome. Springer, Berlin, Germany, pp 75–94
2. Van Eck J (2018) Genome editing and plant transformation of solanaceous food crops. Curr Opin Biotechnol 49:35–41
3. Ron M, Kajala K, Pauluzzi G, Wang D, Reynoso MA, Zumstein K et al (2014) Hairy root transformation using *Agrobacterium rhizogenes* as a tool for exploring cell type-specific gene expression and function using tomato as a model. Plant Physiol 166:455–469
4. Van Eck J, Keen P, Tjahjadi M (2019) *Agrobacterium tumefaciens*-mediated transformation of tomato. In: Kumar S, Barone P and Smith M (Eds) Transgenic plants (methods in molecular biology 1864). Humana Press, New York, NY, pp 225–234
5. Shikata M, Ezura H (2016) Micro-tom tomato as an alternative plant model system: mutant collection and efficient transformation. In: Botella J, Botella M (eds) Plant signal transduction (methods in molecular biology 1363). Humana Press, New York, NY, pp 47–55
6. Garcia D, Narváez-Vásquez J, Orozco-Cárdenas ML (2015) Tomato (*Solanum lycopersicum*). In: Wang K (ed) Agrobacterium protocols (methods in molecular biology 1223). Springer, New York, NY, pp 349–361
7. Lampropoulos A, Sutikovic Z, Wenzl C, Maegele I, Lohmann JU, Forner J (2013) GreenGate—a novel, versatile, and efficient cloning system for plant transgenesis. PLoS One 8:e83043
8. Schiml S, Fauser F, Puchta H (2017) CRISPR/Cas-mediated *in planta* gene targeting. In: Busch W (ed) Plant genomics (methods in molecular biology 1610). Humana Press, New York, NY, pp 3–11
9. Ellul P, Garcia-Sogo B, Pineda B, Ríos G, Roig L, Moreno V (2003) The ploidy level of transgenic plants in *Agrobacterium*-mediated

transformation of tomato cotyledons (*Lycopersicon esculentum* L. mill.) is genotype and procedure dependent. Theor Appl Genet 106:231–238

10. Decaestecker W, Buono RA, Pfeiffer ML, Vangheluwe N, Jourquin J, Karimi M et al (2018) CRISPR-TSKO facilitates efficient cell type-, tissue-, or organ-specific mutagenesis in Arabidopsis. bioRxiv. 474981. https://doi.org/10.1101/474981
11. Popp MW, Maquat LE (2016) Leveraging rules of nonsense-mediated mRNA decay for genome engineering and personalized medicine. Cell 165:1319–1322
12. Zischewski J, Fischer R, Bortesi L (2017) Detection of on-target and off-target mutations generated by CRISPR/Cas9 and other sequence-specific nucleases. Biotechnol Adv 35:95–104
13. Jacobs TB, Martin GB (2016) High-throughput CRISPR vector construction and characterization of DNA modifications by generation of tomato hairy roots. J Vis Exp 110: e53843
14. Fauser F, Schiml S, Puchta H (2014) Both CRISPR/Cas-based nucleases and nickases can be used efficiently for genome engineering in *Arabidopsis thaliana*. Plant J 79:348–359
15. Ritter A, Iñigo S, Fernández-Calvo P, Heyndrickx KS, Dhondt S, Shi H et al (2017) The transcriptional repressor complex FRS7-FRS12 regulates flowering time and growth in *Arabidopsis*. Nat Commun 8:15235
16. Pauwels L, De Clercq R, Goossens J, Iñigo S, Williams C, Ron M et al (2018) A dual sgRNA approach for functional genomics in *Arabidopsis thaliana*. G3 Genes 8:2603–2615
17. Kleinstiver BP, Prew MS, Tsai SQ, Topkar VV, Nguyen NT, Zheng Z et al (2015) Engineered CRISPR-Cas9 nucleases with altered PAM specificities. Nature 523:481–485
18. Houbaert A, Zhang C, Tiwari M, Wang K, de Marcos Serrano A, Savatin DV et al (2018) POLAR-guided signalling complex assembly and localization drive asymmetric cell division. Nature 563:574–578

Chapter 26

Modulation of Activity of a Carotenoid Pathway Through the Use of the TeT-on Regulatory System: Application in the Fungus *Fusarium fujikuroi*

Julia Marente, Pedro Ortega, Javier Pardo-Medina, Javier Avalos, and M. Carmen Limón

Abstract

Carotenoids are widespread pigments in photosynthetic species, but they are also found in nonphotosynthetic microorganisms, such as bacteria and fungi. The amenability of fungi to genetic studies have made some fungal species advantageous models in the study of the genetics and biochemistry of carotenoid biosynthesis, while others have been used for biotechnological carotenoid production. The availability of molecular techniques that allow modulating the expression of target genes is a powerful tool in the manipulation of carotenoid synthesis. An example of an adjustable gene expression is based on the tetracycline-controlled transcriptional activation system, known as Tet-on. We describe here the material and protocols for the construction of a Tet-on regulated gene, its introduction in the filamentous fungus *F. fujikuroi*, and its use to modulate the expression of a negative regulator of carotenoid biosynthesis.

Key words Tet-on, Doxycycline, Neurosporaxanthin, Xanthophyll, *Fusarium fujikuroi*

1 Introduction

Some fungi are important sources of information on the genetics and biochemistry of the genes and enzymes involved in microbial carotenogenesis. Outstanding examples are *Phycomyces blakesleeanus* and *Mucor circinelloides* for the synthesis of β-carotene and *Neurospora crassa* and *Fusarium fujikuroi* for the synthesis of the carboxylic apocarotenoid acid neurosporaxanthin [1, 2]. Other fungi are used for the biotechnological production of some carotenoids, as the filamentous fungus *Blakeslea trispora* for β-carotene or lycopene, and the yeast *Xhanthophyllomyces dendrorhous*, formerly *Phaffia rhodozyma*, for the xanthophyll astaxanthin [3].

The ascomycetous fungus *F. fujikuroi* has been widely used as a model of carotenoid production in fungi [4]. This species produces a mixture of carotenoids, with neurosporaxanthin as the major

Manuel Rodríguez-Concepción and Ralf Welsch (eds.), *Plant and Food Carotenoids: Methods and Protocols*, Methods in Molecular Biology, vol. 2083, https://doi.org/10.1007/978-1-4939-9952-1_26,

product. We formerly described methods for the analysis of this xanthophyll and its precursor carotenoids [5, 6]. Moreover, we have identified all the genes and enzymes involved in this biosynthetic pathway and the molecular mechanisms of its regulation are currently under investigation. The mRNA levels of the structural genes increase in the presence of light [7] or in the absence of nitrogen [8], but they are kept under low levels under standard growth conditions. The main control of carotenogenesis is done by a negative regulator, CarS, whose mutation results in the accumulation of large amounts of mRNA for the structural genes of the pathway and a high carotenoid biosynthetic activity. The gene *carS* codes for a protein of the RING finger (RF) family, with sequence similarity with protein CrgA of *M. circinelloides*, whose mutation results in a similar carotenoid overproducing pattern in this fungus [9, 10]. The degree of similarity between the CarS and CrgA proteins is not very high, as expected for two taxonomically distant fungi, but it covers the most relevant CrgA domains, which include two amino-terminal RF domains, and a LON protease domain [11]. At least one of the RF domains of CrgA is essential for its regulatory function in carotenogenesis, and this seems to be also the case of CarS, as indicates the occurrence of mutations in a RF domain in two independent carotenoid overproducing *carS* mutants [12]. In other proteins, RF finger domains are typical of ubiquitin ligases (E3) which together with ubiquitin conjugating enzymes (E2) mediate ubiquitination of target proteins [13]. This labeling process frequently is a signal for protein degradation but also is a mode of regulation. This suggests that CarS might function as an E3 ubiquitin ligase.

The biological properties of neurosporaxanthin are unknown, and its potential applications in food or feed industries remain to be investigated. The development of new techniques to control the expression of genes related to carotenoid metabolism are a promising approach to modulate carotenoid production of fungi. A method of inducible gene expression that allows transcription to be reversibly turned on or off is known as Tetracycline-Controlled Transcriptional Activation. This method is based on a negative regulatory mechanism governing resistance to tetracycline in Gram-negative bacteria. The regulatory circuit in *Escherichia coli* consists of the repression of the Tn10 tetracycline resistance operon by the binding of the TetR repressor protein to its operator sequence, TetO [14]. Because of its higher stability, the tetracycline analog doxycycline is usually used as the inducer signal under laboratory conditions. In the presence of doxycycline, the antibiotic binds TetR, provoking its dissociation from TetO and allowing the transcription of the operon. TetR was adapted for its use in eukaryotic cells with the opposite regulatory role: it was fused to a transcriptional activation domain from the herpes simplex virus protein 16 (VP16) converting it into a hybrid transactivator, known as rtTA

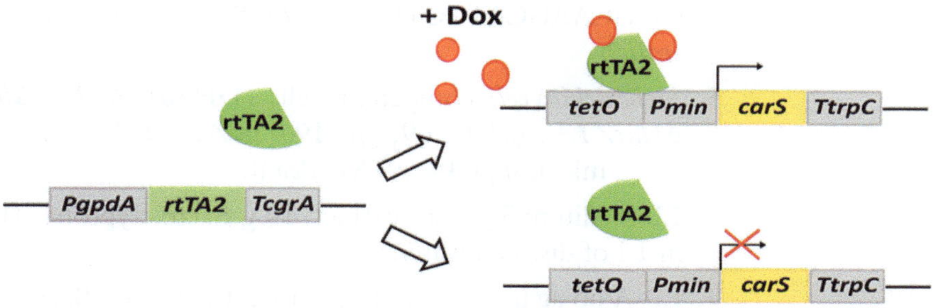

Fig. 1 Control of transcription of *carS* gene by doxycycline using Tet-on system. The Tet-on system consists in two modules: (**a**) Transactivator module: Transactivator rtTA2 is constitutively expressed under the control of *Aspergillus* glyceralhehyde-3P dehydrogenase promoter (*PgpdA*). (**b**) Expression module: A gene of interest is expressed under the doxycycline-dependent Tet-on promoter. In our case, this gene is *carS,* that encodes a down-regulator of carotenoid gene expression. In the presence of Dox (represented by orange circles), rtTA activates the Tet-on promoter and the gene *carS* is transcribed. In the absence of Dox, rtTa is not active and low transcription is expected. Dox: doxycycline

[15]. The binding of rtTA to an eukaryotic promoter allows transcriptional activation, but this is prevented by its interaction with doxycycline. This regulatory circuit, known as Tet-off, was changed to a Tet-on system by further modifications in the rtTA protein, so that it is only able to recognize the target promoter sequence when it is bound to doxycycline (reviewed in Ref. [16]). In addition, the promoter sequence was improved by the addition of several TetO binding elements (*tetO*), resulting in an optimized doxycycline-responsive promoter [17, 18]. Therefore, the use of the Tet-on promoter system allows the control of transcription of any target gene through the addition of doxycycline to the medium (Fig. 1).

We show that the Tet-on system may be used to modulate carotenoid biosynthesis in fungi through the controlled expression of a key regulatory gene. To check the efficiency of this methodology, we have constructed a Tet-on regulated version of *carS* and we have introduced it in a *carS* mutant by transformation. We show that carotenoid production is repressed by the addition of doxycycline to the medium through the activation of the negative regulator CarS. This method may be particularly useful for the controlled expression of genes for bottleneck steps in the carotenoid pathways.

2 Materials

2.1 Construction of Plasmids pPO3 and pPO5

1. Plasmid pLR14 [12], containing gene *carS* from *Fusarium fujikuroi*.
2. Plasmid pGEM®-T easy (Promega) for PCR fragment cloning.
3. Plasmid pVG3 [19].
4. Primers carS-PmeI-1F 5′-CCA*GTTTAAA*CGTACTATACGCACGCAAT-3′ and carS-PmeI-1R: 5′-CGA*GTTTAAA*CGTA

GGGCAAGCATTGTCA-3′. *Pme*I restriction sites are *in italics*.

5. *Escherichia coli* competent cells (DH5α): Φ-80′ Δ*lac*ZM15 Δ(*lacZYA-argF*)U169, *gyrA96 recA1 end*A1 *thi*-1, *hsd*R17 (rk−, mk+), *sup*E44, *rel* A1, *deo* R.
6. LB medium: 5 g yeast extract, 10 g Bacto tryptone, 10 g NaCl in 1 l of distilled water.
7. LB-Amp agar medium: 5 g yeast extract, 10 g Bacto tryptone, 10 g NaCl, and 16 g agar in 1 l of distilled water. Autoclave. Cool until 55 °C. Add 100 μl of 100 mg/ml sodium ampicillin.
8. LB-IPTG-X-Gal-Amp medium: Prepare LB-Amp agar and supplement with 0.5 mM IPTG and 85 μg/ml X-Gal.
9. T4 DNA ligase.
10. Expand High Fidelity PCR System (Roche).
11. Restriction enzymes *Pme*I, *Sac*II, and *Eco*RI.
12. DNA size marker.
13. TAE 50× buffer: 2 M Tris–acetate, pH 8.0, EDTA 0.05 M.
14. Isolate II PCR and Gel Kit (Bioline).
15. NanoDrop ND-1000 spectrophotometer (NanoDrop Technologies, Wilmington, DE, USA).
16. Horizontal electrophoresis system.
17. Power supply.

2.2 Cultivation of Fusarium fujikuroi

Sterilize all media and material in an autoclave at 120 °C for 20 min.

1. *F. fujikuroi* IMI 58289 wild type strain (previously, *Gibberella fujikuroi*), obtained from the Imperial Mycological Institute, Kew, UK.
2. *F. fujikuroi* strain SG39. A superproducer of carotenoids with a mutation in the gene *carS* [19].
3. DG medium [20]: 30 g of glucose, 3 g of $NaNO_3$, 1 g of H_2KPO_4, 0.5 g of $MgSO_4{\cdot}7H_2O$, 0.5 g of KCl, 16 g of agar, and 2 ml of microelement solution in 1 l of distilled water. Autoclave. This is the minimal medium for *F. fujikuroi*.
4. Microelement solution for DG medium: 0.5 mg of HBO_4, 5 mg of $CuSO_4$, 10 mg of $FeCl_3$, 1 mg of $MnCl_2$, 1 mg of $NaMoO_4$, and 100 mg of $ZnSO_4$ in 100 ml of distilled water.
5. Medium EG [21]: 1 g glucose, 1 g yeast extract, 1 g NH_4NO_3, 1 g KH_2PO_4, 0.5 g $SO_4Mg{\cdot}7H_2O$, 16 g agar in 1 l of distilled water. Autoclave. Medium EG was used to obtain spores.
6. Sterile VitraPOR® borosilicate filter crucibles, porosity 1, 15-ml volume, ROBU® (Glasfilter-Geräte GmbH, Hattert, Germany).

7. Cell counting chamber (Neubauer chamber or Bürker chamber).

2.3 Transformation of *Fusarium*

Sterilize all media and material in an autoclave at 120 ° C for 20 min.

1. Darken medium: 15 g corn steep liquor, 30 g sucrose, 2 g $(NH_4)_2SO_4$, and 7 g of $CaCO_3$ in 1 l of distilled water. Autoclave.
2. ICI medium: 10 g glucose, 1 g $(NH_4)_2SO_4$, 5 g KH_2PO_4, 1 g $MgSO_4·7H_2O$, and 2 ml of Geissman microelement solution in 1 l distilled water [21]. Autoclave.
3. Geissman microelement solution: 0.1 g $FeSO_4·7H_2O$; 0.015 g $CuSO_4·5H_2O$; 0.161 g $ZnSO_4·7H_2O$; 0.01 g $MnSO_4·7H_2O$; 0.01 g $(NH_4)_6Mo_7O_{20}·4H_2O$ in 100 ml of distilled water [21].
4. Osmostabilization buffer: 1.2 M KCl, 0.66 M CaCl, pH 5.5. Dissolve 35.97 g of KCl and 2.94 g of $CaCl_2$ in 400 ml of distilled water. Adjust pH to 5.5 and autoclave.
5. Enzyme solution: Mix 200 mg of Lysing enzymes from *Trichoderma harzianum* (Sigma), 150 mg of Driselase from basidiomycetes (Fluca), 15 mg of lyticase (Sigma), and 10 mg of BSA in 50 ml of osmostabilization buffer. Mix with a magnetic stirrer bar during at least 30 min. Sterilize through a 0.42 μm filter.
6. STC 2×: sorbitol 2.4 M, 20 mM Tris–HCl, pH 8, 100 mM $CaCl_2$.
7. 30% PEG Tris–HCl: 30% PEG 8000, 10 mM Tris–HCl, pH 8, 50 mM CaCl.
8. Selective medium: DG medium with 50 mg/l hygromycin (Roche).
9. Regeneration medium: Prepare a 2× regeneration medium containing 100 mg of yeast extract, 100 mg of NZ amine A, and 1.2 g of agar in 50 ml of water. Autoclave. Let it cool until 55 °C and keep it warm in a heat cabinet until preparation of the dishes. Then mix with 50 ml of 1.6 M sucrose solution prewarmed at 42 °C.
10. Regeneration top agar: Prepare regeneration 2× medium replacing agar by the same amount of agarose. Add the same amount of 1.6 M sucrose (2×), mix and maintain the regeneration top agar in a bath at 55 °C. Add 40 μl of 50 mg/ml hygromycin (Roche) per 5 ml of regeneration top agar.
11. Sterile borosilicate filter, porosity 1 and 0, 50-ml volume, ROBU® (Glasfilter-Geräte GmbH).

12. Centrifuge Allegra™ X-22R equipped with a rotor for 50-ml tubes or equivalent.

2.4 qPCR Expression Analyses

1. 0.5 mm zirconia/silica beads.
2. Screw-cap microtubes of 2 ml, without skirted base.
3. RNeasy Plant Mini Kit (QIAGEN).
4. β-mercaptoethanol.
5. DNase I recombinant, RNase-free.
6. Transcriptor First Strand cDNA synthesis kit (Roche).
7. Primers:
   RtFfcarS-1F: 5′-GATACCCGGCGGAAAGGTTA-3′.
   RtFfcarS-1R: 5′-CTGACAGTCCATTTCAGCGC-3′.
   tub-1F: 5′-CCGGTGCTGGAAACAACTG-3′.
   tub-1R: 5′-CGAGGACCTGGTCGACAAGT-3′.
8. LightCycler® 480 SYBR Green I Master 5 × 1 ml (Roche).
9. LightCycler® 480 Instrument (Roche Life Science).
10. Microcentrifuge with a rotor for microcentrifuge tubes (1.5–2 ml) at 12,000 × *g*.

2.5 Carotenoids Extraction

1. Washed sea sand.
2. Screw-cap microtubes of 2 ml without skirted base.
3. Acetone.
4. Lyophilizer.
5. FAST-PREP24 (Biomedicals).
6. Speed vacuum concentrator.
7. UV/visible scanning spectrophotometer.
8. UV cuvette (1-ml volume) and 8-mm high micro cell cuvette (0.1-ml).

2.6 General Equipment

1. Platform shaker.
2. Thermocycler.
3. Vortex.
4. Standard fume hood.

3 Methods

3.1 Construction of a Plasmid with a Carotenoid Regulator Controlled by Tet-on System

1. Perform a DNA extraction of plasmid pLR14 that contains the gene of interest, *carS* [12].
2. Add *Pme*I-restriction sites to *carS* ORF by PCR amplification using primers carS-PmeI-1F and carS-PmeI-1R.

3. Set up a standard PCR reaction mixture containing 50 ng of DNA, 0.2 mM of dNTPs, 1 μM of carS-PmeI-1F and 1 μM carS-PmeI-1R, 1.75 mM of $MgCl_2$, 1× PCR buffer, and 1 U of Expand High Fidelity PCR System.
4. Perform a PCR using a program fitting your primer Tms. For *carS*, program was: Denaturalization at 94 °C for 2 min. 35 cycles of amplification reactions consisting of 94 °C, 20 s; 55 °C, 30 s; 72 °C, 2 min. Final polymerization step of 10 min at 72 °C.
5. Separate DNA PCR fragment in a 0.8% agarose gel electrophoresis (100 V for 45 min).
6. Purify the *carS* ORF fragment. Elute DNA in 10 μl of ultrapure water.
7. Ligate the *Pme*I hanging *carS* fragment with vector pGEM-T easy. Incubate an appropriate DNA amount of insert and 1 μl of vector pGEM-T easy (50 ng/μl) with 1 U of T4 DNA ligase, 5 μl of 2× Rapid Ligation Buffer in a final volume of 10 μl. Incubate overnight at 16 °C or 1 h at room temperature.
8. Incubate competent cells of *E. coli* DH5α with 5 μl of ligation reaction for 30 min in ice. Transform *E. coli* by thermic shock following a standard protocol [22]. Spread transformed bacteria on LB-X-Gal-Amp agar medium and incubate plates overnight at 37 °C.
9. Select several white colonies to screen positives clones. Pick selected colonies to fresh liquid LB-Amp, incubate overnight at 37 °C. Extract plasmid DNA from *E. coli* clones following [22].
10. Digest 1 μg of plasmid with 1 U of *Eco*RI in 1× H buffer. Incubate at 37 °C during 2 h. Check size of fragments by electrophoresis in a 0,8% agarose gel in TAE 1× buffer. The correct clones gave two bands of 3 kb and 2 kb, corresponding to the vector pGEM-T and *carS* gene, respectively. One of the positives clones was named pPO3.
11. Digest pPO3 and pVG3, a plasmid with the Tet-on based system [19], with *Pme*I at 37 °C, to release *carS* ORF from pPO3 (or gene of interest) and to introduce it in pVG3.
12. Separate DNA fragments by electrophoresis in a 0.8% agarose gel in TAE 1× buffer. Include as DNA size marker 5 μl of the 1-kb ladder marker.
13. Cut with a scalpel blade 2 agarose gel slices containing the *carS* ORF from pPO5 vector (2 kb) and the large fragment from plasmid pVG3 (6.7 kb).
14. Purify fragments corresponding to the *carS* ORF from pPO3 (insert) and to pVG3 without the *mluc* (vector) with Isolate II PCR and Gel Kit (Bioline).

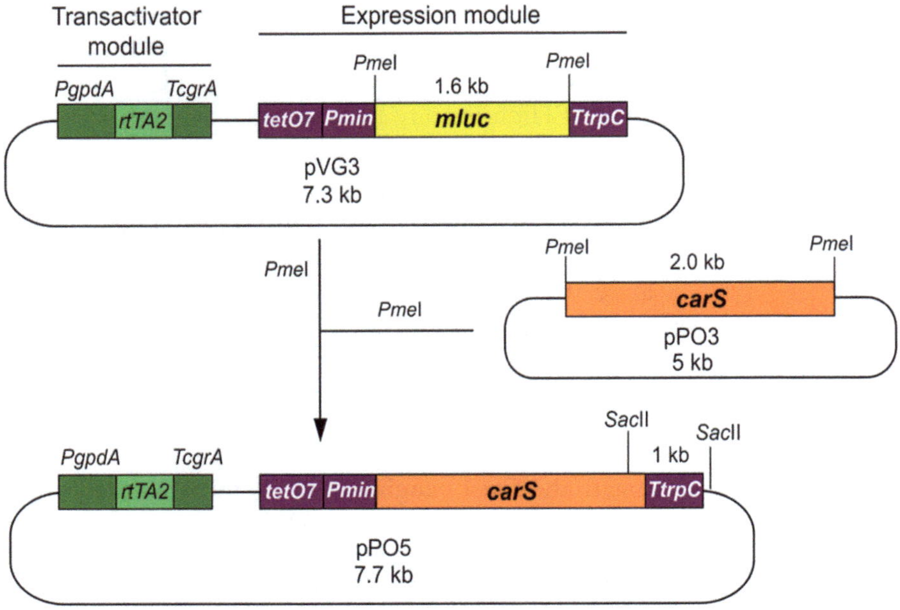

Fig. 2 Construction of a plasmid based on the Tet-on system to control expression of gene in *carS* in *F. fujikuroi*. Plasmid pPO5 was generated by substitution of *mluc* sequence from pVG3, removed by digestion with *Pme*I, by *carS* ORF, obtained from pPO3 digested with the same enzyme. The pPO3 construction was obtained by cloning into pGEM-T easy vector the *carS* ORF sequence with *Pme*I overhangs obtained by PCR amplification from plasmid pLR14

15. Ligate the vector and *carS* to obtain a plasmid with the gene controlled by the Tet-on system. Incubate vector and insert in a 1:3 relation with 5× ligation buffer and 1 U of T4 DNA ligase at 16 °C, overnight. As result, pPO5 was generated by substitution of *mluc* fragment of *Pme*I digested pVG3 with the *carS* from pPO3 digested with *Pme*I (Fig. 2).
16. Transform competent *E. coli* cells with 5 μl of the ligation mixture, as described above, and select on LB-Amp agar.
17. Select a dozen of bacterial colonies resistant to ampicillin. Isolate their plasmid DNA with a miniprep protocol [22]. Digest 1 μg of plasmid DNA with enzyme *Sac*II, in 1× buffer at 37 °C, during 2 h.
18. Check restriction pattern by electrophoresis in a 0.8% agarose gel run at 100 V for 45 min. Restriction with *Sac*II enzyme allows discrimination between both types of insertion. Choose a clone giving two bands of 1 kb and 6.7 kb, corresponding to the correct insertion. The wrong direction produces 2-kb and 5.7-kb bands.

3.2 Transformation of *Fusarium fujikuroi* Protoplasts

The transformation protocols are based on the ones described by [23, 24].

1. Inoculate 50 ml of Darken medium with approximately a 4-mm^2 mycelium plug in a 250-ml flask. Incubate for 3 days, at 30 °C in a shaker at 200 rpm.
2. Take 1 ml of the culture to inoculate 100 ml of fresh ICI medium. Incubate it overnight at 30 °C, and 150 rpm.
3. Filter the young mycelium by vacuum through a sterile filter paper, wash the filtered mycelium with sterile osmostabilization buffer and filter again until there is no liquid in the mycelium. Remove carefully the mycelium from the filter with a sterile spatula, resuspend it in 25 ml of enzyme solution (*see* materials, *see* **Note 1**) and incubate it for 3–4 h at 30 °C at 100–150 rpm (*see* **Note 2**).
4. Separate protoplasts from mycelia filtering through two borosilicate crucibles filters, first use a pore size 0 filter (160–250 μm), collect the filtrate in a sterile Petri dish and pass through pore size 1 filter (100–160 μm). Collect the resulting filtrate in a 50-ml tube.
5. Fill the tube with osmostabilization buffer up to 50 ml and centrifuge it for 15 min at 2000 × *g*.
6. Discard the supernatant and fill again up to 50 ml with osmostabilization buffer. Resuspend the protoplasts and centrifuge them again for 15 min at 2000 × *g* (*see* **Note 3**).
7. Repeat the previous step.
8. Discard the supernatant and resuspend the protoplasts in 500 μl of osmostabilization buffer (*see* **Note 3**).
9. Determine concentration of protoplasts using a Neubauer counting chamber under a microscope.
10. Use 3×10^7 protoplasts for each transformation. Protoplasts can be stored at −80 ° C (*see* **Note 4**). Before storing protoplasts, add 2 μl of DMSO and add PEG to a final concentration of 0.1% per 100 μl of aliquots.
11. Add 10 μg of linear DNA per transformation. In this case, we added 10 μg of pPO5 (plasmid containing the Tet-on cassette) and 7 μg of PAN7–1 [25], which contained a hygromycin-resistance cassette (*see* **Note 5**).
12. Prepare two microcentrifuge tubes with 50 μl STC 2× buffer. One of them will be used both as a control for regeneration and as a control for the selective medium. Add 50 μl of sterile distilled H_2O to the control. Add to the other tube 50 μl of the DNA to transform.
13. Add 100 μl with 3×10^7 protoplasts to each of the tubes.

14. Add 50 μl of 30% PEG-Tris–HCl and mix carefully by inversion.
15. Incubate the tubes in ice, for 20–30 min.
16. While the transformation process is in progress, keep two flasks with 100 ml of 2× regeneration medium at 50–55 °C.
17. Transfer the content of each microcentrifuge tube to a sterile 10-ml tube and add 2 ml of 30% PEG-Tris–HCl to each one. Mix carefully by inversion and incubate for 10 min in ice.
18. Add 4 ml of 1× STC to each tube and mix by inversion.
19. Mix the 6 ml of the transformation tube with 120 ml of regeneration medium (*see* **Note 6**) and prepare five plates spreading approximately 20–25 ml per Petri dish. Similarly, prepare another two Petri dishes with the corresponding amount of protoplasts from the control reaction and carry out appropriate serial dilutions.
20. Check growth on transformation plates on the following day. When a thin mycelial growth is visible, approximately after 19 h of incubation, add top agar with hygromycin to the transformation plates and one of the controls. For this purpose, put 5 ml of regeneration top agar to a 10-ml sterile tube, keep it at 55 °C in a water bath, and add 40 μl of 50 mg/ml hygromycin B. Immediately, lay the top agar to the transformation plates, and incubate at 30 °C. Colonies will appear after 2–3 days. Don't add the antibiotic to one of the control plates to be used as a positive growth control.
21. Pick the colonies and transfer them to selective medium. Transfer again the colonies obtained in the selective medium to EG sporulation medium. Incubate for 7 days and collect spores. Filter them through borosilicate filter with a porosity of 1. Streak spores to isolate colonies originated from a single spore.
22. Repeat twice the single-spore purification procedure (*see* **Note 7**).
23. Keep spores of transformants at −80 °C in 20% glycerol for storage.

3.3 Quantification of Transcript Levels of Carotenoid Genes

3.3.1 Cultivation Conditions

1. Inoculate *F. fujikuroi* wild type and transformants strains on EG agar Petri dishes. Use sterile toothpicks to inoculate in seven symmetrical points on EG medium from a 3-day-old mycelial colony (*see* **Note 8**).
2. Incubate the Petri dishes at 30 °C for 7 days under ~6 W/m^2 white light, ~360 lx. In our case, we used four fluorescent tubes Philips TL-D 18 W/840 at a distance of ca. 60 cm.

3. Add 10 ml of sterile 10% glycerol to each agar culture (*see* **Note 9**). Rub the surface of a sporulated culture with a sterile spatula. Harvest the spores from the dishes with a sterile pipette and pass through a borosilicate (size-1 pore) filter on top of a sterile 250-ml flask (*see* **Note 10**). Transfer the filtered spores to a 50-ml Falcon tube.
4. Centrifuge the spore solution at 2000 × *g* for 20 min at 4 °C. Remove the supernatant and resuspend the pellet of spores in 1 ml of sterile water.
5. Count the spores with a Neubauer chamber.
6. Optimize concentration and time of addition of doxycycline (Dox) for gene induction (*see* **Note 11**).
7. Inoculate 100 ml of DG minimal medium with 10^6 spores, two flasks for each strain, one to induce the Tet-on system and another one as control. Incubate for 3 days in darkness at 30 °C in a shaker at 150 rpm. After 48 h add 10 μg/ml Dox (*see* **Note 11**) to induce the Tet-on system. As controls, strains were cultured without Dox.
8. Filter the mycelia through a sterile filter paper, wash them with sterile water and freeze in liquid nitrogen. Store the samples at −80 °C until use.

3.3.2 RNA Extraction

1. Take approximately 100 μl of 0.5-mm zirconia/silica beads and introduce in a 2-ml microfuge tube with cap. As a reference, fill until reaching the cone of the tube (*see* **Note 12**).
2. Add 450 μl of lysis (RLT) buffer from RNeasy Plant Mini Kit, previously supplemented with β-mercaptoethanol, and add 2–3 mg of frozen mycelium to each 2-ml tube.
3. Grind the mycelium in a FAST-PREP24 in two disruptions pulses of 30 s at 4 m/s.
4. Extract RNA using the RNeasy Plant Mini Kit according to the manufacturer's protocol and quantify RNA (e.g., in a NanoDrop ND-1000 spectrophotometer). Check RNA integrity by agarose gel electrophoresis.

3.3.3 Synthesis of cDNA

1. Add 2.5 μg of RNA in a volume of 7 μl and 1 μl of 10× buffer (included in the kit) to a PCR tube. Add 10 U of DNase I and incubate for 15 min at 25 °C using the thermocycler.
2. Stop the reaction by adding 1 μl of 50 mM EDTA (pH 8.0) and heating at 65 °C for 10 min.
3. Add 1 μl of RNA-free double distilled H_2O and quantify the RNA (e.g., using a NanoDrop ND-1000 spectrophotometer).
4. Use Transcriptor first-strand cDNA synthesis kit for retrotranscription. To the previous 10-μl RNA samples (**steps 1–3**) add 1 μl of anchored-oligo (dT) and 2 μl RNA-free distilled H_2O

(supplied in the kit) and incubate at 65 °C for 10 min. The result is the Template-primer mix.

5. Prepare the RT mix and add it to the Template-primer mix following the supplier's instructions of the Kit.
6. Calculate theoretic cDNA concentration using the equation: Theoretic [cDNA] = [RNA] × Vi × E/Vf. Being [RNA]: RNA concentration after DNAse treatment. Vi: initial volume, E: retrotranscriptase efficiency, and Vf: final volume. In our case [cDNA] theoretic = [RNA] × 10 × 0.7/20.
7. To calculate what volume of water must be added to the sample to get a final concentration of 25 ng/μl use the equation: V = [cDNA] × Vf/ (cc–Vf) = [cDNA] × 20/(25–20).

3.3.4 Analysis by Quantitative RT-PCR

1. In a 1.5-ml microcentrifuge tube, prepare the qPCR mix the LightCycler 480 SYBR Green I Master with the selected primers for *carS* gene. Prepare a primer mix adding 10 μl of 100 μM primer RtFfcarS-1F primer and 10 μl of 100 μM primer RtFfcarS-1F in 80 μl of double distilled H_2O. For the RT mix (*see* **Note 13**), add 2.6 μl of vial 2 (PCR-grade water), 5 μl of vial 1 (Master mix 2×) and 0.4 μl of 10 μM primer mix.

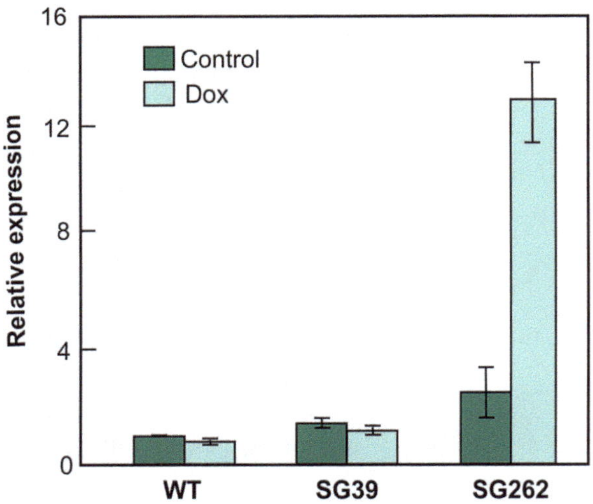

Fig. 3 Expression of *carS* gene controlled by the Tet-on system. Levels of mRNA of the gene *carS* under the control of the Tet-on promoter controlled by doxycycline in strain SG262 and compared to *F. fujikuroi* IMI 58289 wild type (WT), and *carS* mutant (SG39). Strains were cultivated in 100 ml of DG medium in 500-ml flasks during 48 h at 30 °C and 150 rpm in darkness. Tet-on expression was induced with 10 μg/ml of doxycycline (Dox) and incubated for 24 h; non-induced cultures were grown in parallel. RNA was extracted (*see* Subheading 3.3.2), converted to cDNA (*see* Subheading 3.3.3) and analyzed by qPCR (*see* Subheading 3.3.4). β-tubulin gene was used as a reference for relative expression

2. Prepare a similar mix for reference β-tubulin gene with primers tub-2F and tub-2R.
3. Run the qPCR protocol in a LightCycler 480 II real-time instrument. Analyze transcription levels with LightCycler 480. Normalize *carS* (the gene of interest) expression with β-tubulin as a reference gene (*see* **Note 14**). Addition of Dox resulted in an induction of *carS* gene (Fig. 3).

3.4 Quantification of Carotenoids in Transformants with Modulated Expression

3.4.1 Cultivation, Harvesting and Preparation of Mycelia

1. Prepare DG agar and DG-Dox agar (DG with 20 μg/ml Dox) media and lay them in 8.9-cm Petri dishes. Inoculate strains SG39 and SG262 at 7 symmetrical points on both media (*see* **Note 15**).
2. Incubate the Petri dishes at 30 °C, for 7 days under ~6 W/m^2 white light, ~360 lx. In our case, the cultures were illuminated with four fluorescent tubes Philips TL-D 18 W/840 at a distance of ca. 60 cm (*see* **Note 16**).
3. Cut the agar around the mycelial colonies with a clean scalpel blade. Scrap carefully the rest of the agar from the mycelia with the blade. Press the mycelial samples on filter papers to remove excess of water. Freeze in liquid nitrogen, and lyophilize the samples overnight.

3.4.2 Extraction and Quantification of Carotenoids

1. Take a 2-ml screw-cap microtube (*see* **Note 17**). Add washed sea sand (*see* **Note 18**), approximately 20–50 mg of dry mycelia (*see* **Note 19**), and 1 ml acetone (*see* **Notes 20** and **21**).
2. Shake the microtubes containing the samples in a FAST-PREP24 with 2 pulses of 30 s at 6 m/s. This allows extraction of carotenoids.
3. Centrifuge the extraction tubes at maximal speed (>12,000 × *g*) for 2 min and transfer the acetone supernatant to a clean 2-ml microcentrifuge tube (*see* **Note 22**). Dry the samples in a speed vacuum concentrator (*see* **Note 23**).
4. Repeat the extraction procedures (**steps 2–3**) up to total bleaching of the sample, and add the supernatant to the tube of the same sample already dried in the speed vacuum concentrator.
5. Dissolve the dry extract in 1 ml of acetone for carotenoid-rich samples or in 100 μl of acetone for low-content carotenoid samples (*see* **Note 24**). Centrifuge the sample at maximal speed (>12,000 × *g*) for 2 min (*see* **Note 25**).
6. Transfer the supernatant to a clean 1.5-ml microfuge tube. Take the sample to a spectrophotometer UV cuvette and measure the absorbance in the complete spectrum from 350 nm to 650 nm in a spectrophotometer (Fig. 4b) (*see* **Note 26**).

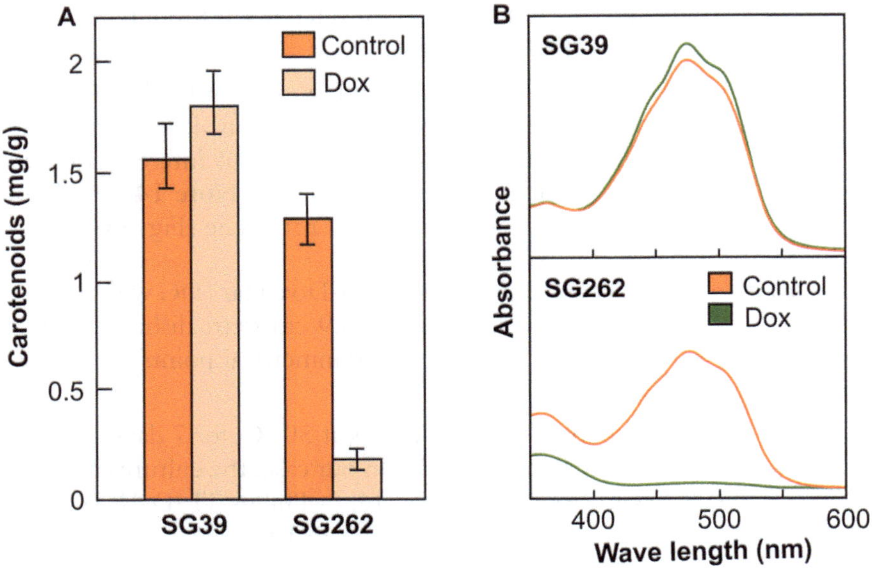

Fig. 4 Modulated production of carotenoids in *F. fujikuroi* by addition of doxycycline. Carotenoids accumulated in the engineered strain SG262 and *carS* mutant SG39 were measured from mycelia cultured on Petri dishes with either DG-Dox agar (DG with 20 μg/ml Dox) or DG agar. Cultures were grown for 7 days at 28 °C under 6 W/ m^2 white light. (**a**) Total carotenoids calculated as described in Subheading 3.4.2. (**b**) Spectra of extracted carotenoids from the *carS* mutant SG39 and the strain SG262 (*carS* gene controlled by Tet-on system). Spectra in green show profile of carotenoids in the presence of Dox (doxycycline) and in orange represent control samples. Spectra were relativized to extracted dry mass

7. Total carotenoid content per dry weight of mycelium is estimated using an overall extinction coefficient (EC) of 200 (1 mg ml^{-1}, 1 cm), using the formula:

Carotenoids (mg/g dry weight) = ([$Abs_{max} - Abs_{650nm}$] × V)/(DW × EC × PL).

V = final volume of solvent with extracted carotenoids in ml. DW = mycelial dry weight in g. EC = absorbance coefficient in mg^{-1} ml cm^{-1}. PL = path length in cm (*see* **Note 27**).

Mutant SG262 produced almost no carotenoids in the presence of Dox (Fig. 4a), and its spectrum analysis shows no peak of carotenoids (Fig. 4b) while the parental strain SG39 produced mainly neurosporaxanthin [4]. The Tet-on system is a useful tool to modulate carotenoid biosynthesis in *F. fujikuroi*.

4 Notes

1. It is recommended to prepare a fresh enzyme solution for protoplast each time because its effectiveness drops if it is frozen at −80 ° C.

2. Check periodically the formation of protoplasts during the incubation of mycelia with enzymes by observing 10-μl aliquots under microscope (200×). Protoplasts, visible as round cells, should start appearing after 30–60 min of incubation.
3. Never use pipette tips to resuspend protoplast pellets. Instead, disaggregate gently the pellet with smooth taps. From this moment on, use cut pipette tips along all the protocol to avoid burst of protoplasts.
4. Competence of protoplasts tends to decrease upon storing at −80 °C.
5. In our case, simultaneous transformation with two plasmids (cotransformation) was done because the selection marker was not included in the plasmid containing the Tet-on cassette.
6. Prepare 1× regeneration medium by mixing 60 ml of warm 2×regeneration medium with 60 ml of warm 1.6 M sucrose. Keep the medium in a bath at 55 °C.
7. Spores (microconidia) are uninucleate. Single-spore purification allows obtaining pure clones in the case of heterokaryotic or unstable transformants.
8. *F. fujikuroi* IMI58289 sporulates poorly. Inoculate at least three Petri dishes for each strain to ensure collecting enough spores.
9. Because of conidia hydrophobicity, use of glycerol is recommended to facilitate suspension of the conidia.
10. If borosilicate filters are not available, the conidia suspension may be passed through a standard Whatman filter paper.
11. The optimal concentration of doxycycline for our growth conditions was formerly determined mixing 50 μl with 10^4 fresh spores of the wild type IMI58289 or the *carS* mutant SG39 and 50 μl of 2× DGpeptone in a 96-well plate. Doxycycline was added at different concentrations (from 10 to 160 μg/ml) after 15 h incubation at 30 °C and growth was checked after 72 h (without shaking). Optimal time for doxycycline additions was also formerly determined for the expression analysis experiments.
12. Wear gloves during RNA extraction and subsequent manipulations to avoid degradation.
13. Calculate the volume of the RT mix considering three replicates for each of the investigated genes (the genes of interest and the reference gene(s)). Add some additional volume to prevent pipetting error.
14. Gene for β-tubulin was used as a reference.

15. The incubation of several colonies in the same Petri dish is recommended because of the slower radial growth of *F. fujikuroi* compared to other fungi.
16. The fluorescent tubes should not be too close to the cultures to avoid excessive heating.
17. Use two 2-ml screw-cap microfuge tubes for the same sample in order to have two replicates.
18. Fill each screw-cap microtube with approximately 100 μl of sand.
19. The recommended amount of dry mycelium depends on its carotenoid content. Approximately 20 mg of dry mycelium are sufficient for samples from strains overproducing carotenoids, and a higher weight (at least 40–50 mg) are needed for the samples with lower carotenoid content, as those of the wild type in the dark.
20. Because of the presence of polar carotenoids in *Fusarium*, acetone is an appropriate solvent for extraction of carotenoids.
21. All manipulations during carotenoid extraction and measurement should be done under green or red light to prevent isomerization and photodegradation of carotenoids.
22. It is important that the supernatant to be dried in the speed vacuum concentrator is totally clean.
23. Do not fill 2-ml microtubes above 1 ml because the sample could spill in the speed vacuum concentrator.
24. If a sample has a very low of carotenoid content, dissolve the sample in only 100 μl of acetone. This requires the availability of microcells for spectrophotometer measuring.
25. If carotenoids were not going to be analyzed immediately after the extraction, extracts should be stored at −20 °C (or lower) until spectrophotometer analysis in order to prevent degradation of carotenoids.
26. If maximal absorbance of the spectrum exceeds the value of 1, dilute the carotenoid extract accordingly to get a lower absorbance.
27. Carotenoids of *F. fujikuroi* usually have a maximal absorption peak between 470 and 480 nm.

Acknowledgments

We thank the Spanish Government (projects BIO2012-39716, BIO2015-69613-R, and CaRed Networks BIO2015-71703-REDT and BIO2017-90877-REDT), and Junta de Andalucía (project CTS-6638) for financial support. J.P-M was granted by

Ministry of Science, Innovation and University and P.O had a Fellowship for Initiation to Research from Plan Propio from University of Seville.

References

1. Avalos J, Nordzieke S, Parra O et al (2017) Carotenoid production by filamentous fungi and yeasts. In: Sibirny AA (ed) Biotechnology of Yeasts and Filamentous Fungi. Springer, Berlin, pp 225–279
2. Avalos J, Díaz-Sánchez V, García-Martínez J et al (2014) Carotenoids. In: Martín JF, García-Estrada C, Zeilinger S (eds) Biosynthesis and molecular genetics of fungal secondary metabolites. Springer, Berlin, pp 149–185
3. Avalos J, Cerdá-Olmedo E (2004) Fungal carotenoid production. In: Arora DK (ed) Handbook of fungal biotechnology, 2nd. edn. Marcel Dekker Inc., New York, pp 367–378
4. Avalos J, Pardo-Medina J, Parra-Rivero O et al (2017) Carotenoid biosynthesis in *Fusarium*. J Fungi 3:1–16
5. Hornero-Méndez D, Limón MC, Avalos J (2018) HPLC analysis of carotenoids in neurosporaxanthin-producing fungi. In: Barreiro C, Barredo J-L (eds) Microbial carotenoids: methods and protocols. Springer, New York, pp 269–281
6. Avalos J, Prado-Cabrero A, Estrada AF (2012) Neurosporaxanthin production by *Neurospora* and *Fusarium*. Methods Mol Biol 898:263–274
7. Avalos J, Estrada AF (2010) Regulation by light in *Fusarium*. Fungal Genet Biol 47:930–938
8. Rodríguez-Ortiz R, Limón MC, Avalos J (2009) Regulation of carotenogenesis and secondary metabolism by nitrogen in wild-type *Fusarium fujikuroi* and carotenoid-overproducing mutants. Appl Environ Microbiol 75:405–413
9. Navarro E, Lorca-Pascual JM, Quiles-Rosillo MD et al (2001) A negative regulator of light-inducible carotenogenesis in *Mucor circinelloides*. Mol Genet Genomics MGG 266:463–470
10. Lorca-Pascual JM, Murcia-Flores L, Garre V et al (2004) The RING-finger domain of the fungal repressor *crgA* is essential for accurate light regulation of carotenogenesis. Mol Microbiol 52:1463–1474
11. Ruger-Herreros M (2016) Participación de la proteína CarS en la regulación de la carotenogénesis y el estrés en *Fusarium fujikuroi*. PhD Thesis, University of Seville
12. Rodríguez-Ortiz R, Limón MC, Avalos J (2013) Functional analysis of the *carS* gene of *Fusarium fujikuroi*. Mol Genet Genomics 288:157–173
13. Metzger MB, Pruneda JN, Klevit RE, Weissman AM (2014) RING-type E3 ligases: master manipulators of E2 ubiquitin conjugating enzymes and ubiquitination. Biochim Biophys Acta 1843:47–60
14. Beck CF, Mutzel R, Muller BJ (1982) A multifunctional gene (*tetR*) controls Tn10-encoded tetracycline resistance. J Bacteriol 150:633–642
15. Urlinger S, Baron U, Thellmann M, Hasan MT, Bujard H, Hillen W (2000) Exploring the sequence space for tetracycline dependent transcriptional activators: novel mutations yield expanded range and sensitivity. Proc Natl Acad Sci U S A 97:7963–7968
16. Das AT, Tenenbaum L, Berhout B (2016) Tet-on systems for doxycycline-inducible gene expression. Curr Gene Ther 16:156–167
17. Gossen M, Bujard H (1992) Tight control of gene expression in mammalian cells by tetracycline-responsive promoters. Proc Natl Acad Sci U S A 89:5547–5551
18. Gossen M, Bonin AL, Freundlieb S, Bujard H (1994) Inducible gene expression systems for higher eukaryotic cells. Curr Opin Biotechnol 5:516–520
19. Meyer V, Wanka F, van Gent J, Arentshorst M, van den Hondel C, Ram A (2011) Fungal gene expression on demand: an inducible, tunable, and metabolism-independent expression system for *Aspergillus niger*. Applied Environ Microbiol 77:2975–2983
20. Avalos J, Casadesús J, Cerdá-Olmedo E (1985) *Gibberella fujikuroi* mutants obtained with UV radiation and N-methyl-N'-nitro-N-nitroso-

guanidine. Appl Environ Microbiol 49:187–191

21. Geissman TA, Verbiscar AJ, Phinney BO, Cragg G (1966) Studies on the biosynthesis of gibberellins from (−)-kaurenoic acid in cultures of *Gibberella fujikuroi*. Phytochemistry 5:933–947
22. Sambrook J, Russell DW (2001) Molecular cloning: a laboratory manual. Cold Spring Harbor Laboratory Press, New York
23. Gafford I, Brown DW, Plattner R, Proctor RH, Qi W, Trail F (2005) Functional analysis of the polyketide synthase genes in the filamentous fungus *Gibberella zeae* (Anamorph *Fusarium graminearum*). Eukaryot Cell 4:1926–1193
24. Twaruschek K, Spörhase P, Michlmayr H, Wiesenberger G, Gerhard A (2018) New plasmids for *Fusarium* transformation allowing positive-negative selection and efficient Cre--*loxP* mediated marker recycling. Front Microbiol 9:1954
25. Punt PJ, Oliver RT, Dingemanse MA, Pouwels PH, van den Hondel CAMJJ (1987) Transformation of *Aspergillus* based on the hygromicyn B resistance marker from *Escherichia coli*. Gene 56:117–124

Part VII

Nutrition and Health

Chapter 27

Intervention Studies in Humans

Begoña Olmedilla-Alonso

Abstract

Intervention studies in humans provide the highest level of evidence in the assessment of the effect of nutrients or bioactive food components on human health /disease. There are different intervention study designs, and their selection depends on the objectives of the study. Here we present the main designs and the factors that should be considered when selecting and developing them.

Key words Human intervention studies, Carotenoids, Provitamin A, Biomarkers

1 Introduction

When evaluating the relationship between the diet, the intake of nutrients or bioactive food components, and human health/disease, intervention studies—those experimental studies where the researcher intervenes at some point throughout the study—are essential tests that provide a direct and high level of evidence needed prior to making decisions in areas such as public health recommendations and effective nutritional intervention programs (*see* **Note 1**) and confirming claims of alleged health-promoting properties of a food or a component [1–4]. Intervention studies on humans provide information that complements that obtained through in vitro digestion assays which, in general, do not fully explain the impact of carotenoid intake in humans but do explain the effect that different food-related factors have [5–7].

Traditionally, intervention studies involving carotenoids have focused on assessing the effect of the intake, at dietary or pharmacological level, of b-carotene on vitamin A status and on reducing the risk of several chronic diseases including some types of cancer. However, in the last decades, there has been growing interest in the diverse biological activities of the major dietary carotenoids and hence in intervention studies on subjects with different chronic diseases or with related risk factors (i.e., cancer, cardiovascular

Manuel Rodríguez-Concepción and Ralf Welsch (eds.), *Plant and Food Carotenoids: Methods and Protocols*,
Methods in Molecular Biology, vol. 2083, https://doi.org/10.1007/978-1-4939-9952-1_27,

disease, diabetes, obesity, macular disease). Moreover, the contribution of dietary provitamin A carotenoids to dietary retinol intake is being questioned based on in vitro and in vivo studies on the three major dietary provitamin A carotenoids (carotene vs. xanthophyll) [8]; thus, more intervention studies with specific provitamin A carotenoids from different food matrices and on subjects with different vitamin A nutritional status should be conducted as these are crucial aspects affecting bioconversion of provitamin A carotenoids into retinol.

Intervention studies designed as randomized controlled trials are the ones considered to provide the most reliable and the strongest evidence in epidemiological research and can generally be considered either "preventive" (risk reduction of certain chronic diseases, that is, lutein and zeaxanthin regarding age-related macular degeneration) or therapeutic (effectiveness of provitamin A carotenoid food sources in preventing or reducing symptoms/mortality due to a vitamin A deficiency).

In brief, the steps in the development of the intervention study are as follows: (1) identifying the objective and design of the intervention study; (2) identifying the target population and define inclusion/exclusion criteria; (3) selection of biomarkers; (4) selecting sampling method and calculate the sample size; (5) selection of data collection methods (quantitative and qualitative); (6) taking out civil liability insurance for research staff involved in the study; (7) submitting the study protocol, participant information sheet and the informed consent form to the ethics committee for approval; (8) recruitment of participants; (9) dietary intervention (food or isolated compounds), using single-dose (usually for the assessment of bioavailability, pharmacokinetic and acute effects) or multiple-dose (medium/long-term trials usually used to evaluate chronic (cumulative) exposure, functional changes or modification of risk factors); (10) dietary/nutritional intervention follow-up and compliance; (11) data analysis and interpretation of results.

2 Materials

2.1 Carotenoid Sources

The carotenoid material used in the intervention studies can be foods containing carotenoids (natural or enriched sources), food supplements or isolated carotenoids at specific doses for a particular study. The use of stable carotenoid isotopes [9] makes it possible to differentiate between administered and endogenous carotenoid but the procedure can be difficult and costly (*see* **Note 2**).

2.2 Food Composition Tables and Databases with Carotenoid Data

In general, food composition tables (FCTs) and databases (FCDBs) did not include carotenoid content and therefore special attention needs to be paid to those showing the individual content of carotenoids (e.g., refs. [10, 11]). These enable us to properly select the foods to be used in intervention studies and also to measure the amounts of carotenoids ingested throughout the study and thus gain insight into their effect on the results of the intervention study (*see* **Note 3**).

When food sources containing carotenoids are used, the different forms of preparation/consumption and, where possible, their bioavailability/bioaccessibility should be considered as these factors affect response and biological activity.

2.3 Participants

Intervention studies must adhere to international principles and applicable standards in each country regarding bioethics, biological safety and data protection while also respecting the overarching principles laid down in the Helsinki Declaration (World Medical Association) [12], in the Council of Europe Convention on Human Rights and Biomedicine [13] and in UNESCO's Universal Declaration on the Human Genome and Human Rights [14] and those required by local ethics committees.

Study participants must be provided with information about the study and must sign an informed consent form before they can be included. Those documents must show that researchers are familiar with and commit to upholding applicable laws regarding personal data protection. In the case of European studies, this serves as a guarantee that the processing, communication and transfer of the personal data of all participants adheres to the provisions of Regulation (EU) 2016/679 of the European Parliament and of the Council of 27 April 2016 and to applicable rules in force in each European Union country [15].

2.4 Biological Samples

In human research studies, biological samples are stored (biobanking) and removed after being used for research purposes. At European level, a set of recommendations has been issued by the Committee of Ministers of the Council of Europe regarding these activities. These recommendations also apply to the use for research purposes of biological materials of human origin stored or previously obtained for a different purpose (i.e., a previous research project) [16].

2.5 Other Considerations

Researchers and staff involved in intervention studies must take out liability insurance before embarking upon a study (*see* **Note 4**).

A detailed and duly justified report on the intervention study must be submitted to the clinical research ethics committee at the hospital or corresponding research center which must authorize said study.

3 Methods

The aim/hypothesis or research question is the first step in a human intervention study and must be clearly identified since it will determine the design, development and interpretation of the results of the study. The aim of the study should be clearly defined, preferably using measurable terms and specifying the study's main variables (at least one dependent and one independent variable). Regarding carotenoids, basically the endpoints must be relevant in the context of vitamin A nutritional status (preventing deficiency) or in that of prevalent chronic diseases or health conditions in which carotenoids dietary intervention would make possible to reduce risk or improve quality of life.

Intervention studies should help produce high-quality information to improve the existing body of knowledge in the field in question (i.e., public health and the taking of health-related decisions). Thus, the first step is to search for the latest information, for example, in PubMed and Cochrane, in order to properly define the objectives and design the study (*see* **Note 5**).

The important points to be considered in the design and implementation of intervention studies on humans are those related with the research question, subject identification, description of the intervention/exposure factors or procedures, outcome definition, reliable markers and appropriate statistical analysis for the study design and type of outcome indicators [17]. In this regard, there are guidelines and checklists of the key elements that should be included in the design, implementation and reporting/interpretation of human intervention studies. These are useful when setting out an intervention study to assess the effect of foods on health/disease (i.e., [17–20].) and specifically for randomized trials, for example, in CONSORT [21] and for nonrandomized studies (e.g., [22]).

3.1 Types of Design

Study design refers to the set of strategies established to achieve the objectives and is affected by the technical and human resources available and by what is ethically feasible in each type of study. The study design includes the description of the intervention and procedures (e.g., protocol, amount of carotenoid supplied, duration of the intervention, handling of drop-outs, and type of masking).

There are different types of design depending on the criteria employed in the classification (e.g., its aim: analytical or descriptive; control of how group assignments are made: observational or experimental). Intervention studies are experimental studies in which subjects are assigned to one of the groups and are then monitored for a period of time (defined on the basis of the aims and the biomarker used) to assess the effectiveness of the intervention, preferably compared to subjects who were given placebos (*see* **Note 6**). In short, they can be randomized (clinical trial, field trial, community trial) or nonrandomized (quasi-experimental intervention study) (*see* **Note 7**). They can be controlled or not controlled and, considering intervention exposure, may be classified as parallel, crossover, factorial, or community. In the crossover design, subjects are assigned to the treatment group and, once the interval for cessation of the residual effect of the carotenoid has passed (wash-out period), they are transferred into the placebo group. Under this design, each individual acts as its own control. The factorial design is used when two or more interventions are performed simultaneously on the same subjects.

Different aspects determine the quality of an intervention study such as randomization (simple, block, stratified, minimized), placebo, control and double-blinding (type of masking). These contribute to obtaining a higher level of evidence in the relationship under examination [4]. A set of criteria must be met in the case of studies to back up nutritional/health claims for foods, that is, those claiming that they contain significant amounts of carotenoids and that these have certain nutritional or physiological effects [1] *see* **Note 8**).

The randomized controlled trial is considered the most rigorous method for determining whether there is a cause–effect relationship between an intervention and outcome as it provides the most reliable and strongest evidence in epidemiological research [2, 4]. However, when choosing a design priority should be given to the one which best meets the objective and gives the study the greatest degree of validity (*see* **Note 9**).

3.2 Selection of Subjects

Subjects should be representative of the target or reference population to which the study results are to be applied (*see* **Note 10**). Participants must meet eligibility criteria (clearly defined inclusion and exclusion criteria), agree to participate and sign the informed consent form once they have been being fully informed about the study.

Eligibility criteria should clearly describe participants so that results can be appropriately interpreted and generalized. Criteria typically include factors such as age, sex, nutritional status, health/disease status, and lifestyle habits.

The number of participants should be sufficiently large to allow valid statistical analysis and a reasonably precise measurement of effect. Sample size must be calculated (i.e., by means of a Power and Sample Size Program, SPSS Sample Power) for a given marker to reflect the final outcome of the study (the number of outcomes or endpoints is a key factor influencing the power of the study) (*see* **Note 11**).

3.3 Identification of Variables (Markers)

Qualitative and/or qualitative variables are the characteristics that are measured in the study and must be clearly identified and defined. Only those that are necessary for the study should be identified. At least one outcome variable must be considered, and also, those variables which could lead to confusion or modify the effect. Here, the time interval between the intervention and the response of the marker must be defined.

Markers of intake, status, or functional effects to assess the effect of a dietary/pharmacological intervention with nutrients or with bioactive compounds, can be biochemical, functional, or clinical measurements. In intervention studies, these are used to assess the endpoints (i.e., surrogate disease endpoints or those related to optimal health or adequate nutritional status), that will eventually be used to support decisions in relation with food, nutrition and health [23–25]. Recent carotenoid exposure may be assessed using analytical markers with relatively "fast turnover" (i.e., postprandial TRL fraction, serum, breast milk) or storage tissue (i.e., adipose tissue, macular pigment) for functional test biomarkers. As for overexposure to carotenoids, markers include clinical signs (i.e., carotenodermia, corneal rings and retinopathy) and biochemical measurements (hypercarotenemia, xanthophyll esters) [25]. It is generally recommended to use more than one biomarker to assess how diet relates to each disease.

Blood tests (serum/plasma) are the most simple and frequently used way to assess carotenoid status and response to an intervention but do not adequately reflect the organism's response to pharmacological doses. Moreover, the carotenoid clearance may differ between subjects and type of carotenoid. More recently, stool samples have been used but with the limitation, inter alia, of possible carotenoid oxidation in the large intestine and retention by microorganisms. In postprandial studies, carotenoid concentrations are assessed in HDL cholesterol or LDL cholesterol and in chylomicrons (useful for the assessment of physiological doses but encountering difficulty in metabolic aspects due to the rapid turnover or transfer among lipoproteins) [25].

An important aspect to be considered in carotenoid intake is the concentration supplied (in both physiological and pharmacological doses) as metabolism (absorption and transport) appears to be different depending on the amount supplied. In addition, when pharmacological doses are administered, safety aspects (related to direct or indirect actions) such as the capacity of provitamin A carotenoids to be converted into retinol should be checked as well [8]. In general, when assessing the effect of food components on the risk of chronic diseases, the biomarkers used should provide information on the stage of the disease in which dietary intervention would be effective for an early diagnosis or a better prognosis of the disease being studied. Therefore, the proper biomarkers must be selected when dealing with each disease (intermediate objectives), and must be methodologically (precision, accuracy, sensitivity, specificity) and biologically (biological plausibility, dose-response ratio, consistency) validated.

Moreover, it should be noted that it is not feasible to perform the direct measurement of the effect of consuming a single food or a single dietary component as these need to be considered in the context of the overall diet. In the case of provitamin A carotenoids, the effects could be due to their conversion to retinol or to their own biological activities.

3.4 Duration of the Study

In addition to the concentration of the bioactive compound supplied, it is also important to consider the time during which it is consumed. Hence, the duration of the study should be sufficient to allow changes in the primary outcome measured but not longer to avoid participant noncompliance or withdrawal. Turnover rates of relevant tissues should also be taken into account.

3.5 Quality, Validity, and Reliability of the Measurements

The equipment and procedures used for biomarker analysis and the training of personnel involved in the study will contribute to the internal and external validity of the intervention study. This includes the calibration of instruments, standardization of procedures, training of researchers, dietary interviews, and anthropometric measurements. In this regard, it is important to check the compliance (adherence to the treatment) of the participants with the intervention assigned by means of biochemical markers or evaluations based on dietary assessment through periodic visits and examinations.

3.6 Data Collection

Interviews of participants, physical examinations (anthropometric data), laboratory procedures (i.e., biomarkers), and other complementary tests (nutritional status assessment—dietary evaluation) must be standardized. Special attention should be paid to factors that could influence the degree of association between intervention and outcome (confounding or interaction factors). If such factors are known, they should be distributed evenly among the groups.

Regardless of the intervention study being conducted, it is always advisable to carry out a dietary assessment in order to know normal (or past) intake patterns and the present intake of the subjects [26] (*see* **Note 12**). This requires the use of FCT with individualized carotenoid data (i.e., [11].). When preparing (or choosing) a questionnaire to assess ingestion, it is important to define which components of the diet that should be the focus of attention, the chemical formats in which they are found (i.e., free-form carotenoids, in ester form as total forms or isomers) and what types of food they are found in. In the case of carotenoids, it is also useful to know what factors may affect the absorption and metabolism of carotenoids such as, for example: tobacco consumption, chronic disease, use of medicines or food supplements. In general, in all dietary assessment questionnaires it is useful to include other general interest data to facilitate the proper interpretation of results (i.e., age, weight, height) and it is preferable to make these measurements in situ rather than to depend on participant responses.

3.7 Data Analysis and Interpretation of Results

Data analysis should be done based on the main expected outcome. The interpretation of results assesses the strength of the effect, alternative explanations such as sampling variation or bias. Interpretation of the results obtained from single or multiple carotenoid intake (short- or long-term studies) is different since they are interpreted in terms of postprandial response/metabolism and of repletion and/or saturation processes, respectively (>1 week) [27]. The degree to which results can be generalized will depend on the representativeness of the sample, statistical accuracy and sensitivity, and whether the study is affected by bias (in the selection of subjects, measurement errors, etc.).

4 Notes

1. When designing effective nutritional intervention programs it is essential to know the bioavailability of each carotenoid from the different types of food sources and from fortified foods.
2. Radioisotope-labeled carotenoids were used but can no longer be used for ethical reasons [28].
3. Foods may also be selected based on the classification of the carotenoid content in fruits and vegetables proposed by Britton and Khachik [29] in four levels, from low (0–0.1 mg/100 g) to very high (>2 mg/100 g).
4. Not all insurance companies offer this type of insurance.
5. Cochrane https://community.cochrane.org/mecir-manual/standards-conduct-new-cochrane-intervention-reviews-c1-c75/performing-review-c24-75. In addition, Cochrane provides search strategies to identify randomized trials in PubMed because not all randomized trials are indexed as RCTs: https://work.cochrane.org/pubmed
6. One must bear in mind that in many cases isolated components may influence different results while in others a single result can be influenced by several components.
7. Randomization is used to ensure that the intervention and control groups are similar in all aspects (including distribution of potential confounding factors except the one being tested), although some confounding variables may emerge randomly.
8. Criteria to be assessed by food studies to obtain a nutrition/health claim should include consistency, strength, quality,

biologically plausibility, specificity, time frame, and dose-response effects [23].

9. The advice of a statistical expert (preferably in the nutrition area) is recommended when selecting a design and also calculating sample size.
10. The term "apparently healthy volunteers" is frequently found in studies which aim to modify risk factors. However, these subjects cannot be considered suitable as a representative sample because some risk factors of the disease being studied can modify metabolism. Therefore a certain number of them (the number depending on the study) should be a prerequisite. Studies with subjects with differing vitamin A nutritional status (deficient and within the normal range) are not comparable when studying the effects of provitamin A carotenoids.
11. Sample size is typically calculated with data from the main dependent variable (outcome) and therefore the following data are needed: mean value and standard deviation, the percentage of variation pursued with the intervention (ideally, to obtain not only statistical significance but also biological/clinical relevance) and the percentage of statistical power assumed. If sample size is not calculated, it would happen that the results (observed effects) obtained do not have enough statistical power to be extrapolated to similar populations.
12. Habitual intake (i.e., accurate individual weight, estimation of foods consumed); past or habitual ingestion (i.e., 24-h recall, dietary history, food frequency consumption). The most frequently used methods are the 24 h dietary recall (3 days) and the food frequency questionnaire.

Acknowledgments

Spanish Carotenoid Network (CaRed) funded by the Spanish MINECO (Ministry of Economy, Industry and Competitiveness) grants BIO2015-71703-REDT and BIO2017-90877-REDT, the European Carotenoid Network (EuroCaroten) funded by the European Commission COST Action CA15136 and Instituto de Salud Carlos III-ERDF grant PI16/01991.

References

1. Reglament EU 1924/2006 Of the European Parliament and of the council of 20 December 2006 on nutrition and health claims made on foods. https://eur-lex.europa.eu/legal-content/en/ALL/?uri=CELEX%3A32006R1924.
2. Commission regulation (EC). No 353/2008 of 18 April 2008 establishing implementing rules for applications for authorisation of health claims as provided for in Article 15 of Regulation (EC) No 1924/2006 of the European Parliament and of the Council

3. European Commission Nutrition and health claims. https://ec.europa.eu/food/safety/labelling_nutrition/claims_en. In: .
4. USDA/HHS commitment to evidence-based process. http://www.nel.gov/default.cfm?library=DGAC.
5. Bohn T, McDougall G, Alegría A, Alminger M, Arrigoni E, Aura A, Brito C, Cilla A, El SN, Karakaya S, Martínez-Cuesta MC, Santos CN (2015) Mind the gap-deficits in our knowledge of aspects impacting the bioavailability of phytochemicals and their metabolites—a position paper focusing on carotenoids and polyphenols. Mol Nutr Food Res 59:1307–1323
6. Granado F, Olmedilla B, Herrero C, Pérez-Sacristán B, Blanco I, Blázquez S (2006) Bioavailability of carotenoids and tocopherols from broccoli: in vivo and in vitro assessment. Exp Biol Med 231(11):1733–1738
7. Granado-Lorencio F, Olmedilla-Alonso B, Herrero-Barbudo C, Pérez-Sacristán B, Blanco-Navarro I, Blázquez-García S (2007) Comparative *in vitro* bioaccesibility of carotenoids from relevant contributors to carotenoid intake. J Agric Food Chem 55(15):6387–6394
8. Rodríguez-Concepción M, Avalos J, Bonet ML, Boronat A, Gomez-Gomez L, Hornero-Mendez D, Limon MC, Meléndez-Martínez AJ, Olmedilla-Alonso B, Palou A, Ribot J, Rodrigo MJ, Zacarias L, Zhu C (2018) A global perspective on carotenoids: metabolism, biotechnology, and benefits for nutrition and health. Prog Lipid Res 70:62–93
9. Sheftel J, Loechl C, Mokhtar N, Tanumihardjo SA (2018) Use of stable isotopes to evaluate bioefficacy of provitamin a carotenoids, vitamin a status, and bioavailability of iron and zinc. Adv Nutr 9:625–636
10. USDA Food Composition Databases. United States Department of Agriculture Agricultural Research Service. https://ndb.nal.usda.gov/ndb/nutrients/index.
11. Dias MG, Olmedilla-Alonso B, Hornero-Méndez D, Mercadante AZ, Osorio C, Vargas-Murga L, Meléndez-Martínez AJ (2018) Comprehensive database of carotenoid contents in ibero-american foods. A valuable tool in the context of functional foods and the establishment of recommended intakes of bioactives. J Agric Food Chem 66:5055–5107
12. WMA (World Medical Association) (2013) Declaration of Helsinki—Ethical principles for medical research involving human subjects. . https://www.wma.net/policies-post/wma-declaration-of-helsinki-ethical-principles-for-medical-research-involving-human-subjects/
13. Council of Europe Convention on Human Rights and Biomedicine. https://www.coe.int/el/web/portal/full-news/-/asset_publisher/Dgh51iCGvfbg/content/the-convention-on-human-rights-and-biomedicine-celebrates-its-20th-anniversary.
14. UNESCO's Universal Declaration on the Human Genome and Human Rights. http://www.unesco.org/new/en/social-and-human-sciences/themes/bioethics/human-genome-and-human-rights/.
15. Regulation 2016/679 of the European Parliament and of the Council of 27 April 2016. https://eur-lex.europa.eu/legal-content/EN/TXT/?uri=celex%3A32016R0679.
16. Council of Europe (2016) Recommendation CM/Rec(2016)6 of the Committee of Ministers to member States on research on biological materials of human origin. . https://search.coe.int/cm/Pages/result_details.aspx?ObjectId=090000168064e8ff
17. Aggett P, Antonine JM, de Vries J, Gallagher A, Hendriks H, Kozianowski G, Meijer G, Richardson D, Rondeau V, Tweedie G, Welch R, Wittwer J (2010) Beyond PASSCLAIM—Guidance to substantiate health claims on foods. In Ates K (ed). ILSI Eur Report Series, 7–19
18. Welch RW, Antonine JM, Berta JL, Bub A, De VJ, Guarner F, Hasselwander FF, Jäkel M, Koletzko BV, Patterson CC, Richelle M, Skarp M, Theis S, Vidry S, Wodside JV (2011) Guidelines for the design, conduct and reporting of human intervention studies to evaluate the health benefits of foods. Br J Nutr 106(2):S3–S15
19. Lucey A, Heneghan C, Kiely ME (2016) Guidance for the design and implementation of human dietary intervention studies for health claim submissions. Nutr Bull 41:378–394
20. USDA-NIH (2018). https://www.nia.nih.gov/research/grants-funding/implementation-policies-human-intervention-studies.
21. CONSORT (Consolidated Standards of Reporting Trials). www.consort-statement.org
22. Deeks JJ, Dinnes J, D'Amico R, Sowden AJ, Sakarovitch C, Song F, Petticrew M, Altman DG (2003) Evaluating non-randomised intervention studies. Health Technol Assesess 7(27)
23. Hawkes C (2004) Nutrition labels and health claims: the global regulatory environment. WHO
24. Combs GF, Trumbo PR, McKinley MC, Milner J, Studenski S, Kimura T, Watkins SM, Raiten DJ (2013) Biomarkers in nutrition: new frontiers in research and application. Ann N Y Acad Sci 1278:1–10

25. Granado-Lorencio F, Blanco-Navarro I, Pérez-Sacristán B, Hernández-Álvarez E (2017) Biomarkers of carotenoid bioavailability. Food Res Int 99:902–916
26. Rutishauser IHE, Black AE (2002) Introducción a la Nutrición Humana. In: Nutr Soc
27. Granado-Lorencio F, Olmedilla-Alonso B (2003) Bioavailability of vitamins. In: Vaquero MP, García-Arias T, Carbajal A, Sánchez-Muñiz FJ (eds) Bioavailability of micronutrients and minor dietary compounds. Metabolic and technological aspets.. Rec Res Dev Nutr. Research Signpost, Kerala, India, pp 19–30
28. Tang G (2010) Bioconversion of dietary provitamin A carotenoids to vitamin A in humans. Am J Clin Nutr 91(suppl):1468S–1473S
29. Britton G, Khachik F (2009) Carotenoids in food. In: Carotenoids, vol. 5, chapter 3. Birkhäuser Verlag Basel

Chapter 28

In Vitro Digestion for Control and Monitoring of Food Effects in Relation to Micellarization Index of Carotenoids

Antonio Pérez-Gálvez and Javier Fontecha

Abstract

The multifactorial system that influences the availability of macronutrients, micronutrients, and phytoconstituents with relevant bioactivities has been assessed by in vitro digestion protocols, which have become an effective technique to support the health-through-food strategy. The measurement of the significance of food structure, food matrix, synergies and competitive effects, processing features, and even some physiological issues has created valuable scientific and technological information, and the development of harmonized protocols. Now it is possible to make further advances by applying this knowledge to obtain data regarding the potential availability of target compound(s) in the food source within a *standard* meal context. This protocol describes the measurement of the micellarization index of carotenoids from dietary rich sources with test meals accounting the effects of high vs. low fat and normo-, hypo-, and hypercaloric content.

Key words In vitro digestion protocol, Micellarization index, Carotenoids, Food effects, Bioaccessibility

1 Introduction

Bioaccessibility, defined as the amount of a compound that is released from its matrix in the gastrointestinal tract, becoming available for absorption, then absorbed by the enterocytes and finally turned into a bioactive form, is the ultimate concept applied to food science and nutrition to enhance the reliability in that the nutritional benefits are achieved. It is a critical issue in the design of food formulations with bioactive ingredients, because it considers the effects of the events taking place during digestion that are significant influencing factors for limiting or enhancing the in vitro bioaccessibility of target compound(s), and with the aim of predicting the outcome in vivo. Hence, straightforward procedures, that is, in vitro digestion protocols have been designed using a decision-making approach related with the kind of compound (s) which bioaccessibility is determined, the question that needs to

Manuel Rodríguez-Concepción and Ralf Welsch (eds.), *Plant and Food Carotenoids: Methods and Protocols*, Methods in Molecular Biology, vol. 2083, https://doi.org/10.1007/978-1-4939-9952-1_28,

be answered, or the mechanism to be figured out [1]. However, the existence of almost tailored procedures for undertaking specific questions made difficult the comparison of data from different laboratories to draw a comprehensive scenario of the lineal combination of critical factors affecting digestibility and absorption of food components. A significant effort has been made by the INFOGEST project [2], which aim was to define a consensus protocol easy to set up in any laboratory, with experimental conditions that should be applied to obtain a satisfactory grade of confidence in the results, to allow their comparison between different laboratories and to establish further improvements in the technique.

In vitro bioaccessibility of carotenoids has been a continuous topic of interest since the onset of this century. The findings obtained with in vivo approaches yielded a priceless legacy to conceive reliable in vitro digestion protocols, which measure the bioaccessibility of carotenoids under different experimental conditions. In the case of carotenoids as in other lipophilic compounds, bioaccessibility is often made equal to micellarization, that is, the technique determines the quantity of ingested carotenoids incorporated into mixed micelles and, consequently, ready for absorption. However, it should be noted that micellarization is not the single element required to measure bioaccessibility of carotenoids, but efficiency of the cellular absorption and first-pass metabolism are contributors to the complete bioaccessibility value. Hence, from the sequential steps involved in the digestion and absorption of carotenoids—release from food matrix, transfer to lipid droplets, emulsification and micellarization, and the subsequent routes of cellular absorption, trafficking of lipids and first-pass metabolism—the in vitro digestion protocols are designed to measure those factors affecting the four initial steps, while different protocols are required to estimate the efficiency of the others. Regarding the sort of factors that influence bioaccessibility of carotenoids, they are grouped within food structure, food matrix and processing features, physiological issues, and genetic and host-related aspects, which meaning has been reviewed recently [3, 4]. Again, the design of in vitro digestion protocols does not allow ascertaining the influence of all those factors. Thus, the impact of genetic and host-related variations is not accessible through this technique, while experimental conditions according to the gastrointestinal capability of different population groups (newborns and infants, the elderly, and patients with certain hepatobiliary disorders) are now included in the design of in vitro digestion protocols only recently [5, 6].

In this milieu, and although a considerable number of in vitro digestion protocols is available, the procedures published by Garrett et al. [7] and Reboul et al. [8] contain the essential knowledge to consistently develop the measurement of the micellarization of carotenoids, or to be tuned with other experimental conditions, as it has been recently discussed [1]. However, these protocols have been applied to single carotenoid food sources (green vegetables and oils, carrots, tomatoes, peppers, oranges, grains, dairy products, food emulsions, etc.) with the objective of gaining information of the factor(s) playing a role in the micellarization of carotenoids. This reductionist approach has been necessary to clearly correlate the observed effect(s) in the efficiency of the micellarization with the examined factor(s), or whether the latter has no influence in the former process. With this strategy a substantial amount of information has been obtained [9] to figure out several weighting factors in the micellarization of carotenoids, including food processing, technological treatments and cooking procedures, application of novel techniques, deposition in the food microstructures, texture, fiber networks and particle size, and amount and kind of lipids (preferably polar lipids). The next move forward could include a holistic assessment of the micellarization of carotenoids in the context of a standard meal or dietary pattern, according to the FDA guidelines for bioavailability and bioequivalence studies [10] and to the harmonized INFOGEST in vitro digestion protocol [2]. Hence, this protocol describes the measurement of the food effect in the micellarization of carotenoids considering normo-, hypo-, and hypercaloric meals to tabulate predictive data of the potential bioaccessibility of carotenoids from dietary food sources, a protocol that could be extended to other lipophilic bioactives and vitamins.

2 Materials

2.1 Standard Meals

1. The standard meals are designed following the guidelines of the report for Human Energy Requirements, which are based on sex BMI and PAL [11] (*see* **Note 1**). For a single meal with a caloric intake of 650–750 kcal (*see* **Note 2**), the macronutrient ratio is established according to the following reference intake ranges: >50% total carbohydrates, 15–30% total fat, and 10–20% total protein (percentage of total energy intake per meal). In addition, the goal of 400 g of fruits and vegetables a day is followed [12]. The design of high or low caloric meals is made applying the same reference intake ranges for macronutrients, and the goal of fruits and vegetables daily intake is followed but increasing or decreasing ca. 30% the total caloric intake in the meal, respectively. Accordingly, the design of high or low-fat meals is made increasing or decreasing the mean reference intake of total fat a 60%, keeping the reference intake of protein in both cases, and decreasing or increasing the reference intake of total carbohydrates, respectively [10]. The food composition tables published by Mataix [13] are taken as reference to calculate the caloric content and macronutrient

Table 1
Description of the meals to account the food effects in the micellarization index of carotenoids

	Caloric content			Fat content	
	Normo	Hypo	Hyper	Low	High
	Pasta (80 g)	Salad (250 g)[c,d]	Salad (320 g)[b,c]	Pasta salad[c,e] (305 g)	Salad (320 g)[b,c]
	Tuna fish (50 g)[a]	Zucchini omelet (egg 55 g, zucchini 200 g, olive oil 5 g)	Bacon (40 g)	York ham (30 g)	Bacon (80)
	Fried tomato (60 g)	Bread (30 g)	Blue cheese (25 g)	Olive oil (8 g)	Egg (88 g)
	Salad (320 g)[b,c3]	Apple (150 g)[c]	Chips potatoes (50 g)	Bread (60 g)	
	Olive oil (5 g)		Bread (50 g)	Watermelon (250 g)[c]	
	Bread (50 g)		Orange juice (100 mL)[c]	Pear (150 g)[c]	
	Orange (200 g)[c]				
Caloric content (kcal)	700	470	900	650	688
Carbohydrates (%)	60	68	52	78	30
Fats (%)	20	19	30	8	50
Proteins (%)	20	13	18	14	20

[a]Fish could be replaced by other meat products that provides the same carbohydrates, fat and protein content
[b]This salad contains iceberg lettuce (100 g), tomato (100 g), onion (30 g), carrot (50 g), pepper (30 g), olive oil (10 g)
[c]This item(s) could be substituted/composed by another rich source of carotenoids
[d]This salad contains iceberg lettuce (150 g), tomato (50 g) onion (30 g), cucumber (15 g), olive oil (5 g)
[e]This salad contains pasta (125 g), tomato (100 g), corn (30 g), carrot (50 g)

composition of the meals. To estimate the carotenoid intake in the designed meals, the European carotenoid database is used [14]. Table 1 includes example meals that we use routinely in our laboratory to measure the food effects in the micellarization index of carotenoids and lipophilic vitamins (*see* **Note 3**).

2.2 Preparing Simulated Gastric Fluids

1. Electrolyte stock solution: 0.625 M KCl, 0.625 M KH_2PO_4, 1.25 M $NaHCO_3$, 2.5 M NaCl, 0.187 M $MgCl_2$ $6H_2O$), 0.625 M $(NH_4)_2CO_3$, and 0.3 M $CaCl_2$ $2H_2O$ (*see* **Note 4**). Transfer each amount of weighted salt to 1 L graduated cylinders with 100 mL water. Make up to 1 L with water and mix. Store at 4 °C.

2. Simulated salivary electrolyte stock solution: Transfer 15.1 mL of 0.625 M KCl, 3.7 mL of 0.625 M KH_2PO_4, 6.8 mL of 1.25 M $NaHCO_3$, 0.5 mL of 0.187 M $MgCl_2$ $6H_2O$, and 0.06 mL of 0.625 M $(NH_4)_2CO_3$ to a 500 mL graduated cylinder with 100 mL water. Make up to 500 mL and mix. Store at −20 °C.
3. Simulated gastric electrolyte stock solution: Transfer 6.9 mL of 0.625 M KCl, 0.9 mL of 0.625 M KH_2PO_4, 12.5 mL of 1.25 M $NaHCO_3$, 11.8 mL 2.5 M NaCl, 0.4 mL of 0.187 M $MgCl_2$ $6H_2O$, and 0.5 mL of 0.625 M $(NH_4)_2CO_3$ to a 500 mL graduated cylinder with 100 mL water. Make up to 500 mL and mix. Store at −20 °C.
4. Simulated intestinal electrolyte stock solution: Transfer 6.8 mL of 0.625 M KCl, 0.8 mL of 0.625 M KH_2PO_4, 42.5 mL of 1.25 M $NaHCO_3$, 9.6 mL 2.5 M NaCl, and 1.1 mL of 0.187 M $MgCl_2$ $6H_2O$ to a 500 mL graduated cylinder with 100 mL water. Make up to 500 mL and mix. Store at −20 °C.
5. Stock solution of 1500 U of α-amylase per mL of simulated salivary electrolyte solution (*see* **Notes 5** and **6**).
6. Stock solution of 25,000 U of porcine pepsin per mL of simulated gastric electrolyte stock solution (*see* **Notes 5** and 7).
7. Stock solution of 800 U of pancreatin from porcine pancreas per mL of simulated intestinal electrolyte stock solution (*see* **Notes 5** and **8**).
8. Stock solution of 0.16 mmol of porcine bile extract per mL of simulated intestinal electrolyte stock solution (*see* **Notes 9** and **10**).
9. Laboratory blender with glass container (1–2 L capacity) with speed control and timer.
10. 100 mL GL45 clean glass laboratory bottles.
11. Magnetic stirrer with 4–6 stirring positions. Cross type PTFE magnetic stir bars 25 mm diameter.
12. Incubator chamber with adjustable temperature control to 0.1 °C and internal power sockets for stirring unit(s).

2.3 Measurement and Adjustment of pH during In Vitro Digestion

1. Adjustment of pH values at the initial and during the gastric and intestinal stages is made with 1 M HCl or 1 M NaOH.
2. The pH measurements should be performed with a pH electrode suitable for difficult samples containing colloids, emulsions, creams, or very viscous samples (*see* **Note 11**).

2.4 Assessment of Lipolysis Efficiency

1. Folch solution, chloroform–methanol 2:1 (v/v), is made fresh each time.
2. 1 M HCl, 1 M NaCl, and H_2O–H_2SO_4 (1:1; v/v).

3. Hexane–diethyl ether–glacial acetic acid 80:20:1 (v/v/v).
4. Silica gel coated TLC plates (10 × 20 cm, 0.25 mm, Si G60).
5. 200 μL syringe for application of sample on TLC plates.
6. TLC heater.
7. Scanner and software for image processing to calculate optical density.

2.5 Isolation of the Micellar Fraction

1. Benchtop centrifuge, temperature 4 °C and equipped with rotors for low and high-speed centrifugation.
2. Centrifuge tubes of 15 mL and 50 mL capacity.
3. Nylon 25 mm (i.d.) × 0.22 μm pore size filters.

2.6 Extraction and Analysis of the Carotenoid Fraction from the Meal Puree and the Micelles Fraction

1. Benchtop centrifuge, temperature 4 °C.
2. Centrifuge tubes of 50 mL capacity.
3. Diethyl ether, hexane, and NaCl (10%, w/v).
4. Nylon 13 mm (i.d.) × 0.22 μm pore size filters.
5. LC-MS/MS platform with UV-visible detector and APCI source coupled to UHR-TOF mass analyzer.
6. C30 analytical column (3 μm particle size, 25 × 0.46 cm i.d.).
7. LC grade solvents for chromatographic separation: methyl-*tert*-butyl ether, methanol and water.
8. Standard stock solutions of astaxanthin, β-carotene, β-cryptoxanthin, canthaxanthin, lutein, lycopene, and violaxanthin, which are prepared at a concentration of 25 mg L^{-1} following the procedures described in [15]. Once the exact concentration is determined, working stock solutions for external calibration curves are prepared at 5 concentration levels ranging from 0.15 to 10.0 mg L^{-1}.

3 Methods/Study Design

3.1 Standard Meals

1. Once the meal is ready to eat, grind it finely in an electric food mixer until a puree is obtained.
2. Keep the puree at 40 °C until the application of the in vitro procedure. Take three portions (5 g) to determine the carotenoid content in the puree.

3.2 Three-Stage In Vitro Digestion Procedure and Isolation of the Micellar Fraction

1. The procedure is based on Minekus et al. [2]. Mix 30 g of the puree with 21 mL of simulated salivary electrolyte stock solution, and place 3 aliquots (17 mL) of the puree in 100 mL GL45 clear glass laboratory bottles (*see* **Note 12**) with a cross type magnetic bar. Mix each aliquot with 1 mL of salivary α-amylase stock solution (1500 U/mL), 50 μL of 0.3 M

$CaCl_2$, and 1.95 mL of water. Incubate the mixture at 37 °C (20 mL) for 10 min or 2 min with magnetic stirring at 1000 rpm.

2. Mix each aliquot (oral bolus) of the preceding step with 15 mL of simulated gastric electrolyte stock solution, 3.2 mL of porcine pepsin solution and 10 μL of 0.3 M $CaCl_2$. Adjust the pH of the mixture to 3.0 with 1 M HCl (*see* **Note 13**) and make the final volume equal to 40 mL. Place the bottle in the incubator chamber at 37 °C for 2 h with magnetic stirring at 1000 rpm.
3. The gastric chyme is mixed with 22 mL of simulated intestinal electrolyte stock solution, 10 mL of pancreatin solution, 5 mL of fresh bile, and 80 μL of 0.3 M $CaCl_2$. Adjust the pH of the mixture to 7.0 with 1 M NaOH (*see* **Note 13**) and make the final volume equal to 80 mL. Place the bottle in the incubator chamber at 37 °C for 2 h, with magnetic stirring at 1000 rpm.
4. The intestinal chyme of each bottle is divided in 40 mL aliquots and transferred to 50 mL centrifuge tubes. Each bottle is washed with two portions of 5 mL of water and added to the centrifuge tubes to make the final volume equal to 50 mL. Isolate the aqueous suspension from the undigested and solid residues by low speed centrifugation at 4000 × *g* for 20 min. Transfer 10 mL aliquots of the aqueous suspension to 15 mL centrifuge tubes and apply the high-speed centrifugation step at 15000 × *g* for 8 min. Isolate the micellar fraction by filtering the clarified suspension through a 0.22 μm Nylon filter.

3.3 Extraction of Carotenoids from the Puree and Micelles

1. The extraction method is based on Ríos et al. [16]. Mix the puree (5 g) or the micellar fraction (5 mL) with diethyl ether (5 mL) and hexane (2 mL) and mix in a vortex for 2 min. Add 5 mL of NaCl (10%, w/v) and mix in a vortex for 2 min. Centrifuge the sample at 4000 × *g* for 5 min.
2. Isolate the organic layer in a rotatory flask and repeat the extraction procedure (avoiding the addition of NaCl solution). After centrifugation, isolate the organic layer again and evaporate the combined organic extracts to dryness in a rotatory evaporator at 25 °C.
3. Dissolve the extract in 0.5 mL of hexane and filter it through a nylon 13 mm (i.d.) × 0.22 μm pore size filter. Store the sample at −20 °C until HPLC-MS/MS analysis (within 1 week).

3.4 LC-MS/MS Analysis

A great variety of LC systems, MS hardware configurations and software tools is available for LC-MS/MS analysis of carotenoids, including column, solvent composition, isocratic or gradient elution, flow rate and postcolumn operation, APCI ionization protocol, parameters and arrangement of the MS analyzer, and conditions for MS/MS ionization and scan of product ions. The following description is the practice routinely used in our laboratory [17].

1. A binary solvent system consisting of methyl-*tert*-butyl ether–methanol–water 85:15:4 (solvent A) and methyl-*tert*-butyl methyl–methanol–water 7:90:3 (solvent B) is used at a flow rate of 1 mL min^{-1}.
2. Carotenoid separation is performed on a C30 analytical column (3 μm particle size, 25 × 0.46 cm i.d.). Injection volume is 20 μL per sample.
3. LC separation is performed using the following gradient setting: hold at 100% A for 10 min, then gradient from 100 to 50% A in 30 min, 50 to 100% B in 10 min, 100% A in 5 min, and finally isocratic 100% A for 5 min for reequilibrating column.
4. A split postcolumn of 0.4 mL min^{-1} is directly introduced on the APCI source, which is operated in positive mode. The *m/z* scan in the UHR-TOF is in the 50–1200 D range. MS data are acquired in broad band Collision Induced Dissociation mode to obtain MS and MS/MS spectra simultaneously.
5. Data evaluation is carried out to identify the carotenoids by using an *in-house* mass database created *ex professo* that contains the monoisotopic masses, elemental composition and, optionally, the retention time and characteristic product ions for 360 carotenes, xanthophylls and xanthophyll esters. The process is represented in Fig. 1. High-resolution mass spectrometry measurements are completed based on mass accuracy and in combination with the isotopic pattern. The characteristics of

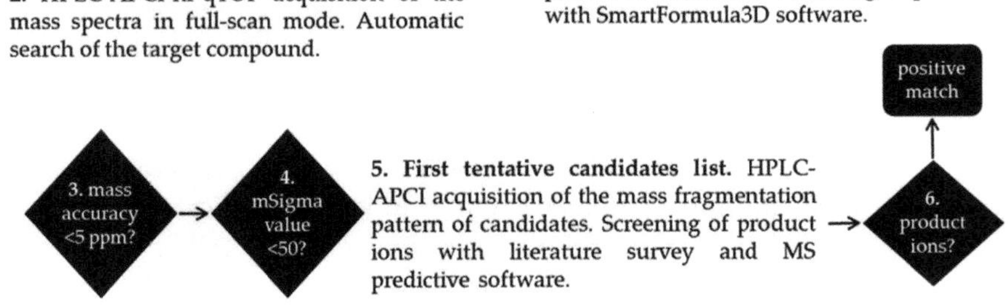

Fig. 1 Scheme of the process for data evaluation after acquisition of HPLC-APCI-qTOF spectra

experimental mass spectrum and the MS^2 data are compared with the data available in the literature for carotenoid identification in food samples, as well as with authentic carotenoid standards isolated from their natural sources. For carotenoid quantification, external calibration curves are applied for individual free carotenoids, while xanthophyll esters are quantified with the calibration curves of their corresponding free xanthophyll counterparts.

3.5 Assessment of Lipolysis Efficiency

As the carotenoids belong to the lipid profile of foods, they should incorporate into mixed micelles to be accessible to intestinal cells for absorption. Micellarization is the outcome of a complex process where several parameters are key players, but particularly the action of digestive secretions that include enzymes and bile salts impacts the efficiency of this stage. Therefore, to obtain reliable results is critical to measure the efficiency of the lipolysis in the chyme obtained after the in vitro digestion protocol.

1. Add 1 g of chyme to a 15 mL centrifuge tube and mix with 5 mL of the Folch solution, cap and mix for 10 min. Mixture could be stored at −20 °C at this point if required.
2. Add 1 mL of 1 M NaCl to the tube and mix for 10 min.
3. Centrifuge for 10 min at 1000 *g* at 4 °C.
4. Discard the aqueous upper layer and transfer the solvent phase (*see* **Note 14**) to a rotatory flask and evaporate to dryness. Add 0.5 mL of chloroform to the dry lipid extract and vortex to dissolve.
5. Spot 200 μL of the lipid extract onto a TLC plate and develop the TLC plate with the elution solvent in the TLC chamber until solvent reaches 4/5 up plate.
6. Remove the TLC plate from the chamber and allow solvents to evaporate at room temperature.
7. Spray the TLC plate with H_2O–H_2SO_4 (1:1, v/v) solution and place it in a plate heater at 180 °C for 15 min. Subsequently, let the TLC plate to reach ambient temperature and scan it. Use an image processing software to analyze one-dimensional images and obtain the grey scale intensity values (*see* **Note 15**).

3.6 Calculations and Statistical Analysis

1. The micellarization index is expressed as percentage value regarding the total initial carotenoid content in the meal and the total carotenoid content in the micellar fraction.
2. A minimum of three replicates per in vitro digestion process is required to determine whether the observed effect (if it is so) is statistically significant.

3. The efficiency of the lipolysis of triacylglycerides is determined by measuring the grey intensity of bands corresponding to the intact triacylglycerides, mono- and diacylglycerides and free fatty acids at the beginning and end of the digestion protocol (*see* **Note 15**).

4 Notes

1. The meals are designed for a 40–49 years old male, 1.70 m high and 65 kg weight, 18.5–25 kg/m^2 (body mass index, BMI) and a moderately activity level (Physical Activity Level, PAL = 1.70–1.99). With this data the caloric requirement is 2450 kcal.
2. The total daily caloric intake (2450 kcal) is distributed 20–25% at breakfast, 15–20% at mid-day/snack, 30% at lunch, 25–30% at dinner.
3. Pasta is cooked in boiling water (1 L per 100 g of pasta) for 12 min. Tuna fish is fried in iron at 120 °C for 3 min. Bacon strips are fried in iron at 120 °C for 3 min.
4. This salt is not added to the simulated electrolyte stock solutions as precipitation may occur.
5. Enzymes solutions need to be used on the same day. Weigh the enzyme carefully because it tends to disperse. The amount of enzyme to be used for the digestion is recalculated according to the activity.
6. We measure the α-amylase enzyme activity with an Alpha-amylase microplate assay kit.
7. We measure the pepsin enzyme activity with a Pepsin microplate assay kit.
8. Pancreatin is a mixture of several digestive enzymes produced by the exocrine cells of the porcine pancreas. However, we measure the lipase enzyme activity to recalculate the amount of pancreatin required.
9. We measure the total bile acid content in the bile extract with an assay kit.
10. Particle size distribution of the bile extract may considerably change among lots as it is depicted in Fig. 2. This may affect the efficiency of the micellarization. Batch 1 is a fine powder, while batch 2 should be ground in a mortar to obtain the same particle size distribution.
11. The pH probe we use in our lab is equipped with temperature sensor and designed for low conductivity samples. Use cleaning solution or mild soap or detergent, and a toothbrush to

Fig. 2 Different particle size distributions of bile extracts purchased from the same dealer

remove oil/fat and chyme residues from the pH bulb after pH measurement.

12. The use of glass bottles is mandatory because intestinal enzymes bind to the plastic materials, so that efficiency of the enzymatic activity is reduced.
13. Determine the amount of 1 M HCl or 1 M NaOH required adjusting the pH value at the gastric and intestinal phase, respectively, in a test experiment so that their addition can be subsequently made quickly. Monitor the pH values at the beginning and intermediate periods of both digestion phases in one of the replicates and adjust the pH value if necessary.
14. Be careful not to disturb the pellet or the solvent phase.
15. We use the free software ImageJ that performs density or grey scale measurement/calibration [18]. Optical density step tablets to calibrate an image are commercially available from Tiffen (Kodak) and Stouffer. The densitometry on bands corresponding to the triacylglycerides, mono- and diacylglycerides, and free fatty acids at the beginning of the digestion phase is the reference value for obtaining the lipolysis efficiency.

Acknowledgments

The financial support of the Ministerio de Economía y Competitividad (MINECO-CICYT, Spanish Government, project AGL2017-87884-R) is gratefully acknowledged. Antonio Pérez-Gálvez is a tenured scientist at the Spanish National Research Council (CSIC).

References

1. Arranz E, Corredig M, Guri A (2016) Designing food delivery systems: challenges related to the in vitro methods employed to determine the fate of bioactives in the gut. Food Funct 7:3319–3336
2. Minekus M, Alminger M, Alvito P, Balance S, Bohn T, Bourlieu C, Carriere F, Boutrou R, Corredig M, Dupont D, Dufour C, Egger L, Golding M, Karakaya S, Kirkhus B, Le Feunteun S, Lesmes U, Macierzanka A, Mackie A, Marze S, McClements DJ, Ménard O, Recio I, Santos CN, Singh RP, Vegarud GE, Wickham MSJ, Weitschies W, Brodkorb A (2014) A standardized static in vitro digestion method suitable for food—an international consensus. Food Funct 5:113–1124
3. Bohn T, McDougall GJ, Alegría A, Alminger M, Arrigoni E, Aura AM, Brito C, Cilla A, El SN, Karakaya S, Martínez-Cuesta MC, Santos CN (2015) Mind the gap-deficits around our knowledge on aspects impacting bioavailability of phytochemicals and their metabolites. Mol Nutr Food Res 59:1307–1323
4. Desmarcherlier C, Borel P (2017) Overview of carotenoid bioavailability determinants: from dietary factors to host genetic variations. Trends Food Sci Nutr 69:270–280
5. Levi CS, Alvito P, Andrés A, Assuncao R, Barberá R, Blanquet-Diot S, Bourlieu C, Brodkorb A, Cilla A, Deglaire A, Denis S, Dupont D, Heredia A, Karakaya S, Giosafatto CVL, Mariniello L, Martins C, Ménard O, El SN, Vegarud GE, Ulleberg E, Lesmes U (2017) Extending in vitro digestion models to specific human populations: perspectives, practical tools and bio-relevant information. Trends Food Sci Technol 60:52–63
6. de Oliveira SC, Deglaire AM, Ménard O, Bellanger A, Rousseau F (2016) Holder pasteurization impacts the proteolysis, lipolysis and disintegration of human milk under in vitro dynamic term newborn digestion. Food Res Int 88:263–275
7. Garrett DA, Failla ML, Sarama RJ (1999) Development of an in vitro digestion method to assess carotenoid bioavailability from meals. J Agric Food Chem 47:4301–4309
8. Reboul E, Richelle M, Perrot E, Desmoulins-Malezet C, Pirisi V, Borel P (2006) Bioaccessibility of carotenoids and vitamin E from their main dietary sources. J Agric Food Chem 54:8749–8755
9. Pérez-Gálvez A (2019) In vitro digestion protocols. The benchmark for estimation of in vivo data. In: Mercadante A (ed) Carotenoid esters in foods. Royal Society of Chemistry, London. (in press)
10. U.S. Food and Drug Administration, Center for Drug Evaluation and Research (2002). Guidance for industry: food-effect bioavailability and fed bioequivalence studies. Office of Training and Communications, Division of Drug Information HFD-240 Rockville Maryland 20857
11. Food and agriculture organization of the United Nations/World Health Organization/United Nations University (2011) Human energy requirements
12. Joint WHO/FAO Expert Consultation on Diet, Nutrition and the Prevention of Chronic Diseases (2002). Diet, Nutrition and the prevention of chronic diseases: report of a joint WHO/FAO expert consultation WHO technical report series 916
13. Mataix-Verdú J (ed) (2011) Tabla de composición de alimentos. Granada, EUG
14. O'Neill ME, Carroll Y, Corridan B, Olmedilla B, Granado F, Blanco I, van den Berg H, Hiniger I, Rousell AM, Chopra M, Southon S, Thurnham DI (2001) A European carotenoid database to assess carotenoid intakes and its use in a five-country comparative study. Br J Nutr 85:499–507
15. Ríos JJ, Xavier AAO, Díaz-Salido E, Arenilla-Vélez I, Jarén-Galán M, Garrido-Fernández J, Aguayo-Maldonado J, Pérez-Gálvez A (2017) Xanthophyll esters are found in human colostrum. Mol Nutr Food Res 61:1700296
16. Britton G, Liaaen-Jensen S, Pfander H (eds) (1995) carotenoids. Volume 1A: isolation and analysis. Birkhäuser Verlag. Baseline
17. Pérez-Gálvez A, Sánchez-García A, Garrido-Fernández J, Ríos JJ (2018) MS tools for a systematic approach in survey for carotenoids and their common metabolites. Arch Biochem Biophys 650:85–92
18. Burger W, Burger MJ (eds) (2010) Digital image processing: an algorithmic introduction using java. Springer, Berlin

Chapter 29

Evaluation of Carotenoids Protection Against Oxidative Stress in the Animal Model *Caenorhabditis elegans*

Patricia Martorell, Silvia Llopis, José Vicente Gil, Salvador Genovés, Daniel Ramón, Lorenzo Zacarías, and María Jesús Rodrigo

Abstract

The nematode *Caenorhabditis elegans* is a versatile and powerful model organism for animal experimental research and, despite being an invertebrate, displays remarkably similar molecular bases and conserved cellular pathways to those of humans. Oxidative stress is an etiological factor that influences numerous diseases, degenerative processes and aging. *C. elegans* has revealed as an opportune and feasible organism to investigate the antioxidant effects of different bioactives or complex food matrices, and a number of protocols have been developed by using different oxidative stressors. Carotenoids are recognized as quenchers and scavengers of reactive oxygen species, and many of their related health benefits attributed in the diet are tightly linked to their antioxidant properties. In this chapter, we report a simple and rapid assay to evaluate the protection capacity of pure carotenoids or complex carotenoid extracts against oxidative stress in the model system *C. elegans*. The protocol describes a representative feeding experiment by adding carotenoids to the nematode growth medium and after an incubation period, the *C. elegans* populations fed with carotenoids are exposed to an acute oxidative stress by using H_2O_2 as oxidative agent. The protection against oxidative stress is evaluated as the survival rate of the nematodes fed with the carotenoid prior to receiving oxidative treatment compared with the survival rate of control nematode population. In order to confirm the carotenoid intake by the nematodes during the feeding experiment a bioassimilation experiment is also reported.

Key words Antioxidant assay, Bioactive ingredients, *Caenorhabditis elegans*, Carotenoids, Experimental animal model, Oxidative stress

1 Introduction

The free-living soil nematode *Caenorhabditis elegans* was adopted in the 1960s as model organism for developmental biology [1] and since then its use as experimental system for many human pathological disorders and as a useful platform to investigate food ingredient or drug bioactive effects, has been increasing and becoming more relevant [2]. *C. elegans* offers many advantages for laboratory manipulation such as small size (1 mm) and transparent body,

Manuel Rodríguez-Concepción and Ralf Welsch (eds.), *Plant and Food Carotenoids: Methods and Protocols*, Methods in Molecular Biology, vol. 2083, https://doi.org/10.1007/978-1-4939-9952-1_29,

differentiate cells and organs, large number of progeny (approximately 300 by self-fertilization), short generation time (2–3 days) and well-defined lifespan (2–3 weeks). More interestingly, its genome is fully sequenced and genetic manipulation tools and mutant collections are available. Despite significant differences between *C. elegans* and animal vertebrates and mammals, 60–80% of *C. elegans* genes have homologues in humans [3]. Another important advantage to consider is that does not require committee approvals for experimentation since it is not legally defined as an animal according to European Union, Canada, or USA regarding its use in laboratory tests [4–6]. One of the most interesting characteristics of this nematode is its use for testing the bioactivity of food ingredients or complex food matrices [2]. In laboratory conditions *C. elegans* is grown on NGM (nematode growth medium) plates which contain *E. coli* bacteria (OP50 strain), the major food for the nematode. The compounds to be tested can be applied or produced by the bacteria that are ingested, directly spread onto the surface of plates, or added to the liquid medium when preparing the plates [7]. The ingredients can be taken up by three different processes: ingestion, through the cuticle, and via exposed sensory neuronal cilia [3]. The ingestion of a compound depends on the nematode satiety and availability in the medium being difficult to control and it is a relatively slow process. Once the compounds enter the intestinal lumen, they absorbed by intestinal cells and distributed throughout the body [3]. Multiple models based on *C. elegans* have been developed to study the effect of food ingredients and drugs in human diseases and degenerative processes such as cancer [8], kidney diseases [9], neurodegenerative disorders including Alzheimer's disease and other tauopathies [10], obesity [2, 11], mitochondrial disorders [12], diabetes [13], or aging [2]. Most of these diseases, as well as the aging process, have in common oxidative stress as an important contributing factor [13].

Several models are available for screening and testing functional ingredients, being murine models and cell cultures the most widely used although, due to their specific characteristics and advantages, are gaining acceptance other models such as zebrafish or the nematode *C. elegans.* Cell cultures with human cells offer, as a main advantage, a system with human biology and genetics, but in a context far from the normal physiological situation with cell lines originated from cancerous tissue growing outside the complete organism. The murine model is the in vivo system most used in preclinical research. Being a mammal is quite similar to humans, so it increases confidence in the extrapolation of the results but, as a counterpart, has great ethical issues and, due to its more complicated handle conditions, it is harder to use it in a routine screenings or in certain experiments as related to aging [14]. On the other hand, the use of *C. elegans* has advantages including rapid screenings with low cost, and could be very useful to obtain valuable information prior to the design of murine model experiments or clinical validations. Therefore, studies using *C. elegans* have significantly contribute to the knowledge into the molecular mechanisms involved in different pathological disorders and responses to stresses by facilitating the generation of results that could hardly been obtained using other more complex biological models.

Oxidative stress is a determinant factor that influences numerous human diseases including diabetes, cardiovascular disease, cancer, neurodegenerative diseases and aging. Oxidative stress is caused by an imbalance between the production of reactive oxygen or nitrogen species and the cellular defense mechanisms that must prevent progressive oxidative damage of cellular components such as DNA, lipids, or proteins. The nematode, despite being an invertebrate animal, is an excellent model to study oxidative stress resistance and aging processes due to evolutionary conserved pathways. The beneficial antioxidant effects of food ingredients can be easily studied in *C. elegans* and conveniently related to defense mechanisms and signaling pathways [11, 15, 16]. The approaches to carry out the study of oxidative stress in *C. elegans* are very diverse, varying the system of generating oxidative stress depending on the objective of the study, the type of oxygen reactive species that need to be generated or the subcellular compartment implicated. It is possible to study the resistance of worms to different oxidant agents such as hydrogen peroxide (H_2O_2), menadione, sodium arsenite, or paraquat, among others [11, 16–18], or by measuring reactive oxygen species (ROS) accumulation [16].

Carotenoids in photosynthetic organisms as well as in other carotenoids-producing organisms play a crucial role protecting the cellular components against ROS acting as quenchers and scavengers [19]. In humans, besides the role of specific carotenoids as provitamin A and their participation in visual function, they are modulators of inflammation-related mechanisms [20–22]. Many of the related health benefits attributed to carotenoids in human diet have been tightly linked to their antioxidant properties as ROS scavengers and quenchers, and also as electrophiles that enhance endogenous antioxidant systems [20]. Therefore, many efforts are focused on the evaluation of carotenoids against oxidative stress by using in vivo and in vitro systems, and *C. elegans* has been shown as a powerful model organism since contains many key components related to oxidative stress networks common to those in mammals [13] This animal system provides a rapid and effective platform to test the bioactivity of carotenoids in pure form, in complex mixtures or in food matrices, and further molecular and biochemical analysis of nematodes fed with carotenoids and upon stress challenge permit the identification of cellular targets and processes, and molecular mechanisms involved in the carotenoid response and protection against oxidative stress. Interestingly, as in other animal models, it has been demonstrated in *C. elegans* a positive connection between lifespan and stress-resistance for a variety of stresses, including oxidative stress [23]. Thus, an enhanced stress resistance may also result in an extended lifespan.

Several studies have used *C. elegans* for testing antioxidant potential of compounds such as selenite [24], cocoa powder [11], and tyrosol [25]. The protective effect of specific carotenoids or food matrices enriched in carotenoids against oxidative stress has been also assessed using *C. elegans* as experimental animal model. In a representative experiment, the nematodes are fed with NGM or NGM supplemented with a certain concentration of the test compound (e.g., carotenoid or carotenoid extract) and subsequently subjected to oxidative stress induced by addition to the nematode culture medium (solid or liquid) of an oxidant stressor [11, 16–18]. In all of the assays there is an increase of ROS levels in the nematode to a point where survival is decreased. After a period of incubation with the stressor, the survival rate of the nematodes fed with the carotenoid prior to receiving oxidative treatment (acute or chronic) is compared with the survival rate of control nematode population. To study the protection capacity of a specific compound (e.g., pure carotenoid or food matrix carotenoid-enriched) it is advisable, if possible, to evaluate a wide range of concentrations of the test-compound in order to perform a dose-response experiment. Carotenoid concentrations below the threshold level to accumulate and reach the specific tissue or cellular targets may not enhance resistance to ROS but elevated doses may also fail due to a prooxidant effect. It should be keep in mind that the results obtained are highly influenced by the type of ROS, rate of oxidative stress exposure, and the target subcellular compartment affected by the specific stressor [26, 27]. Moreover, it is also possible that nematodes sensitive to oxidative stress at one age are resistant to the same agent at a different age. Therefore, when comparing results obtained in different trials and laboratories, all these factors to design the oxidative stress experiment have to be considered and clearly report all conditions and parameters.

By using different oxidant stressors and procedures, the increased resistance to oxidative stress has been reported for several carotenoids such as astaxanthin [28, 29], lutein [30], β-carotene, lycopene, and β-cryptoxanthin [31]. This type of *in vivo* antioxidant assays is not only restricted to evaluate pure carotenoids or mixtures but can also be used for testing the protection of complex food matrices as orange juice. A successful example was reported for sweet oranges with increased levels of β-carotene obtained through metabolic engineering which showed 20% higher antioxidant effect compared to control juice [32].

In this chapter we report in the animal model *C. elegans* a simple and quick assay to evaluate the protection of pure carotenoids or carotenoid extracts against an acute oxidative stress by using H_2O_2 as oxidant agent. The protocol also includes the specific procedure to assess the carotenoid/s intake by the nematode by using an HPLC-PAD system. This protocol can be particularly useful for screening a high number of samples and to compare the protection capacity between them. Furthermore, the procedures can easily be adapted to evaluate the effect of carotenoid supplementation or enrichment in complex food matrices.

2 Materials

Use analytical or, when indicated, HPLC grade reagents, and ultrapure water (MilliQ, prepared by purifying deionized water). Work in the fume cupboard when using organic solvents. For manipulation of bacteria and nematodes work with sterile material and conditions (e.g., laminar flow cabinet). Carotenoids solutions and extracts should be maintained under dim light conditions, and store solutions and dried extracts under nitrogen atmosphere to prevent carotenoid oxidation.

2.1 Caenorhabditis elegans *Cultivation and Manipulation*

1. *C. elegans* strain N2, Bristol (wild-type) is provided by the Caenorhabditis Genetics Center (University of Minnesota, Minneapolis, MN, USA). *C. elegans* is maintained in the laboratory on Nematode Growth Medium (NGM) agar(NaCl, agar, peptone, 5 mg/mL cholesterol, 1 M KPO_4 buffer pH 6.0, 1 M $MgSO_4$), aseptically poured into petri dishes, at 20 °C. Plates are seeded with an *Escherichia coli* OP50 (University of Minnesota, *C. elegans* Genetics Center) bacteria lawn as food source.
2. Transferring nematodes grown on NGM plates is performed using a dissecting stereomicroscope, picking single animals with a worm picker (a platinum wire) which is flamed between transfers to avoid contaminations.

2.2 *Reagents, Solutions, and Buffers*

1. NGM solution and preparation of NGM petri dishes (60 × 15 mm) with *E. coli* OP50. For 1 L of NGM solution, add 0.75 g Bacto peptone, 0.9 g NaCl, 5.1 g Bacto agar, 0.3 mL cholesterol solution (prepared by dissolving 1 g of cholesterol in 200 mL of 100% ethanol, not autoclaved and stored at room temperature). Add distilled water to 292 mL and autoclave at 120 °C, 20 min. After autoclaving, add 0.3 mL of 1 M $CaCl_2$, 0.3 mL of 1 M $MgSO_4$ and 7.5 mL of 1 M K-phosphate (pH 6). Keep it at 60 °C until preparing the dishes. Pour 10 mL of NGM solution into each petri dish and leave at room temperature from several hours to overnight. Make an overnight culture of OP50 with LB and add 50 μL of the OP50 solution to each NGM dish and leave overnight at room temperature.

2. M9 buffer is used to recover, washing and transfer high number of worms, eggs or larvae. For 1 L of buffer, add 3 g KH_2PO_4, 6 g Na_2HPO_4, 5 g NaCl, 1 mL 1 M $MgSO_4$. Sterilize by autoclaving.
3. Basal S medium is used for the oxidative stress. For 1 L, add 5.85 g NaCl, 17 g agar. After autoclave, add 50 mL of 1 M K-phosphate buffer (pH 6) and 1 mL cholesterol (prepared by dissolving 1 g of cholesterol in 200 mL of 100% ethanol, not autoclaved and stored at room temperature).
4. Liquid nitrogen.
5. Hydrogen peroxide 30% stock for acute oxidative stress. Store stock at 4 °C.
6. Nitrogen gas.
7. Micropipettes, Hamilton syringes, 1.5 mL Eppendorf type vials, 2 mL cryovials, 50 mL tubes, QS 1 mL cuvettes and platinum wire (for nematode manipulation).

2.3 Equipment

1. Stereomicroscope equipped with a transmitted light source (e.g., Motic SMZ168) to score animal features like size or motility. View the nematodes directly on plates on the stage of the stereomicroscope.
2. UV-Vis spectrophotometer with cuvette port (e.g., SPECTROstar Nano, BMG Labtech) to determine the concentration of the carotenoid in a solution or extract.
3. Growth chamber at 20 °C.
4. Laminar flow cabinet.
5. Centrifuge (for 1.5 mL vials and 50 mL tubes) and vortex.
6. Gas extraction cabinet to work with organic solvents.

3 Methods

3.1 Preparation of Pure Carotenoid Solution (C) or Carotenoids Extract (CE) to Assay

The first step in this procedure is to obtain a dried aliquot of C or CE that will be dissolve in a suitable solvent and volume immediately before is added to the NGM dishes. The amount of C or CE necessary to perform a representative oxidative stress assay can range between 10 and 50 μg (*see* **Note 1**). The source of the C or CE to be tested can be commercial (*see* **Note 2**) or previously purified from any food or biological source (*see* **Note 3**).

1. Prepare a stock solution of C or CE using appropriate organic solvents (e.g., chloroform, dichloromethane, methanol, or acetone) to reach a final concentration between 100 and 300 ng/mL determined by spectrophotometer (*see* **Note 4**).

2. Take out from the C or CE stock solution the total amount that will be necessary to perform the whole experiment, place in a 1.5 mL vial and completely dried under nitrogen stream (*see* **Note 5**).
3. Store the dried C or CE aliquot at −20 °C until preparing the NGM dishes supplemented with C or CE (*see* **Note 6**).

3.2 Preparation of NGM Petri Dishes with E. coli OP50 Supplemented with C or CE for Feeding Experiments

We recommend preparing the NGM dishes supplemented with C or CE in the same day of the feeding experiment.

1. In exploratory experiments we recommend to perform a dose-response experiment with 5–6 concentrations of C or CE covering a range from 0.005 μg/mL to 0.1 μg/mL (e.g., 0.005, 0.01, 0.025, 0.05, 0.1 μg/mL) (*see* **Note** 7). The NGM dish contains a volume of 10 mL, therefore a total amount of 0.05, 0.1, 0.25, 0.5, and 1 μg of C or CE is added per dish. To that end, dissolve the dried aliquot in the appropriate volume of chloroform–ethanol (ratio 1:15, v/v) and prepare the dilutions.
2. The solution containing the corresponding amount C or CE are directly spread on the agar dish surface in sterile conditions and spread with a Drigalski spatula. To minimize the variability, the same volume of C or CE solution will be added to each dish. In control NGM dishes the same volume of solvent used for C or CE will be added to the dish (*see* **Note 8**).
3. In order to get a comparison of the antioxidant capacity provided by carotenoids respect to other well-recognized antioxidant compounds, we use vitamin C at a final dose of 0.1 μg/mL as positive control in the oxidative stress assays. Commercial vitamin C is prepared immediately before the assay and is added on the agar dish surface.

3.3 Preparation of Nematode Synchronized Population for Feeding Experiments

All assays should be performed with age-synchronized nematodes populations without experienced food deprivation.

1. To obtain a synchronized population of nematodes, place gravid nematodes on 35 mm NGM dishes (control) containing *E. coli* OP50 strain or NGM supplemented with the corresponding C or CE or vitamin C at the corresponding doses. Allow to lay eggs for 2 h.
2. Remove the adults, and incubate the hatched eggs for 4 days at 20 ° C to obtain synchronized young adults (*see* **Note 9**).
3. Place young synchronized adults on NGM dishes (control), supplemented NGM dishes with a specific of C or CE concentration or with vitamin C (0.1 μg/mL) containing *E. coli* OP50 strain. The nematodes are incubated for 7 days at 20 °C. During this incubation period the nematodes uptake the

carotenoid/s from the C or CE supplemented NGM dishes. To check bioassimilation of the C or CE by the nematodes, a specific protocol has been developed and described in Subheading 3.4.

3.4 Acute Oxidative Stress Assay with H_2O_2

It is recommended to perform at least two independent experiments using 100 nematodes per treatment /dose in each experiment. In Fig. 1 a representative scheme of experimental design showing the steps and final outcome is depicted (*see* **Note 10**).

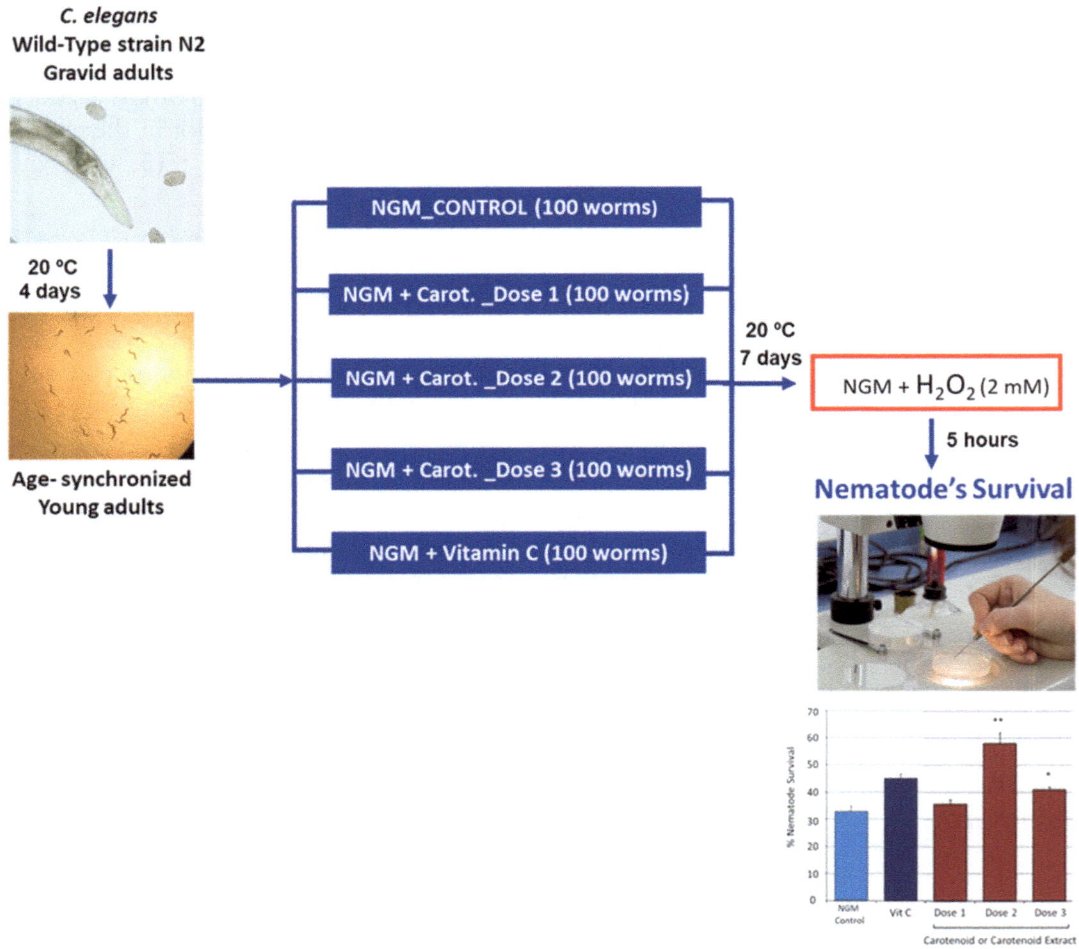

Fig. 1 Flowchart of a representative experiment of carotenoid protection against acute oxidative stress by using *C. elegans* as animal model. Gravid adults are placed in NGM plates (Control) or NGM supplemented with the corresponding carotenoid or carotenoid extract, or vitamin C (positive control). A total of 100 worms per conditions are transferred. Plates are incubated at 20 °C for 7 days. Then, an oxidative stress is provoke by H_2O_2 (2 mM) incubation and nematodes survival is assessed under stereomicroscope

1. Basal S medium plates are prepared as described previously. The H_2O_2 working solution is prepared and added to the medium at final dose of 2 mM (*see* **Note 11**).
2. Nematode adults previously incubated for 7 days at 20 °C on NGM dishes (control), supplemented NGM with a specific of C or CE concentration or with vitamin C (0.1 μg/mL) (*see* **step 3** from Subheading 3.2) are transferred to freshly prepared Basal S medium dishes containing 2 mM H_2O_2 and left for 5 h. This incubation with H_2O_2 provokes an acute oxidative stress.
3. After incubation, the nematodes are scored for viability under the stereomicroscope. The viability of the nematodes is scored by touching them with a platinum wire and considering dead nematodes those that do not longer respond to prodding.
4. Statistically evaluate the viability of nematodes (percentage of survival) between nematodes cultured in control and treatment conditions after oxidative stress by means of one-way analysis of variance (ANOVA) and the Tukey test for comparative analysis.

3.5 Carotenoid Bioassimilation Assay

To confirm C or CE intake by nematodes, a procedure has been developed that determines the presence and composition of the carotenoids in the nematodes upon the feeding experiment by HPLC-PAD. All solvents use in this section are HPLC grade, unless is specified.

1. Obtain a synchronized population of *C. elegans* of 5 days age using the same procedures described in **steps 1** and **2** of Subheading 3.2.
2. Prepare at least 50 dishes of NGM (control) with OP50 and NGM with OP50 supplemented with C or CE at final concentration of 1 μg/mL (*see* **Note 12**).
3. Obtain around 150–200 age-synchronized nematodes per dish and incubate for 24 h at 20 °C.
4. Recover the nematodes from the dish surface by adding 1 mL of M9 buffer to the dish and gently shake it (*see* **Note 13**).
5. Transfer the M9 buffer containing worms from dishes to 50 mL tubes. For each condition (NGM-control and NGM-carotenoid) pool in one single tube the recovered M9 buffer with the nematodes.
6. Centrifuge the 50 mL NGM-control and NGM-carotenoid tubes at 700 × *g*, 2 min, 20 °C. Discard the supernatant.
7. Add 20 mL of M9 buffer to each tube, gently shake, and repeat **step 5**. Repeat this step twice.
8. Add 10 mL of M9 buffer to each tube and incubate for 2 h at 20 °C to facilitate the removal of gut microbiota from the nematodes. Discard the supernatant.

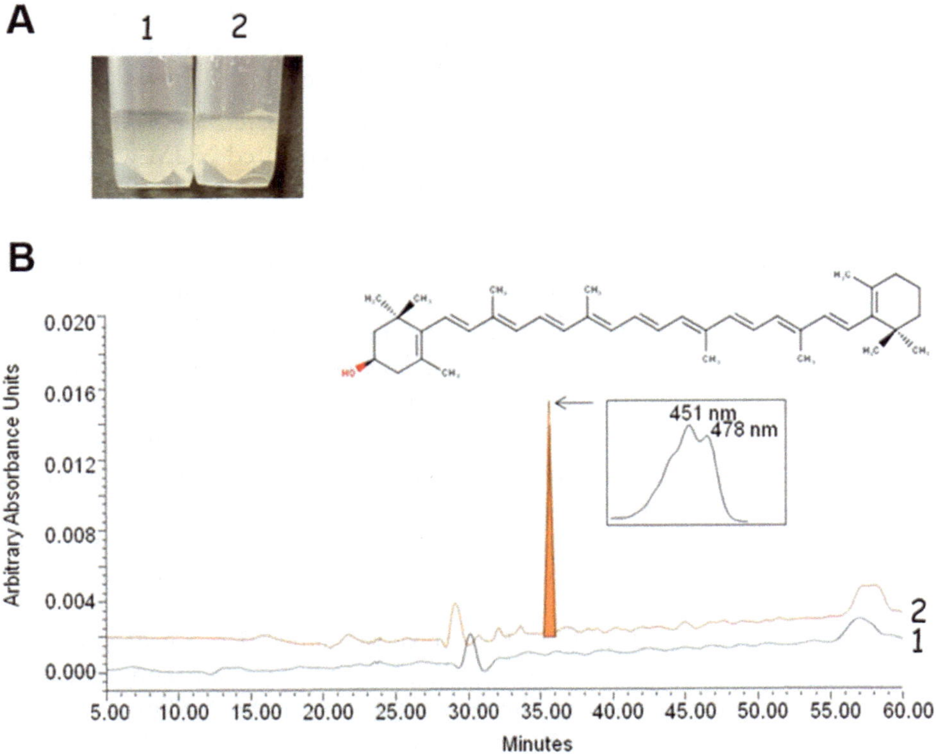

Fig. 2 Analysis of β-cryptoxanthin intake by *C. elegans* in feeding experiments. (**a**) Suspension of washed *C. elegans* grown in control NGM medium (1) and NGM supplemented with β-cryptoxanthin (1 μg/mL) (2). Note the nematode transparent suspension obtained under control condition (1) compared to the yellowish-creamy color of nematodes fed with β-cryptoxanthin (2). (**b**) Max-Plot HPLC-PAD chromatograms from *C. elegans* extracts grown under control conditions (1, *grey line*) and supplemented with β- cryptoxanthin (2, *orange line*). Only one carotenoid peak, marked in orange color, was identified in chromatogram from condition (2) which absorbance spectrum (shown in insert) and retention time matched with β-cryptoxanthin

9. The evacuated and washed nematode pellets are recovered with a micropipette and transferred to 2 mL cryovials. Centrifuge the cryovials at 1500 × *g*, 2 min, 20 °C, to completely remove the supernatant. Take a picture as shown in Fig. 2a (*see* **Notes 6, 14**, and **15**). Immerse the cryovials into liquid nitrogen and store frozen pellets at −80 °C.
10. Defrost the pellets on ice, transfer to 1.5-mL tubes with 0.5 mL of acetone (HPLC grade), and disrupt with micropestle for 2 min.
11. Vortex the disrupted pellets for 30 s and centrifuge for 3 min at 15,700 × *g* at 4 °C. The acetone is recovered, and pellets are reextracted twice with 2 mL acetone. Check for pellet color and if pellet is not translucent or white perform an additional extraction with 1 mL of dichloromethane.

12. Dry the pooled acetone/dichloromethane extracts for each condition under nitrogen stream (*see* **Note 6**).
13. Dissolve the dried residue obtained for each condition in 250 μL dichloromethane, add 250 μL MilliQ water and vortex for 30 s. Recover the organic phases in a 1.5 mL vial, reextract the aqueous phase twice with 250 μL of dichloromethane.
14. Dry the pooled organic phases under a nitrogen stream and store at −20 °C until HPLC-PAD analysis (*see* **Note 6**).
15. Analyze the presence of carotenoids in the nematode pellet extracts by using a HPLC-PAD equipped with a C30 analytical column (4.6 × 250 mm, 5 μm) couple to a C30 guard column (20 × 4.0 mm, 5 μm). In this procedure it is describe the chromatographic conditions and equipment that is routinely used in our laboratory which can resolve most of the common carotenoids present in fruits or fruit-based food. However, these conditions can be modified and adapted depending on the carotenoids composition in the extracts under study.
16. Dissolve the nematode pellet extract in 30 μL of methanol–acetone–chloroform (2:1:3, v/v/v) and centrifuge at 12,300 × *g*, 1 min at 20 °C.
17. Inject 20 μL on the HPLC system using the method described elsewhere [33] with a ternary gradient elution. The initial solvent composition consisted of 90% methanol (MeOH), 5% water and 5% methyl tert-butyl ether (MTBE). The solvent composition changes in a linear fashion to 95% MeOH and 5% MTBE at 12 min. During the next 8 min the solvent composition changes to 86% MeOH and 14% MTBE. After reaching this concentration the solvent is gradually changed to 75% MeOH and 25% MTBE at 30 min. Following 20 min solvent composition changes linearly being, at 50 min, 50% MeOH and 50% MTBE. The final composition is reached at 70 min and consisted of 25% MeOH and 75% MTBE. The initial conditions are established in 5 min and equilibrated for 15 min before the next injection. The flow rate is 1 mL/min, and the column temperature is set to 25 °C. The PAD is set to scan from 250 to 540 nm, and for each elution a Maxplot chromatogram is obtained, which plots each carotenoid peak at its corresponding maximum absorbance wavelength.
18. Compare chromatograms obtained with the pellet extracts of nematodes grown on NGM-control and NGM supplemented with carotenoid, and identify carotenoids analyzing the spectra of the chromatographic peaks (*see* **Notes 16** and **17**). Figure 2b shows chromatograms obtained from *C. elegans* grown in NGM control and NGM supplemented with β-cryptoxanthin at 1 μg/mL. Only one carotenoid peak which spectrum and retention time matched with

β-cryptoxanthin, was identified in the nematodes grown in NGM supplemented with β-cryptoxanthin, confirming the carotenoid uptake by the animal.

4 Notes

1. To avoid variability between trials it is important to be sure that the same batch of test compounds will be used along the experiment. This is particularly determinant when a carotenoid extract is evaluated since may contain a complex and not defined proportion of carotenoids.
2. When possible used the highest purity standard available (>95%) to avoid impurities with other carotenoids.
3. It is possible to assay a novel carotenoid that has been purified by HPLC or TLC.
4. It is important to completely dissolve C or CE. To that end, follow the solvents recommended elsewhere [34] since this can be variable depending of the polarity of your carotenoid, and determine concentration spectrophotometrically as described elsewhere [34]. Be sure that solvents used do not contain antioxidants such as butylated hydroxytoluene which will interfere in the assay.
5. We recommend using a Hamilton syringe to take the aliquot since it is important to accurately measure the volume.
6. Do not store dried carotenoids or extracts for more than 2 weeks. In our experience if sample to evaluate contains lycopene or phytoene do not store for more than 2 days since these carotenes are more unstable.
7. The range of concentrations must be adjusted to each carotenoid/carotenoid extract, and these concentrations corresponds assays performed to β-cryptoxanthin, lycopene and β-carotene in our laboratory. However, we recommend to test in exploratory experiments doses up to 2 μg/mL for pure carotenoids and up to 10 μg/mL for complex extracts.
8. The solvent used to dissolve the C or CE (e.g., chloroform–ethanol) can cause deleterious effects on *C. elegans* performance, therefore the same amount of solvent should be added to control NGM dishes.
9. Nematodes are transferred to new plates using a platinum wire.
10. It is possible to further explore the cellular and molecular mechanism involved in the *C. elegans* response to carotenoid treatment and subsequent oxidative stress by means of transcriptional profile or biochemical (e.g., ROS production) changes in the nematodes. To that end the number of dishes

per condition should be extended to 50 and upon treatment nematodes are collect from the dishes, follow **steps 4–8** from Subheading 3.4, and immediately frozen in liquid nitrogen and store at −80 °C until analysis.

11. The H_2O_2 is added to the autoclaved medium ensuring a temperature below 50 °C.
12. If it has been previously tested the effectiveness of the C or CE in a dose-response experiment, the dishes for the bioassimilation assays can be prepared at the most effective dose. However, in our experience, determination of carotenoids by HPLC-PAD in nematodes fed with pure carotenoids at doses below 0.1, are usually below the detection limits. Therefore, the dose in the dishes can be increase up to 1–5 μg/mL since the only purpose of this experiment is to confirm the carotenoids uptake and not to evaluate the biological response of nematodes.
13. This washing is necessary to remove the *E. coli* OP50 present in the media that could interfere in subsequent determinations.
14. We recommend to take a picture of the nematode pellets in the cryovials prior to freezing. Usually the pellet of nematodes grown in NGM supplemented with carotenoids show a creamy, yellow or pale orange color compared to the translucent aspect of NGM-control. *See* Fig. 2a for an example of a nematode feeding experiment with β-cryptoxanthin.
15. By weighing the cryovials before and after transferring the nematodes, the nematode pellet weight can be determined, and usually is about 100–200 mg.
16. It is recommended to analyze the carotenoid or carotenoid extract use for nematode feeding by using the same chromatographic conditions. This allows a direct comparison facilitating the identification of carotenoids in the nematode extracts.
17. If the nematode pellet is weighed before extraction as suggested in **step 8** (*see* **Note 15**), the relative carotenoid uptake by the nematode can be determined per mg of nematode pellet and compared between experiments or conditions.

Acknowledgments

Financial support by research grant AGL-2015-70218 (Ministerio Economía y Competitividad, Spain) is gratefully acknowledged. MJR and LZ are members of Eurocaroten (COST_Action CA15136) and CaRed (Spanish Carotenoid Network, BIO2015-71703-REDT and BIO2017-90877-REDT).

References

1. Brenner S (1974) The genetics of *Caenorhabditis elegans*. Genetics 77:71–94
2. Shen P, Yue Y, Zheng J, Park Y (2018) *Caenorhabditis elegans*: a convenient *in vivo* model for assessing the impact of food bioactive compounds on obesity, aging, and Alzheimer's disease. Ann Rev Food Sci Technol 9:1–22
3. Kaletta T, Hengartner MO (2006) Finding function in novel targets: *C. elegans* as a model organism. Nat Rev Drug Discov 5:387–398
4. European Parliament and Council of the European Union (2010) Directive 2010/63/EU of the European Parliament and of the Council of 22 September 2010 on the protection of animals used for scientific purposes. Official J Eur Union 276:33–79. 32010L0063
5. U.S. Government Publishing Office (2013) 7 U.S.C., Chapter 54: Transportation, sale, and handling of certain animals. https://www.govinfo.gov/content/pkg/USCODE-2012-title7/pdf/USCODE-2012-title7-chap54.pdf
6. Senate of Canada (2015) BILL S-214: An act to amend the food and drugs act (cruelty-free cosmetics).http://www.parl.ca/DocumentViewer/en/42-1/bill/S-214/third-reading
7. Zheng SQ, Ding AJ, Li GP et al (2013) Drug absorption efficiency in *Caenorhabditis elegans* delivered by different methods. PLoS One 8: e56877
8. Kobet RA, Pan X, Zhang B et al (2014) *Caenorhabditis elegans*: a model system for anticancer drug discovery and therapeutic target identification. Biomol Ther 22:371–383
9. Ganner A, Neumann-Haefelin E (2017) Genetic kidney diseases: *Caenorhabditis elegans* as model system. Cell Tissue Res 369:105–118
10. Pir GJ, Choudhary B, Mandelkow E (2017) *Caenorhabditis elegans* models of tauopathy. FASEB J 31:5137–5148
11. Martorell P, Bataller E, Llopis S et al (2013) A cocoa peptide protects *Caenorhabditis elegans* from oxidative stress and -amyloid peptide toxicity. PLoS One 8:e63283
12. Maglioni S, Ventura N (2016) C. elegans as a model organism for human mitochondrial associated disorders. Mitochondrion 30:117–125
13. Moreno-Arriola E, Cardenas-Rodriguez N, Coballase-Urrutia E et al (2014) *Caenorhabditis elegans*: a useful model for studying metabolic disorders in which oxidative stress is a contributing factor. Oxidative Med Cell Longev 2014:705253. https://doi.org/10.1155/2014/705253
14. Calvo DR, Martorell P, Genovés S et al (2016) Development of novel functional ingredients: need for testing systems and solutions with *Caenorhabditis elegans*. Trends Food Sci Technol 54:197–203
15. Fischer N, Büchter C, Koch K et al (2017) The resveratrol derivatives trans-3,5-dimethoxy-4-fluoro-4′-hydroxystilbene and trans-2,4′,5-trihydroxystilbene decrease oxidative stress and prolong lifespan in *Caenorhabditis elegans*. J Pharm Pharmacol 69:73–81
16. Saier C, Büchter C, Koch K et al (2018) *Polygonum multiflorum* extract exerts antioxidative effects and increases life span and stress resistance in the model organism *Caenorhabditis elegans* via DAF-16 and SIR-2.1. Plants 7:60. https://doi.org/10.3390/plants7030060
17. Sanada Y, Asai S, Ikemoto A et al (2014) Oxidation resistance 1 is essential for protection against oxidative stress and participates in the regulation of aging in *Caenorhabditis elegans*. Free Radic Res 48:919–928
18. Havermann S, Humpf HU, Watjen W (2016) Baicalein modulates stress resistance and life span in *C. elegans* via SKN-1 but not DAF-16. Fitoterapia 113:123–127
19. Rodriguez-Concepcion M, Avalos J, Bonet ML et al (2018) A global perspective on carotenoids: metabolism, biotechnology, and benefits for nutrition and health. Prog. Lipid Res 70:62–93
20. Barros MP, Rodrigo MJ, Zacarias L (2018) Dietary carotenoid roles in redox homeostasis and human health. J Agric Food Chem 66:5733–5740
21. Kaulmann A, Bohn T (2014) Carotenoids, inflammation, and oxidative stress-implications of cellular signaling pathways and relation to chronic disease prevention. Nutr Res 34:907–929
22. Stahl W, Sies H (2005) Bioactivity and protective effects of natural carotenoids. Biochim Biophys Acta Mol basis Dis 1740:101–107
23. Keith SA, Amrit FRG, Ratnappan R et al (2014) The *C. elegans* healthspan and stress-resistance assay toolkit. Methods 68:476–486
24. Li WH, Shi YC, Chang CH et al (2014) Selenite protects *Caenorhabditis elegans* from oxidative stress via DAF-16 and TRXR-1. Mol Nutr Food Res 58(4):863–874
25. Cañuelo A, Gilbert-López B, Pacheco-Liñán P et al (2012) Tyrosol, a main phenol present in extra virgin olive oil, increases lifespan and

stress resistance in *Caenorhabditis elegans*. Mech Ageing Dev 133:563–574

26. Castello PR, Drechsel DA, Patel M (2007) Mitochondria are a major source of paraquat-induced reactive oxygen species production in the brain. J Biol Chem 282:14186–14193
27. Senchuk M, Dues D, Van Raamsdonk J (2017) Measuring oxidative stress in *Caenorhabditis elegans*: paraquat and juglone sensitivity assays. Bio Protoc 7:1–11
28. Liu X, Luo Q, Cao Y et al (2016) Mechanism of different stereoisomeric astaxanthin in resistance to oxidative stress in *Caenorhabditis elegans*. J Food Sci 81:H2280
29. Liu X, Chen X, Liu H et al (2018) Antioxidation and anti-aging activities of astaxanthin geometrical isomers and molecular mechanism involved in *Caenorhabditis elegans*. J Funct Foods 44:127–136
30. Augusti PR, Brasil AVS, Souto C et al (2017) Microcystin-LR exposure induces oxidative damage in *Caenorhabditis elegans*: protective effect of lutein extracted from marigold flowers. Food Chem Toxicol 109:60–67
31. Llopis S, Rodrigo MJ, González N et al (2019) β-Cryptoxanthin reduces body fat and increases oxidative stress response in *Caenorhabditis elegans* model (2019) nutrients 11 (2):232
32. Pons E, Alquézar B, Rodríguez A et al (2014) Metabolic engineering of β-carotene in orange fruit increases its in vivo antioxidant properties. Plant Biotechnol J 12:17–27
33. Alquezar B, Rodrigo MJ, Zacarías L (2008) Regulation of carotenoid biosynthesis during fruit maturation in the red-fleshed orange mutant Cara Cara. Phytochemistry 69:1997–2007
34. Shiedt K, Liaaen-Jensen S (1995) Isolation and analysis. In: Britton G, Liaaen-Jensen, S, Pfander H (ed.) Carotenoids, Volume 1A. Birkhäuser Verlag, Basel, p103–105

Chapter 30

Mouse Models to Study Antiobesogenic Effects of Carotenoids

Joan Ribot, Bojan Stojnic, Andreu Palou, and M. Luisa Bonet

Abstract

Carotenoids entail a vast potential to tackle health problems including obesity and some of its comorbidities. The use of animal models remains necessary, particularly at early stages of research (preclinical) and for advancing in mechanistic aspects of carotenoid action. No single animal model completely mimics human absorption and metabolism of carotenoids, and the best model must be chosen considering the specific application, characteristics of the individual models, and funding and facilities available. Here, we propose three protocols in mice to investigate the potential of a given carotenoid, carotenoid mixture, or carotenoid-rich extract to (a) counteract the development of obesity and prevent the metabolic alterations caused by feeding mice a moderate high-fat diet; (b) improve the metabolic profile of obese animals with metabolic alterations caused by chronic high-fat diet feeding; and (c) act as coadjuvants in weight loss strategies (reversion to a low fat diet) applied to diet-induced obese animals.

Key words Animal models, Obesity, Metabolic syndrome, Carotenoids

1 Introduction

Obesity is defined as the presence of excess body fat and is ultimately caused by a chronic energy imbalance whereby energy intake exceeds energy expenditure [1]. Its prevalence is increasing worldwide in not only industrialized but also developing and transition countries, and nearly three million deaths each year are obesity related [2]. This is thought to be mainly due to changes in the dietary patterns of the world population toward diets richer in calories from fats and sugars, together with physical inactivity [3]. The obesity pandemic has drawn much attention and concern since obesity is strongly associated with an increased risk of developing the metabolic syndrome, a cluster of metabolic derangements characterized by central (visceral) obesity, insulin resistance, dyslipidemia, increased blood pressure, and nonalcoholic fatty liver disease (NAFLD) [4, 5].

Manuel Rodríguez-Concepción and Ralf Welsch (eds.), *Plant and Food Carotenoids: Methods and Protocols*, Methods in Molecular Biology, vol. 2083, https://doi.org/10.1007/978-1-4939-9952-1_30,

In mammals, the amount of fat reserves depends on the interplay between several interconnected processes, including the control of feeding behavior, energy expenditure, the gut environment, partitioning of nutrients among tissues and anabolic versus catabolic pathways, as well as ongoing processes within fat depots determining adipocyte number (e.g., preadipocyte proliferation, adipocyte differentiation, and apoptosis) [1, 6]. Each of these processes involves many signals, receptors, signal transduction cascades, and downstream targets including metabolic enzymes, transport proteins, and regulatory factors. Alterations in key molecular elements of this complex system can favor a positive energy balance, which can ultimately lead to obesity. Such alterations may be inherited or due to environmental exposures, and can at times be very subtle. In addition, multiple sociocultural, socioeconomic, and behavioral factors —often interrelated and many of them still unknown or poorly understood— appear to contribute to the establishment and perpetuation of obese phenotypes.

In order to tackle obesity, a common approach is the use of dietary interventions based on altering diet macronutrient composition or promoting the supplementation with bioactive ingredients which can ultimately target key processes and molecular elements controlling body fat content. In recent years, a novel perspective is emerging in regard to the biological activity of carotenoids and carotenoid-derived metabolites —including, but not restricted to, the vitamin A retinoids— that connects these compounds to the control of body fat reserves, with implications for the management of obesity and obesity-related metabolic disturbances. Evidence for the antiobesity action of specific carotenoids and their derived metabolic products comes so far mainly from cell, animal, and human epidemiological studies. There have been only few human intervention studies in the field to date, but the scenario they depict is encouraging, since beneficial effects have been observed on the accrual of body fat, abdominal fat and related risk parameters following carotenoids consumption. Interestingly, in the case of beta-carotene, these effects have been achieved under mild supplementation, at doses lower than the ones that have caused concern and controversy in large intervention trials in the past (reviewed in [7, 8]). Antiobesity and beneficial metabolic effects of carotenoids can relate to their direct or indirect action on different tissues, including the liver and adipose tissues —which constitute the main reservoirs for carotenoids and vitamin A and whose lipid metabolism is indeed known to be affected by retinoids— as well as on the gut microbiota. At the molecular level, mechanisms of action are emerging that include physical and/or functional interaction of the carotenoid molecules and/or their derivatives with transcription factors of the nuclear receptor superfamily and with proinflammatory and antioxidant cellular signaling pathways.

To date, the most studied carotenoids in relation to an anti-obesogenic action have been beta-carotene, lycopene, cryptoxanthin, astaxanthin, and fucoxanthin, together with beta-carotene-derived retinoids and some other apocarotenoids [7–9]. Nevertheless, there are more than 600 carotenoids in nature and many of them, or their derived apocarotenoids, are possibly capable of interacting with cellular machineries and signaling pathways to elicit beneficial effects against obesity and obesity comorbidities.

It would be ideal to study carotenoid absorption, metabolism and contribution to disease prevention through intervention studies in human subjects. However, human studies face specific safety and ethical issues, they are expensive and often difficult to coordinate, and they do not allow for extensive tissue sampling for mechanistic studies. Because of these limitations, the use of animal models remains necessary, particularly during the early stages of research. Unfortunately, no single animal model completely mimics human absorption and metabolism of carotenoids, and the best model must be chosen considering the specific application, characteristics of the individual models and the funding and facilities available [10].

In mammals including humans, two carotenoid metabolizing enzymes have been identified, denoted as β-carotene-15,15′-dioxygenase (BCO1) and β-carotene-9′,10′-dioxygenase (BCO2). BCO1 converts a limited number of provitamin A carotenoids to retinaldehyde, from which all naturally occurring retinoids can be synthesized by endogenous pathways. Whereas BCO2 displays broad substrate specificity and catalyzes oxidative cleavage at the C9,C10 double bond of a large variety of carotenoids [11] to generate apo-10′-carotenoids (whose biological activity remains largely unknown). The efficiency of carotenoids intestinal cleavage varies greatly across species. For instance, adult rats and mice convert almost all absorbed beta-carotene to vitamin A retinoids in the enterocytes —although they can absorb intact carotenoids, including beta-carotene, when fed diets containing supraphysiological levels— whereas humans, gerbils, and ferrets incorporate more beta-carotene into nascent chylomicrons and deliver it to tissues in its intact form [10]. Thus, the extrapolation of data from rodents to humans must be done with caution. Nevertheless, rodents have been widely used in studies of carotenoid activity in mammals, as they offer advantages over other experimental animal species [10, 12] (see below). Interestingly, we have previously shown [13] that, in overt contrast to adult rats, suckling rats partly absorb dietary beta-carotene intact, even when present at a moderate dose in the diet, which resembles the situation in humans and suggests that suckling rats may be an appropriate animal model to study carotenoids uptake, metabolism and biological activity. In this context, our studies also establish a new potential model for the experimental evaluation of early effects of carotenoids on metabolic programming, including epigenetics, and their long-term consequences, including the susceptibility to obesity later in life [13–15].

Moreover, the use of rodent models is convenient because rodents are extensively used to study the mechanisms by which obesity develops and induces physiological disorders [16–19], whereas only few studies of this type have been conducted in gerbils [20] or in ferrets [21]. Currently, numerous genetic and diet-induced experimental rodent models of obesity are described in the literature. Monogenic animal models of obesity can be useful because their obesity and adiposity are often severe, resulting in a clear distinct phenotype [22]. Nevertheless, environmental or behavioral factors such as excess caloric food intake and physical inactivity are most likely the main contributors to the obesity pandemic, rather than genetic changes [4]. Thus, models with diet-induced obesity have been preferably used in place of monogenic models to make the experimental model as close as possible to the human situation and the purpose of the study [23].

2 Materials

2.1 Animals

4–6 weeks old C57BL/6 J male mice (e.g., from Charles River Laboratories, Barcelona, Spain).

2.2 Diets and Treatments

1. Defined low-fat low-sucrose (LFD) diet with 10% energy as fat and 7% energy as sucrose (e.g., D12450J from Research Diets Inc.).
2. Defined high-fat medium-sucrose (HFD) diet with 45% energy as fat and 17% energy as sucrose (e.g., D12451 from Research Diets Inc.).
3. Olive oil (vehicle).

2.3 Glucose Tolerance Test and Surrogate Insulin Resistance Indexes

1. Glucose.
2. Saline (NaCl 0.9%).
3. Sodium heparinized capillary tubes to collect blood used to obtain plasma.

2.4 Blood and Tissue/Organ Collection

1. Isoflurane as anesthesic.
2. 23 G needle attached to a 1 mL self-prepared heparinized syringes (10–15 IU sodium heparin/syringe mL), for cardiac puncture.
3. Dissection and surgical tools including forceps, spatulas, scissors, and needles.
4. 1.5 mL plastic storage tubs.

5. Liquid nitrogen.
6. 4% paraformaldehyde in 0.1 M phosphate buffer a pH 7.4, to fix fresh tissue biopsies.

3 Methods

3.1 Model to Assess the Capacity of Carotenoids to Counteract the Development of Metabolic Alterations and Obesity Induced by an Obesogenic Diet

Here we propose a protocol in mice to investigate the potential of a given carotenoid, carotenoid mixture, or carotenoid-rich extract to counteract the development of obesity and prevent the metabolic alterations caused by feeding mice a moderate high-fat diet (45% of caloric content) (*see* **Note 1**) for 10 weeks (*see* **Note 2**). If knockout mouse models deficient for carotenoid dioxygenases (*see* **Note 3**) are used, this may allow dissecting the effects of parent carotenoids apart from the effects of derived apocarotenoid products. For instance, as seen in [12], if an antiobesogenic effect of the tested carotenoid is observed in wild-type control mice but is lacking in the matched knockout, an involvement of carotenoid-dioxygenase products could be suspected instead of the parent carotenoid.

1. Perform the experiments with 4–6-week-old (*see* **Note 2**) C57BL/6J (*see* **Note 4**) male (*see* **Note 5**) mice.
2. Animals must be kept confined and handled according to existing ethical guidelines for the use and care of laboratory animals (*see* **Note 6**).
3. Divide mice by weight-matching into four experimental groups and house 3–4 mice per cage in a controlled temperature and humidity room with 12-hour light–dark cycle and free access to food and water (10 animals/group suggested, or 3–4 cages/group). Feed all mice with a defined low-fat low-sucrose (LFD) diet with 10% energy as fat and 7% energy as sucrose.
4. After 1 week (day 0 of the experiment), switch three experimental groups to a defined high-fat medium-sucrose (HFD) diet with 45% energy as fat and 17% energy as sucrose and continue feeding the remaining experimental group with the LFD diet (*see* **Notes 7** and **8**).
5. At day 0, start the treatment. Give the mice of the LFD-fed group and of one HFD-fed group the vehicle (olive oil); these will serve as, respectively, LFD-fed and HFD-fed reference groups. Treat the other two HFD-fed groups with an emulsion of the selected pure carotenoid, mix or extract (*see* **Notes 9** and **10**) at two different doses (*see* **Note 11**). Conduct the treatments daily by oral administration, with the aid of a pipette (20 μL), more or less at the same time during the day. Prepare the carotenoid emulsions every day to prevent degradation.
6. The duration of the nutritional and treatment challenge is suggested to be 10 weeks (*see* **Note 2**).

7. Regularly monitor body weight and food intake (twice per week). Food intake per animal can be estimated on a per-cage basis, from the actual amount of food consumed by the animals, its caloric equivalence, and the number of animals per cage.
8. It is advisable to perform the following analyses and tests in live animals, as in [24]:
   - Body composition using suitable magnetic resonance equipment (e.g., EchoMRI-700TM analyzer).
   - Energy expenditure and respiratory exchange ratio (RER) through indirect calorimetry (e.g., using PhenoMaster Indirect Calorimetry System) and spontaneous locomotor activity through a motion sensor coupled to the indirect calorimetry equipment.
   - A glucose tolerance test and the estimation of the surrogate insulin resistance indexes. For this, intraperitoneally administer a glucose load (2 g per kg of body weight in 100 μL of sterile saline) to mice after a 6-h fast (from 6 am to 12 pm). Collect blood samples (one drop) from the tail before the glucose load (time 0) and at 30, 60, 120, and 180 min thereafter and measure the levels of glucose with a glucometer. In addition, at time 0 collect a blood sample (maximum 200 μL) to measure the levels of insulin and free fatty acids (by specific ELISA and enzymatic assays, respectively) for the estimation of the surrogate insulin resistance indexes.
   - The homeostatic model assessment for insulin resistance (HOMA-IR) score is calculated from fasting insulin and glucose concentrations as follow: HOMA-IR = insulin (in μU/L) × glucose (in mmol/L)/22.5.
   - The revised quantitative insulin sensitivity check index (R-QUICKI) is calculated as follow: R-QUICKI = 1/[log glucose (in mg/dL) + log insulin (in μU/mL) + log NEFA (in mEq/L)].
9. On day 70 (week 10 of treatment), proceed to the euthanization and tissue collection, within the first 3 h of the light cycle (distribute the animals over several days).
10. Anesthetize the animal with isoflurane [25]. For this, place the animal in the anesthetic induction chamber, and set the vaporizer at 4–5% isoflurane. When the animal loses consciousness, remove it from the chamber and place a mask on it, with the vaporizer at 1–2% isoflurane and oxygen at 100% (maintenance).

11. With the anesthetized animal, proceed to the necropsy. For this, place the animal in supine decubitus, fixing it to a surface by the legs.
12. To obtain samples:
    - Blood is obtained by cardiac puncture at the ventricular level. This puncture serves for blood gathering and actual euthanasia. Prepare plasma from collected blood and store at −80 °C until biochemical analysis.
    - Collect the tissues/organs of interest (e.g., white and brown adipose tissue depots, liver) in their entirety and weigh them. After weighing, cut a fragment of fresh biopsies and fix it (in 4% paraformaldehyde, as described in [24]), to be used for morphological and immunohistochemical analysis. Snap-freeze the remaining tissue in liquid nitrogen and store at −80 °C until molecular biology analysis.
    - To characterize the microbiota, collect the caecum, weigh it whole, empty it, collect the faeces and store at −80 °C until analysis. Wash empty caecum with sterile saline and weigh it again to estimate the total content of the caecum.

3.2 Model to Assess the Capacity of Carotenoids to Improve the Metabolic Profile in Animals with Diet-Induced Obesity

Here we propose a protocol in mice to investigate the potential of carotenoids to improve the metabolic profile in obese animals with metabolic alterations caused by feeding a moderate HFD (45% of caloric content) (*see* **Note 1**) for 20 weeks (*see* **Note 2**). As indicated in Subheading 3.1, knockout mouse models deficient for carotenoid-dioxygenases (*see* **Note 3**) can be used to dissect the effects of parent carotenoids apart from the effects of derived apocarotenoid products.

1. Perform the experiments with 4–6 weeks old (*see* **Note 2**) C57BL/6 J (*see* **Note 4**) male (*see* **Note 5**) mice, and proceed as indicated in **steps 1–3** of Subheading 3.1, including the 1 week acclimation to a defined LFD.
2. To obtain diet-induced obese (DIO) mice, switch mice to a defined HFD with 45% energy as fat and 17% energy as sucrose for 20 weeks (*see* **Note 8**). Maintain an experimental group on the LFD diet.
3. To check for the development of obesity and metabolic impairments, regularly monitor body weight, food intake and, if the necessary equipment is available, body composition. It is also advisable to perform by the end of the 20 week HFD challenge a glucose tolerance test and analyze circulating levels of glucose and insulin in the fasting state for the calculation of insulin resistance indexes. After 20 weeks, mice usually display a ~20–25% excess body weight and signs of metabolic syndrome, namely excess body fat, insulin resistance and NAFLD as a result of HFD feeding [26].

4. Divide the (25-week-old) obese C57BL/6J DIO mice (day 0 of the experiment) by weight-matching into three experimental groups and continue to feed animals on the HFD. Maintain the LFD-fed group on the LFD.
5. At day 0, start the treatment. Give the mice of the LFD-fed group and of one HFD-fed group the vehicle (olive oil); these will serve as, respectively, LFD-fed and HFD-fed reference groups. Treat the other two HFD-fed groups with an emulsion of the selected pure carotenoid, mix or extract (*see* **Notes 9** and **10**) at two different doses (*see* **Note 11**). Conduct the treatments and prepare carotenoid emulsions as indicated in **step 5** of Subheading 3.1.
6. Continue the nutritional and treatment challenge for 5 weeks.
7. Regularly monitor body weight and food intake (twice per week). The last week of the experiment, it is advisable to perform analysis and tests in live animals as listed in **step 8** of Subheading 3.1.
8. On day 35 (week 5 of treatment), proceed to the euthanization and tissue collection, as in **steps 9–12** of Subheading 3.1.

3.3 Model to Assess the Potential of Carotenoids to Act as Coadjuvants in Weight Loss Strategies

Here we propose a protocol to investigate the potential of carotenoids to act as coadjuvants in weight loss strategies (e.g., together with a low fat diet) by studying their ability to favor body weight/fat loss and metabolic recovery in DIO mice (*see* **Notes 1** and **2**) upon switching from the obesogenic HFD to a regular LFD. (As indicated in Subheadings 3.1 and 3.2, if knockout mouse models deficient for carotenoid-dioxygenases (*see* **Note 3**) are used, this will allow dissecting the effects of parent carotenoids apart from the effects of derived apocarotenoid products.

1. Perform the experiments with 4–6-week-old (*see* **Note 2**) C57BL/6J (*see* **Note 4**) male (*see* **Note 5**) mice, and proceed to induce DIO and metabolic derangements through HFD (45% energy as fat) feeding for 20 weeks, as described in **steps 1–3** in Subheading 3.2.
2. Distribute the (25-week-old) C57BL/6 J DIO mice (day 0 of the experiment) by weight-matching into four experimental groups. Switch three experimental groups to the LFD and maintain an experimental group on the HFD. Maintain the LFD-fed group on the LFD.
3. At day 0, start the treatment. Give the mice of the LFD-fed group, HFD-fed group and of one reverted HFD-fed group the vehicle (olive oil); these will serve as, respectively, LFD-fed, HFD-fed and reverted HFD-fed reference groups. Treat the other two reverted HFD-fed groups with an emulsion of the selected pure carotenoid, mix or extract (*see* **Notes 9** and **10**) at

two different doses (*see* **Note 11**). Conduct the treatments and prepare carotenoid emulsions as indicated in **step 5** of Subheading 3.1.

4. Continue the nutritional and treatment challenge for 5 weeks.
5. Regularly monitor body weight and food intake (twice per week). The last week of the experiment, it is advisable to perform analysis and tests in live animals as listed in **step 8** of Subheading 3.1.
6. On day 35 (week 5 of treatment), proceed to the euthanization and tissue collection, as in **steps 9–12** of Subheading 3.1.

4 Notes

1. Several types of high-fat diets can be used to induce obesity. Cafeteria diet (CAF)-fed rodents represent a useful model to study human obesity because the CAF is a homemade, palatable, hypercaloric, high-sucrose, and hyperlipidic diet that induces voluntary hyperphagia resulting in increased body weight, abdominal fat, hyperinsulinemia, and hyperglycemia, leading to the metabolic syndrome [27–29]. However, CAF does not seem to be appropriate to study the biological properties of individual food components such as dietary carotenoids, since it is an undefined diet whose ingredients can contain multiple nutrients and nonnutrients compounds. The same can be said about chow diets, which are undefined diets made out of raw components (e.g., ground wheat, corn, or oats, soybean, fish, and vegetable oil, among others). Increasing the amount of certain macronutrients, such as fats, sucrose or both, of a defined diet (made up of refined ingredients, such as cornstarch, sucrose, casein, to name some) is more advisable for our purpose. Defined diets high in both fat and sucrose produce more deleterious effects on animals than diets high in either fat or sucrose alone [30], and, moreover, such high-fat high-sucrose diets more closely mimic most typical human obesogenic diets [30, 31]. Defined obesogenic diets for rodents can be either homemade or obtained commercially.

   The dietary fat source, as well as the percentage of fat supplementation, may result in differences in phenotypes. More pronounced obesity and signs of insulin resistance are observed when the fat source is rich in saturated fat (such as lard) or when supplementation rises to 60% of the caloric intake as fat [32, 33]. Mild obesity challenges (e.g., 40–45% caloric intake as fat instead of 60%) may be better suited to reveal moderate effects of food bioactives on obesity development.

2. The length of the dietary challenge and the time of its onset relative to the age of the animals may affect the macroscopic and molecular obesity-related phenotypes developed. After only 5 days of a (40%) HFD challenge, an impact on white adipose tissue global gene expression profile can already be observed in C57BL/6 J mice [34, 35], which seems to be highly predictable of the long-term changes observed after 12 weeks of HFD challenge. Nevertheless, for specific white adipose tissue genes including *Pparg2*—a master transcription factor for adipogenesis and fat tissue metabolism— and some *Pparg2* transcriptional targets a time-dependent response has been reported [36]. Initially, within the first week of HFD, these genes are downregulated, which may serve to prevent excessive fat deposition; after longer periods of HFD (fourth week onward) they are instead upregulated, which likely contributes to WAT expansion. Moreover, with prolonged HFD, accumulated oxidative damage in tissues including the liver may contribute to aggravate metabolic complications, such as hepatic steatosis [37]. Thus, the duration of the treatment can be extended up to 24 weeks to see more deleterious effects on animals, such as the progression of NAFLD [38]. An early onset of the dietary challenge (e.g., beginning at 4–6 weeks of age) usually results in a more persistent obese phenotype in rodents [39].
3. Mouse models deficient for each of the two carotenoid-dioxygenases have been established. These genetic models allow studying the consequences of the accumulation of carotenoids in mammalian tissues in a tunable fashion [40, 41]. This breakthrough also allows dissecting the effects of parent carotenoids apart from their eventual apocarotenoid-derived signaling products in models (dietary and/or genetic) of diseases epidemiologically associated with carotenoid intake, such as obesity.
4. Most rodents tend to become obese on HFD, but there are variable responses in weight gain, systemic metabolic alterations (i.e., glucose tolerance, insulin resistance, or triglycerides), adipose tissue biology and others parameters depending on the strain [32]. The C57BL/6 J mice are prone to develop obesity when fed an HFD, owing to their strain-specific genetic background, and are one of the most commonly used rodent strain for the study of diet-induced obesity and its comorbidities [18, 42].
5. As general advice, animal studies should be routinely performed in both female and male animals [43]. Moreover, sex-dependent differences exist in lipid handling and partitioning between tissues in response to an HFD [44]. Female rats

show a higher capacity for storing fat in adipose tissue and for oxidizing fatty acids in muscle that could explain their lower tendency to manifest an HFD obesity-linked disorder such as increased hepatic lipid content, though at the expense of increased weight and adiposity gain [44]. In contrast, male rodents are prone to develop obesity-linked disorders, and for this reason have traditionally been more used than females in animal studies investigating antiobesity effects of bioactive compounds. All in all, it could be important to perform the interventions in animals of both sexes, even if full characterization of samples is first conducted in the male animals.

6. The approval of experimental protocols and procedures by the competent Bioethical Committee is mandatory before starting.
7. A relatively low dose of dietary vitamin A is advisable in the experimental diets to maximize intestinal absorption of the supplied carotenoids (especially beta-carotene), since carotenoid absorption can be repressed by preformed vitamin A [45].
8. It is important to replace the HFD in the cage feeders at least twice a week, to avoid rancidity.
9. Pure carotenoids or carotenoids mixtures are usually dissolved in oil. We recommend refined olive oil (not virgin olive oil), because it is largely devoid of polyphenols, contains only trace amounts of carotenoids, and provides good quantities of vitamin E as antioxidant. For some carotenoids, water-soluble, relatively stable formulations (beadlets) can be obtained commercially [46] which contain the selected crystalline carotenoid (s) together with additional components, typically DL-alpha-tocopherol, ascorbyl palmitate, corn oil, fish gelatin, sucrose, and corn starch (e.g., provided by DSM Nutritional Products Ltd.). The same formulation but without the carotenoids can be provided by the manufacturer, to be administered to the control groups. Also, food extracts enriched in a given carotenoid or carotenoid mixture can be used in the animal intervention experiments. When food extracts are used, it is important to characterize their composition and stability.
10. Daily carotenoid administration through pipetting has the advantage that it ensures good control over the actual dosage but it requires animal handling skills and can be time-consuming. Administration of carotenoids can also be carried out through the diet, by mixing the selected pure carotenoid, beadlets, or a food extract enriched in the selected carotenoid (s) with a powdered HFD diet or including them in a pelletized HFD. In the latter case, the experimental diets need to be prepared by cold extrusion so as to protect the carotenoids from heat treatment. When using supplemented diets, the dose

received depends on the animal's food intake, and achieving a constant dosage (in mg per kg of animal per day) might not be feasible. It is recommended that powdered supplemented diets and carotenoid emulsions are prepared every 2 days, to prevent the degradation of carotenoids. It is also important to verify the composition and stability of the food extracts enriched in carotenoids.

11. The dose to be used in animal studies will depend on the carotenoid or carotenoids under study. For beta-carotene, it is prudent not to exceed ~×fourfold the average intake in humans, which is about 4 mg per day in European countries [9]. When establishing the experimental doses, it might also be important to take into account the daily average *total* carotenoids intake (~9.5 to 16 mg per day), as a reference. The body surface area (BSA) normalization method [47] is often used to convert drug doses from animal to human studies, despite recent controversy. Controversy arises because BSA does not take into account physiologic and pharmacokinetic data beyond body surface, and hence may entail oversimplification of dose conversion [48]. In any case, the idea after BSA or other conversion equations is to find a dosage for humans that will allow for therapeutic concentrations of the molecule of interest at the cellular targets, in order to replicate benefits observed in animal models, without toxicity. As an example, a daily oral dose of 2.5 mg of carotenoid per kg of body weight in mice is, based on the BSA equation, equivalent to 12 mg of carotenoids per day for a 60 kg person, which is within the average intake of carotenoids in humans [9]. This dose can be achieved by feeding a diet containing 30 mg of carotenoid per kg of diet, in mice of 30 g of body weight eating 2.5 g of diet per day.

Acknowledgments

BS is the recipient of a "La Caixa" Foundation predoctoral contract at the University of Balearic Islands. The group is a member of the European COST-Action EUROCAROTEN (CA15136; EU Framework Programme Horizon 2020), and the Spanish Network of Excellence CaRed (BIO2015-71703-REDT and BIO2017-90877-REDT; Agencia Estatal de Investigación, MICIU/FEDER, EU). CIBER de Fisiopatología de la Obesidad y Nutrición (CIBERobn) is an initiative of the ISCIII (Spanish Government).

References

1. Palou A, Picó C, Bonet ML (2003) The molecular basis of body weight control. Forum Nutr 56:164–168
2. Scully T (2014) Public health: society at large. Nature 508(7496):S50–S51. https://doi.org/10.1038/508S50a
3. Shao A, Drewnowski A, Willcox DC, Krämer L, Lausted C, Eggersdorfer M, Mathers J, Bell JD, Randolph RK, Witkamp R, Griffiths JC (2017) Optimal nutrition and the ever-changing dietary landscape: a conference report. Eur J Nutr 56(Suppl 1):1–21. https://doi.org/10.1007/s00394-017-1460-9
4. Vasileva LV, Marchev AS, Georgiev MI (2018) Causes and solutions to "globesity": the new fa(s)t alarming global epidemic. Food Chem Toxicol 121:173–193. https://doi.org/10.1016/j.fct.2018.08.071
5. Lumeng CN, Saltiel AR (2011) Inflammatory links between obesity and metabolic disease. J Clin Invest 121(6):2111–2117. https://doi.org/10.1172/JCI57132
6. Isolauri E (2017) Microbiota and obesity. Nestle Nutr Inst Workshop Ser 88:95–105. https://doi.org/10.1159/000455217
7. Bonet ML, Canas JA, Ribot J, Palou A (2015) Carotenoids and their conversion products in the control of adipocyte function, adiposity and obesity. Arch Biochem Biophys 572:112–125. https://doi.org/10.1016/j.abb.2015.02.022
8. Bonet ML, Canas JA, Ribot J, Palou A (2016) Carotenoids in adipose tissue biology and obesity. Subcell Biochem 79:377–414. https://doi.org/10.1007/978-3-319-39126-7_15
9. Rodriguez-Concepcion M, Avalos J, Bonet ML, Boronat A, Gomez-Gomez L, Hornero-Mendez D, Limon MC, Meléndez-Martínez AJ, Olmedilla-Alonso B, Palou A, Ribot J, Rodrigo MJ, Zacarias L, Zhu C (2018) A global perspective on carotenoids: metabolism, biotechnology, and benefits for nutrition and health. Prog Lipid Res 70:62–93. https://doi.org/10.1016/j.plipres.2018.04.004
10. Lee CM, Boileau AC, Boileau TW, Williams AW, Swanson KS, Heintz KA, Erdman JW (1999) Review of animal models in carotenoid research. J Nutr 129(12):2271–2277. https://doi.org/10.1093/jn/129.12.2271
11. Palczewski G, Widjaja-Adhi MA, Amengual J, Golczak M, von Lintig J (2016) Genetic dissection in a mouse model reveals interactions between carotenoids and lipid metabolism. J Lipid Res 57(9):1684–1695. https://doi.org/10.1194/jlr.M069021
12. Amengual J, Gouranton E, van Helden YG, Hessel S, Ribot J, Kramer E, Kiec-Wilk B, Razny U, Lietz G, Wyss A, Dembinska-Kiec A, Palou A, Keijer J, Landrier JF, Bonet ML, von Lintig J (2011) Beta-carotene reduces body adiposity of mice via BCMO1. PLoS One 6(6):e20644. https://doi.org/10.1371/journal.pone.0020644
13. Musinovic H, Bonet ML, Granados N, Amengual J, von Lintig J, Ribot J, Palou A (2014) Beta-carotene during the suckling period is absorbed intact and induces retinoic acid dependent responses similar to preformed vitamin a in intestine and liver, but not adipose tissue of young rats. Mol Nutr Food Res 58 (11):2157–2165. doi:https://doi.org/10.1002/mnfr.201400457
14. Granados N, Amengual J, Ribot J, Musinovic H, Ceresi E, von Lintig J, Palou A, Bonet ML (2013) Vitamin a supplementation in early life affects later response to an obesogenic diet in rats. Int J Obes 37 (9):1169–1176. https://doi.org/10.1038/ijo.2012.190
15. Arreguín A, Ribot J, Mušinović H, von Lintig J, Palou A, Bonet ML (2018) Dietary vitamin a impacts DNA methylation patterns of adipogenesis-related genes in suckling rats. Arch Biochem Biophys 650:75–84. https://doi.org/10.1016/j.abb.2018.05.009
16. Nilsson C, Raun K, Yan FF, Larsen MO, Tang-Christensen M (2012) Laboratory animals as surrogate models of human obesity. Acta Pharmacol Sin 33(2):173–181. https://doi.org/10.1038/aps.2011.203
17. Panchal SK, Poudyal H, Iyer A, Nazer R, Alam MA, Diwan V, Kauter K, Sernia C, Campbell F, Ward L, Gobe G, Fenning A, Brown L (2011) High-carbohydrate, high-fat diet-induced metabolic syndrome and cardiovascular remodeling in rats. J Cardiovasc Pharmacol 57 (5):611–624. https://doi.org/10.1097/FJC.0b013e31821b1379
18. Panchal SK, Brown L (2011) Rodent models for metabolic syndrome research. J Biomed Biotechnol 2011:351982. https://doi.org/10.1155/2011/351982
19. Lai M, Chandrasekera PC, Barnard ND (2014) You are what you eat, or are you? The challenges of translating high-fat-fed rodents to human obesity and diabetes. Nutr Diabetes 4:e135. https://doi.org/10.1038/nutd.2014.30

20. Ventura LL, Fortes NC, Santiago HC, Caliari MV, Gomes MA, Oliveira DR (2017) Obesity-induced diet leads to weight gain, systemic metabolic alterations, adipose tissue inflammation, hepatic steatosis, and oxidative stress in gerbils. PeerJ 5:e2967. https://doi.org/10.7717/peerj.2967
21. Mustonen AM, Puukka M, Rouvinen-Watt K, Aho J, Asikainen J, Nieminen P (2009) Response to fasting in an unnaturally obese carnivore, the captive European polecat Mustela putorius. Exp Biol Med (Maywood) 234 (11):1287–1295. https://doi.org/10.3181/0904-RM-140
22. Hinney A, Vogel CI, Hebebrand J (2010) From monogenic to polygenic obesity: recent advances. Eur Child Adolesc Psychiatry 19 (3):297–310. https://doi.org/10.1007/s00787-010-0096-6
23. Rosini TC, Silva AS (1992) Moraes C (2012) diet-induced obesity: rodent model for the study of obesity-related disorders. Rev Assoc Med Bras 58(3):383–387
24. Petrov PD, Ribot J, Palou A, Bonet ML (2015) Improved metabolic regulation is associated with retinoblastoma protein gene haploinsufficiency in mice. Am J Physiol Endocrinol Metab 308(2):E172–E183. https://doi.org/10.1152/ajpendo.00308.2014
25. Marquardt N, Feja M, Hünigen H, Plendl J, Menken L, Fink H, Bert B (2018) Euthanasia of laboratory mice: are isoflurane and sevoflurane real alternatives to carbon dioxide? PLoS One 13(9):e0203793. https://doi.org/10.1371/journal.pone.0203793
26. Reynés B, Serrano A, Petrov PD, Ribot J, Chetrit C, Martínez-Puig D, Bonet ML, Palou A (2016) Anti-obesity and insulin-sensitising effects of a glycosaminoglycan mix. J Funct Foods 26:350–362. https://doi.org/10.1016/j.jff.2016.07.022
27. Rodriguez E, Ribot J, Rodriguez AM, Palou A (2004) PPAR-gamma 2 expression in response to cafeteria diet: gender- and depot-specific effects. Obes Res 12(9):1455–1463. https://doi.org/10.1038/oby.2004.182
28. Ribot J, Rodríguez AM, Rodríguez E, Palou A (2008) Adiponectin and resistin response in the onset of obesity in male and female rats. Obesity (Silver Spring) 16(4):723–730. https://doi.org/10.1038/oby.2008.113
29. Reynés B, García-Ruiz E, Díaz-Rúa R, Palou A, Oliver P (2014) Reversion to a control balanced diet is able to restore body weight and to recover altered metabolic parameters in adult rats long-term fed on a cafeteria diet. Food Res Int 64:839–848. https://doi.org/10.1016/j.foodres.2014.08.012
30. Sato Mito N, Suzui M, Yoshino H, Kaburagi T, Sato K (2009) Long term effects of high fat and sucrose diets on obesity and lymphocyte proliferation in mice. J Nutr Health Aging 13 (7):602–606
31. Surwit RS, Feinglos MN, Rodin J, Sutherland A, Petro AE, Opara EC, Kuhn CM, Rebuffé-Scrive M (1995) Differential effects of fat and sucrose on the development of obesity and diabetes in C57BL/6J and a/J mice. Metabolism 44(5):645–651
32. Buettner R, Schölmerich J, Bollheimer LC (2007) High-fat diets: modeling the metabolic disorders of human obesity in rodents. Obesity (Silver Spring) 15(4):798–808. https://doi.org/10.1038/oby.2007.608
33. Krishnan S, Cooper JA (2014) Effect of dietary fatty acid composition on substrate utilization and body weight maintenance in humans. Eur J Nutr 53(3):691–710. https://doi.org/10.1007/s00394-013-0638-z
34. Voigt A, Ribot J, Sabater AG, Palou A, Bonet ML, Klaus S (2015) Identification of Mest/Peg1 gene expression as a predictive biomarker of adipose tissue expansion sensitive to dietary anti-obesity interventions. Genes Nutr 10 (5):477. https://doi.org/10.1007/s12263-015-0477-z
35. Voigt A, Agnew K, van Schothorst EM, Keijer J, Klaus S (2013) Short-term, high fat feeding-induced changes in white adipose tissue gene expression are highly predictive for long-term changes. Mol Nutr Food Res 57 (8):1423–1434. https://doi.org/10.1002/mnfr.201200671
36. Margareto J, Gómez-Ambrosi J, Marti A, Martínez JA (2001) Time-dependent effects of a high-energy-yielding diet on the regulation of specific white adipose tissue genes. Biochem Biophys Res Commun 283(1):6–11. https://doi.org/10.1006/bbrc.2001.4733
37. Ciapaite J, van den Broek NM, Te Brinke H, Nicolay K, Jeneson JA, Houten SM, Prompers JJ (2011) Differential effects of short- and long-term high-fat diet feeding on hepatic fatty acid metabolism in rats. Biochim Biophys Acta 1811(7–8):441–451. https://doi.org/10.1016/j.bbalip.2011.05.005
38. Li CC, Liu C, Fu M, Hu KQ, Aizawa K, Takahashi S, Hiroyuki S, Cheng J, von Lintig J, Wang XD (2018) Tomato powder inhibits hepatic Steatosis and inflammation potentially through restoring SIRT1 activity and Adiponectin function independent of carotenoid cleavage enzymes in mice. Mol Nutr Food Res 62(8):e1700738. doi:https://doi.org/10.1002/mnfr.201700738

39. de Castro UG, dos Santos RA, Silva ME, de Lima WG, Campagnole-Santos MJ, Alzamora AC (2013) Age-dependent effect of high-fructose and high-fat diets on lipid metabolism and lipid accumulation in liver and kidney of rats. Lipids Health Dis 12:136. https://doi.org/10.1186/1476-511X-12-136
40. Amengual J, Lobo GP, Golczak M, Li HN, Klimova T, Hoppel CL, Wyss A, Palczewski K, von Lintig J (2011) A mitochondrial enzyme degrades carotenoids and protects against oxidative stress. FASEB J 25 (3):948–959. https://doi.org/10.1096/fj.10-173906
41. Hessel S, Eichinger A, Isken A, Amengual J, Hunzelmann S, Hoeller U, Elste V, Hunziker W, Goralczyk R, Oberhauser V, von Lintig J, Wyss A (2007) CMO1 deficiency abolishes vitamin a production from beta-carotene and alters lipid metabolism in mice. J Biol Chem 282(46):33553–33561. https://doi.org/10.1074/jbc.M706763200
42. West DB, Waguespack J, McCollister S (1995) Dietary obesity in the mouse: interaction of strain with diet composition. Am J Phys 268 (3 Pt 2):R658–R665. https://doi.org/10.1152/ajpregu.1995.268.3.R658
43. Reusch JEB, Kumar TR, Regensteiner JG, Zeitler PS, Participants C (2018) Identifying the critical gaps in research on sex differences in metabolism across the life span. Endocrinology 159(1):9–19. https://doi.org/10.1210/en.2017-03019
44. Priego T, Sanchez J, Pico C, Palou A (2008) Sex-differential expression of metabolism-related genes in response to a high-fat diet. Obesity (Silver Spring) 16(4):819–826. https://doi.org/10.1038/oby.2007.117
45. Lobo GP, Hessel S, Eichinger A, Noy N, Moise AR, Wyss A, Palczewski K, von Lintig J (2010) ISX is a retinoic acid-sensitive gatekeeper that controls intestinal beta,beta-carotene absorption and vitamin a production. FASEB J 24 (6):1656–1666. https://doi.org/10.1096/fj.09-150995
46. Murano I, Morroni M, Zingaretti MC, Oliver P, Sanchez J, Fuster A, Pico C, Palou A, Cinti S (2005) Morphology of ferret subcutaneous adipose tissue after 6-month daily supplementation with oral beta-carotene. Biochim Biophys Acta 1740(2):305–312. https://doi.org/10.1016/j.bbadis.2004.10.012
47. Reagan-Shaw S, Nihal M, Ahmad N (2008) Dose translation from animal to human studies revisited. FASEB J 22(3):659–661. https://doi.org/10.1096/fj.07-9574LSF
48. Blanchard OL, Smoliga JM (2015) Translating dosages from animal models to human clinical trials--revisiting body surface area scaling. FASEB J 29(5):1629–1634. https://doi.org/10.1096/fj.14-269043

INDEX

A

B

C

D

E

F

Manuel Rodríguez-Concepción and Ralf Welsch (eds.), *Plant and Food Carotenoids: Methods and Protocols*, Methods in Molecular Biology, vol. 2083, https://doi.org/10.1007/978-1-4939-9952-1,

MIX
Papier aus verantwortungsvollen Quellen
Paper from responsible sources
FSC® C105338

If you have any concerns about our products,
you can contact us on
ProductSafety@springernature.com

In case Publisher is established outside the EU,
the EU authorized representative is:
Springer Nature Customer Service Center GmbH
Europaplatz 3, 69115 Heidelberg, Germany

Printed by Libri Plureos GmbH
in Hamburg, Germany